Lecture Notes in Mathematics

Edited by A. Dold and B. Eckmann

998

Recent Developments in the Algebraic, Analytical, and Topological Theory of Semigroups

Proceedings of a Conference
Held at Oberwolfach, Germany, May 24–30, 1981

Edited by K. H. Hofmann, H. Jürgensen, and H. J. Weinert

Springer-Verlag
Berlin Heidelberg New York Tokyo 1983

Editors

Karl Heinrich Hofmann
Fachbereich Mathematik, Technische Hochschule Darmstadt
Schloßgartenstr. 7, 6100 Darmstadt, Federal Republic of Germany

Helmut Jürgensen
Fachbereich Informatik, Technische Hochschule Darmstadt
Alexanderstr. 24, 6100 Darmstadt, Federal Republic of Germany

Hanns Joachim Weinert
Institut für Mathematik, Technische Universität Clausthal
Erzstr. 1, 3392 Clausthal-Zellerfeld, Federal Republic of Germany

AMS Subject Classifications (1980): 20 M xx, 16 A 76, 16 A 78, 22 A 15, 22 A 20, 22 A 25, 22 A 26, 22 E 99

ISBN 3-540-12321-0 Springer-Verlag Berlin Heidelberg New York Tokyo
ISBN 0-387-12321-0 Springer-Verlag New York Heidelberg Berlin Tokyo

Printing and binding: Beltz Offsetdruck, Hemsbach/Bergstr.
2146/3140-543210

Vorwort

Vom 24. Mai bis zum 30. Mai 1981 fand am Mathematischen Forschungsinstitut Oberwolfach die zweite Tagung über Halbgruppentheorie statt. Sie stand unter der Leitung von K.H. Hofmann, U.Knauer und H.J. Weinert. Von den 58 Teilnehmern kamen 40 aus 10 europäischen Ländern einschließlich der Sowjetunion, 15 aus Nordamerika und je einer aus Australien, Japan und Kolumbien. Zahlreichen weiteren Interessenten mußte zum Bedauern der Tagungsleitung abgesagt werden.

Insgesamt wurden 48 Vorträge gehalten, an den Vormittagen solche zur algebraischen und kombinatorischen Theorie der Halbgruppen, an den Nachmittagen vorwiegend solche zu topologischen und analytischen Halbgruppen. Dieser Tagungsband enthält zwanzig aus diesen Vorträgen hervorgegangene Arbeiten; weitere Tagungsbeiträge sind oder werden anderweitig veröffentlicht.

Der vorliegende Band gibt einen Einblick in einige Schwerpunktthemen der gegenwärtigen algebraischen, topologischen und analytischen Halbgruppentheorie. In diesem Zusammenhang verweisen wir den Leser auf die zusammenfassenden Vorworte zu den beiden Teilen, in die er aus inhaltlichen Gesichtspunkten gegliedert wurde.

Die Unterzeichneten danken allen Verfassern für ihre Mitarbeit und den Herausgebern der Lecture Notes für ihre Bereitschaft, diesen Tagungsband zu veröffentlichen. Bei der typographischen Gestaltung haben uns neben den Verfassern selbst Frau E. Seitz, Frau E. Steingasser (beide Darmstadt) und Frau E. Weber (Clausthal) geholfen; auch dafür sei an dieser Stelle herzlich gedankt.

K.H. Hofmann, H. Jürgensen, H.J. Weinert
Darmstadt und Clausthal-Zellerfeld, im Januar 1983.

TABLE OF CONTENTS

Recent trends in the theory of semigroups in analysis and topology

Preface to the afternoon papers delivered at the Conference on Semigroups in Oberwolfach 1981 May 24 through May 30

Current research in analytical and topological semigroup theory is represented in these Proceedings through the following main topics:

1. The algebraic and topological structure of compact topological and semitopological semigroups.
2. Harmonic analysis of compact and locally compact semigroups.
3. Differentiable semigroups and Lie theory of semigroups.

The contributions to these topics may be classified as follows:

1.1. Brown, D.R., and J.M.Stepp, Inner points in certain holoidal semigroups, 19 pp.

1.2. Cohen, H., R.J.Koch and J.D.Lawson, Semigroups defined by lattice polynomials, 7 pp.

1.3 Ruppert, W., On structural methods and results in the theory of compact semitopological semigroups, 25 pp.

2.1 Baker, J.W., J.S.Pym and H.L.Vasudeva, Mulitpliers for some measure algebras on compact semilattices, 23 pp.

2.2 Mislove, M., and J.R.Liukkonen, Measure algebras of locally compact semilattices, 16 pp.

2.3 Troallic, J.-P., Semigroupes semitopologiques et présque-périodicité, 13 pp.

3.1 Graham, G. Differentiable semigroups, 71 pp.

3.2 Hofmann, K.H., and J.D.Lawson, Foundations of Lie semigroups, 74 pp.

It is characteristic for the ongoing research in this area that the attribution of some of the articles to one of the three subspecialities which we singled out is ambiguous: Ruppert's article has distinct applications to the theory of weakly almost periodic functions and therefore to the topic of harmonic analysis, and the contribution of Mislove and Liukkonen contains an elegant structural result on locally compact semilattices which is of quite independent interest in category 1 dealing with structural problems.

The viability of research on hard conventional problems about compact semigroups is exemplified through the papers of GROUP 1. The basic problem in the theory of compact topological or semitopological semigroups is to describe the geometry of the underlying space and how it relates to the algebraic structure. For compact topological semigroups, cohomology theory is one of the time honored methods. This line is vividly illustrated by the desire to find sufficient conditions for points of a compact semigroup to be, on one hand, in the interior of the semigroup, or, on the other, to be on its periphery. In the absence of any concrete embeddings into manifolds, one must define the concepts of interiority or peripheriality in terms of cohomology or local cohomology. These ideas show their true power when they are combined with the concept of topological dimension, whose cohomological aspect were pioneered by H. Cohen in his dissertation [1]. The key results known in the middle sixties were that every finite dimensional compact space has inner points, and that the identity of a compact connected finite dimensional monoid is peripheral, unless the monoid is a group (Hofmann, Lawson, Madison, Mostert). The connection between the topological concept of dimension and the algebraic concept of breadth of a semilattice became apparent through the promotion of the theory of locally compact and compact semilattices through Lawson. These proceedings indicate the role played by semilattices in almost all aspects of topological semigroup theory today. The contribution 1.1 by BROWN and STEPP concerns peripherality, dimension, and breadth.

---

[1] On the occasion of H.Cohen's 60 birthday, one evening of the conference was set apart for a review of his contributions to topology and semigroup theory. The review was prepared and deliverd by R.J.Koch.

They show that in a certain type of compact, H-trivial semigroup of cohomological dimension n, an idempotent has to be an inner point if its breadth is n+1, and they use the repertory of cohomological methods all the way from Wallace's Acyclicity Theorem through tools they develop for the purpose.

The explicit structure theory of a compact topological semigroup on a square, i.e. a 2-cell, is amazingly complex. The theory of a 1-cell has been known since the fifties (Clifford, Mostert, Shields, Storey); but a complete structure theory for the semigroups on the 2-cell is so much beyond hope that one must ask the question whether a complete classification is a reasonable problem (and one must probably answer no). However, interesting contributions keep coming. In an effort to investigate at least those multiplications on the 2-cell which are idempotent, in 1.2 , COHEN, KOCH, and LAWSON are led to consider a question which touches upon universal algebra. They determine the lattice ordered semigroup multiplications on the "lattice square" L^2 a distributive lattice L. This turns out to be surprisingly involved even if L has two elements: The authors find exactly 17 multiplications on L^2 in that case. In the general case, they present a fairly comprehensive theory.

The study of compact semitopological semigroups (i.e. semigroups in which multiplication is separately continuous in each of the two variables) is motivated traditionally by the fact that the multiplication in semigroups of operators with the strong or weak operator topology is not continuous, but is separately continuous. Compact semigroups of this kind arise quite naturally in the compactification of topological groups or semigroups constructed for the purpose of classifying all weakly almost periodic functions. Since the work of Glicksberg and de Leeuw on weakly almost periodic compactifications, the theory of compact semitopological semigroups established itself as a parallel branch of compact semigroup theory of which topological compact semigroups forms another branch.[2)]

---

2) The recently studied topic of right or left topological semigroups (Berglund, Junghenn, Hindman, Milnes, Pym) is not represented in these Proceedings even though Pym gave a lecture on an explicit construction of certain left topological semigroups.

Few results carry over from the main body of compact topological semigroup theory to the semitopological case; the structure of the completely simple minimal ideal is one of them, even though the full power of the results on the minimal ideal of a compact topological semigroup does not persist in the semitopological case. For the rest, questions have to be posed in a new spirit for compact semitopological semigroup. In 1.3 , RUPPERT gives an excellent review of some problems in this area to which he has contributed much in recent years. The first deep result on compact semitopological semigroups came from functional analysis via the famous Fixed Point Theorem of Ryll-Nardzweski. The result in question, in essence says that the minimal ideal of a compact semitopological semigroup must be an ideal if its group of units is dense. Ruppert as well as TROALLIC (in 2.3) show different routes of access to this theorem not using (but rahter giving thereby new and independent proofs of) the Ryll-Nardzweski Fixed Point Theorem. The gist of the above theorem could be expressed by saying that the idempotents in the minimal ideal are central. Ruppert further shows that all idempotents in a compact semitopological monoid are central, if the group of units contains a dense continuous homomorphic image N of some locally compact connected group. Moreover, N is then normal in the semigroup (i.e. satisfies sN=Ns for all s). Ruppert further characterizes those locally compact connected groups whose weakly almost periodic compactification is minimal in the sense that it is just the one point compactification. From this point of view it is natural to ask the question of characterizing those locally compact connected groups whose weakly almost periodic compactifications are inverse semigroups or, at the very least, regular semigroups. Ruppert reports a complete answer to this problem. Needless to say that the fine structure theory of locally compact groups and Lie group theory enter significantly.

In the paper 2.2 of GROUP 2 , TROALLIC applies function space methods based on a well-known theorem of Namioka's to prove novel results on compact semitopological semigroups. In particular, he sheds new light on the result mentioned above by giving more general sufficient conditions for the minimal ideal to be a group. He thereby produces a new approach to the Ryll-Nardzewski Fixed Point Theorem and to some classical results on the decomposition of weakly almos periodic functions.

The paper 2.1 by BAKER, PYM, and VASUDEVA discusses convolution measure algebras of a very special type of compact semilattices. The measure algebra of locally compact Lawson semilattices is also the subject of article 2.2 by MISLOVE and LIUKKONEN. The thrust, however, is somewhat different in the two papers. The first one is concerned with a characterisation of multipliers of a convolution measure algebra M in the measure algebra M(S) of certain subsemilatices S of the cube $\mathbb{I}^n$, where $\mathbb{I}$ denotes the unit interval with min-multiplication. The second paper discusses the question of symmetry of the algebra M(S) for a locally compact Lawson semilattice S, which requires a careful study of the Hewitt-Kakutani phenomenon in this case. In the paper 2.1, the case that S has an identity leads to the expected result that a multiplier is uniquely represented by a measure acting under convolution. The real difficulties arise in the absence of an identity. Here multipliers are represented by measures on the containing cube $\mathbb{I}^n$, and only if the set of maximal elements of S is finite can one be sure that the measure can actually be chosen in S. In general, the representing measure is neither unique, nor must have the norm of the multiplier. These facts are illustrated by examples. Curiously enough, the case of breadth 2 makes an exeption, since it is indeed well-behaved in both of these respects. These facets of the problem show that it is not a superficial one.- In the paper by MISLOVE and LUIKKONEN, the class of all locally compact Lawson semilattice is neatly split into two classes. The authors call a locally compact Lawson semilattice S tame if and only if it does not contain an isomorphic copy of the compact semilattice $2^{\mathbb{N}}$. They show that S is tame iff for every compact subset X of S there is a finite subset F with inf X = inf F iff S contains no meet irredundant Cantor set. This is a completely structure theoretical result whose prove requires some subtle arguments. For tame S, the authors now are able to characterize completely the elements of Δ M(S), i.e., the complex algebra homorphisms of M(S) as given by integration of some Borel semicharacter of S. If S is wild, i.e., not tame, then any bounded linear functional on the vector space M(X) of all complex measure on an irredundant Cantor set X in S extends to an element of Δ M(S): Thus M(S) is not symmetric.

The papers of GROUP 3 on differentiable semigroups and Lie theory of semigroups treat a fairly new subject even though its historical roots go right back to Sophus Lie. Semigroups of (local) transformations were in fact discussed in the last three decades of the 19th century under the name of transformation groups. Since the fifties in this century, some attempt at Lie semigroup theory was made from time to time, but was never carried to any appreciable depth. The two articles in these proceedings present for the first time a systematic outline of a theory of differentiable semigroups and the concomitant Lie theory. The article 3.1 by GRAHAM lays the foundations for the calculus on manifolds with generalized with generalized boundary in the necessary generality to deal with the concept of a differentiable semigroup and to treat even the simplest examples. The basic idea here is to introduce the concept of a manifold with generalized boundary which is defined to be a regular space which is locally homeomorphic to a subset of $\mathbb{R}^n$ with dense interior. A function f on such a set to $\mathbb{R}^m$ is called differentiable in a point a of the set if for each $\varepsilon > 0$ there is a $\delta > 0$ and a linear $T:\mathbb{R}^n \to \mathbb{R}^m$ such that for all elements x and y of the domain of f in a δ-neighborhood of a one has $\|f(y)-f(x) - T(y-x)\| < \varepsilon \|y-x\|$. The familiar function f given by $f(x) = x^2 \sin(1/x)$ for $x \neq 0$, else $= 0$, is not differentiable in 0 in this sense. With this concept of differentiability, GRAHAM develops the analysis of strongly C^k-manifolds with generalized boundary with all the necessary paraphernalia to speak of tangent bundles, vector fields, flows and so on. In this context he is then able to define and treat the concept of strongly C^k-differentiable semigroup and to provide the background for a general theory. In particular, he is able to speak about invariant vector fields and associate with a differentiable semigroup a Lie algebra. This is remarkable since traditionally one has the prejudice that a Lie algebra is associated with a group (or, at least, a local group). However, even if the identity of a differentiable semigroup is on an edge, or is even a cusp, there IS still a tangent plane to the manifold with generalized boundary in that point, with which the Lie algebra may be identified. In this fashion, GRAHAM is able to find for each differentiable monoid a local embedding of a neighborhood of the indentity into a Lie group, namely, the one (unique up to local isomorphism), which is associated with the Lie algebra of the monoid. It is historically remarkable, that this theorem, in a way, is contained in Sophus Lie's foundational work on continuous groups; a rigorous proof was presumably not possible with the methods and standards of that period. A Lie monoid S determines, in addition to its Lie algebra L(S) a tangent wedge W(S) giving all the one parameter semigroups of S. This wedge has been the object of intense scrutiny by HOFMANN and LAWSON, and not all aspects of this wedge are fully resolved as of yet. GRAHAM defines an exponential function $\exp: W(S) \to S$ which is a local diffeomorphism onto its image; but immense complications of the theory result from the fact that it will not be generally surjective near 0; in fact W(S) may lack inner points in L(S). If S is a compact differentiable monoid, GRAHAM shows that L(S) has an ideal of codimension 1. If the boundary of a differentiable monoid S is smooth, then the group H(1) of units of S is a Lie group of codimension 1 in S and is open in the boundary. The paper concludes with a sequence of intriguing problems.

In 3.2, HOFMANN and LAWSON develop a general theory of local monoids in a Lie group. In order to present a local theory in a systematic fashion, the fix a Lie algebra L and a sufficiently small zero neighborhood U on which the Campbell-Hausdorff formula yields a convergent series and thus defines a local group structure. They then define local semigroups S in U to contain 0 and to satisfy $S * S \cap U \subseteq S$, where $*$ denotes Campbell-Hausdorff multiplation. They study S in terms of the tangent wedge for S in 0 which they call L(S); one should, however be careful to realize that their L(S) is the W(S) of GRAHAM, while their L corresponds to the L(S) of GRAHAM. The paper 3.2 investigates how a set of vectors generates a local semigroup, or a local semigroup, or a local group and which Lie wedge, respectively Lie algebra arises. The maximal vector subspace of L(S) is shown to be the Lie algebra of the local group of units of S. In a very precise sense it is shown in which way L(S) approximates S locally near 0. It is reported that every proper closed cone in L is the tangent object of a local semigroup; this result will be published elsewhere. Local divisibility is investigated, and various typical examples are explicitly discussed.

In conjunction, the papers 3.1 and 3.2 of group 3 provide the first systematic introduction to a general Lie theory of semigroups. These papers also contain references to current literature in geometric control theory, in which subsemigroups of Lie groups are relevant.

Karl Heinrich Hofmann

MULTIPLIERS FOR SOME MEASURE ALGEBRAS ON COMPACT SEMILATTICES

J.W. Baker, J.S. Pym and H.L. Vasudeva

1. INTRODUCTION

The prototype of representation theorems for multipliers on measure algebras was given by Wendel [15]: the multipliers on the group algebra $L^1(G)$ of a locally compact group G are precisely the operators obtained from convolution by measures on G. Many authors have obtained results of the same kind for measure algebras on other locally compact semigroups. Ideally, each multiplier is realized by a unique measure, possibly not on the original semigroup but on a larger one in which the original semigroup is dense, and the correspondence between multipliers and measures is an isometry. Usually the measure algebras concerned are required to have an approximate identity in some sense. Often there is a problem in deciding exactly which measures on the larger semigroup can act as multipliers.

Thus, for example, Todd [14, §8] takes a semigroup S which possesses a "cancellable element" a and embeds S in a semigroup G in which a is invertible. His algebra consists of all measures absolutely continuous with respect to a suitable measure m on S, and he is able to represent its multipliers isometrically by measures on G. Sleijpen [12] represents multipliers on the natural measure algebra on a "foundation stip" S by measures on a natural completion of S. His representation is isometric, but his characterization of the measures which can be multipliers is in terms of an awkward continuity condition. A similar condition appears in Lahr [6]. He deals with general commutative convolution algebras with "weak approximate identities" and represents multipliers by measures on the Taylor structure semigroup [13] of S. His correspondence is a bicontinuous injection, but not an isometry in general. In the same paper for certain semigroups S he represents multipliers of $\ell^1(S)$ by elements of $\ell^1(\Omega S)$ where ΩS is the multiplier semigroup of S (the definition is in §2 below); S may not be dense in ΩS and the representing measure is not usually uniquely determined by the multiplier, though every element of $\ell^2(\Omega S)$ does yield a multiplier.

The algebras with which we shall be concerned may not contain approximate identities in the senses required (implicitly or explicitly) by most earlier authors, and indeed, it is unlikely that they contain approximate identities in any sense. To make up for this deficiency, they will be fairly concrete: they are compact subsemigroups S of finite products B of compact totally ordered sets with the multiplication min. Harmonic analysis on B has been worked out by Baartz [1] and Newman [11], though we shall not need their results. Our algebras are (almost) arbitrary convolution measure algebras on S (though Theorem 4.4 shows that to some extent this apparent generality is an illusion). Our approach is through the simplest case, in which S has an identity. Here, our results generalize those of Larsen [10] and Todd [14, §2] for the totally ordered case, and of Dhar and Vasudeva [3] and of Johnson and Lahr [5] in the n-dimensional case, particularly in that we allow much more general measure algebras. Our representation is isometric, and we obtain a simple characterization, in measure-theoretic terms, of those measures which can act as multipliers.

For the case in which S has no identity, the basic idea is that each of our semigroups is a union of semigroups which have an identity. We can represent multipliers by measures on B (Theorem 4.6) but not usually by measures on S (Theorem 4.8). The representing measure is not unique (Examples 4.9, 4.11, Remark 5.3), and cannot generally be made isometric (Examples 4.10, 4.12) though curiously it can in the two-dimensional case (Theorem 5.5). The measures which can be multipliers do have a measure theoretic characterization, but this is rather complicated and we merely sketch the ideas (Remark 4.7(ii)). For certain semigroups S, we have $\Omega S = B$ (Lemma 2.6). We give towards the end of §4 results which parallel those of Lahr [6] for ℓ^1 algebras.

2. NOTATION AND PRELIMINARIES

The ordered semigroups S we shall be concerned with are "finite dimensional" in the following sense: they can be embedded in a finite direct product of totally ordered semigroups $I_1,\ldots,I_k$, where each of the latter has multiplication min. Thus, if $x = (x_1,\ldots,x_k)$ and $y = (y_1,\ldots,y_k)$ are elements of S, then

$$xy = (\min\{x_1,y_1\},\ldots, \min\{x_k,y_k\}). \qquad (2.1)$$

Note that $x^2 = x$ for each x. We shall assume that each of $I_1,\ldots,I_k$ is compact and connected in its order topology, and that S is closed in the product topology. Of course, S is a topological semigroup. For $1 \le j \le k$, the maximal element of I_j

is denoted by 1, and the minimal element by 0; we shall assume that $1 \neq 0$. Connectivity of the set I_j implies that, for every $x \in I_j$,

$$x = \sup\{y \in I_j : y < x\}. \tag{2.2}$$

We write $B = I_1 \times \ldots \times I_k$. A <u>box</u> in B is obtained by choosing any $a \in B$ and forming

$$B_a = \{x \in B : x \leq a\}.$$

Then B_a is a subsemigroup of B. It has a semigroup identity, a, and a semigroup zero $0 = (0,\ldots,0)$. In particular, for $1 = (1,\ldots,1)$ we have $B_1 = B$. The semigroup S has a unique minimal element (by Zorn's Lemma and the fact that S is closed under min); we lose nothing if we assume this to be $0 \in B$. There is, in general, no unique maximal element in S, but there are (Zorn's Lemma again) maximal elements, and for each x in S there is such a maximal element a such that $x \leq a$, or equivalently $xa = x$. This last relation says precisely that the set A of maximal elements of S is a <u>set of relative units</u> for S (Hewitt and Zuckerman [4], §7). Obviously

$$S = \bigcup\{B_a \cap S : a \in A\},$$

and S has a semigroup identity if and only if A has exactly one element.

The <u>inside</u> B_a^o of a box B_a is defined to be

$$\{x \in B_a : x_j < a_j \text{ for } 1 \leq j \leq k\};$$

this is the topological interior only when $a_j < 1$ for all j. An <u>upper face</u> F_a^J of B_a is obtained by taking a proper subset J of $\{1,\ldots,k\}$ and writing

$$F_a^J = \{x \in B_a : x_j < a_j \text{ for } j \in J;\ x_j = a_j \text{ for } j \notin J\}. \tag{2.3}$$

The corresponding face of B (= B_1) will simply be written F^J. The inside of B_a together with its upper faces form a partition of B_a. The <u>upper boundary</u> of B_a is $B_a \setminus B_a^o$. We define a canonical map, the <u>explosion</u>, η_a on B_a which projects each face orthogonally into the corresponding parallel face of B; precisely, η_a is the identity on B_a^o, while if $x = (x_1,\ldots,x_k) \in F_a^J$,

$$\eta_a(x) = (y_1,\ldots y_k) \tag{2.4}$$

where $y_j = x_j$ if $j \in J$ (and so $x_j < a_j$), $y_j = 1$ if $j \notin J$ (and so $x_j = a_j$). Note that η_a followed by multiplication by a is just the identity on B_a : $a\eta_a(x) = x$ if $x \in B_a$. On the other hand, $\eta_a(ax) = x$ if and only if x is in the image of η_a, i.e. either $x \in B_a^o$ or for some J, $x \in F^J$ and $x_j < a_j$ for $j \in J$.

For any x in S, write $S_x^o = S \cap B_x^o$, and define the <u>inside</u> of S to be

$$S^o = \bigcup_{x \in S} S_x^o = \bigcup_{x \in A} S_x^o \qquad (2.5)$$

The <u>multiplier semigroup</u> ΩS of S is the set of mappings $f: S \to S$ such that $f(xy) = (fx)y$ for $x,y \in S$, with composition as its semigroup operation. In some cases, ΩS is easy to determine.

2.6 PROPOSITION. *Suppose that $pr_j S = I_j$ for each j (pr_j being the projection of B onto the jth coordinate). Write $u_j = (0,\ldots,0,1,0,\ldots,0)$ where the one occurs in the jth place. Then $\Omega S = B$ if and only if $u_j \in S$ for every $j = 1,2,\ldots,k$.*

PROOF. Suppose $u_j \in S$ for $j = 1,\ldots,k$. Let $f \in \Omega S$ and put $a_j = fu_j$. Since $u_j^2 = u_j$, $u_j a_j = a_j$ and so a_j has all coordinates except the jth equal to 0; let that j^{th} component be $f_j \in I_j$. For any x in S, $(fx)u_j = x(fu_j) = xa_j$. This says that the jth component of fx is $f_j x_j$. This holds for every j, so $fx = (f_1,\ldots,f_k)x$, so $f \in B$.

Conversely, if $\Omega S = B$, given j we choose $x_j \in S$ such that $pr_j x_j = 1$. Then, as $u_j \in \Omega S$,

$$u_j = u_j x_j \in S,$$

and this completes our proof.

By a <u>measure</u> we shall mean a bounded regular Borel measure. The convolution product, $\mu\nu$, of two measures μ,ν is given by the usual formula

$$\mu\nu(\phi) = \int\int \phi(xy)\,d\nu(x)\,d\nu(y)$$

where ϕ is any continuous bounded function. It will be useful to have two simple formulae for certain convolution products. Denote by $\operatorname{supp}\mu$ the support of μ. Then if $\operatorname{supp}\nu \le \operatorname{supp}\mu$ (i.e. $x \le y$ for all x in $\operatorname{supp}\nu$, y in $\operatorname{supp}\mu$) we have

$$\mu\nu = \mu(1)\nu \; ; \qquad (2.7)$$

this follows immediately from the definition since $xy = x$ for x,y as above. For the second formula, we suppose $a = (a_J, a_J')$ where $a_J \in \prod_{j \in J} I_j$ and $a_J' \in \prod_{j \notin J} I_j$, that μ is carried by $\{(x_J, x_J') : x_J \le a_J,\ x_J' \ge a_J'\}$ and ν is carried by

$\{(y_J,y_J') : y_J \geq a_J, y_J' \leq a_J'\}$. Let $\pi_J\mu$ be the projection of μ in the first coordinate, $\pi_J'\mu$ the projection in the second. Then

$$\mu\nu = \pi_J\mu \times \pi_J'\nu \tag{2.8}$$

where the right hand product means the direct product measure on the box B_a. This is easily seen from the defining formula, since

$$\mu\nu = \iint \phi(x_J y_J, x_J' y_J')\, d\mu(x_J,x_J')\, d\nu(y_J,y_J')$$

$$= \iint \phi(x_J,y_J')\, d\mu(x_J,x_J')\, d\nu(y_J,y_J')$$

$$= \iint \phi(x_J,y_J')\, d\pi_J\mu(x_J)\, d\pi_J'\nu(y_J').$$

We shall frequently have cause to use the fundamental result on supports.

2.9 SUPPORT LEMMA. *If μ is carried by the Borel set A (i.e. $|\mu|(S\setminus A) = 0$) and ν is carried by the Borel set B then $\mu\nu$ is carried by AB. If in addition μ and ν are positive then*

$$\text{supp}(\mu\nu) = (\text{supp}\ \mu)\ (\text{supp}\ \nu)$$

In particular notice that if ε_a is the point mass at a and μ is carried by $B\setminus B_a^\circ$ then $\mu\varepsilon_a$ is carried by the upper boundary of B_a.

A convolution measure algebra $\mathcal{M}$ on S is an algebra of measures on S which is complete in the usual measure norm and has the further property that if $\mu \in \mathcal{M}$ and ν is absolutely continuous with respect to the total variation measure $|\mu|$, then $\nu \in \mathcal{M}$ (Taylor[13]). Such an algebra contains the restriction of any of its members to any Borel measurable subset. We assume that $\bigcup\{\text{supp}\ \mu : \mu \in \mathcal{M}\}$ is dense in S. Again to avoid unilluminating special cases, we shall restrict our attention to algebras all carried by the inside of S: $\mu \in \mathcal{M}$ implies

$$|\mu|(S\setminus S^\circ) = 0$$

or equivalently

$$|\mu|(S) = \sup\{|\mu|(K) : K \subseteq S^\circ \text{ and } K \text{ is compact}\}.$$

As a direct consequence of this last formula, given $\varepsilon > 0$ we can find $a_1,\ldots,a_n$ in A such that

$$|\mu|(S\setminus \bigcup_{j=1}^{n} B_{a_j}) < \varepsilon, \tag{2.10}$$

the property we actually need. We shall also assume that supp $\mathcal{M}$ = S; as supp $\mathcal{M}$ is already a semigroup, this represents no real loss.

If S has an identity e, then $\mathcal{M}$ has an approximate identity of norm 1. To construct one, simply take a basic set {U} of neighbourhoods of e directed by reverse inclusion, and for each U take any measure δ_U in $\mathcal{M}$ carried by U with $\delta_U \geq 0$ and $\|\delta_U\| = 1$. Standard arguments (which use the fact that $\mathcal{M}$ is carried by S^o) show that $\{\delta_U\}$ is as required. Of course, the algebra $\mathcal{M}(S)$ of all measures on S has an identity, viz. ε_e.

We now establish a condition on S for $\mathcal{M}(S)$ to have an identity, say ε. (This is a special case of the theorem of Lardy [8], but we shall give the easy proof in our situation.) Let a be a maximal element of S, and write $\varepsilon = k\varepsilon_a + \mu$, where k is a real number and μ is carried by $S\backslash\{a\}$. Then

$$\varepsilon_a = \varepsilon\varepsilon_a = k\varepsilon_a + \mu\varepsilon_a.$$

The support lemma (2.9) can be used to prove that $\mu\varepsilon_a(\{a\}) = 0$, and we conclude that k = 1. This argument holds for any maximal element of S, so that, if $a_1,\dots,a_n$ are n of them,

$$\varepsilon = \varepsilon_{a_1} + \dots + \varepsilon_{a_n} + \nu$$

where ν is singular to each ε_{a_i}. Therefore

$$\|\varepsilon\| = \|\varepsilon_{a_1}\| + \dots + \|\varepsilon_{a_n}\| + \|\nu\| \geq n.$$

We conclude that S has at most a finite number of maximal elements. Conversely, if S has a finite set $a_1,\dots,a_n$ of maximal elements, then

$$\varepsilon = \sum_i \varepsilon_{a_i} - \sum_{i<j} \varepsilon_{a_i a_j} + \sum_{i<j<h} \varepsilon_{a_i a_j a_h} - \dots \tag{2.11}$$

is, as Hewitt and Zuckerman point out ([4; Theorem 7.5]), easily seen to be an identity for $\mathcal{M}(S)$.

Hewitt and Zuckerman also remark that there can be a great deal of cancellation in this formula. We point out that ε has no mass inside S. Indeed, let μ be the restriction of ε to a box B_x^o, where $x \in S$. Then

$$\varepsilon_x = \varepsilon\varepsilon_x = (\varepsilon-\mu)\varepsilon_x + \mu\varepsilon_x.$$

Now $(\varepsilon-\mu)$ is carried by $B\setminus B_x^o$, so $(\varepsilon-\mu)\varepsilon_x$ has no mass in B_x^o, by the remark after (2.9); moreover, from (2.7) we see that $\mu\varepsilon_x = \mu$. If we now restrict both sides of the last equation to B_x^o, we see that $\mu = 0$.

On the other hand, if S has more than one maximal element, ε is not a positive measure, and $\|\varepsilon\| \geq 3$, as is easily deduced from (2.11).

A multiplier on an algebra is a multiplier on the multiplicative semigroup of the algebra. The algebras $\mathcal{M}$ we consider are commutative and semisimple (see [1] or [11]) and the multipliers on them are automatically linear and continuous (see the first chapter of [9]). The obvious question is whether each multiplier on $\mathcal{M}$ can be represented by a measure on S; the answer is that this is true if and only if S has a finite number of maximal elements. As the proof of this result is simplified by some technical work, we delay it until Theorem 4.8.

3. THE CASE IN WHICH S HAS AN IDENTITY

3.1 THEOREM. *We assume the notation and conventions of §2. Let S have an identity a. Let M be a multiplier on $\mathcal{M}$. Then there is a unique measure μ on S such that M is simply convolution by μ. Moreover*

(i) the restriction μ^o of μ to S^o is in $\mathcal{M}$; and

(ii) if $\mathcal{M}_U$ denotes the restriction of $\mathcal{M}$ to a set U and μ^J denotes the restriction of μ to the face F_a^J of B_a (so that μ^J is carried by $S \cap F_a^J$), we have, for each proper subset J of $\{1,\ldots,k\}$,

$$\mu^J \in \bigcap\{\mathrm{proj}_J\ \mathcal{M}_U :\ U \text{ is an open nhbd of } F_a^J\}$$

where proj_J is the projection of B_a onto F_a^J.

Conversely, if μ is a measure on B_a which satisfies (i) and (ii), then convolution by μ is a multiplier on $\mathcal{M}$.

Finally, in the above situation $\|\mu\| = \|M\|$, and if M is a positive operator, then μ is a positive measure.

PROOF. We continue with the notation of §2. The net $(M\delta_U)$, where (δ_U) is the approximate identity for $\mathcal{M}$, is contained in the weak* compact ball of radius $\|M\|$ in the full algebra $\mathcal{M}(S)$. It therefore has a convergent subnet, say $M\delta_{U_i} \to \mu$ (weak*); note that $\|\mu\| \leq \|M\|$. As multiplication in closed balls of $\mathcal{M}(S)$ is weak* continuous, we have for $\nu \in \mathcal{M}$

$$M\nu = \lim \delta_{U_i}(M\nu) = \lim (M\delta_{U_i})\nu = \mu\nu$$

as required. Note also from this formula that $\|M\| \leq \|\mu\|$. Hence $\|M\| = \|\mu\|$. To prove uniqueness, note that if μ and μ' both yield M, then $\mu\delta_U = M\delta_U = \mu'\delta_U$ for every U, whence $\mu = \mu'$ on taking the limit. If M is positive, then because δ_U is positive for each U, μ is also positive.

It remains to establish (i) and (ii) and their converse. The proof is inductive and establishes the result and its converse for each face in turn.

We begin by considering μ^o. Take any x in $B_a^o \cap S = S^o$, and form $V_x = \prod_1^k\,]x_j,a_j]$; this is an open neighbourhood of a in B_a. Write $\delta_x = \delta_{V_x}$ for brevity, and observe that $\mu\delta_x \in \mathcal{M}$, and hence the restriction of $\mu\delta_x$ to B_x^o is in $\mathcal{M}$. We claim this restriction is just μ_x, where μ_x is the restriction of μ to B_x^o. Indeed by (2.7), $\mu_x\delta_x = \mu_x$. Moreover, by the support lemma (2.9), $(\mu-\mu_x)\delta_x$ is carried by $(B_a \setminus B_x^o).V_x \subseteq B_a \setminus B_x^o$. This establishes our claim. Now let $x \nearrow a$ in $S^o = S \cap B_a^o$ (this is possible because $\mathcal{M}$ is carried by S^o and supp$\mathcal{M}$ is dense in S) to see that the restriction μ^o of μ to S^o is in $\mathcal{M}$.

Conversely, any measure μ^o in $\mathcal{M}$ defines a multiplier on $\mathcal{M}$ by convolution.

We now continue with the proof of the direct part of the theorem. We now know that $\mu - \mu^o$ defines a multiplier by convolution, and moreover $\mu - \mu^o$ has no mass inside B_a. We turn our attention to a (k-1) dimensional upper face of B_a, and for definiteness take $J = \{1,2,\ldots,k-1\}$. Write $F_a^J(x) = \{y \in F_a^J : y_j \leq x_j, j \in J\}$, let μ_x^J be the restriction of μ^J to $F_a^J(x)$, and denote by π_J the projection onto $I_1 \times \ldots \times I_{k-1}$. Then the use of the support lemma shows that on the set

$$\pi_J F_a^J(x) \times\,]x_k,a_k] = \{(y_j) : y_j \leq x_j \text{ for } 1 \leq j \leq k-1 \text{ and } x_k < y_k \leq a_k\}$$

the restrictions of the measures $(\mu-\mu^o)\delta_x$ and $\mu_x^J\delta_x$ coincide; but by (2.8) the latter is the direct product

$$\pi_J\mu_x^J \times \mathrm{proj}_k\, \delta_x$$

where proj_k is the projection onto the kth coordinate. Any restriction of $(\mu-\mu^o)\delta_x$ is, of course, in $\mathcal{M}$. If U is an open neighbourhood of F_a^J, then by taking x_k large enough (and fixing $x_1 < a_1,\ldots,x_{k-1} < a_{k-1}$) we can ensure that

$$\pi_J F_a^J(x) \times\,]x_k,a_k] \subseteq U$$

(remember that $F_a^J(x)$ is compact), and hence we conclude that $\pi_J\mu_x^J \times \mathrm{proj}_k\delta_x \in \mathcal{M}_U$. As U is arbitrary, $\mu_x^J \in \bigcap_U \mathrm{proj}_J\mathcal{M}_U$. Finally, as this intersection is a closed set, we can let $x \nearrow a$ to get $\mu^J \in \bigcap_U \mathrm{proj}_J\mathcal{M}_U$.

The next step is to prove the converse to the last assertion: given $\mu^J \in \bigcap_U \mathrm{proj}_J \mathcal{M}_U$, we prove that μ^J is a multiplier for $\mathcal{M}$. We take first an arbitrary $z \in B_a^o \cap S$ and we let ν be any measure in $\mathcal{M}$ carried by B_z^o. Then we take any $x \in B_a^o \cap S$ with $x \geq z$ and define μ_x^J as above. We shall prove that $\mu_x^J \nu \in \mathcal{M}$. Taking limits first over x, then over z will show that $\mu^J \nu \in \mathcal{M}$ for all $\nu \in \mathcal{M}$, and the conclusion will follow.

We begin by choosing $\lambda \in \mathcal{M}$ which is carried by

$$\pi_J F^J(x) \times]x_k, a_k]$$

and is such that $\mathrm{proj}_J \lambda = \mu_x^J$; the possibility of this follows quickly from our hypothesis. Now note from (2.8) that

$$\lambda \varepsilon_x = \pi_J \mu_x^J \times \mathrm{proj}_k \varepsilon_x$$

and (using (2.8) again) that this is also the value of $\mu_x^J \varepsilon_x$. From (2.7), $\varepsilon_x \nu = \nu$. Hence we have

$$\mu_x^J \nu = \mu_x^J \varepsilon_x \nu = \lambda \varepsilon_x \nu = \lambda \nu \in \mathcal{M},$$

as required.

We now return again to the direct part of the proof. We now know that $\mu - (\mu^o + \mu^J)$ is a multiplier. We choose another (k-1) dimensional face, and proceed as before. When these are exhausted, we pass to (k-2) dimensional faces, and so on. After treating the (unique) 0-dimensional face $\{a\}$, the theorem is proved.

3.2 REMARKS. (i) The technique of subtracting the measures already dealt with in the above proof is merely for (we hope) clarity. It avoids an argument resting heavily on epsilons which would assert that the effect of a measure on an r-dimensional face was negligible when considering an s-dimensional face for $r \neq s$.

(ii) We show how Theorem 3.1 generalizes work of earlier authors. Larsen [10] considers $B = S = I$ the unit interval (with max rather than min, but the reader can make the change), and takes $\mathcal{M}$ to be the usual space L^1 of Lebesgue integrable functions. Clearly, $B \setminus S^o$ is just $\{1\}$, and $\{1\}$ is the only lower face of B. Thus, a multiplier is the sum of a measure in $\mathcal{M} = L^1$ and a measure carried by $\{1\}$, i.e. a multiple of ε. The set U of 3.1(ii) is just an interval $]u,1]$ with $u < 1$, and the projection of integrable functions on this to the set $\{1\}$ consists of all measures on $\{1\}$ (i.e. all multiples of ε) so we have found all multipliers. Similar remarks apply to [14;§2] on adjoining an identity to the semigroups considered there.

In [3], $B = S = I \times I$ (again, the authors use max), and $\mathscr{M}$ consists of Lebesgue integrable functions on S. Put $I_1 = [0,1[\times \{1\}$, $I_2 = \{1\} \times [0,1[$. The projection of the restriction of $\mathscr{M}$ to a neighbourhood U of I_1 (or I_2) onto I_1 (resp. I_2) gives all Lebesgue integrable functions on I_1 (resp. I_2). The projection of $\mathscr{M}$ restricted to a neighbourhood of 1 onto {1} gives all multiples of ε_1. Thus, multiplier measures are exactly those of the form

$$\mu = \mu^0 + \mu^1 + \mu^2 + \alpha\varepsilon_1$$

where $\mu^0 \in \mathscr{M}$, μ^1 (resp. μ^2) is Lebesgue integrable on I_1 (resp. I_2) and α is a real number. This is one of the characterizations in [3]. In [5], this same result is obtained for $B = S = I^k$ for any positive integer k, but is expressed in terms of tensor products.

(iii) We might add here that S need not be the whole of B. For example, with I the unit interval as above, we could take $B = I \times I$, $S = \{(x,y) \in B : \alpha(1-x) \leq 1-y \leq \beta(1-x)\}$ for some fixed α,β with $0 < \alpha < \beta$ (that is, the intersection of a cone with vertex at (1,1) with B) and $\mathscr{M}$ any algebra on $S^o = S\setminus\{1\}$. Then $U = \{(x,y) : (1-y) < \alpha(1-x)\}$ is a neighbourhood of I_1 which does not meet S, so $\mathscr{M}_U = \emptyset$; thus, a multiplier measure has no mass on I_1. Similarly, it has no mass on I_2. Thus, it is the sum of a multiple of ε_1 and a measure in $\mathscr{M}$.

(iv) Although the condition imposed upon $I_1,\ldots,I_k$ is restrictive, S can be fairly arbitrary and, in particular, need not be connected. For example, take I to be the unit interval and put $B = I \times I$, as above. One could then take S to be $T \times T$ where T is either (a) the sequence $\{0,\frac{1}{2},\frac{3}{4},\ldots,1\}$, or (b) the Cantor ternary set, with $\mathscr{M}$ any convolution measure algebra carried by $(T\setminus\{0\}) \times (T\setminus\{0\})$ whose support is $T \times T$.

4. THE GENERAL CASE

We approach the general case by reducing it to that considered in §3. To do this we must ensure that there are enough semigroups of the form S_x to which Theorem 3.1 is applicable. To this end we make the following hypothesis.

4.1. HYPOTHESIS. *There is a subsemigroup* E *of* S *such that* $x \in \mathrm{cls}\, S_x^o$ *for each* x *in* E *and* $\bigcup_{x \in E} S_x^o$ *is dense in* S.

It is, in fact, enough to have a *set* E with this property since if $x \in \mathrm{cls}\, S_x^o$ and $y \in \mathrm{cls}\, S_y^o$ then $xy \in \mathrm{cls}\, S_{xy}^o$. We denote the restriction of $\mathcal{M}$ to S_x^o by $\mathcal{M}_x$. Then our assumption that $\mathrm{supp}\,\mathcal{M} = S$ implies that $\mathrm{supp}\,\mathcal{M}_x = \mathrm{cls}\, S_x^o$ for any x in E. We are also assured by 4.1 that the linear span of $\bigcup_{x \in E} \mathcal{M}_x$ is dense in $\mathcal{M}$.

We are grateful to the referee for pointing out the error in our original draft which led us to this hypothesis. We shall show in §6 that many semigroups with measure algebras which satisfy the conditions of §2 do satisfy (4.1).

4.2. LEMMA. *Let* $\mathcal{M}_x$ *denote the restriction of* $\mathcal{M}$ *to* S_x^o. *Then the restriction* M_x *of a multiplier* M *on* $\mathcal{M}$ *to* $\mathcal{M}_x$ *is a multiplier on* $\mathcal{M}_x$.

PROOF. All we need to show is that M maps $\mathcal{M}_x$ into $\mathcal{M}_x$. Let (δ_U) be an approximate identity for $\mathcal{M}_x$. Then, if $\mu \in \mathcal{M}_x$,

$$M\mu = M \lim_U \delta_U \mu = \lim_U (M\delta_U)\mu \in \mathcal{M}_x$$

since $M\delta_U$ is carried by S, μ is carried by S_x^o, S_x^o is an ideal in S so that $M\delta_U$ is carried by S_x^o (use the support lemma (2.9)), and $\mathcal{M}_x$ is closed.

The theorem which allows us to make the reduction now follows.

4.3. THEOREM. *Suppose that, for each* $x \in S$, *we have a multiplier* M_x *on* $\mathcal{M}_x$. *The following statements are equivalent:*

(i) *there is a multiplier* M *on* $\mathcal{M}$ *whose restriction to* $\mathcal{M}_x$ *for each* $x \in S$ *is* M_x;

(ii) *if* $x, y \in E$ *with* $y \leq x$ *then* M_y *is the restriction of* M_x *to* $\mathcal{M}_y$, *and* $\sup \|M_x\| < \infty$.

Under the conditions of (i) *or* (ii), $\|M\| = \sup \|M_x\|$.

PROOF. Obviously (i) implies (ii). If we are given (ii) then it is straightforward to define a multiplier M on the linear span of $\bigcup_{x \in E} \mathcal{M}_x$. To deduce (i), it is enough to prove that M is bounded.

Let ν be in the linear span of $\bigcup_{x \in E} \mathcal{M}_x$, so there are $x_1, \ldots, x_n$ such that $\nu \in \mathcal{M}_{x_1} + \ldots + \mathcal{M}_{x_n}$, and ν is carried by $S_{x_1} \cup \ldots \cup S_{x_n}$. Write ν_i for the restriction of ν to $S_{x_i} \setminus (S_{x_1} \cup \ldots \cup S_{x_{i-1}})$ $(1 \le i \le n)$. Then $\nu_1, \ldots, \nu_n$ are carried by disjoint sets, so $\|\nu\| = \|\nu_1\| + \ldots + \|\nu_n\|$. Let $K = \sup \|M_x\|$. Then

$$\|M_\nu\| = \|M_{x_1}\nu_1 + \ldots + M_{x_n}\nu_n\| \le \|M_{x_1}\nu_1\| + \ldots + \|M_{x_n}\nu_n\| \le K\|\nu\|.$$

This proves (i), and incidentally shows that $\|M\| \le K$. To prove that $K \le \|M\|$ we only have to make the simple observation that $\|M_x\| \le \|M\|$ for each x.

Our next result, though of some interest in itself, plays an important part in our exposition. In the proof of Theorem 3.1, the question which measures could act as multipliers on $\mathcal{M}$ played a crucial role. The answer to this question is not so simple in the general case and we shall not burden the reader with all the complexities involved (see 4.7(ii) below), but the following theorem enables us to avoid the point in our proof that multipliers can be represented by measures.

4.4 THEOREM. Let M be a multiplier on $\mathcal{M}$. Then M has a unique extension as a multiplier to the whole of $\mathcal{M}(S^O)$ (i.e. the algebra of all measures inside S). This extension has norm $\|M\|$.

PROOF. On each μ_x $(x \in E)$, M is given by a unique measure μ_x carried by S_x. Each μ_x defines, by convolution, a unique multiplier M_x on $\mathcal{M}(S_x^O)$ of norm $\|\mu_x\|$. The family (M_x) satisfies the conditions of Theorem 4.3(ii), and so defines a unique multiplier on the closure of the span of $\bigcup_x \mathcal{M}(S_x^O)$, i.e. $\mathcal{M}(S^O)$, of norm $\sup\|\mu_x\| = \|M\|$. This multiplier clearly extends M.

The existence of the extension enables us to determine the measures μ_x easily in certain circumstances.

4.5 COROLLORY. Let the extension of M to $\mathcal{M}(S^O)$ be again denoted by M. Then, if $x \in E \cap S^O$, we have $\mu_x = M\varepsilon_x$.

PROOF. If $x \in E \cap S^O$, then $\varepsilon_x \in \mathcal{M}(S^O)$. If $\nu \in \mathcal{M}(S_x^O)$, then $\varepsilon_x\nu = \nu$, so

$$\mu_x\nu = M\nu = (M\varepsilon_x)\nu.$$

We now come to our main theorem. Notice that no mention is made of the uniqueness of the representing measure; this is because uniqueness does not hold (see Examples 4.9, 4.11 below).

4.6 THEOREM. Let M be a multiplier on $\mathcal{M}$. Then there is a measure μ on B such that $M\nu = \mu\nu$ ($\nu \in \mathcal{M}$). The restriction of μ to S^o is in $\mathcal{M}$. Moreover, μ can be chosen so that it is carried by S^o and the upper faces of B.

PROOF. The actual procedure of the following construction does enable us to find representing measures in some cases, so we shall include one or two details which are not strictly necessary.

Let $S = \bigcup_{x \in E} S_x$. On each S_x, M is realized by a measure μ_x of total mass not exceeding $\|M\|$. The uniqueness part of Theorem 3.1 shows that on the intersection $S_x^o \cap S_y^o = S_{xy}^o$, the measures μ_x and μ_y coincide. Thus, we can define a measure μ^o on S^o whose restriction to S_x^o is μ_x restricted to S_x^o (i.e. μ_x^o) for each x. As $\mu_x^o \in \mathcal{M}$ for each x (Theorem 3.1 again), $\mu^o \in \mathcal{M}$.

Following the tactics adopted to prove Theorem 3.1, we now consider the multiplier $M - \mu^o$. We regard this as acting on $\mathcal{M}(S^o)$, thanks to Theorem 4.4. We fix a (k-1) dimensional upper face of B, say F^J. Now $M - \mu^o$ is represented on $\mathcal{M}(S_x^o)$ by a measure μ_x which has restriction μ_x^J to the face F_x^J of B_x parallel to F^J. The explosion η_x (see §2) maps η_x^J to a measure on F^J. Moreover, the relationship $x\eta_x(y) = y$ for $y \in F_x^J$ (see after (2.4)) yields $\varepsilon_x\eta_x(\mu_x^J) = \mu_x^J$.

We want to piece together all the measures $\eta_x(\mu_x^J)$ on F^J to form the measure μ^J. For this to be justified, all we need do is check that if $\eta_x(F_x^J)$ and $\eta_y(F_y^J)$ intersect, then $\eta_x(\mu_x^J)$ and $\eta_y(\mu_y^J)$ coincide on the intersection. Now for any x,y, $\eta_x(F_x^J)$ and $\eta_y(F_y^J)$ certainly do intersect, and the intersection is just $\eta_{xy}(F^J_{xy})$. It will be enough for us to establish that the restriction of $\eta_x(\mu_x^J)$ to $\eta_{xy}(F^J_{xy})$ is $\eta_{xy}(\mu^J_{xy})$.

We introduce some more notation. As $B_{xy} \subseteq B_x$, there is an explosion map from B_{xy} to B_x; call it η^x_{xy}. Obviously, $\eta_{xy} = \eta_x\eta^x_{xy}$. As above, we have $\varepsilon_{xy}\eta^x_{xy}(\mu^J_{xy}) = \mu^J_{xy}$. Now the measure on B_{xy} which represents $M - \mu^o$ is in fact $\varepsilon_{xy}\mu_x$ (compare Corollary 4.5), so that μ^J_{xy} is the restriction of $\varepsilon_{xy}\mu_x$ to F^J_{xy}. Now it is not hard to see that the only points of F_x^J which multiplication by xy sends to F^J_{xy} are those in $\eta_{xy}(F^J_{xy})$, and from this we conclude that $\varepsilon_{xy}\mu_x$ restricted to F^J_{xy} is just the convolution product of ε_{xy} with the restriction of μ_x^J to $\eta^x_{xy}(F^J_{xy})$. Transferring this statement to F^J using η_{xy} gives that $\eta_{xy}(\mu^J_{xy})$ is the restriction of $\eta_x(\mu_x^J)$ to $\eta_{xy}(F^J_{xy})$, as required.

We have now found our measure μ^J on F^J, and the next step is to show that μ^J is a multiplier on our algebra. Recall that the latter is now $\mathcal{M}(S^o)$. Now for each x, ε_x acts as the identity operator on $\mathcal{M}_x(S^o)$, and so μ^J acts as $\mu^J\varepsilon_x$. But $\mu^J\varepsilon_x$ is carried by the upper boundary of B_x, and has norm $\|\mu^J\varepsilon_x\| \le \|\mu^J\|$ (using Theorem 3.1). From Theorem 4.3 we see that μ^J acts as a multiplier on $\mathcal{M}(S^o)$.

We now proceed inductively as before. First consider $M - (\mu^o + \mu^J)$ as a multiplier on $\mathcal{M}(S^o)$. Then treat all the (k-1) dimensional faces in turn, before passing to the (k-2) dimensional faces, and so on. After the zero dimensional case (the vertex $\mathbf{1}$), the proof is finished.

4.7 <u>REMARKS</u>. (i) There is now more point in the procedure which removes μ^J at each stage. For $x \in S^o$, and $K \subseteq J$, the product xF^J contains F_x^K, so that if μ^J were not removed, the measure $\varepsilon_x\mu^J$ could seriously modify μ_x^K.

(ii) We shall now discuss briefly how the measures on B which act as multipliers on an algebra $\mathcal{M}$ carried by S^o might be characterized. For special algebras, there might be simple ad hoc methods available; for example, if S^o is an ideal in B, any measure on B is a multiplier on $\mathcal{M}(S^o)$, and again a measure μ which is a multiplier on $\ell^1(S^o)$ will be in $\ell^1(B)$ for most semigroups S. One easy characterization is to note that μ is a multiplier if and only if $\mu\varepsilon_x$ is a multiplier for $\mathcal{M}_x$ for every x in E and to rely on Theorem 3.1 for a characterization of the measures $\mu\varepsilon_x$, but this does not seem very satisfactory.

The principle expressed in Theorem 3.1 was that a measure on a face was a multiplier if and only if it was approximately given by a measure in $\mathcal{M}$ carried by a neighbourhood of that face. In the present, more general, situation we cannot make a similar assertion for whole faces, but must proceed locally within faces. Thus, the construction followed in the proof of the last theorem does not necessarily produce measures whose supports cover the whole of F^J; they are in fact all carried by $\bigcup_{x \in E} \eta_x(F_x^J) = G^J$, say. Moreover, a point $y \in G^J$ may not be in S. (An example of a semigroup for which these statements are true can be found in Example 4.11 below.) However, for $y \in G^J$, $\{z \in S : z \ge y\}$ is a non-empty, compact subsemigroup of S, and it therefore has a set H of maximal elements. Let ν be a measure on G^J. For a neighbourhood W of y, let ν_W be its restriction to W. For $z \in H$, the measure $\nu_W\varepsilon_z$ is carried by Wz. The requirement is that $\nu_W\varepsilon_z$ should be approximable on a neighbourhood of Wz by a measure in $\mathcal{M}$. The complications involved in making this precise do not seem worthwhile.

The question arises of why we have chosen to represent M by a measure on B rather than on S itself. The answer is given by the following theorem.

4.8. <u>THEOREM</u>. It is possible to represent every multiplier on $\mathcal{M}$ by a measure on S if and only if the set of maximal elements of S is finite.

<u>PROOF</u>. If the identity multiplier is a measure on S, then $\mathcal{M}(S)$ has an identity (use (4.4) and the weak* continuity of convolution in $\mathcal{M}(S)$) so that S has a finite set of maximal elements (§2). Conversely, if $\mathcal{M}(S)$ has an identity ε, then there is a net (ν_i) in $\mathcal{M}$, bounded by $\|\varepsilon\|$, such that $\nu_i \to \varepsilon$ (weak*) (this is a simple bipolar argument - the unit ball of $\mathcal{M}$ is weak* dense in the unit ball of $\mathcal{M}(S)$). Then, if M is any multiplier, $(M\nu_i)$ is a bounded net, and so, by passing to a subnet, we may assume that $M\nu_i \to \mu$ (weak*) for some $\mu \in \mathcal{M}(S)$. It is straightforward to prove that μ realizes M.

Theorem 4.6 recalls results of Todd [14] and Sleijpen [12] in that multipliers are represented by measures on semigroups larger than the original one. However, the cases differ in that our semigroups S may not have cancellable elements (Todd) nor is B a completion of S (Sleijpen). Both Todd and Sleijpen find unique representing measures, and the correspondences between multipliers and measures is isometric. These features are not generally present in our situation, which is therefore nearer that of Lahr [7].

4.9 <u>EXAMPLE</u>. <u>The representation of a multiplier by a measure need not be unique, nor need the measure have the same norm as the multiplier.</u>

<u>CONSTRUCTION AND PROOF</u>. Take any semigroup S with a finite number of, but at least two, maximal elements. The identity multiplier is represented both by the identity ε of $\mathcal{M}(S)$ and by the measure ε_1 on B. Moreover $\|\varepsilon\| \geq 3$ (see the end of §2).

In this example, it will be noticed that there is a representing measure for the identity, namely ε_1, which does have the correct norm. Our next example will show that this is not always the case. Indeed, we shall show more generally that there may be no representing measure with the correct norm on any semigroup which contains S, and not just on B. It is a curiosity that in two dimensions positive results hold; we devote §5 to this case. In example 4.9 it is also seen that the representing measure ε_1 which has minimal norm is carried by the boundary of B (indeed, it is the measure produced by the construction of Theorem 4.6); we shall see in Lemma 5.1 that this need not be the case, even in two dimensions.

4.10 EXAMPLE. There is a semigroup S, a convolution measure algebra $\mathscr{M}$ on S^o, and a positive multiplier M on $\mathscr{M}$ such that M is represented by no positive measure on any semigroup T which contains S. Moreover, if μ represents M, $\|\mu\| > \|M\|$.

CONSTRUCTION AND PROOF. We take $I_1 = I_2 = I_3 = [0,1]$, so that B is the usual unit cube in $\mathbb{R}^3$. We write $C =]\frac{1}{2},1] \times]\frac{1}{2},1] \times]\frac{1}{2},1]$ and put $S = B \setminus C$. The maximal elements of S are

$$a_1 = (\tfrac{1}{2},1,1),\ a_2 = (1,\tfrac{1}{2},1),\ a_3 = (1,1,\tfrac{1}{2})$$

so that $S = B_{a_1} \cup B_{a_2} \cup B_{a_3}$.

We shall define a measure λ which is carried by the "inner edges" of C; these are the lines

$$L_1 = \{(t,\tfrac{1}{2},\tfrac{1}{2}) : \tfrac{1}{2} < t \le 1\},\ L_2 = \{(\tfrac{1}{2},t,\tfrac{1}{2}) : \tfrac{1}{2} < t \le 1\},\ L_3 = \{(\tfrac{1}{2},\tfrac{1}{2},t) : \tfrac{1}{2} < t \le 1\}.$$

On each L_i we put the measure λ_i which is just linear Lebesgue measure; thus $\lambda_i \ge 0$ has total mass $\frac{1}{2}$. At the point $P = (\frac{1}{2},\frac{1}{2},\frac{1}{2})$ we put a point mass of size $-\frac{1}{2}$. We write

$$\lambda = \lambda_1 + \lambda_2 + \lambda_3 - \tfrac{1}{2}\varepsilon_P.$$

The multiplier M is to be convolution with λ on a suitable algebra $\mathscr{M}$. In view of Theorem 4.4 we might as well take $\mathscr{M} = \mathscr{M}(S^o)$ (though we could choose certai: smaller algebras instead). To discover how M acts, we take our cue from Theorem 4.3, and consider the semigroups $S_{a_i} = B_{a_i}$. If ν is carried by $B^o_{a_i}$, then

$$M\nu = \lambda\nu = (\lambda\varepsilon_{a_i})\nu$$

as ε_{a_i} is the identity on $\mathscr{M}(B^o_{a_i})$. Now convolution with ε_{a_i} just projects outside B_{a_i} onto the faces of B_{a_i}. If we take i = 1 to fix ideas, we see that L_1 is projected to $(\frac{1}{2},\frac{1}{2},\frac{1}{2})$, so that λ_1 is projected to $\frac{1}{2}\varepsilon_P$, and that λ_2, λ_3 and ε_P are left alone, as they are carried by B_{a_1}. Hence

$$\lambda\varepsilon_{a_1} = \tfrac{1}{2}\varepsilon_P + \lambda_2 + \lambda_3 - \tfrac{1}{2}\varepsilon_P = \lambda_2 + \lambda_3.$$

Thus M_{a_1} (the restriction of M to $\mathscr{M}(S^o_{a_1})$) is positive and $\|M_{a_1}\| = \|\lambda_2 + \lambda_3\| = 1$.

From Theorem 4.3 we see that $\|M\| = 1$, and by the methods of its proof we see that M is positive.

For later use we observe that, if O is the minimal element of S, and so the zero of the semigroup (and also the minimal element of B - see §2), then

$$\|M\varepsilon_0\| = \|\lambda\varepsilon_0\| = \|\lambda(1)\varepsilon_0\| = \|\varepsilon_0\| = 1 = \|M\|.$$

Now let T be a semigroup which contains S and let a measure μ on T represent M. By the support lemma (2.9), we may assume that S is an ideal in T. Assume $\mu \geq 0$.

The uniqueness part of Theorem 3.1 applied to B_{a_1} shows that $\varepsilon_{a_1}\mu = \lambda_2 + \lambda_3$. As μ is positive, we conclude that $\|\mu\| = \|\lambda_2 + \lambda_3\| = 1$, and from (2.7) that $a_1 \cdot \operatorname{supp}\mu = L_2 \cup L_3$. We write $D = \operatorname{supp}\mu$ and

$$D_i = \{x \in D : a_1 x \in L_i\} \quad (i = 2,3).$$

Thus, $D = D_2 \cup D_3$. Since ε_{a_1} maps the restriction of μ to D_i to the measure λ_i, we see that $\mu(D_2) = \mu(D_3) = \frac{1}{2}$. Also, $a_1 \cdot (D_2 \cap D_3) = \{P\}$, so that $\mu(D_2 \cap D_3) = 0$. In the same way $a_2 D = L_3 \cup L_1$, and we can write $D = C_3 \cup C_1$ with $a_2 C_i = L_i$ $(i = 3,1)$, etc.

Now suppose, if possible, that $C_3 \neq D_3$, say $C_3 \setminus D_3 \neq \emptyset$. As D_3 is relatively closed in C_3 (the continuous inverse image of a compact set), $\mu(C_3 \setminus D_3) > 0$. We conclude first that there is $x \in C_3 \setminus D_3$ such that $a_2 x \in L_3 \setminus P$, and secondly that $\mu(D_3 \setminus C_3) > 0$ (for both C_3 and D_3 have measure $\frac{1}{2}$). Thus, there is $y \in D_3 \setminus C_3$ with $a_1 y \in L_3 \setminus P$. We therefore have

$$a_1 a_2 xy = (a_2 x)\ (a_1 y) \in L_3 \setminus P$$

(for multiplication in L_3 is just min in a linearly ordered set). On the other hand, since $x \notin D_3$, $x \in D_2$ and hence $a_1 x \in L_2$, while similarly $y \in C_1$ and $a_2 y \in L_1$. Hence

$$a_1 a_2 xy = (a_1 x)\ (a_2 y) \in L_2 L_1 = P.$$

This contradiction shows that $C_3 = D_3$.

We next need to consider the third minimal element a_3, and to write $D = G_1 \cup G_2$ with $a_3 G_i = L_i$ for $i = 1,2$. Then, by what we have already proved

$$C_3 = D_3,\ D_2 = G_2,\ G_1 = C_1.$$

Now, D_3 meets C_1 in $C_1 \cap C_3$, a set of measure zero, and D_2 meets G_1 in $G_2 \cap G_1$, a set of measure zero. Thus $G_1 = C_1$ meets $D = D_2 \cup D_3$ in a set of measure zero. But G_1 is a subset of D which has measure $\frac{1}{2}$.

This contradiction shows that the assumption $\mu \geq 0$ is untenable.

To obtain the conclusion about the norm, we recall that we stated earlier in this proof that $\|M\| = \|M\varepsilon_0\|$; if μ represents M, then $\|M\varepsilon_0\| = \|\mu\varepsilon_0\| = \|\mu(1)\varepsilon_0\| = |\mu(1)|$; and $|\mu(1)| < \|\mu\|$ because μ is not positive.

4.11 EXAMPLE. There is a semigroup $T \subseteq B$ and a multiplier M' on a convolution measure algebra on T^0 which is represented by two distinct measures on the upper boundary of B.

CONSTRUCTION AND PROOF. We take B as in Example 4.10. We write

$$b_1 = (1,\tfrac{1}{2},\tfrac{1}{2}),\ b_2 = (\tfrac{1}{2},1,\tfrac{1}{2}),\ b_3 = (\tfrac{1}{2},\tfrac{1}{2},1),$$

and $T = B_{b_1} \cup B_{b_2} \cup B_{b_3}$. Then T is a cube of side $\frac{1}{2}$ placed in the bottom corner of B with three equal-sized cubes stuck to its three inside faces. Obviously $T \subseteq S$, where S is as in Example 4.10, and the line L_1 of Example 4.10 is actually part of T. The multiplier M' is to be that given by convolution by the measure λ_1 of Example 4.10 on the algebra $\mathcal{M}(T^0)$.

If we carry out the procedure of the proof of Theorem 4.6 for M' we obtain a representing measure on the upper boundary of B which is the sum of the following:

linear Lebesgue measure on the line $\{(t,1,1) : \frac{1}{2} \leq t \leq 1\}$;

point masses of $\frac{1}{2}$ at $(1,\frac{1}{2},1)$ and $(1,1,\frac{1}{2})$;

a point mass of -1 at $(1,1,1)$.

(The simplest way to see that this gives M' is to consider the convolution by ε_{b_i} onto each cube B_{b_i}.)

To obtain a second representation, we observe that λ_1 also acts by convolution as a multiplier M'' on $\mathcal{M}(S^0)$. Obviously the restriction of M'' to $\mathcal{M}(T^0)$ is M', so that a measure which represents M'' also represnts M'. We apply the procedure of Theorem 4.6 to M'', and obtain a measure which is the sum of:

linear Lebesgue measure on $\{(t,\frac{1}{2},1) : \frac{1}{2} \leq t \leq 1\}$ and $\{(t,1,\frac{1}{2}) : 0 \leq t \leq 1\}$;

minus linear Lebesgue measure on $\{(t,1,1) : \frac{1}{2} \leq t \leq 1\}$;

point masses of $\frac{1}{2}$ at $(1,\frac{1}{2},\frac{1}{2})$ and $(1,1,1)$;

point masses of $-\frac{1}{2}$ at $(1,\frac{1}{2},1)$ and $(1,1,\frac{1}{2})$.

These two representing measures are clearly distinct.

In [7], Lahr identifies (algebraically and topologically, but not isometrically) the multiplier algebra of $\ell^1(S)$, for certain semigroups S, with the quotient $\ell^1(\Omega S)/\text{Ann}\ \ell^1(S)$, where ΩS is the multiplier semigroup of S (see §2) and Ann $\ell^1(S)$ is the annihilator of $\ell^1(S)$ in $\ell^1(\Omega S)$. A parallel result can be established in our situation. Given a convolution measure algebra $\mathcal{M}$ on S^o, write

$$\mathcal{M}_B = \{\mu : \mu \text{ is carried by B and } \mu\mathcal{M} \subseteq \mathcal{M}\}$$

that is, $\mathcal{M}_B$ is just the set of measures on B which act as multipliers. Since every multiplier on $\mathcal{M}$ has at least one representation by a measure on B, algebraically we have that $\mathcal{M}_B/\text{Ann}\,\mathcal{M}$ is isomorphic to the algebra of multipliers on $\mathcal{M}$. By the closed graph theorem, this isomorphism is bicontinuous. Must it be an isometry? The negative answer is given by the following strengthening of Example 4.10.

4.12 EXAMPLE. In Example 4.10, there is $\varepsilon > 0$ such that for each μ which represents M, $\|\mu\| \geq \|M\| + \varepsilon$.

PROOF. Suppose this is not true. Then for each n we can find μ_n representing M such that $\|\mu_n\| < \|M\| + \frac{1}{n}$. Now μ_n can have no mass inside S, that is, μ_n is carried by $T \setminus S^o$. This set is compact. Thus, (μ_n) has a weak* cluster point μ carried by $T \setminus S^o$. Since $\|\mu\| \leq \|M\| + \frac{1}{n}$ for each n, we have $\|\mu\| = \|M\|$. As convolution is weak* continuous on bounded sets, μ represents M. Hence (as in the proof of Example 4.10)

$$\|\mu\| = \|M\| = \|M\varepsilon_0\| = |\mu(1)|,$$

so that μ is positive, which contradicts Example 4.10.

5. THE TWO-DIMENSIONAL CASE

Example 4.10 was three dimensional. The aim of this short section is to point out that this is necessary: in two dimensions, every positive multiplier can be represented by a positive measure.

First observe that, in two dimensions, if S has a finite set of maximal elements and we arrange these in terms of decreasing first coordinate

$$pr_1 a_1 > pr_1 a_2 > \ldots > pr_1 a_n,$$

then the second coordinates must increase:

$$\mathrm{pr}_2 a_1 < \mathrm{pr}_2 a_2 < \ldots < \mathrm{pr}_2 a_n.$$

Thus, S has a kind of "saw-tooth" edge.

We deal first with the simplest case (after that of §2) in which S has two minimal elements.

5.1 LEMMA. Let $B = I_1 \times I_2$. Let $a_1 = (1,v)$, $a_2 = (u,1)$, $S = B_{a_1} \cup B_{a_2}$. Let M be a positive multiplier on a convolution algebra $\mathcal{M}$ on S^o. Then there is a positive measure μ on B which represents M. Moreover, $\|\mu\| = \|M\|$.

PROOF. Write $b = a_1a_2 = (u,v)$. Since $\mathcal{M}$ has an approximate identity, there is a measure λ on S which represents M (Theorem 4.7). Write λ as the sum of λ_o carried by S^o, λ_1 carried by the line from b to a_1, λ_2 carried by the line from b to a_2 and a point mass $k\varepsilon_b$. The measure which represents M on the rectangle B_{a_1} is

$$\lambda\varepsilon_{a_1} = \lambda_o\varepsilon_{a_1} + \lambda_1\varepsilon_{a_1} + \lambda_2\varepsilon_{a_1} + k\varepsilon_n\varepsilon_{a_1}$$

$$= \lambda_o\varepsilon_{a_1} + \lambda_1 + (\lambda_2(1) + k)\varepsilon_b.$$

By Theorem 3.1, this is positive, so $\lambda_o\varepsilon_{a_1} \geq 0, \lambda_1 \geq 0$, and $\lambda_2(1) + k \geq 0$. Similarly, $\lambda_o\varepsilon_{a_2} \geq 0$ (whence $\lambda_o \geq 0$), $\lambda_2 \geq 0$, and $\lambda_1(1) + k \geq 0$. Thus, if $k \geq 0$, λ is already a positive measure representing M.

Suppose $k = -h \leq 0$. Then we take measures σ_i with $0 \leq \sigma_i \leq \lambda_i$ and $\sigma_i(1) = h$, for $i = 1,2$. We define μ to be the sum of

λ_o in S^o;

$\lambda_1 - \sigma_1$ on the line from a_1 to b;

$\lambda_2 - \sigma_2$ on the line from a_2 to b;

$({}^1/_h)(\sigma_1 \times \sigma_2)$ on the rectangle with corners 1, a_1,b,a_2.

This is clearly positive. Moreover, its projection onto B_{a_i} for $i = 1,2$ coincides with that of λ, so the multiplier it defines is M.

The norm equality is true more generally.

5.2 LEMMA. If a positive multiplier M is represented by a positive measure μ, then $\|M\| = \|\mu\|$.

PROOF. Let O be the zero of S. We may assume M is acting on $\mathcal{M}(S^o)$ (by 4.4) which contains ε_O. Hence

$$\|\mu\| = \mu(1) = \|\mu(1)\varepsilon_O\| = \|\mu\varepsilon_O\| = \|M\varepsilon_O\| \le \|M\|.$$

On the other hand, it is always true that $\|M\| \le \|\mu\|$.

5.3 REMARK. There is no uniqueness in Lemma 5.1. This is easily seen, since the choice of the measures σ_i is arbitrary.

5.4 COROLLARY. *The conclusion of Lemma 5.1 holds if S has a finite set of maximal elements.*

PROOF. Let the maximal elements be $a_1, a_2, \ldots, a_n$, arranged as above. Put $c_2 = \max\{a_1, a_2\}$. We apply Lemma 5.1 to the rectangle B_{c_2} with its subsemigroup $B_{a_1} \cup B_{a_2}$. We find a positive measure μ_2 on B_{c_2} which represents M on $B_{a_1} \cup B_{a_2}$; note that $\|\mu_2\| \le \|M\|$ (not equality, as we are dealing with a restriction of M). We extend M to $\mathcal{M}(B^o_{c_2})$ by defining it to be convolution by μ_2. By Theorem 4.2, M is now defined on $B_{c_2} \cup S$, and the norm of the extension remains $\|M\|$.

We now write $c_3 = \max\{c_2, a_3\}$. We apply what we have just proved to B_{c_3}. Proceeding in this way establishes the result.

5.5 THEOREM. *Let $B = I_1 \times I_2$, let S be a subsemigroup of B, and let $\mathcal{M}$ be a convolution measure algebra on S^o. Any positive multiplier M on $\mathcal{M}$ can be represented by a positive measure μ on B, and $\|\mu\| = \|M\|$.*

PROOF. Let A be any finite set of maximal elements of S. Then the restriction of M to $\bigcup_{x \in A} B_x$ can be represented by a positive measure μ_A on B, and $\|\mu_A\| \le \|M\|$, from 5.4. The bounded net (μ_A) has a weak*-cluster point in $\mathcal{M}(B)$, and now standard arguments will complete the proof.

Theorem 5.5 can be extended to assert that every multiplier can be represented by a measure of the correct norm. The technique is to write the multiplier as a linear combination of positive multipliers. When there are two maximal elements, this is done by writing the representing measure λ on S in terms of $\lambda_0, \lambda_1, \lambda_2$ and k as in the proof of Lemma 5.1, by observing from Theorem 4.3 that

$$\|\lambda\| = \max\{\|\lambda_0\| + \|\lambda_1\| + |\lambda_2(1)+k|, \|\lambda_0\| + \|\lambda_2\| + |\lambda_1(1)+k|\},$$

by decomposing $\lambda_0, \lambda_1, \lambda_2$ into a combination of positive measures, and then by writing k as a sum of other numbers in such a way that the norms of the positive multipliers obtained are as small as possible. We do not give the details.

6. HYPOTHESIS 4.1

We are able to show that hypothesis 4.1 is satisfied if we assume metrisability. This of course means that each interval is isomorphic to the unit interval.

6.1 PROPOSITION. Let $I_1, \ldots, I_k$ be unit intervals and suppose that S and $\mathcal{M}$ satisfy the conditions of §2. Then 4.1 is automatically satisfied.

PROOF. By our earlier assumptions, $\bigcup_{x \in A} S_x^o$ is dense in S. It will be enough to show that $\{x \in \text{cls } A : x \in \text{cls } S_x^o\}$ is dense in cls A.

For each positive integer n, let $B(x, 1/n)$ be the open ball with centre x and radius $1/n$, and put

$$U_n = \{x \in \text{cls } A : B(x, 1/n) \cap S_x^o = \emptyset\}.$$

Then U_n is closed, for suppose that $x_r \to x$ and $x \notin U_n$. Then there is y in $B(x, 1/n) \cap S_x^o$. But then, for sufficiently large r, y is also in $B(x_r, 1/n) \cap S_{x_r}^o$, and so $x_r \notin U_n$. Secondly U_n has empty interior relative to cls A. For let W be a neighbourhood of x in cls A. Then we can assume that $x \in A$. By the assumption on S there is a sequence (x_r) in A and $y_r \in S_{x_r}^o$ for all r such that $y_r \to x$. Since $y_r \leq x_r$ for all r, we have $z \geq \lim y_r = x$ for each cluster point z of (x_r). As $z \in S$ and x is maximal in S, $z = x$ so that $x_r \to x$. So x_r is eventually in W. But the distance from x_r to y_r must converge to zero, so $y_r \in B(x_r, 1/n)$ eventually. Since $y_r \in S_{x_r}^o$, we conclude that $x_r \notin U_n$ for sufficiently large r.

We now apply the Baire category theorem to cls A to deduce that $\{x \in \text{cls } A : x \in \text{cls } S_x^o\} = \text{cls } A \setminus \bigcup_{n=1}^{\infty} U_n$ is dense in cls A.

ACKNOWLEDGEMENT. We would like to express our gratitude to the referee whose careful reading of the paper has saved us from several errors.

REFERENCES

[1] Baartz, A.P., The measure algebra of a locally compact semigroup, Pacific J.Math., 21(1967) 199-214.

[2] Berglund J.F., and K.H. Hofmann, Compact semitopological semigroups and weakly almost periodic functions, Lecture Notes in Mathematics 42, Springer, Berlin, 1967.

[3] Dhar R.K., and H.L. Vasudeva, Characterisations of multipliers of $L_1(R)$, to appear.

[4] Hewitt E., and H.S. Zuckerman, The ℓ_1-algebra of a commutative semigroup, Trans.Amer.Math.Soc. 83(1956) 70-97.

[5] Johnson D.L., and C.D. Lahr, Multipliers of L^1algebras with order convolution, Publ.Math.Debrecen, 28(1981) 153-161.

[6] Lahr C.D., Multipliers for certain convolution measure algebras, Trans.Amer.Math.Soc., 185(1973) 165-181.

[7] Lahr C.D., Multipliers for ℓ_1-algebras with approximate identities, Proc.Amer.Math.Soc. 42(1974) 501-506.

[8] Lardy L.J., On the identity in a measure algebra, Proc.Amer.Math.Soc. 19(1968) 807-810.

[9] Larsen R., An introduction to the theory of multipliers, Springer, Berlin, 1971.

[10] Larsen R., The multipliers of $L_1([0,1])$ with order convolution, Publ.Math.Debrecen, 23(1976)239-248.

[11] Newman S.E., Measure algebras on idempotent semigroups, Pacific J.Math. 31 (1969) 161-169.

[12] Sleijpen G.L.G., L-Multipliers for foundation semigroups with identity element, Proc.London Math.Soc., (3) 39(1979) 299-330.

[13] Taylor J.L., The structure of convolution measure algebras, Trans.Amer. Math.Soc. 119(1965) 150-166.

[14] Todd D.G., Multipliers of certain convolution algebras over locally compact semigroups, Math.Proc.Cambridge Phil.Soc. 87(1980) 51-59.

[15] Wendel J.G., Left centralizers and isomorphisms of group algebras, Pacific J.Math. 2(1952) 251-261.

[16] Gierz, G., et al, A Compendium of Continuous Lattices, Springer-Verlag, Berlin, Heidelberg, New York (1980), 371 pp.

[17] Lawson, J.D., J.R. Liukkonen, and M. Mislove, Measure algebras of semilattices with finite breadth,Pacific J. Math., 69 (1977), 125-139.

J.W. BAKER, Department of Pure Mathematics, The University, Sheffield S3 7RH, England.

J.S. PYM, Department of Pure Mathematics, The University, Sheffield S3 7RH, England.

H.L. VASUDEVA, Department of Mathematics, Panjab University, Chandigarh, India 160014.

INNER POINTS IN CERTAIN HOLOIDAL SEMIGROUPS

D.R. Brown and J.W. Stepp

1. Introduction. A semigroup is uniquely divisible if each element has a unique *nth* root for every positive integer n. Compact, uniquely divisible topological semigroups, particularly commutative ones, have attracted the attention of several researchers, and a modest bibliography has resulted. For papers prior to 1970, see the listing in [4]; others include [5], [7], [11], and [14]. The most illustrious subclass of the UDC semigroups, topological semilattices, has, of course, a much more extensive library when it is considered on its own merits. No attempt will be made here to document this; one may consult [9] to obtain an idea of the tremendous scope of results involving semilattices theory that have occurred during the past decade.

Among compact semilattices, a particularly well behaved family is comprised of those semilattices in which, for each $x \in S$, $M(x) = \{y: x \leq y\}$ is a connected subset of S. These objects, known also as chainwise connected semilattices, had their genesis in [2]. They are known to be dimensionally stable [17] (definitions follow in the next section), and to satisfy the inequalities codimension(S) $\leq$ breadth(S) $\leq$ codim(S) + 1 [18, 19], where the first inequality becomes equality when S has an identity. Recently, [6], the authors showed that, when the second inequality becomes equality, then every element of maximal breadth must be an inner point of S (in the cohomological sense). The question thus arises: what family of compact, UDC semigroups is the "right" subclass to represent an extension of the chainwise connected semilattices, in the sense that

1. it includes the class of chainwise conneceted semilattices; and
2. a large measure of the theory above, with necessary interpretations, continues to hold in this class?

In this paper, we show that one such class is the collection of compact UDC semigroups S such that 1. $M(e) = \{x: e \in xS\}$ is connected for each idempotent $e \in S$; and 2. S is holoidal.

A standard example of a chainwise connected semilattice without identity is furnished by $\{(x_1,\ldots,x_n) \in [0,1]^n: x_1x_2\ldots x_n = 0\}$, the "lower" faces of the unit cube, where multiplication is "min" in each coordinate. The same set supplies motivation for our extension, if multiplication is regarded as the usual real product in each coordinate. Variety is provided by replacing $[0,1]$ in any coordinates by an arbitrary UDC semigroup on an arc.

<u>2. Preliminaries</u>. Following the usual convention, the letter E will be used exclusively to represent the set of idempotents of a semigroup; it will be subscripted by the name of the semigroup only when the possibility of confusion arises. A semigroup is holoidal if all of its maximal subgroups are singletons. In this case, if the semigroup S is also compact and commutative, then the relation $a \leq b$ if and only if $a \in bS$ is antisymmetric and transitive; if also $S = ES$, then this relation becomes a partial order with closed graph [12]. Let

$L(x) = \{y: y \le x\}$, $M(x) = \{y: x \le y\}$, $L(A) = \cup\{L(x): x \in A\}$, $M(A) = \cup\{M(x): x \in A\}$. If A is a closed subset of S, and S is compact, then $L(A)$ and $M(A)$ are also closed. If A is an open set, then $M(A)$ is open, although $L(A)$ need not be. However, in one important case, $L(A)$ is open (see lemma 3.3). Clearly $L(A) = AS$, but no such algebraic characterization exists for $M(A)$. With the hypotheses given above on S, E is a compact semilattice, and $(\le)$ restricted to E is precisely the usual semilattice order.

Uniquely divisible, commutative semigroups will be abbreviated UDC semigroups. If S is a compact, holoidal UDC semigroup, and $x \in S$, then the net $\{x^{1/n}\}$ converges to an idempotent e such that $ex = x$; indeed, e is the minimal idempotent satisfying $x \le e$ [13]. In a similar manner, the maximal idempotent f below x is the limit of the net $\{x^n\}$. If x is not an idempotent, then $\{x^{p/q}$: p/q positive rational$\}$ is a subsemigroup of S; its closure is known to be topologically isomorphic to the interval $[0,1]$ with usual real multiplication [10], in such a way that 1 corresponds to e above and 0 to f. This fact permits positive irrational exponents to have meaning; if x^0 is defined to be e, then x^z makes sense for all non-negative z. The function $(x,z) \to x^z$ is known to be jointly continuous [3].

The additional condition in use in this paper, that $M(e)$ is connected for every idempotent e, assures the connectivity of S. Moreover, if e, $f \in E$, $f \le e$, then there exists a "thread" between f and e; that is, a semigroup defined on a topological arc with identity e and zero f. Since uniqueness of roots prevents the existence of nilpotent elements in threads, we eschew this notation for the less colorful but more descriptive expression "arc UDC subsemigroup". It is easy to see that, in compact holoidal UDC semigroups in which $M(e)$ is connected, $e \in E$, we have $x \leq y$ if and only if there exists an arc UDC subsemigroup T such that $x \in yT$. It should be noted that, from the remarks above, the condition $S = ES$ is automatically satisfied by compact UDC semigroups.

The cohomology theory of Alexander-Spanier-Wallace [21] is used throughout. In particular, the strong excision theorem from this theory establishes isomorphism between the groups of the compact pair $(A \cup B, A)$ and $(B, A \cap B)$. We will need the generalized homotopy theorem, which permits any connected set in place of the unit interval. The Vietoris-Begle mapping theorem will be used in the following sense: if $f: X \to Y$ is a closed map such that $f^{-1}(y)$ is acyclic for all y, then f^* is an isomorphism in all dimensions. A space is acyclic if all of its (reduced) groups are trivial. A space has codimension $\leq n$ provided $i^*: H^n(A) \to H^n(X)$ is surjective for every closed subset A of X, where i is the inclusion of A into X. Thus, an acyclic space has codimension $\leq n$ if its closed subsets all have trivial n-cohomology. A point x is an inner point if there exists an open set U containing x such that, for every open set V containing x, with $V \subseteq U$, there exists an interger p such that $i^*: H^p(S, S\backslash V) \to H^p(S, S\backslash U)$ is not the zero homomorphism [20]. We will make use of Wallace's acyclicity theorem sufficiently often to warrant its restatement here.

<u>Wallace's Acyclicity Theorem [24]</u>: Let X and Y be compact Hausdorff spaces, and let M be a set to set function which assigns to each closed

set A in X a closed connected set $M(A)$ in Y and with the properties:

a. $M(A_1 \cup A_2) = M(A_1) \cup M(A_2)$, and there exists A_3 such that $M(A_1) \cap M(A_2) = M(A_3)$;

b. if $x \in X$ and if W is an open set about $M(x)$ then there is an open set U about x with $M(\overline{U}) \subseteq W$;

c. $H^p(M(x)) = 0$, all $p \geq 0$, all $x \in X$.

Then $H^p(M(A)) = 0$, all $p \geq 0$, all closed sets $A \subseteq X$.

One of the principal corollaries of this theorem is that, if S is a commutative continuum semigroup with 0 and $S = ES$, then S is acyclic [25].

The breadth of a semilattice S, denoted $Br(S)$, is the smallest positive integer n such that any finite subset F of S has a subset F_1 of at most n elements such that $\inf(F_1) = \inf(F)$. Points of breadth n are those points possessing an irreducible representation as a product of n elements.

A semigroup S of finite codimension is dimensionally stable in a class C of semigroups if every homomorphic image of S which belongs to class C has codimension less than or equal to that of S.

The following lemmas are all proved by straightforward applications of the exact sequence theorem for pairs and the Mayer-Vietoris exact sequence. Proofs of all four may be found in [6].

Lemma 2.1. Let X be an acyclic space. Then $x \in X$ is an inner point of X if and only if there exists an open set U containing x such that for each open set V containing x, $V \subseteq U$, the homomorphism induced by the inclusion map $j: X\backslash U \to X\backslash V$ is nonzero in some dimension.

Lemma 2.2. Let A be a closed subset of X, a compact Hausdorff space. Let $\operatorname{codim}(X) = n$, $H^n(X) = 0 = H^{n-1}(A)$. If is an inner point in dimension $n - 1$ relative to A, then x is an inner point relative to X in the same dimension.

Lemma 2.3. Let $\{A_i\}$, $i = 1,\ldots,n$ be a collection of acyclic sets such that $\cap\{A_j: j \in F\}$ is acyclic for every nonempty $F \subseteq \{1,\ldots,n\}$. Then $\cup\{A_i: i = 1,\ldots,n\}$ is acyclic.

Lemma 2.4. Let $\{A_i\}$ be a collection of distinct compact acyclic sets such that $\cap\{A_j: j \in F\}$ is acyclic and nonempty for each F properly contained in $\{1,\ldots,n + 1\}$, but $\cap\{A_i: i = 1,\ldots,n + 1\} = \square$. Then $H^{n-1}(A_1 \cup \ldots \cup A_{n+1}) \neq 0$.

3. Breadth and Codimension. Throughout the balance of this paper, the statement "S is a UDC semigroup" will be understood to include the assumption that S is compact.

Lemma 3.1. Let S be a holoidal UDC semigroup in which $M(e)$ is connected for each $e \in E$. Let $f: S \to T$ be a continuous homomorphism, where T is also a UDC semigroup. Then, for any finite subset $E_1 \subseteq E$, $f^{-1}(t) \cap E_1S$ is acyclic for each $t \in T$.

Proof. It suffices to consider E_1S as the domain of f. We argue inductively. If $|E| = 1$, then S has an identity. Fix $t \in T$, let $E(t) = \{e \in E_S: f(e)t = t\}$. Let e_0 be the zero of $E(t)$. Clearly, the generalized homotopy theorem yields that the cohomology groups of

$f^{-1}(t)$ are isomorphic to those of $e_0f^{-1}(t)$. We complete the argument for this case by showing that the latter set is contractible within itself. Fix $a \in e_0f^{-1}(t)$; for $x \in e_0f^{-1}(t)$ and $z \in [0,1]$, let $g(x,z) = x^z a^{1-z}$. Then g is jointly continuous, [3], $g(x,1) = x$, and $g(x,0) = a$. Moreover, because T is uniquely divisible, the range of g is contained in $e_0f^{-1}(t)$. Hence $e_0f^{-1}(t)$ and $f^{-1}(t)$ are acyclic.

Assume the inductive hypothesis and let $E_1 = \{e_1,\ldots,e_n\}$. Let $f_i^{-1}(t) = e_if^{-1}(t)$, and decompose $f^{-1}(t)$ as $\cup\{f_i^{-1}(t)\colon i = 1,\ldots,n-1\} \cup f_n^{-1}(t)$. By direct calculation, $f_i^{-1}(t) \cap f_n^{-1}(t) = e_ie_nf^{-1}(t)$. Since $\{e_ie_n\colon i = 1,\ldots,n-1\}$ consists of at most $n - 1$ idempotents, the inductive hypothesis gives $\cup\{f_i^{-1}(t) \cap f_n^{-1}(t)\colon i = 1,\ldots,n-1\}$ the acyclic property. The Mayer-Vietoris exact sequence proves the acyclicity of $f^{-1}(t)$, using the decomposition mentioned above.

If T is not required to be uniquely divisible, then g defined in the proof above may not stay within $f^{-1}(t)$. Examples similar to those constructed in [8] show that, in this case, the map f may not be acyclic.

Also, it is conjectural that f is an acyclic map without the additional requirement that E_1 be finite. For example, if $f^{-1}(t)$ is locally connected metric and finite dimensional in the covering sense, then the map $(x,y) \to (x^{\frac{1}{2}}y^{\frac{1}{2}}, x^{\frac{1}{2}}y^{\frac{1}{2}})$ is a symmetric retaction of $f^{-1}(t) \times f^{-1}(t)$ onto its diagonal; by an unpublished, but well known, theorem of J.L. Kelley, this suffices to make $f^{-1}(t)$ an absolute retract, hence acyclic.

Corollary 3.2. Let S be a holoidal UCD semigroup in which $M(e)$ is connected for each $e \in E$. Suppose also that S has finitely many maximal idempotents. Then S is dimensionally stable in the class of UDC semigroups.

Proof. Let A be a closed subset of $f(S)$, f_1 the restriction of f to $f^{-1}(A)$. By the Vietoris-Begle mapping theorem, both f and f_1 induce isomorphisms in all cohomology dimensions. Since the homomorphism induced by injecting $f^{-1}(A)$ into S is surjective at the level of the codimension of S, it follows that the homomorphism induced by injecting A into T is also onto in this dimension.

Lemma 3.3. Let S_i, $i = 1,\ldots,n$ be a family of not necessarily metric arcs, and let S be their cartesian product. Representing S_i as $[z_i,e_i]$, let $B = \{(x_i) \in S\colon x_j = z_j$ for some $j\}$, $T_j = \{(x_i) \in S\colon x_j = e_j\}$, $T = \cup\{T_j\colon j = 1,\ldots,n\}$. Let X be a proper closed subset of S, $T \cup B \subseteq X$. Then $H^{n-1}(X) \neq 0$.

Proof. If each S_i is metric, this result dates back to A.D. Wallace, [23]. Note $T_1,\ldots,T_n$, B satisfy the hypothesis of Lemma 2.4, so that $H^{n-1}(T \cup B) \neq 0$. Choose $(x_i) \in S\backslash X$, and, for each i, let $[a_i,b_i]$ be an interval neighborhood of x_i such that $\Pi\{[a_i,b_i]\colon i = 1,\ldots,n\} \subseteq S\backslash X$. If f_i is the evident retraction of $[z_i,e_i]$ onto $[a_i,b_i]$, then $f = \Pi f_i$ satisfies $f^{-1}(t)$ is acyclic for each $t \in \Pi[a_i,b_i]$. Let F be the boundary of $\Pi[a_i,b_i]$; observe that g, the restriction of f to $T \cup B$, is onto F and also satisfies $g^{-1}(t)$ is acyclic for each $t \in F$. By application of the Vietoris-Begle theorem, $H^{n-1}(T \cup B) \cong H^{n-1}(F) \cong H^{n-1}(f^{-1}(F))$. Finally $T \cup B \subseteq X \subseteq f^{-1}(F)$ and $g = hk$, where k is the inclusion of $T \cup B$ into $f^{-1}(F)$ and h is f restricted to $f^{-1}(F)$. It now follows that the homomorphism induced by the inclusion of $T \cup B$ into X is surjective, which completes the proof.

If, in addition to being an arc, each S_i in the lemma above is also a UDC semigroup, then so, of course, is S. Since the structure of such semigroups is completely understood, [12], it is clear that the H-order on each S_i is the same as the order induced on S_i by considering it as a topological lattice in the cut point order topology. More importantly, this equivalence of orders continues to hold in S. Thus all results connecting topology and order in S are valid, regardless of the actual multiplication in S. In particular, if U is an open set in S, then $L(U)$ and $M(U)$ are both also open [1]. Note also that, if n is the index cardinal in the product construction of S, then any product of more than n elements in S being equal to 0 implies that all but n of these elements may be discarded without losing 0 as product. This follows from the fact that in a UDC semigroup on an arc there can be no zero divisors. These remarks highlight the princiapl alterations necessary to adapt the proof of Lemma 3.2 of [18] to the following lemma. We include a proof in the spirit of self-containment.

Lemma 3.4. Let R be a holoidal UDC semigroup, S the cartesian product of n UDC semigroups S_i, each defined on an arc. Let h be a continuous surjective homomorphism of S onto R. If $h^{-1}(h(0)) = 0$, then $\operatorname{codim}(R) > n - 1$.

Proof. Assume $\operatorname{codim}(R) \le n - 1$. Let B and T be as described in 3.3, and decompose B as $\cup B_j$, where $B_j = \{(x_i): x_j = z_j\}$. Note B_j is isomorphic to $\Pi\{S_i: i \neq j\}$. Set $A = h^{-1}h(T \cup B)$; by lemma 3.1,

$h^{-1}(r)$ is acyclic for each $r \in R$, hence $H^{n-1}(A)$ is isomorphic to $H^{n-1}(h(T \cup B))$. Since R is acyclic and $\operatorname{codim}(R) \leq n - 1$, we have $H^{n-1}(h(T \cup B)) = 0$, whence $H^{n-1}(A) = 0$, also. Since $T \cup B \subseteq A$, by lemma 3.3 we have $A = S$, so that $h(T \cup B) = R$. By hypothesis, $0 = (z_i) \notin T$ forces $h(0) \notin h(T)$, whence $0 \in U = h^{-1}(R \backslash h(T))$ and $h(U) \subseteq h(B)$. Let $C_j = h^{-1}h(B_j) \cap U$; by the Baire theorem, some C_k has non-empty interior. From earlier remarks, $L(C_k^o)$ is open and contains 0. Since $h^{-1}h(B_k)$ is an ideal, $h^{-1}h(B_k) = L(h^{-1}h(B_k))$. Thus we choose an element (a_i) in $h^{-1}h(B_k)$ such that $a_i \neq z_i$ for $i = 1,\ldots,n$. Let $x(j)$ be the element in $h^{-1}h(B_k)$ such that $x(j)_i = a_i$, $i \neq j$, $x(j)_j = z_j$, $j = 1,\ldots,n$. Choose $y(j)$ in B_k such that $h(y(j)) = h(x(j))$, $j = 1,\ldots n$. We now have $0 = (z_i) = x(1)x(2)\ldots x(n)$, whence $y(1)y(2)\ldots y(n) \in h^{-1}h(0) = 0$. Invoking an earlier remark, at least one $y(j)$ may be discarded from this product, say $y(2)\ldots y(n) = 0$. Then $x(2)\ldots x(n) \in h^{-1}h(0)$, and this product is not equal to 0, since its first coordinate is $a_1^{n-1} \neq z_1$. This contradiction establishes the lemma.

<u>Theorem 3.5.</u> Let S be a holoidal UDC semigroup of codimension n. Assume $M(e)$ is connected for each $e \in E$. Then $\operatorname{breadth}(E) \leq n + 1$, and, if S has an identity, then $\operatorname{breadth}(E) \leq n$.

<u>Proof.</u> We again borrow liberally from the comparable theorem in [18]. Suppose first that S has an identity, 1. Let $z = c_1c_2\ldots c_m$ be an element of maximal breadth in E, where this product is an irreducible representation of z. Since $M(c_i)$ is connected for each c_i, there exist arc UDC subsemigroups S_i with zero c_i and identity 1 [15].

Let $W = \Pi\{S_i : i = 1,\ldots,m\}$, and $m: W \to S$ be defined by $m((x_i)) = x_1x_2\ldots x_m$. Since $m^{-1}(z)$ is a closed subset of W, we may assume $(c_1,c_2,\ldots,c_m)$ is maximal in this set, rechoosing the c_i's in the representation of z, if necessary. If W is assigned coordinatewise operations, then m is a continuous homomorphism, and $m^{-1}m(0) = 0$. By application of lemma 3.4, $\operatorname{codim}(m(W)) > m - 1$. Hence $n = \operatorname{codim}(S) \geq \operatorname{codim}(m(W)) > m - 1$, so that $m \leq n$ as claimed. Abandoning the condition that S have an identity, again let $z = c_1c_2\ldots c_m$ be an element of maximal breadth. For $i = 1,\ldots,m - 1$, let $d_i = c_ic_m$; then $\{d_1,\ldots,d_{m-1}\}$ is a set of $m - 1$ distinct (by the irreducible representation of z idempotents in the holoidal UDC semigroup c_mS with identity c_m. By the first half of the proof, $m - 1 \leq n$, or, equivalently, $m \leq n + 1$.

Since a holoidal UDC semigroup of arbitrary finite codimension may have as few as two idempotents, an identity and a zero, it is impossible to bound the breadth of E below by any function of the codimension of S, in contrast with the chainwise connected semilattice case.

4. Breadth and Inner Points. Throughout this section, S will be a holoidal UDC semigroup in which $M(e)$ is connected for each $e \in E$. We continue to borrow liberally, this time from our own work, [6].

Lemma 4.1. If S has an identity, then $M(x)$ is acyclic for every $x \in S$.

Proof. If x is an idempotent, then $M(x)$ is a continuum semigroup with zero, commuting idempotents, and the property that every element has an idempotent identity; by Wallace's acyclicity theorem, $M(x)$ is acyclic in this case [25]. If x is not an idempotent, let e be the identity and $M(e)$ is connected, the action of $M(e)$ on $M(x)$ by translation maps back into $M(x)$ and generates a (generalized) homotopy, so that the cohomology groups of $M(x)$ are isomorphic to those of $M(x) \cap eS$. The latter set contracts to e under the mapping $g(t,z) = t^z$, where $z \in [0,1]$, [3]. Hence $M(x)$ is again acyclic.

Lemma 4.2. If S has an identity, and V is an open ideal in S, then $S \backslash V$ is acyclic.

Proof. Clearly, $M(S \backslash V) = S \backslash V$. It suffices to show that the M relation satisfies the hypothesis of Wallace's Acyclicity Theorem. Since $M(x)$ is connected for each x (lemma 4.1), and S has an identity, $M(A)$ is connected for every closed $A \subseteq S \backslash V$. The partial order properties of the H-relation guarantee $M(A_1 \cup A_2) = M(A_1) \cup M(A_2)$ and $M(A_1) \cap M(A_2) = M(M(A_1) \cap M(A_2))$. Moreover, $M(A_1) \cap M(A_2)$ is a closed set, by the remarks in section 2. Finally, let $x \in S \backslash V$, and let W be an open set containing $M(x)$; we may assume W is open in S. By standard compactness arguments, using the continuity of multiplication, there exists an open set U_1 containing $M(x)$ such that $S(S \backslash W) \subseteq S \backslash U_1$. For any $t \in S \backslash W$, $St \subseteq S \backslash U_1$, so that $t \notin M(U_1)$. It follows that $M(U_1) \subseteq W$. Since $M(x)$ is closed and S is normal, there exists an open set $U \supseteq M(x)$, $\overline{U} \subseteq U_1$. By lemma 4.1, $M(x)$ is acyclic for each $x \in S \backslash V$, so that $M(A)$ is acyclic for each closed set A in $S \backslash V$. In particular, $M(S \backslash V) = S \backslash V$ is acyclic.

Lemma 4.3. Suppose $\operatorname{codim}(S) = n$, and z is an element of breadth $n + 1$, with irreducible representation $z = x_1 x_2 \ldots x_{n+1}$; $z, x_i \in E$. Let

$y_i = x_1x_2\ldots\hat{x}_i\ldots x_{n+1}$, where $\hat{x}$ means "delete x". Then, for any proper subset $I \subset \{1,\ldots,n+1\}$, $\cap\{M(y_i): i \in I\} \neq \square$, whereas $\cap\{M(y_i): i = 1,\ldots,n+1\} = \square$.

Proof. If $j \notin I$, then $x_j \in \cap\{M(y_i): i \neq j\} \subseteq \{M(y_i): i \in I\}$. Now suppose $a \in \cap\{M(y_i): i = 1,\ldots,n+1\}$. Let $\{a^{1/n}\}$ converge to $e \in E$, and let $b_i = ex_i$, $i = 1,\ldots,n+1$. By the irreducible representation of z, the b_i's are distinct elements, and $z = ez = b_1b_2\ldots b_{n+1}$. Since eS has an identity, breadth$(eE) = n$ (theorem 3.5). Hence we may discard one b_i from this representation of z, say $z = b_1b_2\ldots b_n$. Then $z = e(x_1\ldots x_n) = ey_{n+1} = y_{n+1} = x_1\ldots x_n$, in contradiction of the assumption that z is an element of breadth $n + 1$.

Lemma 4.4. With all notation as in the previous lemma, let Γ be a non-empty subset of $\{2,\ldots,n+1\}$. Let $w = \Pi\{x_j: j \notin \Gamma, j \geq 2\}$. Then $\cap\{x_1M(y_i): i \in \Gamma\} \cap x_1wM(y_1)$ is connected.

Proof. This set is a compact subsemigroup; let f represent its zero. Note also that x_1w is an identity for this semigroup. It suffices to produce a connected subset of this semigroup containing x_1w and f. Let $N = \{t \in wM(y_1): x_1t = f\}$. Note that N is a closed subsemigroup of $wM(y_1)$; let f_1 be its zero. Since $wM(y_1)$ is a holoidal UDC subsemigroup of S, there exists an arc UDC subsemigroup D of $wM(y_1)$ having w as its identity and f_1 as its zero. It follows that x_1D is an arc UDC subsemigroup of $x_1wM(y_1)$ with identity x_1w and zero $x_1f_1 = f$. Moreover, since $M(f)$ is an increasing set, x_1D is contained in $M(f)$, also.

Theorem 4.5. With all notation as in lemma 4.3, the element z is an inner point of S.

<u>Proof</u>. The procedure is parallel to that used in [6]. By lemma 2.2, it suffices to prove that z is inner in a smaller semigroup. Fix i between 1 and n, and let C_{ij} be any arc UDC semigroup between the idempotents x_i and $x_i x_j$; let $A_i = C_{i1}C_{i2}\dots C_{in}$, where C_{ii} may be regarded as $\{x_i\}$. Note A_i contains all finite products of the x_j's which contain x_i as a factor. The subsemigroup in which we work is the one generated by $\{A_1,\dots,A_n\}$. It is easily verified that this subsemigroup is compact, uniquely divisible, and has $M(e)$ connected for each idempotent e within it. Since it contains all x_i's, it contains z (indeed, as its zero) and hence has breadth $n + 1$. By theorem 3.5, it therefore has codimension n; for the sake of simplicity, we name it S and perform all further calculations within it. Recall $y_i = x_1x_2\dots\hat{x}_i\dots x_{n+1}$, let $B = \cup\{M(y_i): i = 1,\dots,n + 1\}$. By lemmas 2.4 and 4.3, $H^{n-1}(B) \neq 0$. Let $U = S\backslash B$; we show U satisfies the hypothesis of lemma 2.1. Let W be an open set containing z, $W \subseteq U$. Let V be the union of all ideals contained in W; V is known to be open [16] and is certainly an ideal. It suffices to show that $j^*:H^{n-1}(S\backslash V) \to H^{n-1}(S\backslash U)$ is not trivial. For ease of notation, let $e = x_1$ henceforth. Let $\lambda_e:S \to eS$ by $\lambda_e(t) = et$, $\lambda_1 = \lambda_e|S\backslash V$ and $\lambda = \lambda_e|B$. The following diagram is independent of path:

$$\begin{array}{ccc} H^{n-1}(e(S\backslash V)) & \xrightarrow{\lambda_1^*} & H^{n-1}(S\backslash V) \\ \Big\downarrow j_1^* & & \Big\downarrow j^* \\ H^{n-1}(eB) & \xrightarrow{\lambda^*} & H^{n-1}(B) \end{array}$$

In order to have $j^* \neq 0$, it suffices to show that λ^* is an isomorphism and $j_1^* \neq 0$. We use the Vietoris-Begle theorem again to accomplish the former. Fix $x \in eB$; we show $\lambda^{-1}(x)$ is an acyclic subset of B. Clearly $\lambda^{-1}(x) \cap M(y_i) = \lambda_e^{-1}(x) \cap M(y_i)$. Let $\Lambda \subseteq \{2,\ldots,n+1\}$ be defined by $i \in \Lambda$ if and only if $\lambda^{-1}(x) \cap M(y_i) \neq \Box$. Thus $\lambda^{-1}(x) = [\lambda^{-1}(x) \cap M(y_1)] \cup [\cup \{\lambda^{-1}(x) \cap M(y_i): i \in \Lambda\}]$.

The second bracketed set is actually equal to $\lambda^{-1}(x) \cap (\cap \{M(y_j): j \in \Lambda\})$, for, suppose $t \in \lambda^{-1}(x) \cap M(y_i)$ for some $i \in \Lambda$. For any $j \in \Lambda$, let $s \in \lambda^{-1}(x) \cap M(y_j)$; then $et = x = es$. Since $e = x_1$ and $j \neq 1$, $e \in M(y_j)$ also, so that $x = ex \in M(y_j)$. Thus $t \in M(x) \subseteq M(y_j)$, as claimed. An argument in the same vein shows that, if the second bracketed set is non-empty, then it contains the first bracketed set. In any case, $\lambda^{-1}(x)$ is contained in either $M(y_1)$ or in $\cap\{M(y_j): j \in \Lambda\}$, both of which are UDC subsemigroups satisfying the hypothesis of lemma 3.1. Regarding λ as a map from this semigroup into eS, we have that $\lambda^{-1}(x)$ is acyclic. By the Vietoris-Begle theorem, λ^* is an isomorphism.

The set $e(S\backslash V) = [(S\backslash V) \cap eS] \cup eB$. This is a consequence of our construction of S. Let $et \in e(S\backslash V)$, $t \in S\backslash V$. It $t \in eS$, then $et = t$; otherwise, we may write $t = t_1 \ldots t_m$, where each t_k belongs to an arc UDC semigroup C_{ij} with identity x_i, $i \neq 1$, and zero $x_i x_j$. if $x_j = e$, then $et_k = ex_i \in eM(y_1)$; if $x_j \neq e$, then $y_1 \leq x_i x_j \leq t_k$, so that again $et_k \in eM(y_1)$. Thus $et = (et_1)\ldots(et_m) \in eM(y_1) \subseteq eB$. The reverse inclusion is immediate.

The morphism j_1^* is proved surjective by showing that $H^n(e(S\backslash V), eB) = 0$ and appealing to the exact sequence theorem for pairs. By the Stong Excision theorem, using the decomposition of $e(S\backslash V)$ derived in the previous paragraph, it suffices to show $H^n((S\backslash V) \cap eS, (S\backslash V) \cap eB) = 0$; this is accomplished by showing each factor is acyclic, and once more invoking the exactness of pairs. The first factor is acyclic by direct application of lemma 4.2. More care must be used in the treatment of the second factor. If $n = 1$, then $(S\setminus V) \cap eB = (S\setminus V) \cap e \cdot \{e, x_2\} = \{e\}$. Thus we may assume that $n \geq 2$. For $2 \leq j \leq n + 1$, define:

$$S_j = \cup\{M(y_k) \cap eS: \; k \neq 1, j\},$$

$$T_j = (S\setminus V) \cap ex_j M(y_1),$$

$$J_j = S_j \cup T_j$$

By direct argument, $(S\setminus V) \cap eB = \cup\{J_j: 2 \leq j \leq n + 1\}$. We show that the family of J_j's satisfy the hypothesis of lemma 2.3.

Each set in the union comprising S_j is a connected UDC subsemigroup of eS, hence acyclic. Any finite intersection of these sets is of the form $M(f) \cap eS$, and is thus also acyclic. (Here, f is the zero of the subsemigroup determined by the intersection.). By lemma 2.3, it follows that S_j is acyclic.

On the other hand, T_j is acyclic by lemma 4.2. Also, $S_j \cap T_j = \bigcup_{k \neq i, j} \{M(y_k) \cap ex_j M(y_1)\}$, since $S_j \subseteq S\setminus V$. Each set in this union is a UDC semigroup with identity ex_j and zero element y_k. By lemma 4.4, such a subsemigroup is connected and hence is acyclic. A finite intersection of sets in this union is of the form $M(f) \cap ex_j M(y_1)$ for some idempotent f, $ex_j \in M(f)$. Again by lemma 4.4, such a subsemigroup is connected and therefore acyclic. Thus, by lemma 2.3, each J_j is acyclic.

Next, for any non-empty subset Λ of $\{2, \ldots, n + 1\}$; define $J_\Lambda = \cap\{J_j: j \in \Lambda\}$, $S_\Lambda = \cap\{S_j: j \in \Lambda\}$, $T_\Lambda = \cap\{T_j: j \in \Lambda\}$. Fix one such Λ; if $\Gamma \subseteq \Lambda$, let $F_\Gamma = S_\Gamma \cap T_{\Lambda\setminus\Gamma}$, where it is understood that $F_\Lambda = S_\Lambda$, and $F_{\square} = T_{\Lambda}$. It is immediate that $J_{\Lambda} = \cup\{F_{\Gamma}: \Gamma \subseteq \Lambda\}$, and that $F_{\Gamma} \cap F_{\Omega} = F_{\Gamma\cup\Omega}$ for any subsets $\Gamma,\Omega \subseteq \Lambda$. We show that each F_{Γ} is acyclic and that F_{Λ} is non-empty; lemma 2.3 will then apply once more to prove J_{Λ} is acyclic. By calculation, the set

$$F_{\Gamma} = \cap\{eM(y_i): i \in \Gamma\} \cap ewM(y_1)\ , \quad \text{where} \quad w = \Pi\{x_j: j \in \Lambda\backslash\Gamma\};$$

by lemma 4.4, this set is a connected UDC subsemigroup of eS and is therefore acyclic. Note $e \in F_{\Lambda}$, so that F_{Γ} is not empty for any $\Gamma \subseteq \Lambda$. Therefore J_{Λ} is acyclic, whence the family $\{J_j: 2 \leq j \leq n + 1\}$ satisfies the hypothesis of lemma 2.3, from which it follows at last that $(S\backslash V) \cap eB$ is acyclic. This proves $j_1^* \neq 0$, so that $j^* \neq 0$, which completes the proof of the theorem.

REFERENCES

1. L.W. Anderson, "On the breadth and codimension of a topological lattice," Pacific J. Math. 9(1959), 327-333.

2. D.R. Brown, "Topological semilattices on the two cell," Pacific J. Math. 15(1965), 36-46.

3. _________ and M. Friedberg, "Representation theorems for uniquely divisible semigroups," Duke Math. J. 35(1968), 341-352.

4. _________________________, "A survey of compact divisible commutative semigroups," Semigroup Forum 1(1970), 143-161.

5. _________________________, "Linear representations of certain compact semigroups," Trans. Amer. Math. Soc. 160(1971), 453-465.

6. D.R. Brown and J.W. Stepp, "Inner points and breadth in certain compact semilattices," Proc. Amer. Math. Soc. (1982).

7. M. Friedberg, "Almost periodic functions, compactifications, and faces of finite dimensional cones," Math. Zeit. 176(1981), 53-61.

8. ___________, "Some examples of clans," Semigroup Forum 4(1972), 156-164.

9. G. Gierz et al, "A compendium of continuous lattices," Heidelberg, 1980.

10. J.A. Hildebrant, "On compact unithetic semigroups," Pacific J. Math. 21(1967), 265-273.

11. ______________ and J.D. Lawson, "On semigroup compactifications,"

12. K.H. Hofmann and P.S. Mostert, "Elements of compact semigroups," Columbus, Ohio, 1966.

13. A.L. Hudson, "Some semigroups on the two cell," Proc. Amer. Math. Soc. 10(1959), 648-655.

14. K. Keimel, "Lokal kompakte Kegelhalbgruppen und deren Einbettung in topologische Vektorraume," Math. Zeit. 99(1967), 7-25.

15. R.J. Koch, "Arcs in partially ordered spaces," Pacific J. Math. 9(1959), 723-728.

16. ________ and A.D. Wallace, "Maximal ideals in compact semigroups," Duke Math. J. 21(1954), 681-686.

17. J.D. Lawson, "Vietoris mappings and embeddings of topological semilattices," U. Tenn. dissertation, 1967.

18. ________, "The relation of breadth and codimension in topological semilattices," Duke Math. J. 37(1970), 207-212.

19. ________, "The relation of breadth and codimension in topological semilattices II," Duke Math. J. 38(1972), 555-559.

20. ________ and B.L. Madison, "Peripheral and inner points," Fund. Math. 69(1970), 253-266.

21. E.H. Spanier, "Algebraic topology," New York, N.Y., 1966.

22. J.W. Stepp, "Semilattices which are embeddable in a product of min intervals," Proc. Amer. Math. Soc. 28(1971), 81-86.

23. A.D. Wallace, "An outline for algebraic topology I," Tulane University lecture notes, 1949.

24. ________, "A theorem on Acyclicity," Bull. Amer. Math. Soc. 67(1961), 123-124.

25. ________, "Acyclicity of compact connected semigroups," Fund. Math. 50(1961), 99-105.

Department of Mathematics
University of Houston
Houston, Texas 77004

SEMIGROUPS DEFINED BY LATTICE POLYNOMIALS

by

H. Cohen, R.J. Koch, and J.D. Lawson

In a recent paper [2] Cohen and Koch exhibit several semigroups whose multiplication is described by some lattice polynomials. More specifically, their semigroups on the two-cell have multiplications defined by $(x_1,y_1) \cdot (x_2,y_2) = (P(x_1,y_1,x_2,y_2), Q(x_1,y_1,x_2,y_2))$ for certain polynomials P and Q on the interval lattice. Using "+" for join and juxtaposition for meet, for any lattice L and for any positive integer n we can formally define a lattice polynomial $P:L^n \to L$ by

$P(x_1,x_2, \ldots, x_n) = \sum c_{i_1 i_2 \ldots i_n} x_1^{i_1} x_2^{i_2} \ldots x_n^{i_n}$, where each i_j is either 0 or 1. For example if n=2, $P(x_1,x_2) = c_{00} + c_{10}x_1 + c_{01}x_2 + c_{11}x_1x_2$ [which means, of course, $c_{00} \vee (c_{10} \wedge x_1) \vee (c_{01} \wedge x_2) \vee (c_{11} \wedge x_1 \wedge x_2)$].

This paper is an outgrowth of attempts to answer the question: For which lattice polynomials does a definition like the one above result in an associative multiplication? It is easily seen than not all polynomials will work. For example, letting $P(x_1,y_1,x_2,y_2) = y_1$ and $Q(x_1,y_1,x_2,y_2,) = x_2$ provides a counterexample. Actually the problem investigated here is: For which lattice polynomials P and Q on any distributive lattice L does the definition $(x_1y_1) \cdot (x_2,y_2) = (P(x_1,y_1,x_2,y_2), Q(x_1,y_1,x_2,y_2))$ Make L × L an idempotent semigroup with identity?

1. Preliminaries:

Definition: A partially-ordered semigroup is a semigroup with a partial order $\le$ such that $a \le b$ and $x \le y$ implies $a \cdot x \le b \cdot y$. A lattice-ordered semigroup is a partially ordered semigroup where the partial order is a lattice order.

Lemma I. *Let* $(S, \cdot, \le)$ *be a lattice-ordered idempotent semigroup. Then for all* $x, y \in S$, *we have* $x \wedge y \le x \cdot y \le x \vee y$.

Proof: Since $(x \wedge y) \le x$ and $(x \wedge y) \le y$, $(x \wedge y) = (x \wedge y) \cdot (x \wedge y) \le x \cdot y$. Similarly, since $x \le x \vee y$ and $y \le x \vee y$, we get $x \cdot y \le x \vee y$.

Lemma 2. *Let* $(S, \cdot, \le)$ *be lattice-ordered idempotent semigroup with largest element 1 and smallest element 0; then 1(resp. 0) is a identity element for* $(S, \cdot)$ *if and only if the multiplication in* S *is given by the meet (resp. join) operation.*

Proof: Let 1 be the identity. For $x, y \in s$ we have $x \wedge y \le x \cdot y$ by Lemma 1. But $x \le 1 \Longrightarrow x \cdot y \le y$, and $y \le 1 \Longrightarrow x \cdot y \le x$. Therefore, $x \cdot y \le x \wedge y$. The converse is obvious, and the 0 case follows by a dual argument.

2. Now let L be a distributive lattice and suppose that S is a semigroup consisting of the elements of $L \times L$ with multiplication "$\cdot$" defined by $(s,t) \cdot (x,y) = (P(s,t,x,y), Q(s,t,x,y))$ where P and Q are lattice polynomials.

The most general form for a lattice polynomial in four variables is

$$(1)\quad P(s,t,x,y) = c_0 + c_1 s + c_2 t + c_3 x + c_4 y + c_5 st + c_6 sx + c_7 sy + c_8 tx + c_9 ty + c_{10} xy + c_{11} stx + c_{12} sty + c_{13} sxy + c_{14} txy + c_{15} stxy.$$

S, of course, inherits the lattice order from $L \times L$ and, with respect to this order, is a lattice-ordered semigroup since lattice polynomials are order-preserving functions.

We show first that if the lattice fails to have a maximal or minimal element, the multiplication is easily determined.

Suppose, therefore, that L has no minimal element (a dual argument will obtain if there is no maximal element), and that (a,b) is the identity for S. Using the fact that P has the form

(1), let x be less than the meet of $\{c_0, c_1, \ldots c_{15}, a, b\}$ and $y=a$. Then $P(a,b,x,y) = x$ precludes the existence of $c_0, c_1, c_2, c_4, c_5, c_7, c_9$, and c_{12}, while $P(x,y,a,b) = x$ eliminates c_3, c_5, c_{10}, and c_{14}. Thus $P(s,t,x,y)$ is reduced to $c_6sx + c_{11}stx + c_{13}sxy + c_{15}stxy$. Now if $x > a$, $P(a,b,x,y)$ cannot equal x. Thus a must be a maximal element, and since a similar argument shows b must also be maximal, applying lemma 2 gives the following result.

Theorem 1. *If* L *has no 0 [resp. 1], then it must have 1 [resp. 0] and multiplication in* S *is* $(s,t) \cdot (x,y) = (s \wedge x, t \wedge y)$ [*resp.* $s \vee x, t \vee y$].

From now on we assume that L has 0 and 1. By lemma 1 we have $s \wedge x \leq P(s,t,x,y) \leq s \vee x$ and $t \wedge y \leq Q(s,t,x,y) \leq t \vee y$. If we let $s = x = 0$ and $t = y = 1$, we get $P(0,1,0,1) = c_0 + c_2 + c_4 + c_9 = 0$; so $c_0 = c_2 = c_4 = c_9 = 0$. Also, since $P(s,t,x,y) \geq sx$, we may w.l.o.g. take $c_6 = 1$ whereupon the terms containing c_{11}, c_{13} and c_{15} are redundant and we may omit them. We have thus shown that

(2) $$P(s,t,x,y) = c_1s + c_3x + c_5st + sx + c_7sy + c_8tx + c_{10}xy + c_{12}sty + c_{14}txy.$$

Similarly we get

(3) $$Q(s,t,x,u) = d_2t + d_4y + d_5st + d_7sy + d_8tx + ty + d_{10}xy + d_{11}stx + d_{13}sxy.$$

We note that for any choice of the constants c_i and d_i in (2) and (3), the resulting multiplication is idempotent.

3. The case L = **2**. Before trying to finish the general problem, it is instructive to look at the special case when L is the two element lattice, **2**, consisting of 0 and 1. Since in this case there are only four elements in S, and, since if the identity is either (0,0) or (1,1), the semigroup is determined by lemma 2, we need only consider the case when (1,0) is the identity. (A dual calculation will handle the case when (0,1) is the identity.)

We wish to determine all multiplications on S = **2** × **2** with (1,0) as identity. Using (1,0) as a left identity

$P(1,0,x,y) = c_1 + c_3x + x + c_7y + c_{10}xy = x$. Letting $x = 0$ and $y = 1$ shows $c_1 = c_7 = 0$. Also from (3) $Q(1,0,x,y) = d_4y + d_7y + d_{10}xy + d_{13}xy = y$ and $x = 0, y = 1$ shows $d_4 + d_7 = 1$. Since (1,0) is a right identity, $P(x,y,1,0) = c_3 + c_5xy + x + c_8y = x$; so that $c_3 = c_8 = 0$. Also $Q(x,y,1,0) = d_2y + d_5xy + d_8y + d_{11}xy = y$; so $d_2 + d_8 = 1$. Incorporating this information we have:

(4) $P(s,t,x,y) = c_5st + sx + c_{10}xy + c_{12}sty + c_{14}txy$ and

(5) $Q(s,t,x,y) = d_2t + d_4y + d_5st + sy + tx + ty + d_{10}xy$.

Since there are only four elements in our semigroup and (1,0) is the identity, we may use (4) and (5) to calculate the remainder of the multiplication table. We get

TABLE 1

	(0,0)	(0,1)	(1,1)
(0,0)	(0,0)	$(0,d_4)$	$(c_{10},d_4 + d_{10})$
(0,1)	$(0,d_2)$	(0,1)	$(c_{10} + c_{14}, 1)$
(1,)	$(c_5,d_2 + d_5)$	$(c_5 + c_{10},1)$	(1,1)

At this point we may use the associative law to test which c's and d's are admissible (or have the computer check them for us). We prefer, however, to use some properties of semigroups to shorten the work.

In particular, the semigroup must contain a minimal ideal K, which must consist of 1, 2, or 3 elements. Moreover, the elements of K must all be left zeroes or all right zeroes. If we suppose K has three left zeroes, we see from the table above that d_4, c_{10}, d_{10} and c_{14} must all be zero while d_2 and c_5 are 1. Thus (4) and (5) reduce to $P = st + sx$ and $Q = t + sy$ respectively. We have just calculated the first of what turns out to be seventeen possibilities. We omit the calculations and list the results.

A) K has three elements and

1) left zeroes, $P = st + sx$; $Q = t + sy$

2) right zeroes, $P = sx + xy$; $Q = y + tx$

B) K has two elements and

3) (1,1) and (0,0) are left zeroes $P = sx + st + txy$
$Q = sy + tx + ty + st$

4) (1,1) and (0,0) are right zeroes $P = sx + xy + sty$
$Q = sy + tx + ty + st$

5) (1,1) and (0,1) are left zeroes $P = sx + st$; $Q = t + y$

6) (1,1) and (0,1) are right zeroes $P = sx + xy$; $Q = t + y$

7) (0,1) and (0,0) are left zeroes $P = sy$; $Q = t + sy$

8) (0,1) and (0,0) are right zeroes $P = sx$; $Q = y + tx$

C) K has only one element (i.e, semigroup has a zero)

9) (0,1) is zero; $P = sx$; $Q = t+ y$

10) (0,0) is zero; $P = sx + sty$; $Q = sy + tx + ty$

11) (0,0) is zero; $P = sx + txy$; $Q = sy + tx + ty$

12) (0,0) is zero; $P = sx$; $Q = sy + tx + ty$

13) (0,0) is zero; $P = sx + sty + txy$; $Q = tx + ty + sy$

14) (1,1) is zero; $P = sx + st + xy$; $Q = sy + t + xy$

15) (1,1) is zero; $P = sx + st + xy$; $Q = tx + y + st$

16) (1,1) is zero; $P = sx + st + xy$; $Q = t + y$

17) (1,1) is zero; $P = sx + st + xy$; $Q = sy + tx + st + ty + xy$

4. We turn now to the general case. That is L is an arbitrary distributive lattice with least element 0 and greatest element 1. We consider the problem of finding all lattice polynomial pairs (P,Q) such that the multiplication on L × L given by $(s,t) \cdot (x,y) = (P(s,t,x,y), Q(s,t,x,y))$ is associative, idempotent and has (1,0) for an identity.

Now each of these conditions can be expressed as polynomial equations in the lattice. For example, the fact that (1,0) is the identity is equivalent to

$$P(1,0,x,y) = P(x,y,1,0) = x \quad \text{and}$$
$$Q(1,0,x,y) = Q(x,y,1,0) = y.$$

The arguments given earlier to reduce equations (2) and (3) to

(4) and (5) still hold. We know, however, that even in the case L = **2** that arbitrary values of the constants will not, in general, result in admissible multiplications. We seek, therefore, relations among the constants that will produce the seventeen cases when L = **2**. Unfortunately, we know of no simple procedure to follow. We have by careful scrutiny, coupled with trial and error and arduous calculations * found several sets of relationships, of which the following are perhaps the most concise.

i) $d_2 + d_5 = d_2 + c_5$

ii) $d_4 + d_{10} = d_4 + c_{10}$

iii) $c_{14}(d_2 + d_4) \leq c_{10} \leq d_4 + c_5 + c_{12}$

iv) $c_{12}(d_2 + d_4) \leq c_5 \leq d_2 + c_{10} + c_{14}$

(where, of course, sums and products are the lattice operations).

It can easily be seen that these relationships hold in all seventeen cases. Moreover, one can check that for any combination of 0's and 1's satisfying i) through iv) one of the seventeen cases results. These relationships, therefore are equivalent to the polynomial equations giving rise to idempotent semigroups with (1,0) as identity on **2** × **2**.

Now, as is well known [e.g. 1], if one set of inequalities is equivalent to another set for a given lattice, they are equivalent for the variety that the lattice generates (i.e. all lattices obtained by taking products, sublattices, and quotients). Moreover, as is also well known, the lattice **2** generates all distributive lattices. Thus, we have:

Theorem 2: *For any distributive lattice L with 0 and 1, S = L × L will be an idempotent semigroup with (1,0) as identity if and only if:*

$$P = c_5 st + sx + c_{10}xy + c_{12}sty + c_{14}txy \quad \textit{and}$$

$$Q = d_2t + d_4y + d_5st + sy + tx + ty + d_{10}xy \quad \textit{and}$$

i), ii), iii) and iv) hold.

---

* We are grateful to Fred J. Sullivan who checked some of these calculations on the computer.

5. We have just seen how to determine the polynomials P and Q if (1,0) is the identity element. It can be shown that if L is a chain, then the identity element must be (0,0), (0,1), (1,0), or (1,1). The (0,0) and (1,1) cases are trivial by lemma 2, the (1,0) case was done above and the (0,1) case can be handled dually. If, however, L is not a chain, other elements may serve as the identity. We have made no effort to investigate this case, but offer the following example to show that this case may indeed arise.

Example: Let L be the four element lattice

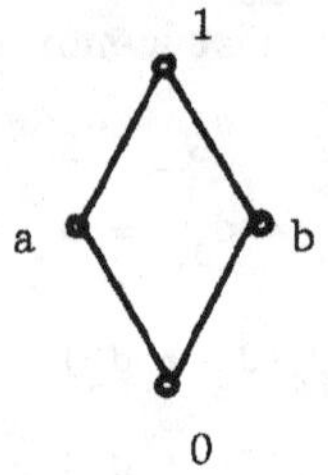

Let $P(s,t,x,y) = bs + bx + sx$

$Q(s,t,x,y) = t + y$

Then $(a,0) \cdot (x,y) = (ab + ax, y) = ((b + a)x, y) = (x,y)$

and $(s,t) \cdot (a,0) = (bs + ba + sa, t) = ((b+a)\ s, t) = (s,t)$

So (a,0) is the identity.

The associative law may be checked directly

$$(s,t)\ [(s,y)\ (z,w)] = (s,t)\ (bx + bz + xz, y + w)$$
$$= (bs + bx + bz, t + y + w)$$

while $[(s,t)\ (x,y)]\ (z,w) = (bs + bx + sx, t + y)\ (z,w)$

$$= (bs + bx + bz, t + y + w).$$

REFERENCES

1. Birkhoff, G. Lattice Theory, Amer. Math. Soc. Providence 3rd Ed. (1967).

2. Cohen, H. and Koch, R. J. Idempotent Semigroups Generated by Threads I, Semigroup Forum 23 (1981) 247-254.

THE UNIVERSITY OF MASSACHUSETTS
LOUISIANA STATE UNIVERSITY

DIFFERENTIABLE SEMIGROUPS

George E. Graham[1]

TABLE OF CONTENTS

---

[1] Portions of this paper are contained in the author's dissertation written at the University of Houston.

INTRODUCTION

The purpose of this paper is to present a cohesive Lie theory for semigroups in which differentiability of the semigroup multiplication is a central feature. This is accomplished by utilizing the constructions of modern differential geometry to give an intrinsic definition of a differentiable semigroup. We show that once this definition is properly formulated, one is able to extend many of the techniques of Lie group theory to the semigroup case. In particular, we define a Lie algebra for each monoid with a 3-times-differentiable multiplication function. The main contribution of this paper is the development of the elementary theory of differentiable semigroups in the form of a variety of results on cancellation, reversibility, subsemigroups and subgroups, and one-parameter semigroups.

The fundamental theory of Lie transformation groups, presented by Sophus Lie over 100 years ago, remains a vital element of both mathematics and physics and is becoming increasingly important in engineering as well (e.g. geometric control theory.) It is somewhat surprising therefore that the development of a corresponding theory of Lie transformation semigroups has been lacking, especially when one considers, as A. D. Wallace was known to observe, that many of the phenomena of the physical world are inherently irreversible and for this reason are perhaps better modeled by semigroups than groups. A major obstruction to the establishment of such a theory has been the difficulty of identifying an appropriate notion of a Lie semigroup. Within the semigroup literature there are a variety of results of a Lie-theoretic nature, but until the works of Graham [G, G_1] and Hofmann and Lawson [H-L] (these proceedings), there was no systematic approach to a general theory of Lie semigroups. The status of Lie semigroup theory up to the present time has already been discussed in detail by Hofmann and Lawson and is also discussed by Hofmann in [H].

There now appear to be at least two avenues to the development of a cohesive Lie theory of semigroups. First, there is the approach of Hofmann and Lawson. They have developed the basic theory of subsemigroups of Lie groups in a systematic manner. With this approach one is able to take advantage of the special relationship of Lie groups and Lie algebras, but even in this setting one immediately encounters difficult problems of a fundamental nature. As they have shown, considerable care must be exercised in defining Lie semigroups and Lie wedges if one is to extend the functorial correspondence between Lie groups and Lie algebras.

Our approach to Lie semigroup theory is to generalize the notion of a differentiable manifold with boundary in order to give an intrinsic definition of a differentiable semigroup. Although we do not postulate the existence of a containing group, there is sufficient geometric structure available for a fairly substantial Lie theory of differentiable semigroups. This elementary Lie theory is established in Part II of this paper. These results may be used to show that a finite-dimensional differentiable monoid is locally embeddable (near 1) in any Lie group having an isomorphic Lie algebra (cf. [G] or [G_2]). Although we will not prove this result here, we use it in section II.6 to unify the two approaches to Lie semigroup theory.

The first problem one encounters in attempting to identify a workable notion of a differentiable semigroup is the nature of the underlying space. If we define a differentiable semigroup to be a semigroup S such that the multiplication $m: S \times S \to S$ is a differentiable map, then the underlying space of S must be of an appropriate geometric type. Requiring S to be a manifold or manifold with smooth boundary eliminates many interesting cases. For instance, a cone in $\mathbb{R}^n$ is (among other things) a subsemigroup of the additive group of $\mathbb{R}^n$. We would like to include cones with differentiable semigroups but cones may contain a vertex or a corner at the identity. As another example, consider the subsemigroup of the Heisenberg group described by Hofmann and Lawson in [H-L]. Specifically, the Heisenberg group is the group G of matrices of the form

$$\begin{pmatrix} 1 & x & z \\ 0 & 1 & y \\ 0 & 0 & 1 \end{pmatrix}$$

such that x, y, z are real. This group may be identified with $\mathbb{R}^3$ (as a manifold) in the obvious way. Let S be the subsemigroup of G defined by $x, y \geq 0$ and $z \in [0, xy]$. When identified with a subset of $\mathbb{R}^3$, S is the region in the first octant below the graph of $z = xy$. Thus S has a cusp at the identity. Since multiplication in S is the restriction of matrix multiplication, S is a reasonable candidate for a differentiable monoid. This example illustrates the need to allow corners, cusps and possibly other irregularities in the boundary.

In this paper we give an intrinsic description of a differentiable manifold with generalized boundary and establish the requisite differential geometry of such manifolds. Once this is done, the development of the Lie theory of differentiable semigroups proceeds in a straightforward manner.

PART I. DIFFERENTIABLE MANIFOLDS WITH GENERALIZED BOUNDARY

I.1. CALCULUS IN BANACH SPACES:

The underlying space of a differentiable semigroup is a differentiable manifold with generalized boundary, which we shall define in this part. We show that by strengthening the definition of the derivative of a function at a point, one is able to extend the differential calculus and geometry of manifolds without boundary to include these spaces. Our approach to manifold theory is essentially that of Lang [L]. Other references for Part I are Dieudonné [D], for calculus in Banach spaces, and Spivak [S] and Graham [G] for manifolds. We shall use the following notation:

E, F	Banach spaces
A°	the topological interior of a subset A of a topological space
A^*	the topological closure of A
$B_\delta(a)$	the open ball of radius δ centered at a
L(E,F)	the Banach space of continuous linear maps from E to F
$L_k(E,F)$	the Banach space of continuous k-multilinear maps from E^k to F
$L_k^s(E,F)$	the subspace of symmetric maps in $L_k(E,F)$

A subset A of a topological space X is an <u>admissible subset of</u> X if A has dense interior in X. When no confusion seems possible as to the containing space, we shall say that A is an <u>admissible set</u>. We note that admissibility is a transitive relation. In particular, a relatively open subset of an admissible subset of X is itself an admissible subset of X. Admissible subsets of a Banach space E are the local models of E-manifolds.

Let E be a Banach space. An E-manifold with generalized boundary, or simply an E-manifold, is a regular topological space M such that if $p \in M$, then there is an open set $U \subset M$ about p and a homeomorphism φ from U onto an admissible subset of E. It is easily shown that an E-manifold is a T_1 space and hence is a Hausdorff space. An n-manifold is an $\mathbb{R}^n$-manifold with a countable base for its topology. It follows that an n-manifold is normal and paracompact.

For many local questions, an E-manifold may be assumed to be an admissible subset of E. For example, differentiability of maps between manifolds is defined in terms of the differentiability of maps $f: A \subset E \to F$ where A is an admissible set. We now consider two formulations of the concept of differentiability in this setting.

Let $f: A \subset E \to F$ and let $a \in A$. A linear map $T \in L(E,F)$ is a Fréchet derivative of f at a if for each $\varepsilon > 0$, there is a $\delta > 0$ such that

$$|f(x) - f(a) - T(x-a)| \le \varepsilon|x - a|$$

$$\text{whenever } x \in B_\delta(a) \cap A .$$

If a is an isolated point of A, then any $T \in L(E,F)$ is a Fréchet derivative of f at a, but if $a \in A^\circ$, then there is at most one such T. When f has only one Fréchet derivative at x, we denote it by $f'(a)$ and say that $f'(a)$ exists.

Let $U \subset E$ be an open set and let $f: U \to F$. Then f is continuously Fréchet differentiable, or simply C^1, if $f'(x)$ exists for each $x \in U$ and the map $f': U \to L(E,F)$ is continuous. Inductively, f is k-times continuously Fréchet differentiable, or C^k, if f is C^1 and f' is C^{k-1}. Finally, f is C^∞ if f is C^k for each positive integer k. If E and F are finite-dimensional, then f is a C^k map if and only if each component function f_i has continuous partial derivatives through order k. In this case the matrix of $f'(x)$ is the (Jacobian) matrix of partial derivatives $\left(\frac{\partial f_i}{\partial x_j}\right)$ at x. In general, the $j^{\underline{th}}$ derivative of a C^k map f $(j \leq k)$ is the map $f^{(j)} = (f^{(j-1)})': U \to L_j(E,F)$. It follows that $f^{(j)}(x) \in L_j^s(E,F)$ for each $x \in U$ (cf. [D, p. 175]).

Let $f: A \subset E \to F$. Then f is C^k differentiable if for each $a \in A$, there is an open set $U \subset E$ about a and a C^k map $g: U \to F$ such that g agrees with f on $U \cap A$.

NOTE. Defining C^k differentiability as in the previous paragraph allows one to extend the usual differential calculus to include functions on arbitrary domains. We now show that by utilizing the strong derivative one may obtain such a differential calculus without requiring the existence of local C^k extensions of the functions.

Let $f: A \subset E \to F$ and let $a \in A$. A linear map $T \in L(E,F)$ is a strong derivative of f at a if for each $\varepsilon > 0$, there is a $\delta > 0$ such that

$$|f(y) - f(x) - T(y-x)| \leq \varepsilon|y - x| \quad \text{whenever } x, y \in B_\delta(a) \cap A. \tag{1.1}$$

Strong derivatives need not be unique in general but are unique when $a \in (\overset{\circ}{A})^*$. If f has a unique strong derivative at a, we denote it by $df(a)$ and say that $df(a)$ exists.

A strong derivative is a Fréchet derivative but not conversely. For example, the function $f: \mathbb{R} \to \mathbb{R}$, defined by $f(x) = x^2\sin(x^{-1})$ for $x \neq 0$ and $f(0) = 0$, is Fréchet differentiable at each point but not strongly differentiable at 0. The problem here is that f' is not continuous at 0. Specifically, if $U \subset E$ is an open set and $f: U \to F$ is C^1, then $f'(x)$ is a strong derivative by the Mean-Value Theorem. The converse holds as well.

LEMMA 1.2. *Let* $A \subset E$ *be an admissible set and let* $f: A \to F$ *be a map. If* $df(x)$ *exists for each* $x \in A$, *then* $df: A \to L(E,F)$ *is continuous.*

PROOF. See Graham [G_1, Lemma 2.3].

NOTE. Lemma 1.2 implies that C^1 differentiability is equivalent to strong differentiability on open sets. This fact has tended to obscure the role of strong differentiability in the development of the differential calculus, although the condition (1.1) of strong differentiability has been recognized by several authors (e.g. Bartle [B], Graves [Gr], Leach [Le], Nashed [N], and Nijenhuis [Nj]) as being crucial to the proof of inverse and implicit mapping theorems. Strong differentiability was investigated as early as 1892 by Peano [Pn], but only recently has the concept begun to receive a systematic treatment (cf. [G], [G_1], [Le], [N], [Nj].) The theorems of this section help to illustrate the role of strong differentiability in the differential calculus.

Let $A \subset E$ be an admissible set and let $f: A \to F$. Then f is *strongly differentiable*, or C^1_s, if $df(x)$ exists for each $x \in A$. Inductively, f is *k-times strongly differentiable*, or C^k_s, if f is C^1 and df is C^{k-1}_s. Finally, f is C^∞_s if f is C^k_s for all positive integers k. If f is C^k_s, then the jth *strong derivative* of f $(j \leq k)$ is the map $d^j f = d(d^{j-1}f): A \to L_j(E,F)$.

If $U \subset E$ is an open set, then $f: U \to F$ is C^k if and only if f is C^k_s. If $A \to E$ is an admissible set and $f: A \to F$ is a C^k_s map, then $f|A^\circ$ is a C^k map. It follows that $d^j f(x) = f^{(j)}(x)$ for each $x \in A^\circ$ and hence $d^j f(x)$ is a symmetric map for each

$x \in A$. It also follows that a C^k map on an admissible set is a C^k_s map, but the converse does not hold unless the domain satisfies additional geometric conditions. The relationship between these two notions of differentiability is summarized by the following theorem.

THEOREM 1.3. *Let* $A \subset E$ *be an admissible set and let* $f: A \to F$.

1. *If* f *is* C^k, *then* f *is* C^k_s.
2. *If* A *is an open set, then* f *is* C^k *if and only if* f *is* C^k_s.
3. *If* E *and* F *are finite dimensional and* A *is either a closed convex set or the intersection of an open set and a closed convex set, then* f *is* C^k *if and only if* f *is* C^k_s.
4. *If* E *and* F *are finite-dimensional and* A *is closed, then* f *is* C^1 *if and only if* f *is* C^1_s.
5. *There is a closed admissible set* $B \subset \mathbb{R}$ *and a* C^2_s *function* $f: B \to \mathbb{R}$ *such that* f *is not* C^2.

PROOF. See Graham [G_1].

NOTE. From the preceding theorem one sees that the class of C^k_s functions properly includes the class of C^k functions. We consider the concept of C^k_s differentiability to be more intrinsic than C^k differentiability in the sense that the existence of local C^k extensions of the functions is not required. Moreover, one may still obtain a fairly complete differential calculus in this setting since the condition (1.1) of strong differentiability is in essence a form of the Mean-Value Theorem.

In the remainder of this section, we present without proof the elements of a strong differential calculus. Proofs of these theorems may be found in $[G_1]$.

THEOREM 1.4. *Let* $f\colon A \subset E \to F$ *be a map and let* $T \in L(E,F)$ *be a linear map such that* (1.1) *holds for some* $a \in A$. *Then*

$$|f(y) - f(x)| \le (|T| + \varepsilon)|y - x|$$

$$\text{whenever } x, y \in B_\delta(a) \cap A.$$

In particular, f *satisfies a Lipschitz condition on a neighborhood of* a.

NOTE. A finite Cartesian product of Banach spaces is again a Banach space with any of the equivalent norms: the Euclidean norm, the sum norm, or the sup norm (cf. [D]). Since strong differentiability is invariant under change to an equivalent norm, one may choose the most convenient norm available.

THEOREM 1.5. *Let* $A \subset E$ *be an admissible set and let* $f\colon A \to F_1 \times \dots \times F_n$. *Then* f *is* C_s^k *if and only if each component function* $f_i\colon A \to F_i$, $i = 1, \dots, n$, *is* C_s^k, *and in this case*

$$(df(x))_i = df_i(x) \text{ for each } x \in A, \quad i = 1, \dots, n.$$

THEOREM 1.6 (Chain Rule). *Let* $A \subset E$ *and* $B \subset F$ *be admissible sets and let* $f\colon A \to B$ *and* $g\colon B \to G$ *be* C_s^k *maps, where* G *is a Banach space. Then* $g \circ f$ *is a* C_s^k *map and*

$$d(g \circ f)(x) = dg(f(x)) \circ df(x) \text{ for each } x \in A.$$

THEOREM 1.7 (Inverse Function Theorem). *Let* $A \subset E$ *be an admissible set and let* $f\colon A \subset F$ *be a* C_s^k *map. If* $df(a)$ *is an isomorphism onto* F *for some* $a \in A$, *then there is an open set* U *about* a *such that* $f|U \cap A$ *is a homeomorphism onto the admissible set* $f(U \cap A)$, $f|U \cap A^{\circ}$ *is an open map, and if* $g = (f|U \cap A)^{-1}$, *then* g *is* C^k_s *and*

$$dg(f(x)) = df(x)^{-1} \quad \text{for each} \quad x \in U \cap A.$$

A linear map $T \in L(E,F)$ *splits in* F if the range of T is closed and complemented in F.

THEOREM 1.8. *Let* $A \subset E$ *be admissible, let* $f: A \to F$ *be a* C^k_s *map, and let* $a \in A$. *Suppose that* $df(a)$ *is one-to-one and that* $df(a)$ *splits in* F. *Let* N *be a closed complement to the range of* $df(a)$. *Then there is an open set* $U \subset E$ *about* a *and a* C^k_s *map* $g: f(U \cap A) + N \to A$ *such that* $f|U \cap A$ *is a homeomorphism onto* $f(U \cap A)$, *the set* $f(U \cap A) + N$ *is an admissible subset of* F, *and*

$$g(f(x)) = x \quad \text{and} \quad g^{-1}(x) = f(x) + N$$

$$\text{whenever} \quad x \in U \cap A.$$

Let $A \subset E$ be admissible and let $f: A \to F$ be a C^k_s map. Then f is an *immersion* if for each $x \in A$ $df(x)$ is one-to-one and the range of $df(x)$ splits in F. The map f is an *embedding* if f is an immersion and a homeomorphism onto its image. Finally, f is a *local embedding* if for each $a \in A$ there is an open set $U \subset E$ about a such that $f|U \cap A$ is an embedding. Theorem 1.8 implies that an immersion is a local embedding.

Let $A \subset E$ and $B \subset F$ be admissible sets and let $f: A \times B \to G$ be a map where G is a Banach space. For each $a \in A$ let $f^a: B \to G$ be the map $f^a(b) = f(a,b)$. For each $b \in B$, define $f_b: A \to G$ by $f_b(a) = f(a,b)$.

THEOREM 1.9 (*Product Rule*). *Let* $A \subset E$ *and* $B \subset F$ *be admissible sets and let* $f: A \times B \to G$ *be a* C^k_s *map* *Then the maps* f^a *and* f_b *are* C^k_s *maps for each* $a \in A$ *and each* $b \in B$ *and*

$$df(a,b)(v,w) = df_b(a)(v) + df^a(b)(w)$$

for each $v \in E$ and each $w \in F$.

THEOREM 1.10. (Parameterized Mapping Theorem). Let $A \subset E$ and $B \subset F$ be admissible sets, let $f: A \times B \to F$ be a C^k_s map, and let $(a,b) \in A \times B$. Suppose that $df^a(b)$ is an isomorphism onto F. Then there are open sets $U \subset E$ about a and $V \subset F$ about b such that:

(i) for each $x \in U \cap A$, the map $f^x|V \cap B$ is an embedding and $f^x(V \cap B)$ is admissible in F, and

(ii) if $V_0 \subset F$ is an open set with $c \in V_0 \cap V$ for some $c \in B^o$, then there is an $\varepsilon > 0$ and an open set $W \subset F$ about $f(a,c)$ such that $W \subset f^x(V_0)$ for each $x \in B_\varepsilon(a) \cap A$.

I.2. DIFFERENTIABLE STRUCTURES: An atlas for an E-manifold M is a collection $\mathcal{A}$ of functions satisfying:

(i) each $\varphi \in \mathcal{A}$ is a homeomorphism from an open subset $\operatorname{dom}\varphi$ of M onto an an admissible subset $\operatorname{im}\varphi$ of E, and

(ii) $M = \bigcup \operatorname{dom}\varphi$ $(\varphi \in \mathcal{A})$.

If φ and ψ are members of the atlas $\mathcal{A}$, then the domain $\varphi(\operatorname{dom}\varphi \cap \operatorname{dom}\psi)$ of $\psi\circ\varphi^{-1}$ is an open subset of $\operatorname{im}\varphi$ and hence is an admissible subset of E. An atlas $\mathcal{A}$ for M is a C^k_s atlas if $\psi\circ\varphi^{-1}$ is a C^k_s map for each $\varphi, \psi \in \mathcal{A}$. Let $\mathcal{A}$ be a C^k_s atlas for M and let $\mathcal{D}$ be the collection of all homeomorphisms ψ such that $\operatorname{dom}\psi$ is an open subset of M, $\operatorname{im}\psi$ is an admissible subset of E, and $\psi\circ\varphi^{-1}$ and $\varphi\circ\psi^{-1}$ are C^k_s maps for each $\varphi \in \mathcal{A}$. Then

$\mathcal{D}$ is the unique maximal C_s^k atlas for M containing A. A C_s^k differentiable structure for M is a maximal C_s^k atlas. If A is a C_s^k atlas for M, then members of the unique differentiable structure containing A are called A-charts or simply charts. A chart at $p \in M$ is a chart containing p in its domain. A C_s^k manifold (with generalized boundary) is a pair $(M,\mathcal{D})$ where M is an E-manifold for some E and $\mathcal{D}$ is a C_s^k differentiable structure for M. When no confusion seems possible $(M,\mathcal{D})$ is contracted to M.

Let M and N be C_s^k manifolds and let $f: M \to N$. Then f is C_s^r differentiable $(r \le k)$ if f is continuous and $\psi \circ f \circ \varphi^{-1}$ is C_s^r differentiable for each chart φ on M and each chart ψ on N.

NOTE. Use of a maximal atlas facilitates the verification of C_s^k differentiability of a map $f: M \to N$ in the following sense: If B is any subatlas of the differentiable structure of M, then f is C_s^k if and only if $f \circ \varphi^{-1}$ is C_s^k for each $\varphi \in B$.

Let $A \subset E$ be an admissible set and let $i: A \to E$ be the inclusion map. Then $\{i\}$ is a C_s^∞ atlas for A, where A carries the relative topology. Thus each admissible subset of E is a C_s^∞ E-manifold.

Let M be a C_s^k manifold and let $p \in M$. If for some chart φ at p, $\varphi(p)$ is an interior point of $\operatorname{im}\varphi$, then $\psi(p)$ is an interior point of $\operatorname{im}\psi$ for each chart ψ at p by the Inverse Function Theorem. The boundary of M, denoted by ∂M, is the set of all $q \in M$ such that $\varphi(q)$ is not in $(\operatorname{im}\varphi)^\circ$ for each chart φ at q. If $\partial M = \emptyset$, then M is a manifold without boundary. In any case, $M \setminus \partial M$ is an open dense subset of M and is a manifold without boundary.

A half-space in E is a set of the form $T^{-1}([0,\infty))$ where $T \in L(E,\mathbb{R})$. A C_s^k E-manifold $(M,\mathcal{D})$ is a manifold with smooth boundary if there is an atlas $A \subset \mathcal{D}$ such that $\operatorname{im}\varphi$ is an open subset of a half-space in E for each $\varphi \in A$.

NOTE. In a similar way one may define a C^k manifold. Theorem 1.3 implies that a C_s^k manifold without boundary is a C^k manifold without boundary and conversely. It also follows that a finite-dimensional C_s^k manifold with smooth boundary is a C^k manifold with smooth boundary and conversely.

Let M be a C_s^k E-manifold and let N be a C_s^k F-manifold. For each chart φ on M and ψ on N, define $\varphi \times \psi$ by $(\varphi \times \psi)(p,q) = (\varphi(p), \psi(q)) \in E \times F$. Then $\operatorname{dom}(\varphi \times \psi)$ is open in $M \times N$ with the product topology and $\operatorname{im}(\varphi \times \psi)$ is an admissible subset of $E \times F$. Since $\varphi \times \psi$ is a homeomorphism, $M \times N$ is an $E \times F$-manifold. It follows that the collection of all such maps $\varphi \times \psi$ is a C_s^k atlas for $M \times N$. The C_s^k differentiable structure generated by this atlas is called the product structure for $M \times N$. In a similar way, one may show that any finite Cartesian product of C_s^k manifolds is a C_s^k manifold. We note however that the product of manifolds with smooth (nonempty) boundary does not have a smooth boundary with the product structure (it has corners).

I.3. THE TANGENT BUNDLE: An important ingredient of the geometry of manifolds is the tangent bundle, which consists of all the tangent spaces at points. The essential idea in the construction of the tangent space of M at a point $p \in M$ is to use a chart φ at p to identify vectors at p with vectors at $\varphi(p)$. We add that if E is a Banach space and $q \in E$, then the tangent space of E at q is E.

Let M be a C_s^k E-manifold and let $p \in M$. If φ and ψ are charts at p and if $v, w \in E$, then (φ,v) is p-*equivalent* to (ψ,w) if $d(\psi\circ\varphi^{-1})(\varphi(p))v = w$. It is easily verified (by the Chain Rule and Inverse Function Theorem) that p-equivalence is an equivalence relation. Let T_pM denote the set of equivalence classes $[(\varphi,v)]_p$ where φ is a chart at p and $v \in E$. The map $\hat{\varphi}_p: E \to T_pM$, defined by $\hat{\varphi}_p(v) = [(\varphi,v)]_p$, is a bijection of sets. Moreover, if T_pM is given the unique vector space structure such that $\hat{\varphi}_p$ is an isomorphism of vector spaces and if ψ is a chart at p, then $\hat{\psi}_p$ is an isomorphism. The *tangent space* of M at p is the set T_pM with the unique vector space structure such that $\hat{\varphi}_p$ is an isomorphism for each chart φ at p.

Let M and N be C_s^k manifolds and let $f: M \to N$ be a C_s^k map. If $p \in M$, then the *(strong) derivative* of f at p is the map $df(p): T_pM \to T_{f(p)}N$ defined by $df(p) = \hat{\psi}_{f(p)} \circ d(\psi\circ f\circ\varphi^{-1})(\varphi(p)) \circ (\hat{\varphi}_p)^{-1}$, where φ is a chart at p and ψ is a chart at $f(p)$. The definition of $df(p)$ is independent of the choice of charts φ and ψ by the Chain Rule (Theorem 1.6.)

THEOREM 3.1 (Chain Rule). Let M, N, *and* P *be* C_s^k *manifolds and let* $f: M \to N$ *and* $g: N \to P$ *be* C_s^k *maps. Then* $g\circ f$ *is a* C_s^k *map and*

$$d(g\circ f)(x) = dg(f(x)) \circ df(x), \quad \textit{for each } x \in M .$$

PROOF. Transfer to Banach spaces using charts and apply Theorem 1.6. □

NOTE. Let $A \subset E$ be an admissible set. Then $\{i\}$ is a C_s^∞ atlas for A, where $i: A \to E$ is inclusion. Thus $\hat{i}_a: E \to T_aE$ is an isomorphism for all $a \in A$, and we may identify T_aE with E.

Let M be a C_s^k E-manifold. A curve in M is a continuous map $c: I \subset \mathbb{R} \to M$ where I is an interval. If c is a C_s^k curve in M, let $c'(t) = dc(t)(1) \in T_{c(t)}M$. Let $p \in M$ and let φ be a chart at p. Then $d\varphi(p) = (\hat{\varphi}_p)^{-1}$. If $p \notin \partial M$ and $v = \hat{\varphi}_p(w) \in T_pM$, then there is an $\varepsilon > 0$ such that $\varphi(p) + tw$ is in the interior of $\operatorname{im}\varphi$ for all $t \in (-\varepsilon,\varepsilon)$. If $c(t) = \varphi^{-1}(\varphi(p) + tw)$ for $t \in (-\varepsilon,\varepsilon)$, then c is a C_s^k curve with $c(0) = p$ and $c'(0) = v$. Thus at nonboundary points p of M tangent vectors may be thought of as tangents to curves in M through p. Tangent vectors at points of ∂M need not be tangent to curves in M unless ∂M is smooth.

Let M be a C_s^k E-manifold with smooth boundary and let $p \in \partial M$. Let φ be a chart at p such that $\operatorname{im}\varphi$ is an open subset of $T^{-1}([0,\infty))$ where $T \in L(E,\mathbb{R})$ is nontrivial. Moreover, we may assume that $\varphi(p) = 0$. Now the kernel of T is a closed subspace of E of codimension 1. Let $T_p\partial M$ be the $\hat{\varphi}_p$ image of the kernel of T. Then $T_p\partial M$ is a linear subspace of T_pM of codimension 1. A vector $v \in T_pM$ is tangent to the boundary if $v \in T_p\partial M$, or equivalently, if $T(d\varphi(p)v) = 0$. A vector $v \in T_pM$ is inward pointing if $T(d\varphi(p)v) > 0$ and v is outward pointing if $-v$ is inward pointing. If v is tangent to the boundary, then $c(t) = \varphi^{-1}(td\varphi(p)v)$ is defined for $t \in (-\varepsilon,\varepsilon)$ for some $\varepsilon > 0$, $c'(0) = v$, and $c(t) \in \partial M$. If v is inward pointing, then there is a C_s^k curve $c: [0,\varepsilon) \to M$ for some $\varepsilon > 0$ such that $c(0) = p$, $c'(0) = v$, and $c(t) \notin \partial M$ for $t > 0$. In a similar way, if v is outward pointing, then there is a C_s^k curve $c: (-\varepsilon,0] \to M$ for some $\varepsilon > 0$ such that $c(0) = p$, $c'(0) = v$, and $c(t) \notin \partial M$ for $t < 0$. Thus each vector in T_pM is tangent to a curve in M.

<u>NOTE</u>. Let $f: M \to N$ be a C_s^k map and let $p \in M$. If $c: I \to M$ is a C_s^k curve with $c(0) = p$, then $f \circ c$ is a C_s^k curve and $(f \circ c)'(0) = df(p)(c'(0))$.

We turn now to the construction of the tangent bundle of a manifold. Let M be a C_s^k E-manifold and let $TM = \{(p,v): p \in M$ and $v \in T_pM\}$. For each chart $\varphi: U \subset M \to A \subset E$, define $T\varphi$ from $TU = \{(p,v): p \in U$ and $v \in T_pM\}$ onto $A \times E$ by $T\varphi(p,v) = (\varphi(p), d\varphi(p)v) = (\varphi(p), (\hat{\varphi}_p)^{-1}v)$. The map $T\varphi$ is a bijection of sets. Moreover, the collection of sets of the form $(T\varphi)^{-1}(W)$, where φ is a chart on M and W is an open subset of $E \times E$, is a base for a topology on TM. With this topology, the collection of all maps $T\varphi$, for φ a chart on M, is a C_s^{k-1} atlas for TM as an $E \times E$-manifold. The <u>tangent bundle</u> of M is the C_s^k map $\pi = \pi_M: TM \to M$ defined by $\pi(p,v) = p$. A standard abuse of terminology, which we shall use on occasion, is to call TM the tangent bundle of M and suppress mention of the map π.

Let $\pi: TM \to M$ be the tangent bundle of the C_s^k E-manifold M. For each $p \in M$, $\pi^{-1}(p) = T_pM$, which is isomorphic to E as a vector space. Moreover, T_pM is a topological vector space with its relative topology as a subset of TM. If φ is a chart at p, then the topology of T_pM is the norm topology for the unique norm on T_pM such that $\hat{\varphi}_p$ is an isometry. Use of a different chart ψ may produce a different norm, but the two norms will be equivalent and hence produce the same topology. Thus T_pM is equivalent to E as a topological vector space.

<u>NOTE</u>. The tangent bundle is an example of a vector bundle, which we shall not define. If $A \subset E$ is an admissible set, then

A is a C_s^∞ manifold and the tangent bundle of A may be identified with $\pi_1: A \times E \to A$. A vector bundle of this type (i.e. a product) is called a trivial bundle. The tangent bundle of a manifold is, in some sense, locally equivalent to a trivial bundle via the maps $T\varphi$. However, it is well-known (cf. [S, p. 3-9]) that the tangent bundle of the two-sphere S^2 is not equivalent to the trivial bundle $\pi_1: S^2 \times \mathbb{R}^2 \to \mathbb{R}^2$.

Let M be a C_s^k E-manifold and let N be a C_s^k F-manifold. Then the collection of all maps of the form $\varphi \times \psi$, where φ is a chart on M and ψ is a chart on N, is a C_s^k atlas on $M \times N$ as an $E \times F$-manifold. For each $(p,q) \in M \times N$, $T_{(p,q)}M \times N$ is isomorphic to $T_pM \times T_qN$ by the map $(\hat{\varphi}_p \times \hat{\varphi}_q) \circ (\widehat{(\varphi \times \psi)}_{(p,q)})^{-1}$.

Let M be a C_s^k manifold. For each $p \in M$, let $T_p^*M = L(T_pM, \mathbb{R})$. Let $T^*M = \{(p,\lambda): p \in M$ and $\lambda \in T_p^*M\}$, and let $\pi: T^*M \to M$ be the map $\pi(p,\lambda) = p$. By a method similar to that of TM, T^*M can be given the structure of a C_s^{k-1} manifold. The <u>dual bundle</u> of M is the C_s^{k-1} map $\pi: T^*M \to M$. If M and N are C_s^k manifolds and $f: M \to N$ is a C_s^k map, then $d^*f(p): T_{f(p)}^*N \to T_p^*M$ is the map $(df(p))^*$.

<u>I.4.</u> <u>SUBMANIFOLDS AND MAPPINGS</u>: Let M and N be C_s^k manifolds and let $f: M \to N$ be a C_s^k map. Then f is an <u>immersion</u> if for each $p \in M$, $df(p)$ is one-to-one and the image of $df(p)$ is closed and complemented in $T_{f(p)}N$. Equivalently, f is an immersion if for each $p \in M$, $df(p)$ has a continuous left inverse (i.e. a map $S \in L(T_{f(p)}N, T_pM)$ such that $S \circ df(p) = 1$). The map f is an <u>embedding</u> if f is an immersion and a homeomorphism onto its image (with the relative topology). Also, f is a <u>diffeomorphism</u> if f is an embedding, $f(M) = N$, and $df(p): T_pM \to T_{f(p)}N$ is an isomorphism

for each $p \in M$.

Let M be a C_s^k manifold. A subset P of M is an immersed submanifold of M if P can be given the structure of a C_s^k manifold in such a way that the inclusion map $i: P \to M$ is a C_s^k immersion. A subset P of M is an embedded submanifold, or simply a submanifold, of M if P can be given the structure of a C_s^k manifold in such a way that the inclusion map is a C_s^k embedding.

NOTE. An embedded submanifold carries the relative topology while an immersed submanifold may not (e.g. a dense wind in a torus). Also an admissible subset A of M is a submanifold of M; the collection of maps of the form $\varphi|A$ where φ is a chart on M is a C_s^k atlas for A. In this case $TA = \cup\{T_pM: p \in A\}$.

THEOREM 4.1 (Inverse Function Theorem). *Let M and N be C_s^k manifolds, let $f: M \to N$ be a C_s^k map and let $p \in M$. If $df(p)$ is an isomorphism onto $T_{f(p)}N$, then there is an open set U about p such that $f|U$ is a diffeomorphism onto the admissible subset $f(U)$ of N, $g = (f|U)^{-1}$ is a C_s^k map, and*

$$dg(f(x)) = (df(x))^{-1} \quad \text{for each } x \in U.$$

PROOF. Transfer to Banach spaces using charts and apply Theorem 1.7. □

NOTE. Theorem 1.8 implies that an immersion is a local embedding. The Inverse Function Theorem implies that the inverse of a C_s^k diffeomorphism is a C_s^k diffeomorphism. Also, if $f: P \to M$ is a one-to-one C_s^k immersion [embedding], then $f(P)$ is an immersed [embedded] submanifold with the unique C_s^k structure that makes $f: P \to f(P)$ a diffeomorphism.

Let M be a C_s^k E-manifold with smooth boundary and let $p \in \partial M$. Then there is a chart φ at p such that $\text{im}\,\varphi$ is an open subset of $T^{-1}[0,\infty)$ for some nontrivial $T \in L(E,\mathbb{R})$. Now $K = T^{-1}(0)$ is closed and complemented in E, and $\varphi^{-1}|K \cap \operatorname{im}\varphi$ is a homeomorphism onto an open neighborhood of p in ∂M. Thus ∂M is an embedded submanifold of M, and $T_p\partial M$ is a subspace of T_pM of codimension 1.

PROPOSITION 4.2. *Let $f\colon M \to N$ be a C^k_s map of C^k_s manifolds and let $P \subset N$ be an immersed submanifold such that $f(M) \subset P$. If f is continuous as a map into P, then f is C^k_s as a map into P.*

PROOF. Let $p \in M$, let φ be a chart at $f(p)$ of the E-manifold P, and let ψ be a chart at $f(p)$ of the F-manifold N. We assume that $\operatorname{dom}\varphi \subset \operatorname{dom}\psi \cap P$ so that $h = \psi \circ i \circ \varphi^{-1}\colon \operatorname{im}\varphi \subset E \to F$ is a C^k_s map where $i\colon P \to N$ is inclusion. Then $dh(\varphi(f(p)))$ is one-to-one and splits in F. Let K be a closed complement to the range of $dh(\varphi(f(p)))$. By Theorem 1.8 there is an open set $U \subset E$ about $\varphi(f(p))$ and a C^k_s map $g\colon h(U \cap \operatorname{im}\varphi) + K \to \operatorname{im}\varphi$ such that $h|U \operatorname{im}\varphi$ is a homeomorphism, $h(U \cap \operatorname{im}\varphi) + K$ is an admissible subset of F, and $g(h(x)) = x$ and $g^{-1}(x) = h(x) + K$ whenever $x \in U \cap \operatorname{im}\varphi$. Now let $W = f^{-1}(\varphi^{-1}(U \cap \operatorname{im}\varphi))$. Since f is continuous as a map into P, W is an open set about p. If $w \in W$, then $\varphi(f(w)) \in U \cap \operatorname{im}\varphi$ and $h(\varphi(f(w))) = \psi(f(x)) \in h(U \cap \operatorname{im}\varphi) \subset h(U \cap \operatorname{im}\varphi) + K$. Hence $f = \varphi^{-1} \circ g \circ \psi \circ f$ on W and since f is C^k_s as a map into N, f is C^k_s as a map into P by the Chain Rule. □

NOTE. If in the above theorem P is an embedded submanifold, then f is continuous as a map into P and hence f is C^k_s as a map into P.

THEOREM 4.3 (Open Mapping Theorem). *Let M and N be C^k_s manifolds and let $f\colon M \to N$ be a C^k_s map. If $df(p)$ maps onto*

$T_{f(p)}N$ <u>for some</u> $p \in M$, <u>then there is an open set</u> U <u>about</u> p <u>such that</u> $f|U \setminus \partial U$ <u>is an open map and</u> $f(U)$ <u>is an admissible subset of</u> N.

<u>PROOF</u>. By using charts we may assume that $M \subset E$ and $N \subset F$. For each $\varepsilon > 0$, there is a $\delta = \delta(\varepsilon) > 0$ such that $|f(y) - f(x) - df(p)(y-x)| \leq \varepsilon|y-x|$ whenever $x, y \in B_\delta(p) \cap M$. Thus we may pick ε small enough to satisfy Graves' approximate derivative condition in Theorem 1 of [Gr]. Let $\delta = \delta(\varepsilon)$ and let $U = B_\delta(p) \cap M$. Then $U \setminus \partial U = U^\circ$ is an open subset of E on which the approximate derivative condition holds. Since $df(p)$ maps onto F, $f|U^\circ$ is an open map by Theorem 1 of [Gr]. The set $f(U)$ is admissible since $f(U^\circ)$ is dense in $f(U)$.

The following theorem is a straightforward generalization of a theorem of Holmes [H1].

<u>THEOREM 4.4</u>. <u>Let</u> M <u>be a connected</u> C^k_s <u>manifold and let</u> $f: M \to M$ <u>be a</u> C^k_s <u>map such that</u> $f^2 = f \circ f = f$. <u>Then</u> $f(M)$ <u>is an embedded submanifold of</u> M <u>and</u> f <u>is a</u> C^k_s <u>map as a map into</u> $f(M)$.

<u>PROOF</u>. Let $p = f(p) \in f(M)$ and let $\varphi : U \to E$ be a chart at p such that $\varphi(p) = 0$. Since f is continuous, $f^{-1}(U \cap f(M)) = f^{-1}(U)$ is an open subset of M about p. Let $V = U \cap f^{-1}(U)$ so that $V \subset U$ is an open set about p and $f(V) \subset V$. Then we may assume that V is an admissible subset of E and $p = 0 \in E$. Let $T = df(p)$. Now $f^2 = f$ implies that $T^2 = T$ and that $F = \text{im } T$ and $K = \text{im}(1-T) = \text{kernel of } T$ are closed complementary subspaces of E.

Let $h = Tf + (1-T)(1-f): V \to E$. Then $dh(p) = T^2 + (1-T)^2 = 1$, the identity on E. Also, $d(Tf)(p) = T^2 = T$. An application of the Inverse Function Theorem and the Open Mapping Theorem yields a $\delta > 0$ such that if $V_\delta = B_\delta(0) \cap V$, then $h|V_\delta$ is a C^k_s diffeomorphism onto the admissible subset $h(V_\delta)$ of E and $Tf|V^\circ_\delta$ is an open map as a map into F. If $x = f(x) \in f(V)$, then $h(x) = Tf(x) = T(x) \in F$ and thus $h|f(V) = T|f(V)$. Let $W = V_\delta \cap f(V) = B_\delta(0)$ $B_\delta(0) \cap f(V)$ which is an open set about p in $f(V)$. Now $T(W) = Tf(B_\delta(0) \cap f^{-1}(B_\delta(0)) \cap V)$. Since Tf is continuous and $Tf|V^\circ_\delta$ is an open map, it follows that $T(W)$ is an admissible subset of F. Let $\psi = h|W$. Then $\psi^{-1}\colon T(W) \to W$ is a C^k_s embedding as a map into E. Thus $\{\psi\}$ is a C^k_s atlas for $W = B_\delta(0) \cap F(V)$ as an F-manifold and the inclusion map $i\colon W \to V \subset E$ is a C^k_s embedding since $i \circ \psi^{-1}$ is. Moreover, if $q \in W$ and $\eta\colon \mathrm{dom}\,\eta \subset W \to G = \mathrm{im}\, df(q)$ is a local homeomorphism derived in the same manner as ψ, then $\eta \circ \psi^{-1}$ is a C^k_s map since $\psi^{-1}\colon T(W) \to E$ is a C^k_s map and η is the restriction of a C^k_s map on an open subset of V. It follows that $G = \mathrm{im}\, df(q)$ is isomorphic to $F = \mathrm{im}\, df(p)$ for all $q \in W$. Now let $A = \{q \in f(V)\colon \mathrm{im}\, df(q) \cong \mathrm{im}\, df(p)\}$. The preceding argument shows that A is open and closed in $f(M)$ and since M is connected, $A = f(M)$. Thus if $q \in f(M)$ and $\eta\colon \mathrm{dom}\,\eta \subset f(M) \to G = \mathrm{im}\, df(q)$ is a chart derived in the same manner as ψ, then we may, by composing with a linear isomorphism if necessary, assume that $G = F$. Hence there is a C^k_s atlas for $f(M)$ as an F-manifold so that $f(M)$ is an embedded submanifold of M. □

Let M and N be a C^k_s manifolds and let $f\colon M \times N \to N$ be a C^k_s map. For each $a \in M$, $f^a\colon N \to N$ is defined by $f^a(b) = f(a,b)$. In a similar manner $f_b\colon M \to N$ is defined by $f_b(a) = f(a,b)$ for each $b \in N$. Theorem 1.9 implies that f^a and f_b are C^k_s maps and that $df(a,b)(v,w) = df_b(a)v + df^a(b)w$ for each $v \in T_aM$ and

and $w \in T_bB$.

THEOREM 4.5 (*Parameterized Mapping and Implicit Mapping Theorems*). *Let* M *and* N *be* C_s^k *manifolds and let* $f: M \to N$ *be a* C_s^k *map. Suppose that* $df^a(b)$ *is an isomorphism for some* $(a,b) \in M \times N$. *Then:*

1. *there are open sets* $U_0 \subset M$ *about* a *and* $V_0 \subset N$ *about* b *such that for each* $x \in U_0$, $f^x(V_0)$ *is admissible in* N *and* $f^x|V_0$ *is a* C_s^k *diffeomorphism onto* $f^x(V_0)$ *and*
2. *if* $b \notin \partial N$, *then there are open sets* $U \subset M$ *about* a, $V \subset N$ *about* b, *and* $W \subset N$ *about* $f(a,b)$ *and a* C_s^k *map* $g: U \times W \to V$ *such that for each* $x \in U$, $f^x(V)$ *is an admissible subset of* N *containing* W, $f^x|V$ *is a* C_s^k *diffeomorphism onto* $f^x(V)$, *and*

$$f(x, g(x,w)) = w \text{ for each } (x,w) \in U \times W.$$

PROOF. Transfer to Banach spaces using charts and apply Theorem 1.10. Existence of the map g is provided by Theorem 2.11 of $[G_1]$.

I.5. VECTOR FIELDS: Let M be a C_s^k manifold and let $\pi: TM \to M$ be its tangent bundle. A *vector field on* M is a section of π, i.e. a map $X: M \to TM$ such that $\pi \circ X = 1_M$. If X is a vector field on M and $p \in M$, then $X(p)$ is also denoted by X_p. Thus $\pi(X_p) = p$ for each $p \in M$. If $A \subset E$ is an admissible set, then A is a C_s^∞ E-manifold and the tangent bundle of A may be identified with $\pi_1: A \times E \to A$. If $Y = Y_1 \times Y_2: A \to A \times E$ is a vector field on A, then $Y_1(a) = a$ for all $a \in A$. Thus Y is completely determined by $Y_2: A \to E$. The *principal part* of a vector field Y on A is the function Y_2. A vector field X on M is a C_s^r vector field if $X: M \to TM$ is a C_s^r map. Since TM is a C_s^{k-1}

manifold, $r \le k-1$. If Y is a vector field on $A \subset E$, then Y is C_s^r if and only if Y_2 is a C_s^r map.

Let M and N be C_s^k manifolds, let $f\colon M \to N$ be a C_s^k diffeomorphism, and let X be a C_s^{k-1} vector field on M. Then $Tf(X)$ is the C_s^{k-1} vector field on N defined by $Tf(X)(p) = df(f^{-1}(p))(X(f^{-1}(p)))$ for each $p \in N$. If $\varphi\colon U \to E$ is a chart on M, then $T\varphi(X|U)$ is a C_s^{k-1} vector field on im φ.

Let M be a C_s^k manifold and for each $r \le k-1$, let $V^r(M)$ be the set of all C_s^r vector fields on M. If $X, Y \in V^r(M)$ and $t \in \mathbb{R}$, then $X + tY$, defined by $(X + tY)_p = X_p + tY_p \in T_pM$, is in $V^r(M)$. Thus $V^r(M)$ is a vector space. Now let $F^r(M)$ be the set of all real-valued C_s^r maps on M and let $F^0(M)$ be the set of all continuous real-valued maps on M. Then $F^r(M)$ is a commutative ring with identity $(0 \le r \le k)$. For each $X \in V^{k-1}(M)$ and each $f \in F^k(M)$, define $Xf\colon M \to \mathbb{R}$ by $Xf(p) = df(p)(X_p)$. Thus X may be thought of as a map from $F^k(M)$ into $F^{k-1}(M) \subset F^0(M)$. Moreover, X is a <u>derivation</u> in the sense that $X(f + tg) = Xf + t(Xg)$ and $X(fg) = f(Xg) + (Xf)g$ for all $f, g \in F^k(M)$ and all $t \in \mathbb{R}$. If $A \subset M$ is an admissible set, then $X|A\colon F^k(A) \to F^{k-1}(A)$ is a derivation which we also denote by X.

<u>PROPOSITION 5.1</u>. *Let M be a C_s^k manifold and let $X \in V^{k-1}(M)$. If for each open set $U \subset M$, $X\colon F^k(U) \to F^{k-1}(U)$ is the 0 map, then $X_p = 0$ for each $p \in M$.*

<u>PROOF</u>. Suppose $X_p \ne 0$. Let $\varphi : U \to E$ be a chart at p and let $S \in L(E, \mathbb{R})$ be a linear map such that $S(d\varphi(p)(X_p)) \ne 0$. Then $S \circ \varphi \in F^k(U)$ and $X(S \circ \varphi)(p) = d(S \circ \varphi)(p)(X_p) = S \circ d\varphi(p)(X_p) \ne 0$ which is a contradiction. Thus $X_p = 0$ for all $p \in M$. □

<u>NOTE</u>. Proposition 5.1 implies that if X and Y induce the

the same derivations then $X = Y$ since $X - Y$ is the zero vector field.

PROPOSITION 5.2. Let M be a C_s^k manifold $(k \geq 2)$ and let $X, Y \in V^{k-1}(M)$. Then there is a unique C_s^{k-2} vector field $[X,Y]$ on M such that for each open set U and each $f \in F^k(U)$

$$[X,Y]f = X(Yf) - Y(Xf).$$

PROOF. Proposition 5.1 implies that $[X,Y]$ is unique if it exists. We establish to existence by using local representatives of X and Y. Let $\varphi: U \to A \subset E$ be a chart on M. Then $T\varphi(X)$ and $T\varphi(Y)$ are C_s^{k-1} vector fields on A. Let X_2 and Y_2 denote the principal parts of these vector fields. Let $[T\varphi(X), T\varphi(Y)]_2(a)$ $= dY_2(a)(X_2(a)) - dX_2(a)(Y_2(a))$ for each $a \in A$. If $f \in F_s^k(A)$, then $[T\varphi(X), T\varphi(Y)](f)(a) = df(a)([T\varphi(X), T\varphi(Y)]_2(a) =$ $df(a) \circ dY_2(a)(X_2(a)) - df(a) \circ dX_2(a)(Y_2(a)) =$
$= d^2f(a)(Y_2(a), X_2(a)) + df(a) \circ dY_2(a)(X_2(a)) - d^2f(a)(X_2(a), y_2(a)$ $- df(a) \circ dX_2(a)(Y_2(a)) = d(T\varphi(Y)(f))(a)(X_2(a)) - d(T\varphi(X)(f))(a)$ $- d(T\varphi(X)(f))(a)(Y_2(a)) = T\varphi(X)(T\ (Y)(f))(a) - T\varphi(Y)(T\varphi(X)(f))(a)$. Here we have used the fact that $d^2f(a)$ is a symmetric map. Thus we see that $[T\varphi(X), T\varphi(Y)]$ has the desired effect on functions. Now $T(\varphi^{-1})$ may be used to define $[X,Y]$ on U. It is straightforward but tedious to verify that $[X,Y]$ is well-defined (i.e. independent of the choice of charts.) Additional details of this verification may be found in [G]. A similar computation shows that $[X,Y]$ has the desired effect on functions. Clearly, $[X,Y]$ is C_s^{k-2} differentiable. □

PROPOSITION 5.3. Let M be a C_s^k manifold $(k \geq 3)$, let $X, Y, Z \in V^{k-1}(M)$, and let $t \in \mathbb{R}$. Then:

1. $[X, Y+tZ] = [X,Y] + t[X,Z]$,

2. $[Y,X] = -[X,Y]$, (Anti-commutativity)

3. $[X, [Y,Z]] + [Y, [Z,X]] + [Z, [X,Y]] = 0$. (Jacobi identity)

PROOF. The proof of this proposition follows by applying the vector fields on each side of the equality to a function f to obtain an equality. □

NOTE. Proposition 5.3 implies that the bracket operation $[\cdot,\cdot]$ is a bilinear map.

A Lie algebra is a vector space L together with a bilinear, anti-commutative bracket operation $[\cdot,\cdot]: L \times L \to L$ which satisfies the Jacobi identity. If M is a C_s^∞ manifold, then $V^\infty(M)$ is a Lie algebra by Proposition 5.3. If L and K are Lie algebras, then $f: L \to K$ is a Lie algebra homomorphism if f is linear and $[f(X), f(Y)] = f([X,Y])$ for each $X, Y \in L$.

EXAMPLE 5.4. Let $M = \mathbb{R}^3$ and let X, Y, Z be the C^∞ vector fields with principal parts $X_2(x,y,z) = (1,0,0)$, $Y_2(x,y,z) = (0,1,x)$ and $Z_2(x,y,z) = (0,0,1)$. If $f: M \to \mathbb{R}$ is a C^∞ map, then $Xf(p) = df(p)(1,0,0) = \frac{\partial f}{\partial x}(p)$. Thus we may identify X with the differential operator $\frac{\partial}{\partial x}$. In a similar manner $Y = \frac{\partial}{\partial y} + x\frac{\partial}{\partial z}$ and $Z = \frac{\partial}{\partial z}$. Now $[X,Y]f = X(Yf) - Y(Xf) = X(\frac{\partial f}{\partial y} + x\frac{\partial f}{\partial z}) - Y(\frac{\partial f}{\partial x})$ $= \frac{\partial^2 f}{\partial x\partial y} + \frac{\partial f}{\partial z} + x\frac{\partial^2 f}{\partial x\partial z} - \frac{\partial^2 f}{\partial y\partial x} - x\frac{\partial^2 f}{\partial z\partial x} = \frac{\partial f}{\partial z} = Zf$ by equivalence of mixed partials. Hence $[X,Y] = Z$. It follows that the linear span of $\{X,Y,Z\}$ in $V^\infty(M)$ is a 3-dimensional Lie algebra since $[X,Z] = [Y,Z] = 0$. In fact this Lie algebra is isomorphic to the Lie algebra of the Heisenberg group described in the introduction to this paper.

Let M and N be C_s^k manifolds and let $f: M \to N$ be a C_s^k

map. If f is a diffeomorphism, then each vector field X on M induces a vector field $Tf(X)$ on N. If f is not a diffeomorphism this may be impossible. A vector field X on M is f-related to a vector field Y on N if $df(p)(X_p) = Y_{f(p)}$ for each $p \in M$.

PROPOSITION 5.5. Let M and N be C_s^k manifolds $(k \geq 2)$ and let $f: M \to N$ be a C_s^k map. Suppose that $X_1, X_2 \in V^{k-1}(M)$ and $Y_1, Y_2 \in V^{k-1}(N)$ and that X_i is f-related to Y_i for $i = 1,2$. Then $[X_1,X_2]$ is f-related to $[Y_1,Y_2]$.

PROOF. Since f-relatedness is defined pointwise, we may transfer to Banach spaces using charts. The proof then follows by computation (cf. [G, Theorem 3.22]). □

I.6. INTEGRATION: Let M be a C_s^k manifold and let X be a vector field on M. An integral curve of X is a C_s^1 curve $\alpha: J \to M$ where $J \subset \mathbb{R}$ is an interval such that $\alpha'(t) = X(\alpha(t))$ for each $t \in J$. An integral curve of X with initial condition $p \in M$ is an integral curve $\alpha: J \to M$ of X such that $0 \in J$ and $\alpha(0) = p$. If X is a C_s^r vector field, then α is C_s^{r+1} by the Chain Rule.

Let $A \subset E$ be an admissible set. Then A is a C_s^∞ manifold and TA may be identified with $\pi_1: A \times E \to A$. A vector field X on A is completely determined by its principal part $X_2 = \pi_2 \circ X$ (cf. §I.5). A vector field X on A satisfies a Lipschitz condition on $U \subset A$ if there is a constant $K > 0$ such that $|X(a) - X(b)| \leq K|a-b|$ for all $a, b \in U$. Equivalently, the vector field X on A satisfies a Lipschitz condition on $U \subset A$ if there is a constant $M > 0$ such that $|X_2(a) - X_2(b)| \leq M|a-b|$ for all $a, b \in U$. In a similar manner, X is C_s^k if and only if X_2 is C_s^k. If X is a C_s^1 vector field on A and $p \in A$, then X satisfies a Lipschitz condition on a neighborhood U of p by Theorem 1.4.

A vector field X on A need not have an integral curve with initial condition p if $p \in \partial A = A \setminus A^\circ$. However, uniqueness of integral curves is established in the next two theorems.

THEOREM 6.1. Let $A \subset E$ and let X be a vector field on A which satisfies a Lipschitz condition on A. If $\alpha: I \to A$ and $\beta: J \to A$ are integral curves of X with the same initial condition, then $\alpha = \beta$ on $I \cap J$.

PROOF. If $I \cap J = \{0\}$, the result holds since $\alpha(0) = \beta(0)$. If $I \cap J$ contains an interval of the form $(-\delta, 0]$ for $\delta > 0$, then $c(t) = \alpha(-t)$ and $d(t) = \beta(-t)$ are integral curves of $-X$ defined on $[0,\delta)$. Thus we may assume that $I \cap J$ contains an interval of the form $[0,\delta)$ for $\delta > 0$.

We show now that $\alpha = \beta$ on $[0,\varepsilon]$ for some $0 < \varepsilon < \delta$. Suppose not and let $0 < \varepsilon < \delta$. Then there is a $t \in (0,\varepsilon]$ such that $\alpha(t) \neq \beta(t)$. Let $M > 0$ such that $|X_2(a) - X_2(b)| \leq M|a-b|$ for all $a, b \in A$. Let $f(s) = \alpha(s) - \beta(s)$ for all $s \in [0,\delta)$. Then f is C^1_s and $f'(s) = \alpha'(s) - \beta'(s) = X_2(\alpha(s)) - X_2(\beta(s))$ for each $s \in [0,\delta)$. Thus $|f'(s)| \leq M|f(s)|$ for all $s \in [0,\delta)$. Since f' is continuous, there is an $s \in (0,t]$ such that $|f'(r)| \leq |f'(s)|$ for all $r \in [0,t]$. Then $|f(s)| = |\int_0^s f'(r)dr| \leq \int_0^s |f'(r)|dr \leq s|f'(s)|$, and $s^{-1}|f(s)| \leq |f'(s)| \leq M|f(s)|$. Hence for each $\varepsilon > 0$ there is an $s > 0$ with $s < \varepsilon$ such that $M \geq s^{-1} \geq \varepsilon^{-1}$. Since this is impossible, there is an $\varepsilon > 0$ with $\varepsilon < \delta$ such that $\alpha = \beta$ on $[0,\varepsilon]$.

We now show that $\alpha = \beta$ on $I \cap J \cap [0,\infty)$. Let $K = \{t \geq 0: \alpha = \beta \text{ on } [0,t]\}$. By the previous paragraph there is an $\varepsilon > 0$ such that $[0,\varepsilon] \subset K$. Suppose there is a $t > \varepsilon$ in $I \cap J$ such that $\alpha(t) \neq \beta(t)$. Then t is an upper bound of K.

Let T be the least upper bound of K. It follows that for each $s \in [0,T)$, $s \in K$ and hence $\alpha(T) = \beta(T)$. Now let $\gamma(t) = \alpha(T+t)$ and $\eta(t) = \beta(T+t)$ for $t \in I \cap J - T$. Then γ and η are integral curves of X with initial condition $\alpha(T) = \beta(T)$ defined on the interval $[0, t-T]$. Again by the previous paragraph, there is a $\rho > 0$ such that $\gamma = \eta$ on $[0,\rho]$. It follows that $\alpha = \beta$ on $[0, T+\rho]$ and that T is not an upper bound of K. Hence $\alpha = \beta$ on $I \cap J \cap [0,\infty)$.

Now if $I \cap J \cap (-\infty,0] \neq \{0\}$, then $\alpha(t) = \beta(t)$ on this set since $c(t) = \alpha(-t)$ and $d(t) = \beta(-t)$ are equal on $(-I) \cap (-J) \cap [0,\infty)$. It follows that $\alpha = \beta$ on $I \cap J$. □

<u>NOTE</u>. If $\alpha\colon I \to A$ and $\beta\colon J \to A$ are integral curves of X with $\alpha(t_0) = \beta(t_0)$ for some $t_0 \in I \cap J$, then $\alpha = \beta$ on $I \cap J$ since $c(t) = \alpha(t+t_0)$ and $d(t) = \beta(t+t_0)$ are integral curves of X with the same initial condition.

An integral curve of a vector field X on a manifold M is <u>maximal</u> if it is not properly contained (as a set of ordered pairs) in any integral curve of X. A simple maximality argument shows that each integral curve of X is contained in a maximal integral curve of X.

<u>THEOREM 6.2</u>. *Let X be a C^1_s vector field on a C^k_s manifold M. If $\alpha\colon I \to M$ and $\beta\colon J \to M$ are maximal integral curves of X with the same initial condition then $I = J$ and $\alpha = \beta$.*

<u>PROOF</u>. Suppose that $\alpha(t) \neq \beta(t)$ for some $t > 0$, and let $K = \{s > 0\colon \alpha = \beta \text{ on } [0,s]\}$. Then t is an upper bound for K. Let L be the least upper bound of K. It follows that $\alpha = \beta$ on $[0,L]$ and that for each $\varepsilon > 0$, there is an $s \in (L, L+\varepsilon)$ such that $\alpha(s) \neq \beta(s)$. Now let φ be a chart at $\alpha(L) = \beta(L)$. Then $\varphi\circ\alpha$ and

$\varphi \circ \beta$ are integral curves of $T\varphi(X)$ and $\varphi(\alpha(L)) = \varphi(\beta(L))$. By Theorem 6.1, $\varphi \circ \alpha = \varphi \circ \beta$ on their common domain. But there is an $s > L$ such that $\alpha(s) \neq \beta(s)$ and both are contained in $\mathrm{dom}\,\varphi$. Since this is impossible, $\alpha(t) = \beta(t)$ for each $t \geq 0$ in $I \cap J$. By a similar argument for $t \leq 0$, it follows that $\alpha = \beta$ on $I \cap J$.

Suppose now that $I \neq J$. We assume that $I \setminus J \neq \emptyset$. Let $\gamma(t) = \beta(t)$ for $t \in J$ and let $\gamma(t) = \alpha(t)$ for $t \in I \setminus J$. Then $J \cup (I \setminus J) = J \cup I$ is an interval containing 0. If $I \cap J$ is an interval (i.e. nondegenerate), then γ is clearly C^1_s since α and β are. If $I \cap J = \{0\}$, then α and β have one-sided derivatives at 0 and they are both equal to $X(\alpha(0))$. It follows that γ is C^1_s on $I \cup J$ and is an integral curve of X. Since γ extends β, $\beta = \gamma$ by maximality and hence $I \cup J \subset J$. In a similar way $J \subset I$ and thus $I = J$. □

<u>NOTE</u>. The proof of Theorem 6.2 shows that if $\alpha\colon [0,\varepsilon) \to M$ and $\beta\colon (-\delta,0] \to M$ are integral curves of X with $\alpha(0) = \beta(0)$, then $\alpha \cup \beta$ is an integral curve of X. It also follows that if α and β are maximal integral curves of the same vector field and $\alpha(t) = \beta(t)$ for some $t \in \mathrm{dom}\,\alpha \cap \mathrm{dom}\,\beta$, then $\alpha = \beta$. We turn now to the question of existence of integral curves.

<u>THEOREM 6.3. Let</u> M <u>be a</u> C^k_s <u>manifold</u> $(k \geq 2)$ <u>and let</u> X <u>be a</u> C^1_s <u>vector field on</u> M. <u>If</u> $p \in M \setminus \partial M$ <u>then there is an</u> $\varepsilon > 0$ <u>and a unique integral curve</u> α <u>of</u> X <u>on</u> $(-\varepsilon,\varepsilon)$ <u>such that</u> $\alpha(0) = p$.

<u>PROOF</u>. The vector field $X|M \setminus \partial M$ is a C^1 vector field on the manifold without boundary $M \setminus \partial M$. The theorem follows by the local existence and uniqueness theorem for solutions of differential equations in Banach spaces (cf. [L, p. 63].) □

NOTE. If $\alpha: I \to M$ is a maximal integral curve of X and a is an endpoint of I, then $\alpha(a) \in \partial M$ since $\alpha(a) \notin \partial M$ implies that α can be extended to an interval containing a in its interior. Also if $p \in \partial M$ and $X_p = 0$, then $\alpha(t) = p$ for $t \in \mathbb{R}$ is the unique integral curve of X with initial condition p.

THEOREM 6.4. *Let* X *be a* C_s^1 *vector field on the* C_s^k *manifold* M $(k \geq 2)$ *and let* $p \in \partial M$. *Suppose that there is an integral curve* $\alpha: (a,b) \to M$ *of* X *such that* $p \in (\text{im}\,\alpha)^*$. *Then there is an integral curve of* X *with initial condition* p.

PROOF. Since $p \in (\text{im}\,\alpha)^*$, there is a sequence $\{t_n\}$ in (a,b) such that $\{\alpha(t_n)\}$ converges to p. If $\{t_n\}$ clusters to $t \in (a,b)$, then $\alpha(t) = p \in \text{im}\,\alpha$. By reparameterizing if necessary, we may have $\alpha(0) = p$. Now $\{t_n\}$ must cluster somewhere in $[a,b]$. Without loss of generality, we may assume that $\{t_n\}$ converges to b. Thus for each open set U about p and for each $\varepsilon > 0$, there is a $t \in (b-\varepsilon, b)$ such that $\alpha(t) \in U$.

We now show that for each open set U about p, there is a $t_0 \in (a,b)$ such that $\alpha([t_0,b)) \subset U$. Let U be an open set about p and let φ be a chart at p. Since a manifold is a regular space, there is an open set W about p such that $W^* \subset U \cap \text{dom}\,\varphi$. Now $\varphi(W)$ is open in $\text{im}\,\varphi$. Hence there is an $\varepsilon > 0$ such that $B_\varepsilon(\varphi(p))^* \cap \text{im}\,\varphi \subset \varphi(W)$. Let $V = \varphi^{-1}(B_\varepsilon(\varphi(p)))$. Then $V \subset W$ and $V^* \subset U \cap \text{im}\,\varphi$. Now $T\varphi(X)$ is a C_s^1 vector field on $\text{im}\,\varphi$. We may assume then that ε has been chosen sufficiently small that there is an $M > 0$ such that $|T\varphi(X)_x| \leq M$ whenever $x \in \varphi(V)$. By the preceding paragraph, there is a $t_0 \in \left(b - \frac{\varepsilon}{4M}, b\right)$ such that $\varphi(\alpha(t_0)) \in B_\delta(\varphi(p))$ where $\delta = \frac{\varepsilon}{2}$. Let $K = \{t \in [t_0,b):$ $\alpha([t_0,t]) \subset V\}$. Suppose $K \neq [t_0,b)$ and let L be the least upper bound of K so that $L < b$. Then $\alpha([t_0,L)) \subset V$ and hence

$\alpha(L) \in V^{*} \setminus V$. However, for each $t \in [t_0,b)$, $\varphi(\alpha(t)) - \varphi(\alpha(t_0)) = \int_{t_0}^{t} (\varphi\circ\alpha)^1(s)ds = \int_{t_0}^{t} T\varphi(X)(\varphi(\alpha(s)))ds$. It follows that $|\varphi(\alpha(t)) - \varphi(\alpha(t_0))| \leq M\cdot|t-t_0| \leq \frac{\varepsilon}{4}$. Thus $\varphi(\alpha(t)) \in B_\rho(\varphi(p))$ where $\rho = 3\varepsilon/4$. Then $\varphi(\alpha(L)) \in B_\varepsilon(\varphi(p))$ and $\alpha(L) \in V$. This is a contradiction and so $\alpha([t_0,b)) \subset V \subset U$.

By the previous paragraph, α can be extended to a continuous curve β on $(a,b]$ by defining $\beta(b) = p$. Since $\alpha'(t) = X(\alpha(t))$ approaches X_p as t approaches b, α' can be extended to a continuous function η on $(a,b]$ with $\eta(b) = X_p$. By using the local representatives as above, we may assume that β and η have values in a Banach space E. Let $\gamma(t) = \int_{t_0}^{t} \eta(s)ds$ for $t \in [t_0,b]$. Then γ is C_s^1 on $[t_0,b]$ and $\gamma'(t) = \eta'(t) = \alpha'(t)$ for $t \in [t_0,b)$. It follows that $\gamma(t) = \alpha(t) = \beta(t)$ for $t \in [t_0,b)$ and hence $\gamma(b) = \beta(b)$ since both are continuous. Then $\delta(t) = \beta(t+b)$ is an integral curve of X with initial condition p. □

Let M be a manifold. A <u>local flow</u> on M is a continuous map $\pi: D \subset \mathbb{R} \times M \to M$ where D is an open set such that for each $p \in M$, $\{t \in \mathbb{R} \mid (t,p) \in D\}$ is an open interval about 0, $\pi(0,p) = p$, and if (t,p), $(s, \pi(t,p))$, and $(s+t,p)$ are in D, then $\pi(s, \pi(t,p)) = \pi(s+t, p)$. A <u>flow</u> on M is a local flow with domain $\mathbb{R} \times M$. If π is a C_s^k flow on M, then for each $t \in \mathbb{R}$, $\pi^t: M \to M$ is a C_s^k diffeomorphism of M with inverse π^{-t}. If π is a C_s^1 local flow on M, then π_p is a C_s^1 curve in M with $\pi_p(0) = p$. Define a vector field X_π on M by $X_\pi(p) = \pi_p'(0)$. Each curve π_p is an integral curve of X with initial condition p, since $\pi_p(s+t) = \pi_{\pi(t,p)}(s)$ for sufficiently small s and t. If X is a vector field on M and π is a local flow on M, then π is a <u>local flow</u> for X if $X = X_\pi$.

<u>THEOREM 6.5</u>. *Let X be a C^k_s vector field on the C^{k+1}_s manifold M. Then there is a C^k local flow for $X|M\setminus\partial M$ on $M\setminus\partial M$.*

<u>PROOF</u>. Since X is C^k_s, $X|M\setminus\partial M$ is C^k. The theorem follows from Theorem IV.2.5 of [L]. □

Let $\mathbb{H} = [0,\infty)$. A *semiflow* on a manifold M is a continuous function $\pi: \mathbb{H}\times M \to M$ such that for each $p \in M$ and all $s, t \in \mathbb{H}$, $\pi(0,p) = p$ and $\pi(s, \pi(t,p)) = \pi(s+t,p)$. If π is a C^k_s semiflow on M, then for each $t \geq 0$ $\pi^t: M \to M$ is a C^k_s map, and for each $p \in M$ $\pi_p: \mathbb{H} \to M$ is a C^k_s curve. As above we may define X_π on M by $X_\pi(p) = \pi'_p(0)$ for each $p \in M$. A semiflow π is a *semiflow for* X on M if $X = X_\pi$. A vector field X on M is *semi-integrable* if for each $p \in M$ there is an integral curve $\alpha: \mathbb{H} \to M$ of X with initial condition p. It is an open problem as to whether the C^k_s semiflows on M are precisely the semiflows for C^k_s semi-integrable vector fields on M.

<u>THEOREM 6.6</u>. *Let M be a C^{k+1}_s manifold and let π be a C^k_s semiflow on M such that X_π is a C^k_s vector field on M. Then for each $t \geq 0$, $\pi^t: M \to \pi^t(M) \subset M$ is a C^k_s diffeomorphism onto the admissible subset $\pi^t(M)$ of M.*

PROOF. Suppose that $\pi^t(p) = \pi^t(q)$ for some $p, q \in M$. Then the curves π_p and π_q are integral curves of X and $\pi_p(t) = \pi_q(t)$. By Theorem 6.2, $p = \pi_p(0) = \pi_q(0) = q$. Thus each map π^t is one-to-one on M.

We now show that $d\pi^t(p): T_pM \to T_{\pi(t,p)}M$ is an isomorphism for each $t \geq 0$ and each $p \in M$. Let $p \in M$. Since π^0 is the identity on M, there is by Theorem 4.5 an $\varepsilon > 0$ and an open set U about p such that if $t \in [0,\varepsilon)$, then $\pi^t|U$ is a diffeomorphism. In particular, $d\pi^t(p)$ is an isomorphism for each $t \in [0,\varepsilon)$.

Let $K = \{t \geq 0:\ d\pi^s(p)$ is an isomorphism for each $s \in [0,t]\}$. If $K \neq \mathbb{H}$, then there is a least upper bound T of K. Again by Theorem 4.5 there is a $\delta > 0$ and an open set V about $q = \pi(T,p)$ such that if $t \in [0,\delta)$ then $\pi^t|V$ is a diffeomorphism. Continuity of π implies that there is a $t \in K$ such that $\pi(t,p) \in V$ and $T - t \in [0,\delta)$. Now $\pi^T = \pi^{T-t} \circ \pi^t$ and by the chain rule $d\pi^T(p) = d\pi^{T-t}(\pi(t,p)) \circ d\pi^t(p)$. Thus $d\pi^T(p)$ is an isomorphism, and if $s \in [0,\delta)$ then $d\pi^{T+s}(p) = d\pi^s(q) \circ d\pi^T(p)$ is an isomorphism. Hence K is not bounded above and $d\pi^t(p)$ is an isomorphism for each $t \geq 0$ and each $p \in M$.

It remains to show that π^t is an open map for each $t \geq 0$. The open mapping theorem implies that $\pi^t|M \setminus \partial M$ is an open map into $M \setminus \partial M$ and therefore $\pi^t(M)$ is an admissible subset of M. The parameterized mapping theorem implies that for each $p \in M$ and each $t \geq 0$ there is an open set U about p such that $\pi^t|U$ is a homeomorphism onto $\pi^t(U)$. However, it is not immediately clear that $\pi^t(U)$ is open in $\pi^t(M)$. We now proceed to show that $(\pi^t)^{-1}$ is continuous for each $t \geq 0$.

We claim that for each $p \in M$, there is an $\eta > 0$ such that $(\pi^t)^{-1}$ is continuous on a neighborhood of $\pi(t,p)$ in $\pi^t(M)$ whenever $t \in [0,\eta)$. Let $p \in M$ and let φ be a chart at p. For each $\varepsilon > 0$ let $U_\varepsilon = \varphi^{-1}(B_\varepsilon(\varphi(p)))$. The local representative $T\varphi(X)$ of $X = X_\pi$ is C^k_s and hence locally bounded. Thus there is a $\delta > 0$ and an $L > 1$ such that if $y \in U_\delta$ then $|T\varphi(X)_{\varphi(y)}| < L$. By the parameterized mapping theorem we may assume that δ has been chosen small enough that $\pi^t|U_\delta$ is a homeomorphism onto $\pi^t(U_\delta)$ for each $t \in [0,\delta)$. Since π is continuous, there is a $\rho > 0$ with $\rho > \frac{\delta}{4L}$ such that $\pi^t(U_\rho) \subset U_\delta$ whenever

$t \in [0,\rho)$. In the same way there is an $\eta > 0$ with $\eta < \rho$ such that $\pi^t(U_\eta) \subset U_\rho$ whenever $t \in [0,\eta)$. Let $t \in [0,\eta)$. We claim that $(\pi^t)^{-1}$ is continuous on a neighborhood of $\pi(t,p)$. Let $\gamma = \frac{\delta}{4}$, let $V = \varphi^{-1}(B_\gamma(\varphi(\pi(t,p))))$, and let $y = \pi^t(q) \in V \cap \pi^t(M)$. Note that $\pi(t,p) \in U_\rho$ and that $V \subset U_{\frac{1}{2}\delta}$. Now $\pi_q: [0,t] \to M$ is an integral curve of X. Define $\alpha: [0,t] \to M$ by $\alpha(s) = \pi_q(t-s)$. Then α is an integral curve of $-X$ with $\alpha(0) = y$ and $\alpha(t) = q$. For s sufficiently small, $\alpha(s) \in U_\delta$ so that we may identify $\alpha(s)$ with $\varphi(s)$. Then $|\alpha(s)-y| = |\alpha(s) - \alpha(0)| = |\int_0^s \alpha'(r)dr|$ $\leq |s|L < tL < \frac{\delta}{4}$. Thus for small values of s, $|\alpha(s)-p| \leq$ $|\alpha(s)-y| + |y-\pi(t,p)| < \frac{3\delta}{4}$, when transferred to $\operatorname{im} \varphi$ by φ. It follows that $\alpha(s) \in U_\delta$ for all $s \in [0,t]$ and in particular $q = (\pi^t)^{-1}(y) = \alpha(t) \in U_\delta$. Hence $(\pi^t)^{-1}(V) \subset U_\delta$. Since $\pi^t|U_\delta$ is a homeomorphism onto $\pi^t(U_\delta)$, $(\pi^t)^{-1}$ is continuous on $V \cap \pi^t(M)$.

Now let $K = \{t \geq 0 : (\pi^s)^{-1}$ is continuous at $\pi(s,p)$ whenever $s \in [0,t]\}$. If $K \neq \mathbb{H}$, then K has a least upper bound $T > 0$. Let $\eta > 0$ such that $(\pi^t)^{-1}$ is continuous at $\pi(t,q)$ whenever $t \in [0,\eta)$ where $q = \pi(T,p)$. Let $t = \frac{1}{2}\eta$, let $y = \pi(t,q)$, and let $V \subset M$ be an open set about y such that $(\pi^t)^{-1}$ is continuous on $V \cap \pi^t(M)$. Let U be an open set about q such that $\pi^t(U) \subset V$ and let $\varepsilon > 0$ with $\varepsilon < t$ such that $\pi^{T+s}(p) \in U$ whenever $|s| < \varepsilon$. Let $r = \frac{\varepsilon}{2}$ and let $s \in [0,r)$. Then $t-r-s > 0$. Now let $\{x_\alpha\} = \{\pi^{T+s}(y_\alpha)\}$ be a net in $U \cap \pi^{T+s}(M)$ converging to $\pi^{T+s}(p)$. We claim that $\{y_\alpha\}$ converges to p. Now $\pi^{t-r-s}(x_\alpha)$ converges to $\pi^{t-r-s}(\pi^{T+s}(p)) = \pi^t(\pi^{T-r}(p))$. But $\pi^{t-r-s}(x_\alpha) =$ $\pi^{t-r-s}(\pi^{T+s}(y_\alpha)) = \pi^{T+t-r}(y_\alpha) = \pi^t(\pi^{T-r}(y_\alpha))$. Since $\pi^{T-r}(p) \in U$, $\pi^t(\pi^{T-r}(p)) \in V \cap \pi^t(M)$. It follows from the continuity of $(\pi^t)^{-1}$ on $V \cap \pi^t(M)$ that $\{\pi^{T-r}(y_\alpha)\}$ converges to $\pi^{T-r}(p)$. Since

$T-r \in K$, $\{y_\alpha\}$ converges to p. Thus $T + s \in K$ for $s \in [0,r)$ and T cannot be an upper bound of K. Hence, $(\pi^t)^{-1}$ is continuous at $\pi(t,p)$ for all $t \geq 0$. Therefore, π^t is a homemmorphism from M onto $\pi^t(M)$.

Finally, since π^t is a homeomorphism and a local C_s^k diffeomorphism, it follows that π^t is a C_s^k diffeomorphism onto the admissible set $\pi^t(M)$. □

EXAMPLE 6.7. Let X and Y be the vector fields of example 5.4, namely $X = \frac{\partial}{\partial x}$ and $Y = \frac{\partial}{\partial y} + x\frac{\partial}{\partial z}$ on $\mathbb{R}^3$. The flow for X is given by $\pi(t,x,y,z) = (x+t,y,z)$, and the flow for Y is given by $(s,x,y,z) = (x,y+s,z+xs)$. Each map π^t and φ^s is a C^∞ diffeomorphism of $\mathbb{R}^3$ for all $t, s \in \mathbb{R}$. The subgroup G of the group of all C^∞ diffeomorphisms of $\mathbb{R}^3$ generated by $\{\pi^t: t \in \mathbb{R}\} \cup \{\varphi^s: s \in \mathbb{R}\}$ is isomorphic to the Heisenberg group described in the introduction. The subsemigroup S of G corresponding to the previously described subsemigroup of the Heisenberg group is the semigroup generated by $\{\pi^t: t \geq 0\} \cup \{\varphi^s: s \geq 0\}$. The semigroup S acts on $\mathbb{R}^3$ in a natural manner and the orbit of a point $p \in \mathbb{R}^3$ is precisely the set of points which can be reached by piecewise C^∞ curves, starting at p, which are integral curves of X or Y on subintervals.

PART II. DIFFERENTIABLE SEMIGROUPS

II.1. DEFINITION OF A C_s^k SEMIGROUP: A C_s^k semigroup $(1 \leq k \leq \infty)$ is a semigroup S on a C_s^k manifold (with generalized boundary) such that the multiplication $m: S \times S \to S$ is a C_s^k map, where $S \times S$ carries the product C_s^k structure. A C_s^k monoid is a C_s^k semigroup with a two-sided identity element, usually denoted by 1.

A C^k_s group is a C^k_s semigroup which is algebraically a group. A <u>Lie group</u> is a finite-dimensional C^∞_s group.

<u>NOTE</u>: If S is a C^k_s semigroup, then S is a topological semigroup since a manifold is a Hausdorff space. If $a \in S$, then <u>left translation by</u> a, defined by $\lambda_a: S \to S: b \to ab$, is a C^k_s map by Theorem I.1.9. In a similar way <u>right translation by</u> a, denoted by ρ_a, is a C^k_s map. If G is a C^k_s group, then λ_g is a C^k_s diffeomorphism with inverse $\lambda_{g^{-1}}$. It follows that $\partial G = \emptyset$. In this case C^k_s differentiability is equivalent to C^k differentiability. Moreover it is shown in Proposition 2.2 that if G is a C^k_s group, then inversion is a C^k_s diffeomorphism of G.

Let G be a Lie group and let S be a subsemigroup of G with dense interior. Then S is a C^∞_s semigroup. Thus cones in $\mathbb{R}^n$ are C^∞_s monoids as are ray subsemigroups of Lie groups (cf. §II.4 or [H-L].) In particular the Lie semigroups of Hofmann and Lawson [H-L] (this volume) are C^∞_s monoids. These examples illustrate the diversity of geometric structure that can occur. For instance in the subsemigroup of the Heisenberg group described in the introduction to this paper the identity of the semigroup lies at a cusp of the boundary. The presence of the cusp at the identity indicates the possible deficiencies of utilizing partial or directional derivatives for the description of a differentiable monoid.

A C^k_s semigroup need not be a subsemigroup of a group. For example the semigroup $M_n(\mathbb{R})$ of all $n \times n$ real matrices is a C^∞_s monoid on a manifold without boundary. In general the Banach space $L(E,E)$ of endomorphisms of the Banach space E is a C^∞_s monoid under composition of operators. In these examples the subset of linear isomorphisms is an open dense subgroup of the semigroup. Other examples may be found with the aid of the following proposition.

PROPOSITION 1.1. *Let T be a C_s^k semigroup and let S be a subsemigroup of T. If S is a C_s^k submanifold of T, then S is a C_s^k semigroup. In particular, if S is an admissible subset of T, then S is a C_s^k semigroup.*

PROOF. Since S is a submanifold of T, the inclusion map $i: S \to T$ is an embedding. Thus $m|S \times S : S \times S \to S$ is continuous. By Theorem I.4.2 $m|S \times S$ is a C_s^k map. Finally, if S is an admissible subset of T, then S is a submanifold of T with the relative C_s^k structure. □

Examples of connected one-dimensional C_s^∞ monoids include the circle group T, the real line under multiplication or addition, the interval [0,1] under multiplication, and intervals of the form [a,1] or (a,1] for $a \in [-1,0]$ under multiplication. It is an open problem as to whether this list contains all connected one-dimensional C_s^∞ monoids.

The closed unit disc D in the complex plane is a compact connected commutative C_s^∞ monoid under complex multiplication. The space of D is a manifold with smooth boundary and ∂D is the circle group T. Since a finite Cartesian product of C_s^k semigroups is a C_s^k semigroup, semigroups of the form $[-1,1]^p \times D^q$ and $[-1,1]^p \times D^q \times T^r$ are compact, connected commutative C_s^∞ monoids.

An example of a noncommutative compact connected C_s^∞ monoid is the affine triangle. Specifically, let T be the semigroup of all real matrices of the form $\begin{pmatrix} x & y \\ 0 & 1 \end{pmatrix}$. Then T is a C^∞ monoid on a manifold without boundary. The affine triangle is the subsemigroup S of T defined by $x, y \geq 0$ and $x+y \leq 1$. The minimal ideal of S is the subset of S defined by $x = 0$ and $y \in [0,1]$.

All examples considered thus far may be realized as connected, admissible submonoids of a C^∞ monoid on a manifold without boundary. It is unknown if all connected C_s^∞ monoids arise in this manner. We also note that all compact examples have faithful matrix representations and that it is unknown if this is always the case. Of course, a compact connected Lie group has a faithful representation as unitary matrices by the Peter-Weyl Theorem.

As a final example we show that any C_s^k manifold supports the structure of a C_s^k semigroup. Let M be a C_s^k manifold and give M the left trivial multiplication, i.e. $pq = p$ for $p, q \in M$. Then M is a C_s^k semigroup. For each $p \in M$, $p^2 = p$, λ_p is a trivial map and ρ_p is the identity map. The minimal ideal of the affine triangle has left trivial multiplication. This example illustrates the potential for pathology in the absence of a two-sided identity.

PROPOSITION 1.2. *If S is a locally compact C_s^k semigroup, then S is finite-dimensional.*

PROOF. Suppose that the space of S is an E-manifold. The set $S \setminus \partial S$ is open and an E-manifold without boundary. If $p \in S \setminus \partial S$ and φ is a chart at p, then there is an open set U about p such that U^* is compact and contained in $\operatorname{dom} \varphi \cap (S \setminus \partial S)$. Then $\varphi(U)$ is open in E and $\varphi(U^*) = \varphi(U)^*$ is compact. It follows that E is locally compact and hence finite-dimensional. Thus S is finite-dimensional. □

NOTE. The proof of Proposition 1.2 shows that a locally compact manifold is finite-dimensional. The converse does not hold since admissible subsets of $\mathbb{R}^n$ need not be locally compact.

II.2. MORPHISMS AND SUBOBJECTS: If S is a C^k_s semigroup, then the tangent space $T_{(a,b)}S \times S$ may be identified with $T_aS \times T_bS$.

THEOREM 2.1 (PRODUCT RULE). *Let (S,m) be a C^k_s semigroup and let $a, b \in S$. Then*

$$dm(a,b)(v,s) = d\rho_b(a)(v) + d\lambda_a(b)(w)$$

for each $(v,w) \in T_aS \times T_bS$.

PROOF. This theorem follows immediately from Theorem I.1.9. □

PROPOSITION 2.2. *Let G be a C^k_s group and let $\theta: G \to G$ be the inversion map. Then θ is a C^k_s map.*

PROOF. Define $f: G \times G \to G \times G$ by $f(x,y) = (x, xy)$. Then f is a one-to-one C^k_s map and $f^{-1}(z,w) = (z, z^{-1}w)$. Now $df(1,1)(v,w) = (v, v+w)$ by the Product Rule and it follows that $df(1,1)$ is an isomorphism. Thus f^{-1} is C^k_s on a neighborhood of $(1,1)$ and therefore θ is C^k_s on a neighborhood U of 1 since $\theta(z) = z^{-1} = \pi_2(f^{-1}(z,1))$. Note also that $\theta(z) = pp^{-1}z^{-1} = p(zp)^{-1} = \lambda_p \circ \theta \circ \rho_p(z)$ for any $p \in G$. If $q \in G$, then θ is C^k_s on Uq since $\theta = \lambda_p \circ \theta \circ \rho_p$ for $p = q^{-1}$ and θ is C^k_s on $U = \rho_p(Uq)$. It follows that θ is a C^k_s diffeomorphism of G. □

Let S and T be monoids. A *local homomorphism* from S to T is a continuous map $f: U \to T$, where $U \subset S$ is an open set about 1, such that $f(1) = 1$ and $f(xy) = f(x)f(y)$ whenever $x, y, xy \in U$. A *homomorphism* from S to T is a local homomorphism with domain S. If S and T are C^k_s monoids, then f is a [*local*] C^k_s *homomorphism* if f is a [local] homomorphism and a C^k_s map, The collection of C^k_s monoids is a category with morphisms the C^k_s homomorphisms. This category is closed under finite Cartesian products by Theorem I.1.5.

Homomorphisms and C_s^k homomorphisms of C_s^k semigroups are defined in the obvious way. Homormophisms between monoids will be assumed to satisfy the condition $f(1) = 1$ unless it is stated that f is a homomorphism of semigroups.

Let S be a C_s^k semigroup and let T be a subsemigroup of S. Then T is a C_s^k subsemigroup of S if T can be given the structure of a C_s^k manifold in such a way that: (i) T is a C_s^k semigroup, and (ii) the inclusion map $i: T \to S$ is a C_s^k immersion. In particular, a C_s^k subsemigroup is an immersed C_s^k submanifold. The semigroup T is an embedded C_s^k subsemigroup of S if T is a C_s^k subsemigroup of S and $i: T \to S$ is an embedding. Finally, if S and T are monoids, then T is a C_s^k submonoid of S if T is a C_s^k subsemigroup of S and the identity of T is the identity of S. Embedded C_s^k submonoids are defined analogously.

NOTE: If G is a Lie group and H is a C_s^∞ subgroup, then H is a Lie subgroup of G (i.e. H is a Lie group and $i: H \to G$ is an immersion) by Proposition 2.2. A Lie subgroup need not be an embedded submanifold (e.g. a dense wind in a torus). However, a closed subgroup of G is an embedded Lie subgroup of G (cf. [S]). A closed submonoid of a C_s^k monoid need not be an embedded C_s^k submonoid. For example, let T be the submonoid of the affine triangle defined by $x = 0$ or $y = 0$ and let p be the element of T with $x = y = 0$. In the relative topology, T is an arc and p is an interior point of the arc. Now ρ_p is the constant p map on the subarc $[p,1]$ from p to 1 and ρ_p is the identity on the complement of $[p,1]$ in T. Hence ρ_p cannot be differentiable at p and T is not a C_s^k subsemigroup with the relative topology. However, T can be given the structure of a C_s^k (immersed) submonoid by topologizing T so that the components of T are $(p,1]$ and its complement.

Differentiable subsemigroups of a C_s^k semigroup S may be found with the aid of Theorem I.4.4. We shall use the following notations. The set of idempotents of S is the set $E = E(S) = \{e \in S: e^2 = e\}$. If $e \in E(S)$, then $eS = \{ex: x \in S\} = \{x \in S: ex = x\}$, $Se = \{x \in S: xe = x\}$, and $eSe = eS \cap Se = \{x \in S: ex = xe = x\}$. The maximal subgroup of eSe containing e is denoted by $H(e)$. If S is compact, then the minimal ideal of S is denoted by $M(S)$.

<u>NOTE</u>. If $e \in E(S)$, then $\lambda_e \circ \lambda_e = \lambda_e$ and $\rho_e \circ \rho_e = \rho_e$, i.e. λ_e and ρ_e are idempotent maps. If $x \in eS$, then $ex = x$ and $d\lambda_e(x): T_xS \to T_xS$. By the chain rule, $d\lambda_e(x)$ is an idempotent (i.e. a projection) for each $x \in eS$. In a similar way $d\rho_e(y)$ is an idempotent for each $y \in Se$. Now associativity of multiplication implies that $\lambda_x \circ \rho_y = \rho_y \circ \lambda_x$ for all $x, y \in S$. In particular, if $z \in eSe$, then $d\lambda_e(z)$ and $d\rho_e(z)$ are commuting projections since $d\lambda_e(z) \circ d\rho_e(z) = d(\lambda_e \circ \rho_e)(z) = d(\rho_e \circ \lambda_e)(z) = d\rho_e(z) \circ d\lambda_e(z)$. Thus $\operatorname{im} d(\lambda_e \circ \rho_e)(z) = \operatorname{im} d\lambda_e(z) \cap \operatorname{im} d\rho_e(z)$ for each $z \in eSe$. The map $1 - d\lambda_e(z)$: $T_zS \to T_zS$ is projection onto the kernel of $d\lambda_e(z)$, which is a closed complement of $\operatorname{im} d\lambda_e(z)$ in T_zS. Hence the tangent space T_zS for $z \in eSe$ may be decomposed into the sum of four subspaces of the form $V \cap W$ where V is either the image or kernel of $d\lambda_e(z)$ and W is either the image or kernel of $d\rho_e(z)$. By deleting any trivial summands, we obtain a direct sum decomposition

of T_zS.

The following theorem and Theorem 2.6 are straightforward generalizations of a theorem of Holmes [H1].

THEOREM 2.3. *Let* S *be a connected* C^k_s *semigroup and let* $e \in E$. *Then* eS, Se, *and* eSe *are embedded* C^k_s *semigroups. If* $x \in eS$, $y \in Se$, *and* $z \in eSe$, *then* $d\lambda_e(x)$ *is projection onto* $T_x(eS)$, $d\rho_e(y)$ *is projection onto* $T_y(Se)$, *and* $d\lambda_e(z) \circ d\rho_e(z) = d\rho_e(z) \circ d\lambda_e(z)$ *is projection onto* $T_z(eSe)$.

PROOF. Since $(\lambda_e)^2 = \lambda_e$ and $\lambda_e(S) = eS$, eS is an embedded C^k_s submanifold of S by Theorem I.4.4. Since eS carries the relative topology, $m | eS \times eS$ is continuous as a map into eS. By Proposition I.4.2, eS is a C^k_s semigroup. In a similar way, $Se = \rho_e(S)$ and $eSe = (\lambda_e \circ \rho_e)(S)$ are embedded C^k_s subsemigroups of S. Now by the proof of Theorem I.4.4, $T_x(eS) = \operatorname{im} d\lambda_e(x)$ for each $x \in eS$ and similarly for Se and eSe. The remainder of the proof follows from the comments preceding the statement of the theorem. □

NOTE. If $e^2 = e$, then $\rho_e | eS$ is an idempotent homomorphism with image eSe. Thus eSe is an embedded C^k_s subsemigroup of eS.

COROLLARY 2.4. *Let* S *be a connected* C^k_s *monoid and let* $e = e^2$ *be a nonidentity idempotent. Then* eS *has empty interior and* $d\lambda_e(e)$ *is projection onto the proper subspace* $T_e(eS)$ *of* T_eS.

PROOF. If eS has interior, then eS has dense interior since eS is a submanifold. Now eS is clearly closed and we claim that eS is open. Let $x \in eS$. Since $\lambda_e | eS$ is the identity, $d\lambda_e(x)$ is the identity. Therefore there is an open set

$U \subset S$ about x such that $\lambda_e|U$ is one-to-one. Let $V \subset S$ be an open set about x such that $eV \subset U$. If $y \in V$, then ey and y are in U and $e(ey) = ey$. Thus $ey = y$ and $V \subset eS$. Hence $eS = S$. But $e \cdot 1 = 1 = e$. Since $e \neq 1$, eS cannot have interior. Now $d\lambda_e(e)$ is projection onto $T_e(eS)$ by Theorem 2.3. If $d\lambda_e(e)$ is the identity, then λ_e is a diffeomorphism on a neighborhood of e. In this case, eS must have interior. Thus $d\lambda_e(e)$ is not the identity map and $T_e(eS)$ is a proper subspace of T_eS. □

NOTE. If S has dimension n and $e^2 = e \neq 1$, then the rank of $d\lambda_e(e)$ must be less than n.

LEMMA 2.5. *Let S be a C_s^k monoid. Then there is a neighborhood U of 1 such that if $x \in U \setminus \{1\}$, then $x^n \notin U$ for some positive integer n.*

PROOF. Let $\varphi: W \to E$ be a chart at 1. Without loss of generality, we may assume that W is an admissible subset of a Banach space E and that $1_S = 0_E \in W$. Let V be an open neighborhood of 1 such that $V^2 \subset W$ and identify V with $\varphi(V)$. Now let $g(x) = x^2$ for $x \in V$ so that $g(V) \subset W$. Then $dg(0)(v) = 2v$ for each $v \in E$ by the Product Rule. Let $U \subset V \cap B_1(0)$ be a relatively open neighborhood of 0 such that

$$\frac{|g(x) - g(y) - dg(0)(x-y)|}{|x - y|} < \frac{1}{2}$$

whenever $x, y \in U$. Then

$$\frac{|x^2 - 2x|}{|x|} = \frac{|g(x) - g(0) - dg(0)(x)|}{|x|} < \frac{1}{2} .$$

Since $\frac{|2x|}{|x|} = 2$, it follows that $\frac{|x^2|}{|x|} > \frac{3}{2}$ or $|x^2| > \frac{3}{2}|x|$ for each $x \in U \setminus \{0\}$. If $x \in U \setminus \{0\}$ and $x^n \in U$ for each n, then $|x^{2n}| > (\frac{3}{2})^n |x|$ for each n which is impossible since $U \subset B_1(0)$. □

THEOREM 2.6. <u>Let</u> S <u>be a compact connected</u> C_s^k <u>monoid and let</u> $e^2 = e \in M(S)$. <u>Then</u> $eS \cap E$ <u>and</u> $Se \cap E$ <u>are embedded</u> C_s^k <u>subsemigroups of</u> S <u>and</u> eSe <u>is a Lie group</u>. <u>If</u> $x \in eS \cap E$ <u>and</u> $y \in Se \cap E$, <u>then</u> $T_x(eS \cap E) = \operatorname{im} d\lambda_e(x) \cap \ker d\rho_e(x)$ <u>and</u> $T_y(Se \cap E) = \operatorname{im} d\rho_e(y) \cap \ker d\rho_e(y)$.

PROOF. It is well known that if S is compact and $e^2 = e \in M(S)$, then $H(e) = eSe$ (cf. [H-M, p. 16].) Thus eSe is a compact connected C_s^k group and inversion is a C_s^k map by Proposition 2.2. By Lemma 2.5 eSe has no small subgroups and therefore is a Lie group (cf. [M-Z]) with respect to some C^∞ structure on eSe. We note that if S is a C_s^∞ monoid, then the Lie group C^∞ structure of eSe is unique and hence must be the C^∞ structure given by Theorem 2.3.

We now show that $eS \cap E$ is an embedded C_s^k subsemigroup of S. Define $p: S \to S$ by $p(x) = \theta(exe)x$ where θ is the inversion map on the group eSe. The map p is a C_s^k map by the chain rule. A straightforward calculation shows that $p^2 = p$ and $p(S) = eS \cap E$. Thus $eS \cap E$ is an embedded C_s^k subsemigroup of S.

By the proof of Theorem I.4.4, $T_e(eS \cap E) = \operatorname{im} dp(e)$. The chain rule and product rule imply that $dp(e) = d\lambda_e(e) \circ (1 - d\rho_e(e))$. Thus $\operatorname{im} dp(e) = \operatorname{im} d\lambda_e(e) \cap \operatorname{im}(1 - d\rho_e(e)) = \operatorname{im} d\lambda_e(e) \cap \ker d\rho_e(e)$. Now $eS \cap E$ has right trivial multiplication (cf. [H-M, pp. 13,16]), and if $x \in eS \cap E$, then $eS \cap E = xS \cap E$. It follows that $T_x(eS \cap E) = \operatorname{im} d\lambda_x(x) \cap \ker d\rho_x(x)$ for each $x \in eS \cap E$. Moreover, $ex = x = x^2$ implies that $d\lambda_e(x) \circ d\lambda_x(x) = d\lambda_x(x)$ and hence $\operatorname{im} d\lambda_x(x) \subset \operatorname{im} d\lambda_e(x)$. Dually, $xe = e$ implies that $d\lambda_x(x) \circ d\lambda_e(x) = d\lambda_e(x)$ and hence $\operatorname{im} d\lambda_e(x) \subset \operatorname{im} d\lambda_x(x)$. By similar computations of $d\rho_e(f)$ and $d\rho_f(f)$, we see that $\ker d\rho_e(f) = \ker d\rho_f(f)$. Thus

$T_x(eS \cap E) = \text{im } d\lambda_e(x) \cap \ker d\rho_e(x)$. The case for $Se \cap E$ is handled analogously. □

COROLLARY 2.7. *Let S be a C^k_s semigroup and let $e \in E$. Then there is an open set U about e such that if $x \in (eSe \cap U) \setminus \{e\}$, then $x^n \notin U$ for some positive integer n.*

PROOF. The semigroup eSe is a C^k_s semigroup with identity e by Theorem 2.3. Hence there is an open set $V \subset eSe$ about e such that $x \in V \setminus \{e\}$ implies $x^n \notin V$ for some n by Lemma 2.5. Since eSe carries the relative topology, $V = U \cap eSe$ for some open set $U \subset S$. □

COROLLARY 2.8. *If S is a locally compact C^k_s semigroup and G is a closed connected subgroup of S, then G is a Lie group.*

PROOF. Let e be the identity of G so that $G \subset eSe$. Let $U \subset S$ be an open set about e as in Corollary 2.7. Then $U \cap G$ contains no subgroup other than $\{e\}$. Since G is closed, G is locally compact. It follows from a theorem of Ellis [E] that inversion is continuous and that G is a topological group. Thus G is a locally compact connected topological group with no small subgroup and hence is a Lie group (cf. [M-Z].) □

NOTE. It will be shown in Proposition 6.6 that if S is a C^∞_s semigroup, then the inclusion map $i: G \to S$ is a C^∞ embedding. We also note that a locally compact C^k_s semigroup is finite-dimensional by Proposition 1.2.

LEMMA 2.9. *Let S be a C^k_s semigroup and let $e \in E$. Then there is an open set U about e such that $U \cap Se \cap E$ has left trivial multiplication.*

PROOF. Let $\varphi: W_0 \to E$ be a chart at e and let W be an open set about e such that $W^2 \subset W_0$. Without loss of generality, we may assume that $W_0 \subset E$ and that $e = 0 \in E$. Let V be a relatively open set about e such that if $(x,y), (p,q) \in V \times V$ and $(x,y) \neq (p,q)$ then

$$(*) \qquad \frac{|xy - pq - dm(e,e)(x-p,\ y-q)|}{|(x-p,\ y-q)|} < \frac{1}{2},$$

where $E \times E$ carries the sum norm $|(x,y)| = |x| + |y|$. Let $f \in V \cap Se \cap E$ with $f \neq e$. Then $fe = f = f^2$ and by (*)

$$\frac{|f^2 - fe - dm(e,e)(0,f-e)|}{|f-e|} = \frac{|dm(e,e)(0,f-e)|}{|f-e|} < \frac{1}{2}.$$

It also follows from (*) that

$$\frac{|ef - e^2 - dm(e,e)(0,f-e)|}{|f-e|} < \frac{1}{2}.$$

Then

$$\frac{|ef|}{|f|} = \frac{|ef - e|}{|f - e|} \leq \frac{|ef - e - dm(e,e)(0,f-e)|}{|f - e|} + \frac{|dm(e,e)(0,f - e)|}{|f - e|}$$

$$< 1$$

Thus $|ef| < |f|$ for each $f \in V \cap Se \cap E$ with $f \neq e$. Let U be a relatively open set about e such that $U^2 \subset V$ and let $f \in U \cap Se \cap E$. Then $(ef)e = e(fe) = ef$, $(ef)(ef) = eff = ef$ and hence $ef \in V \cap Se \cap E$. If $ef \neq e$, then $|ef| = |e(ef)| < |ef|$ which is impossible. Thus $ef = e$ for each $f \in U \cap Se \cap E$.

Now let $f, \hat{f} \in U \cap Se \cap E$. Then $f\hat{f} = (fe)\hat{f} = f(e\hat{f}) = fe = f$ and $U \cap Se \cap E$ has left trivial multiplication.

PROPOSITION 2.10. Let S be a C_s^k semigroup and let $e \in E$. Then the component of $Se \cap E$ containing e has left trivial multiplication.

PROOF. Let $A = \{f \in Se \cap E: ef = e\}$. Then $e \in A$ and A is clearly closed. Also A has left trivial multiplication since $f, \hat{f} \in A$ implies $f\hat{f} = (fe)\hat{f} = f(e\hat{f}) = fe = f$. We claim that A is open. Let $f \in A$. By Lemma 2.9 there is an open set U about f such that $U \cap Sf \cap E$ is left trivial. Now let $\hat{f} \in V \cap Se \cap E$. Then $\hat{f}f = (\hat{f}e)f = \hat{f}(ef) = \hat{f}e = \hat{f}$ so $\hat{f} \in V \cap Sf \cap E$. Thus $e\hat{f} = (ef)\hat{f} = e(f\hat{f}) = ef = e$ and $\hat{f} \in A$. Since $f \in V \cap Se \cap E \subset A$, A is open and closed and hence contains the component of $Se \cap E$ containing e. □

NOTE. There is a dual result for $eS \cap E$.

II.3. THE LIE ALGEBRA: Let S be a C^k_s semigroup. A vector field X on S is right-invariant if

$$d\rho_b(a)(X_a) = X_{ab}$$
$$\text{for each } a, b \in S.$$

Equivalently, a vector field X is right-invariant if S is ρ_b-related to itself for all $b \in S$ (cf. §I.5). A vector field X is left-invariant if X is λ_a-related to itself for all $a \in S$. The collection of right-invariant vector fields of S is denoted by $L(S)$. It is straightforward to verify that $L(S)$ is a vector subspace of the vector space of all vector fields on S.

NOTE. The right-invariant vector fields of S are of particular significance when considering differentiable left actions of S on a manifold (cf. [G] or [G_2].) It is for this reason that we focus on right rather than left invariance of vector fields.

PROPOSITION 3.1. *Let S be a C^k_s monoid. Then $L(S)$ is linearly isomorphic to the tangent space T_1S.*

PROOF. For each $v \in T_1S$, let $X = X(v)$ be the vector field defined by $X_a = d\rho_a(1)(v)$. If $a, b \in S$, then $X_{ab} = d\rho_{ab}(1)(v)$ $= d\rho_b(a)(d\rho_a(1)(v)) = d\rho_b(a)(X_a)$ and thus X is right invariant. Also $X(v+w) = X(v) + X(w)$ for each $v, w \in T_1S$ by the linearity of $d\rho_a(1)$ for each $a \in S$. Thus the map from T_1S into $L(S)$ which sends v to $X(v)$ is a linear map. Now if $Y \in L(S)$, then $Y = X(Y_1)$, i.e. $Y_a = d\rho_a(1)(Y_1)$ for all $a \in S$. Hence the map that sends v to $X(v)$ is a linear isomorphism. □

NOTE. Since T_1S can be given the structure of a Banach space, $L(S)$ may be given the structure of a Banach space.

PROPOSITION 3.2. *Let S be a C_s^k monoid and let $X \in L(S)$. Then X is continuous, and if $k > 1$, then X is C_s^{k-1} differentiable.*

PROOF. If $X \in L(S)$, then $X_a = d\rho_a(1)(X_1) = dm(1,a)(X_1,0)$ by the product rule. Since dm is continuous, X is continuous. If $k > 1$, then dm is C_s^{k-1} and thus X is C_s^{k-1} by the chain rule. □

THEOREM 3.3. *Let S be a C_s^k monoid with $k > 2$. If $X, Y \in L(S)$, then $[X,Y] \in L(S)$.*

PROOF. If $X, Y \in L(S)$, then X and Y are ρ_a-related to themselves for all $a \in S$. By Theorem I.5.5, $[X,Y]$ is ρ_a-related to itself for all $a \in S$ and hence $[X,Y]$ is right-invariant. □

If S is a C_s^k monoid with $k \geq 3$, then the *Lie algebra of* S is the Lie algebra $L(S)$ under Lie bracket of vector fields. The fact that $L(S)$ is a Lie algebra follows from Theorem 3.3 and Theorem I.5.3. In a similar way the collection of left-invariant vector fields of S is a Lie algebra. It is shown in Proposition 6.7 that these algebras are isomorphic.

Let S and T be C_s^k monoids and let $f: S \to T$ be a C_s^k

local homomorphisms. Define $L(f): L(S) \to L(T)$ by

$$L(f)(X)_p = d\rho_p(1)(df(1)(X_1))$$
$$\text{for } X \in L(S),\ p \in T.$$

Note that $L(f)(X)$ is the unique $Y \in L(T)$ such that $Y_1 = df(1)(X_1)$.

<u>THEOREM 3.4. Let</u> S <u>and</u> T <u>be</u> C_s^k <u>monoids with</u> $k \geq 3$ <u>and let</u> $f: S \to T$ <u>be a</u> C_s^k <u>local homomorphism. Then</u> $L(f)$ <u>is a Lie algebra homomorphism.</u>

<u>PROOF</u>. Clearly $L(f)$ is a linear map. Since f is a local homomorphism, $f \circ \rho_a = \rho_{f(a)} \circ f$ on a neighborhood of 1 for each $a \in S$ sufficiently near 1. If $X \in L(S)$ and $a \in \text{dom } f$, then $df(a)(X_a) = df(a)(d\rho_a(1)(X_1) = d(f \circ \rho_a)(1)(X_1) = d(\rho_{f(a)} \circ f)(1)(X_1)$ $= d\rho_{f(a)}(1)(df(1)X_1) = L(f)(X)_{f(a)}$. Thus each $X \in L(S)$ is f-related to $L(f)(X)$. Hence if $X, Y \in L(S)$, $[X,Y]$ is f-related to $[L(f)X, L(f)Y]$ by Theorem I.5.5. By a similar computation to the one above, it follows that $L(f)([X,Y]) = [L(f)X, L(f)Y]$ and $L(f)$ is a Lie algebra homomorphism. □

<u>NOTE</u>. In categorical terms, L is a functor from the category of C_s^k monoids $(k \geq 2)$ into the category of Lie algebras. The verification of this fact is straightforward.

<u>PROPOSITION 3.5. Let</u> S <u>be a</u> C_s^k <u>monoid</u> $(k \geq 3)$ <u>and let</u> $p \in S$. <u>Then</u> $K_p = \{X \in L(S): X_p = 0\}$ <u>is a subalgebra of</u> $L(S)$.

<u>PROOF</u>. Suppose $X_p = Y_p = 0$ for $X, Y \in L(S)$. By using a chart at p, we may assume that X and Y are the principal parts of their local representatives. In this case, the principal part of $[X,Y]_p = dY(p)(X_p) - dX(p)(Y_p) = 0$. Thus $[X,Y] \in \{Z \in L(S): Z_p = 0\}$. □

<u>NOTE</u>. If $X \in L(S)$, then $X_p = 0$ if and only if $X_1 \in \ker d\rho_p(1)$.

In a similar way, if $\tilde{L}(S)$ is the collection of left invariant vector fields of S, then $\tilde{L}(S)$ is a Lie algebra, and $\{X \in \tilde{L}(S): X_p = 0\}$ is a subalgebra. We also note that if e is a right zero for S, i.e. $xe = e$ for all $x \in S$, then ρ_e is a constant and hence $X_e = 0$ for all $X \in L(S)$.

<u>II.4. ONE-PARAMETER SEMIGROUPS AND RAY SEMIGROUPS</u>: Let S be a monoid. A [local] <u>one-parameter subsemigroup of</u> S is a [local] homomorphism from $\mathbb{H} = ([0,\infty), +)$ into S. A [<u>local</u>] <u>one-parameter submonoid of</u> S is a [local] one-parameter subsemigroup α of S such that $\alpha(0) = 1_S$. A [<u>local</u>] <u>one-parameter subgroup of</u> S is a [local] homomorphism from the additive real numbers into S.

<u>NOTE</u>. It will be shown in Proposition 6.2 that if S is a finite-dimensional C_s^∞ monoid, then each one-parameter subsemigroup or subgroup of S is a C_s^∞ map. Also, if α is a local one-parameter subsemigroup [subgroup] of S, then there is a one-parameter subsemigroup [subgroup] $\hat{\alpha}$ of S such that $\alpha = \hat{\alpha} \mid \text{dom}\,\alpha$ (cf. [H-M, p. 99]).

Let S be a C_s^k monoid and let α be a C_s^k one-parameter submonoid of S. Then α defines a C_s^k semiflow (cf. I.6) on S by $\pi(t,x) = x\alpha(t)$, since $\alpha(s)\alpha(t) = \alpha(s+t) = \alpha(t)\alpha(s)$ for all $s, t \geq 0$. Thus $\pi^t = \rho_{\alpha(t)}: S \to S\alpha(t)$ is a C_s^k diffeomorphism by Theorem I.6.6. Thus we have the following:

<u>THEOREM 4.1. Let</u> S <u>be a</u> C_s^k <u>monoid and let</u> α <u>be a</u> C_s^k <u>one-parameter submonoid of</u> S. <u>Then</u> $\lambda_{\alpha(t)}$ <u>and</u> $\rho_{\alpha(t)}$ <u>are</u> C_s^k <u>diffeomorphisms (into</u> S). <u>In particular,</u> $d\lambda_{\alpha(t)}(x)$ <u>and</u> $d\rho_{\alpha(t)}(x)$ <u>are isomorphisms for each</u> $t \geq 0$ <u>and each</u> $x \in S$.

<u>COROLLARY 4.2. Let</u> S <u>be a</u> C_s^k <u>monoid and let</u> α <u>be a</u> C_s^k <u>one-parameter submonoid of</u> S <u>such that</u> $\alpha(t) \in S \setminus \partial S$ <u>for some</u>

$t > 0$. Then $\alpha(s) \in S \setminus \partial S$ for all $s \geq t$.

PROOF. Let $s > t$ and let U be an open set about $\alpha(t)$ such that $U \cap \partial S = \emptyset$ and $\rho_{\alpha(s-t)}|U$ is a C^k_s diffeomorphism. Then $\alpha(s) \in U_{\alpha(s-t)}$ and $U_{\alpha(s-t)} \cap \partial S \neq \emptyset$. Hence $\alpha(s) \notin \partial S$. □

Let S be a C^k_s monoid and let α be a C^k_s one-parameter submonoid of S. Then $\alpha'(0) \in T_1S$ and there is a unique $X \in L(S)$ such that $X_1 = \alpha'(0)$. Then α is an integral curve of X since $\alpha'(t) = d\alpha(t)(1) = d\rho(t)(d\rho_t(0)(1)) = d(\alpha \circ \rho_t)(0)(1) = d(\rho_{\alpha(t)} \circ \alpha)(0)(1) = d\rho_{\alpha(t)}(1)(\alpha'(0)) = X_{\alpha(t)}$. Since X is ρ_a-related to itself for all $a \in S$, $\rho_a \circ \alpha$ is an integral curve of X for each $a \in S$. The constant 1 map from $\mathbb{H}$ into S is a one-parameter submonoid of S. This may be the only one-parameter submonoid of S as in the next example.

EXAMPLE 4.3. Let $S = \{(x,y) \in \mathbb{R}^2 : x > 0 \text{ and } y \in (0,x^2)\}$ together with $(0,0)$. Then S is a C^∞_s submonoid of the additive plane. The image of a non-trivial one-parameter submonoid of the plane is a ray from the origin and S contains none of these rays. Hence the only one-parameter submonoid of S is the trivial one.

DEFINITION 4.4. A ray semigroup is a C^∞_s monoid S such that S is generated by the set of all elements of S of the form $\alpha(t)$ where α is a C^∞_s submonoid of S and $t \geq 0$.

NOTE. If S is a ray semigroup and $\alpha(t)$ is a generator of S, then $\rho_{\alpha(t)}$ and $\lambda_{\alpha(t)}$ are one-to-one maps by Theorem 4.1. Thus if $x \in S$, then λ_x and ρ_x are one-to-one maps. Hence if $x, y, z \in S$ and $xy = xz$ or $yx = zx$, then $y = z$, i.e. S is cancellative. Note also that S is (arc-wise) connected.

Hofmann and Lawson in [H-L] defined ray semigroups as (isomorphic to) subsemigroups of Lie groups. Our definition is equivalent to theirs in this case. However, cancellation alone is insufficient to guarantee group embeddability of a monoid. It is an open problem as to where every ray semigroup embeds in a Lie group, although we establish some partial results in this direction in [G_2] (see also §II.6).

II.5. CANCELLATION AND REVERSIBILITY: Let S be a semigroup and let $U \subset S$. Then the multiplication of S is cancellative on U if $x, y, z \in U$ and $xy = xz$ or $yx = zx$ implies $y = z$. The semigroup S is cancellative if multiplication is cancellative on S. Equivalently, S is cancellative if λ_x and ρ_x are one-to-one maps for each $x \in S$.

We have already seen that ray semigroups are cancellative. Not every C_s^∞ monoid is cancellative (e.g. $M_n(\mathbb{R})$), but every C_s^k monoid is cancellative on a neighborhood of 1.

THEOREM 5.1. Let S be a C_s^k monoid and let $p \in S$. Then there are open sets U about 1 and V about p such that $\lambda_x|V$ and $\rho_x|V$ are diffeomorphisms (into S) for each $x \in U$. In particular the multiplication of S is cancellative on a neighborhood of 1.

PROOF. This theorem follows immediately by applying Theorem I.4.5 and its dual to $m: S \times S \to S$. □

If S is a monoid, the group of units of S is the maximal subgroup $H(1)$ of S with identity 1. Mostert and Shields showed in [M-S_1] that if the identity of a monoid has a Euclidean neighborhood, then $H(1)$ is an open subgroup and a Lie group. We now extend this result to the infinite-dimensional (smooth) case.

THEOREM 5.2. Let S be a C_s^k monoid such that $1 \notin \partial S$. Then $H(1)$ is open and a C_s^k group. If S is also compact and connected, *then* S *is a Lie group.*

PROOF. Let U_0 be an open set about 1 such that $U \cap \partial S = \emptyset$ and such that S is cancellative on U_0. Let U be an open set about 1 such that $U^2 \subset U_0$. By Theorem I.4.5, there are open sets W and V about 1 such that $W \subset xU$ for each $x \in V$. Let $x \in U \cap W$. Then there is a $y \in U$ such that $xy = 1$. Moreover, $(yx)(yx) = yx \in U^2 \subset U_0$ and hence $yx = 1$. Thus $U \cap W \subset H(1)$. Since $H(1)$ is homogeneous, $H(1)$ is an open subgroup. Clearly multiplication is C_s^k when restricted to $H(1)$.

Now suppose that S is compact and connected. Then $H(1)$ is open since $1 \notin \partial S$ and $H(1)$ is closed since S is compact. Hence $H(1) = S$. By Corollary 2.8 S is a Lie group. We note that S must be finite-dimensional in this case. □

NOTE. If S is a C_s^k monoid and $1 \in \partial S$, then $H(1) \subset \partial S$ since a group is a homogeneous space. In this case $H(1)$ must have empty interior. However, it is not unreasonable to expect that S has an open submonoid which embeds in a group, although we are unable to prove this in general.

Let S be a semigroup. Then S is *right reversible* if $Sx \cap Sy \neq \emptyset$ whenever $x, y \in S$. Equivalently, S is right reversible if $\operatorname{im} \rho_x \cap \operatorname{im} \rho_y \neq \emptyset$ whenever $x, y \in S$. The semigroup S is *left reversible* if $yS \cap xS \neq \emptyset$ whenever $x, y \in S$.

PROPOSITION 5.3. Let S *be a* C_s^k *semigroup, let* $p \in S \setminus \partial S$, *and let* $q \in S$. *Suppose that* $d\rho_q(p)$ *is an isomorphism. If* U *is an open set about* p, *then there are open sets* V *about* q *and* W *about* pq *such that* $W \subset Ur$ *whenever* $r \in V$.

PROOF. Since $p \notin \partial S$ we may assume that $U \cap \partial S \neq \emptyset$. The proposition follows immediately from Theorem I.4.5. □

NOTE. If S is a C_s^k monoid, then $d\rho_1(p)$ is the identity map for each $p \in S$. If $p \in S \setminus \partial S$ and U is an open set about p, then there is an open set V about 1 such that $p \in Ur$ for each $r \in V$. In particular, $Ur \cap Us \neq \emptyset$ for $r, s \in V$ so that S is, in a sense, locally right reversible.

Let S be a C_s^k semigroup. We shall call an element x of S a pseudo-unit if λ_x and ρ_x are C_s^k embeddings and $d\lambda_x(y)$ and $d\rho_x(y)$ are isomorphisms for each $y \in S$. We denote the set of pseudo-units by $P(S)$. If $x \in P(S)$, then $\lambda_x, \rho_x | S \setminus \partial S$ are open maps into $S \setminus \partial S$ and hence xS and Sx are admissible subsets of S. Thus $x \in P(S)$ if and only if λ_x and ρ_x are diffeomorphisms onto admissible subsets of S. The chain rule implies that $P(S)$ is a subsemigroup of S and if S has an identity, then $H(1) \subset P(S)$. It follows that $P(S) \setminus \partial S$ is an ideal of $P(S)$.

PROPOSITION 5.4. *Let S be a ray semigroup. Then $P(S) = S$.*

PROOF. If α is a generator of S (i.e. a C_s^k one-parameter submonoid of S), then $\lambda_{\alpha(t)}$ and $\rho_{\alpha(t)}$ are C_s^k diffeomorphisms into S. Thus $\alpha(t) \in P(S)$ for all $t \geq 0$. It follows that $P(S) = S$. □

PROPOSITION 5.5. *Let S be a compact C_s^k monoid. Then $P(S)$ is an open submonoid of S.*

PROOF. Let $p \in P(S)$. We show first that there is an open set W about p such that if $x \in W$ and $y \in S$, then $d\lambda_x(y)$ and $d\rho_x(y)$ are isomorphisms. For each $y \in S$, there are open sets U_y about p and V_y about y such that $\lambda_x | V_y$ and $\rho_x | V_y$ are C_s^k diffeomorphisms into S, by Theorem I.4.5. Let $\{(U_y, V_y): y \in S\}$ be a collection of such pairs so that $\{V_y: y \in S\}$ is an open cover of S and let $\mathcal{V} = \{V_i: i = 1, \ldots, n\}$ be an open subcover. Let

$\mathcal{U} = \{U_i: i = 1, \ldots, n\}$ be the corresponding collection of open sets about p. Let $x \in W = \cap\mathcal{U}$ and let $y \in S$. Then $y \in V_i$ for some i and $x \in U_i$. Hence, $d\lambda_x(y)$ and $d\rho_x(y)$ are isomorphisms for all $y \in S$.

Suppose $P(S)$ is not open. Then there is a $p \in P(S)$ and a net $\{x_\alpha: \alpha \in D\}$ converging to p such that $x_\alpha \notin P(S)$. By the result of the previous paragraph, we may assume that there are nets $\{y_\alpha: \alpha \in D\}$ and $\{z_\alpha: \alpha \in D\}$ such that $y_\alpha \neq z_\alpha$ and $x_\alpha y_\alpha = x_\alpha z_\alpha$ or $y_\alpha x_\alpha = z_\alpha x_\alpha$ for all $\alpha \in D$. Since S is compact, $\{y_\alpha\}$ must cluster to some $y \in S$ and $\{z_\alpha\}$ must cluster to some $z \in S$. By taking subnets we may assume that $y = \lim y_\alpha$, $z = \lim z_\alpha$, and $x_\alpha y_\alpha = x_\alpha z_\alpha$ for all $\alpha \in D$. Then $py = \lim x_\alpha y_\alpha = \lim x_\alpha z_\alpha = pz$ and since $p \in P(S)$, $y = z$. Now since $d\lambda_p(y)$ is an isomorphism, there are, by Theorem I.4.5, open sets U about p and V about $y = z$ such that $\lambda_x|V$ is one-to-one whenever $x \in U$. But for α sufficiently large, $x_\alpha \in U$, $y_\alpha \in V$, $z_\alpha \in V$, and $x_\alpha y_\alpha = x_\alpha z_\alpha$. This is a contradiction. It follows that $P(S)$ is an open set. □

PROPOSITION 5.6. *Let S be a connected C^k_s semigroup such that $P(S)$ is open and nonempty. If $e \in P(S)^*$ is a left [resp., right] zero for $P(S)$, then e is a left [resp., right] zero for S.*

PROOF. If e is a left zero for $P(S)$, then $ex = x$ for all $x \in P(S)$. Since $P(S)$ is open, $d\lambda_e(x) = 0$ for all $x \in P(S)$. By continuity, $d\lambda_e(e) = 0$. If $eS \neq \{e\}$, then $T_e(eS)$ is nontrivial and $d\lambda_e(e) \neq 0$ by Theorem 2.3. Thus $eS = \{e\}$ and e is a left zero for S. □

II.6. GROUP EMBEDDABILITY: In this section we examine consequences of the local Lie group embedding theorem which is stated below. This theorem is proved in [G] and [G_2].

THEOREM 6.1. Let S be a finite-dimensional C_s^∞ monoid, let G be a Lie group, and let $\Phi: L(S) \to L(G)$ be an isomorphism of Lie algebras. Then there is a C_s^∞ diffeomorphism and local isomorphism f from S onto an admissible subset of G such that $L(f) = \Phi$.

NOTE. Since every finite-dimensional Lie algebra is isomorphic to the Lie algebra of a Lie group, it follows that each C_s^∞ monoid is locally embeddable in a Lie group. We shall call a local isomorphism such as provided by Theorem 6.1 a local Lie group embedding of S. The word isomorphism in Theorem 6.1 may be replaced throughout by the word homomorphism. To see this, let $\Phi: L(S) \to L(G)$ be a Lie algebra homomorphism onto $L(G)$. Let H be a Lie group with $L(H) \cong L(S)$. Then S is locally embeddable in H and Φ may be thought of as a homomorphism from $L(H)$ onto $L(G)$. In this case there is a local homomorphism (of groups) from H onto a local subgroup of G. By restricting the maps as necessary we obtain the desired result.

We now state without proof some properties of Lie groups. Proofs may be found in [S]. Let G be a Lie group. For each $X \in L(G)$, the integral curve α of X with $\alpha(0) = 1$ is a local one-parameter subgroup of G. Since this local one-parameter subgroup can be extended to all of $\mathbb{R}$, it follows that for each $X \in L(G)$, there is a unique one-parameter subgroup α such that $\alpha'(0) = X_1$. The exponential map is the map $\exp: L(G) \to G$ defined by $\exp(X) = \alpha(1)$ where $\alpha'(0) = X_1$. The map exp is C^∞ and a local diffeomorphism on a neighborhood of 0. If $\alpha'(0) = X_1$, then $\alpha(t) = \exp(tX)$ for all $t \in \mathbb{R}$. It follows that $\exp|\langle X\rangle$ is a homomorphism where $\langle X\rangle$ is the linear span of X (i.e. all scalar multiples of X.) The map exp is not a homomorphism unless G is commutative. A Lie algebra L is <u>commutative</u> if and only if $[X,Y] = 0$ for all $X, Y \in L$. A Lie group is commutative if and only if $L(G)$ is commutative. If $f: G \to H$ is a continuous homomorphism of Lie groups, then f is C^∞ and $\exp \circ L(f) = f \circ \exp$. In particular, each one-parameter subgroup of a Lie group is a C^∞ function.

<u>COROLLARY 6.2</u>. <u>Let α be a one-parameter subsemigroup of a finite-dimensional C_s^∞ monoid</u> S. <u>Then α is C_s^∞ differentiable.</u>

<u>PROOF</u>. Let $e = \alpha(0)$. Then $e^2 = e$ and $\alpha(t) \in eSe$ for all $t \geq 0$. Since eSe is a C_s^∞ monoid, we may assume that α is a one-parameter submonoid of S, i.e. $\alpha(0) = 1$.

Let f be a local C_s^∞ embedding of S into a Lie group G. Then $f \circ \alpha$ is a local one-parameter submonoid of G which can be extended to a one-parameter subgroup of G. It follows that $f \circ \alpha$ is a C_s^∞ map. Since f is a local C_s^∞ diffeomorphism $\alpha = f^{-1} \circ f \circ \alpha$ is C_s^k on $[0,\varepsilon)$ for some $\varepsilon > 0$. Now for $t \geq \varepsilon/2$, $\alpha(s) = \alpha(t - \frac{\varepsilon}{2})\alpha(s - t + \frac{\varepsilon}{2})$ when $|s-t| < \varepsilon/2$. If $T = t - \frac{\varepsilon}{2}$, then $\alpha = \lambda_{\alpha(T)} \circ \alpha \circ \lambda_{-T}$ on $(t - \frac{\varepsilon}{2}, t + \frac{\varepsilon}{2})$. Since λ_{-T} is C^∞ and α is C_s^∞ on $[0,\varepsilon)$, α is C^∞ on a neighborhood of t. Thus α is C_s^∞. □

Let S be a C_s^∞ monoid and let $W(S) = \{X \in L(S)\colon X_1 = \alpha'(0)$ for some one-parameter submonoid of $S\}$. Define $\exp\colon W(S) \to S$ by $\exp(X) = \alpha(1)$ where $\alpha'(0) = X_1$. It follows that $\alpha(t) = \exp(tX)$ for all $t \geq 0$ and hence $W(S)$ is closed under multiplication by nonnegative scalars.

<u>COROLLARY 6.3</u>. <u>Let</u> S <u>be a finite-dimensional ray semigroup</u>. <u>Then</u> $W(S)$ <u>generates</u> $L(S)$.

<u>PROOF</u>. Let $f\colon U \subset S \to G$ be a local embedding of S in G

with $f(U)$ an admissible subset of G. Let R be the ray subsemigroup of G generated by $L(f)(W(S))$. Then $f(U) \subset R$. By a theorem of Jurdjevic and Sussman [J-S], R has interior if and only if $L(f)(W(S))$ generates $L(G)$. Since $L(f)$ is an isomorphism and R has interior, $W(S)$ generates $L(S)$. □

A <u>wedge</u> is a subset of a vector space which is closed under addition and multiplication by nonnegative scalars.

Let S be a C_s^k monoid. Then $L(S)$ is isomorphic to T_1S as a vector space. Since T_1S can be given the structure of a Banach space, so can $L(S)$. If S is finite-dimensional, then all norms on $L(S)$ which make addition continuous are equivalent and in this case there is a norm such that $|[X,Y]| \leq |X| \cdot |Y|$ for all $X, Y \in L(S)$.

<u>COROLLARY 6.4</u>. <u>Let</u> S <u>be a locally compact</u> C_s^∞ <u>monoid</u>. <u>Then</u> $W(S)$ <u>is a closed wedge in</u> $L(S)$ <u>and</u> $\exp: W(S) \to S$ <u>is a</u> C_s^∞ <u>diffeomorphism into</u> S <u>on a neighborhood of</u> 0.

<u>PROOF</u>. Since S is locally compact, S is finite-dimensional. Let $f: U \subset S \to G$ be a local Lie group embedding of S. Let $V \subset S$ be an open set about 1 such that V^* is compact and contained in U. Then $f(V^*) = f(V)^*$ is a compact local submonoid of G. Let $\varepsilon > 0$ be such that $\exp|B_\varepsilon(0)$ is a diffeomorphism and $\exp(B_\varepsilon(0)) \cap f(U) \subset f(V)$. Suppose now that $X = L(f)(Y)$ for some $Y \in W(S)$. Then there is a one-parameter submonoid β of S such that $\beta'(0) = Y$. The map $f \circ \beta$ is a local one-parameter submonoid of $f(U)$ and $(f \circ \beta)'(0) = X_1$. In a similar manner, if γ is a local one-parameter submonoid of $f(U)$ with $\gamma'(0) = X_1$, then $f^{-1} \circ \gamma$ is a local one-parameter submonoid of S which can be extended to all of $\mathbb{H}$. If $Y_1 = (f^{-1} \circ \gamma)'(0)$, then $L(f)(Y) = X$.

It follows then that $X \in L(f)(W(S))$ if and only if X_1 is tangent to a local one-parameter submonoid of $f(U)$.

We show now that $L(f)(W(S))$ is closed. Let $\{X^\alpha\}$ be a net in $L(f)(W(S))$ converging to $X \in L(G)$. There is a $\delta > 0$ such that $tX \in B_\varepsilon(0)$ whenever $|t| < \delta$. Since exp is continuous, $\{\exp tX^\alpha\}$ converges to $\exp tX$ for all t and in particular for $|t| < \delta$. If $t \in [0,\delta)$ then eventually $tX^\alpha \in B_\varepsilon(0)$ and $\exp(tX^\alpha) \in f(V)$. Thus $\exp tX \in f(V)^* \subset f(U)$ and $X \in L(f)(W(S))$. Hence $W(S)$ is closed since $L(f)^{-1}$ is continuous.

We now claim that $W(S)$ is a wedge. Closure of $W(S)$ under multiplication by nonnegative scalars is clear. Let X and Y be in $L(f)(W(S))$. In a Lie group, $\exp(X+Y) = \lim_n [\exp(\frac{1}{n}X) \exp(\frac{1}{n}Y)]^n$ and $\exp(tX)\exp(tY) = \exp(tX + tY + o(t^2))$ for small t where $t^{-2}o(t^2)$ is bounded for small t. If X and Y are in $B_{\frac{1}{3}\varepsilon}(0) \cap L(f)(W(S))$, then $\exp(\frac{1}{n}x) \exp(\frac{1}{n}Y) = \exp(Z)$ where $Z \in B_{\frac{1}{n}\varepsilon}(0)$ for n sufficiently large. It follows that $[\exp(\frac{1}{n}X) \exp(\frac{1}{n}Y)]^n = \exp(nZ) \in f(V)$ and that $\exp(X+Y) \in f(U)$. Hence $W(S)$ is a wedge.

Now $f \circ \exp = \exp \circ L(f)$ on $L(f)^{-1}(B_\varepsilon(0))$ in $W(S)$. Since f^{-1} is C_s^∞, it follows that $\exp\colon W(S) \to S$ is a C_s^∞ diffeomorphism (into S) on a neighborhood of $0 \in W(S)$. □

<u>NOTE</u>. The wedge $W(S)$ may have dimension less than the dimension of S even if S is a ray semigroup (e.g. subsemigroups of the Heisenberg group.) In this case $\exp(W(S))$ will have empty interior in S.

We also note that if the semigroup S in Corollary 6.4 is a subsemigroup of a Lie group, then our wedge $W(S)$ is the Lie wedge of S defined by Hofmann and Lawson [H-L]. In general, if S is a C_s^∞ monoid, then we may identify a Lie wedge (in the sense of Hofmann and Lawson) for S by embedding a local submonoid of S in a Lie group G. In this case, it may not be possible to define an exponential map on the whole Lie wedge of S. However, we may define a local exponential map from an admissible subset of $L(S)$ about 0 onto a neighborhood of 1 by restricting the exponential map of G appropriately.

COROLLARY 6.5. *Let S be a finite-dimensional ray semigroup. Then S is commutative if and only if $L(S)$ is commutative.*

PROOF. Let $f: U \subset S \to G$ be a C_s^∞ local embedding such that $L(f)$ is an isomorphism onto $L(G)$. Let $V \subset S$ be an open set about 1 such that $V^2 \subset U$. If S is commutative, then $f(x)f(y) = f(y)f(x)$ for all $x, y \in V$. Now $A = f(V)[f(V)]^{-1}$ is a neighborhood of 1 in G. It follows that if $g, h \in A$, then $gh = hg$. Thus the identity component of G is commutative and hence $L(G) \cong L(S)$ is commutative. If $L(S)$ is commutative, then G is commutative and S is commutative on U. Since a ray semigroup is locally generated, it follows that S is commutative. □

NOTE. If S is a commutative ray semigroup (of arbitrary dimension), then exp is a homomorphism. To see this let $X, Y \in W(S)$ and let α and β be C_s^∞ one-parameter subsemigroups such that $\alpha'(0) = X_1$ and $\beta'(0) = Y_1$. Let $\gamma(t) = \alpha(t)\beta(t)$ for all $t \geq 0$. Then γ is a C_s^∞ one-parameter submonoid and $\gamma'(0) = X_1 + Y_1$ by the Product Rule. It follows that $\exp(X+Y) = (\exp X)(\exp Y)$ and that $\exp: W(S) \to S$ is a homomorphism. In a similar manner, exp maps onto S. Thus S is the (algebraic) homomorphic image of a wedge in a Banach space.

COROLLARY 6.6. *Let* G *be a closed connected subgroup of a locally compact connected* C_s^∞ *monoid* S. *Then* G *is a Lie group and the inclusion map* $i: G \to S$ *is a* C_s^∞ *embedding.*

PROOF. By Corollary 2.8, G is a Lie group in the relative topology. Now if e is the identity of G, then $G \subset eSe$, which is a closed embedded C_s^∞ subsemigroup of S. Thus we may assume that $e = 1$.

Let $f: U \subset S \to H$ be a C_s^∞ local embedding of S in a Lie group H. Then $f|U \cap G$ is a continuous local homomorphism from G into H. The proof that a continuous homomorphism of Lie groups is C^∞ extends to continuous local homomorphisms (cf. [S, Theorem 10.12].) It follows that $f|U \cap G$ is C^∞ as a map into H. Since $f(U \cap G) \subset f(U)$, $f|U \cap G$ is C_s^∞ as a map into $f(S)$. Then $i = f^{-1} \circ f$ on $U \cap G$ and thus i is C_s^∞ on a neighborhood of 1. If $g \in G$, then $\lambda_g^{-1}: G \to G$ is C^∞ and hence $i = \lambda_g \circ i \circ \lambda_{g^{-1}}$ is C_s^∞ on a neighborhood of g. □

Let S be a C_s^∞ monoid. For each $v \in T_1S$, let $\tilde{v}$ be the vector field on S defined by $\tilde{v}_x = d\lambda_x(1)(v)$, for all $x \in S$. Then $\tilde{v}$ is C_s^∞ and left-invariant. Let $\tilde{L}(S)$ be the collection of left-invariant vector fields of S. By the dual of Theorem 3.3, $\tilde{L}(S)$ is a Lie algebra which is isomorphic to T_1S as a vector space.

COROLLARY 6.7. *Let* S *be a finite-dimensional* C_s^∞ *monoid. Then* $L(S)$ *and* $\tilde{L}(S)$ *are isomorphic Lie algebras.*

PROOF. Let $f: U \subset S \to G$ be a local Lie group embedding so that $L(f): L(S) \to L(G)$ is an isomorphism. In a similar way, $\tilde{L}(f): \tilde{L}(S) \to \tilde{L}(G)$ is an isomorphism, where $\tilde{L}(f)(\tilde{X})_g = d\lambda_g(1)$ $d\lambda_g(1)(df(1)(\tilde{X}_1))$ for all $g \in G$. Thus it suffices to show that $L(G)$ and $\tilde{L}(G)$ are isomorphic.

Let $\theta: G \to G$ denote inversion, i.e. $\theta(x) = x^{-1}$ for all $x \in G$. If $X \in L(G)$, then there is a one-parameter subgroup α such that $\alpha'(0) = X_1$, and α is an integral curve of X with initial condition 1. If $x \in G$, then $\rho_x \circ \alpha$ is an integral curve of X with initial condition x since X is right-invariant. Now $(\theta \circ \rho_x \circ \alpha)(t) = (\alpha(t)x)^{-1} = x^{-1}\alpha(-t)$ and $(\theta \circ \rho_x \circ\alpha)^1(t) = d\lambda_{x^{-1}}(1)(-\tilde{X}_1) = (-\tilde{X}_1)(\theta(x))$. It follows that X is θ-related to $-\tilde{X}_1$ for all $X \in L(G)$. By Theorem I.5.5, the map which sends X to $-\tilde{X}_1$ is a Lie algebra homomorphism.

In a similar manner, this map is an isomorphism. □

<u>NOTE</u>. If G is a matrix Lie group, then $L(G)$ and $\tilde{L}(G)$ can be realized as Lie algebras of matrices, by identifying each of these with T_1G. For example, if G is the Heisenberg group of all matrices of the form

$$\begin{pmatrix} 1 & x & z \\ 0 & 1 & y \\ 0 & 0 & 1 \end{pmatrix}$$

for x, y, z real, then T_1G is the vector space of matrices of the form

$$\begin{pmatrix} 0 & x & z \\ 0 & 0 & y \\ 0 & 0 & 0 \end{pmatrix}$$

for x, y, z real since these are precisely the matrices which occur as tangents to curves in G at 1. Now if $A \in G$, then $d\lambda_A(1) = A$, i.e. $d\lambda_A(1)(B) = AB$, since λ_A is a linear map. If we identify T_1G with $L(S)$ and with $\tilde{L}(S)$, then $[A,B] = BA-AB$ in $L(S)$ and $[A,B] = AB - BA$ in $\tilde{L}(S)$.

It was shown in Proposition 3.5 that $K_p = \{X \in L(S): X_p = 0\}$ is a subalgebra of $L(S)$. In a similar manner, $\tilde{K}_p = \{\tilde{X} \in \tilde{L}(S): \tilde{X}_p=0$

is a subalgebra of $\tilde{L}(S)$. Since $L(S)$ is isomorphic to $\tilde{L}(S)$ by the map which sends X to $-\tilde{X}_1$, it follows that $\{X \in L(S): d\lambda_p(1)(X_1) = 0\}$ is a subalgebra of $L(S)$ when S is finite-dimensional.

Although $L(S)$ and $\tilde{L}(S)$ are isomorphic Lie algebras, their geometric properties as sets of vector fields may be quite different. For example, let S be the affine triangle. If $e \in M(S)$, then $M(S) = Se \cap E$, which is left-trivial, and e is a left zero for S. Thus $\tilde{X}_e = 0$ for all $\tilde{X} \in \tilde{L}(S)$. However, $M(S) = Se$ is one dimensional and $T_e(Se) = \operatorname{im} d\lambda_e(1)$. Hence, K_e is one-dimensional and in particular there are vector fields $X \in L(S)$ with $X_e \neq 0$.

The proof of the following theorem is based on a method introduced by Horne in [Hr] for obtaining a matrix representation of a differentiable semigroup with a right or left zero.

<u>THEOREM 6.8</u>. <u>Let S be a nontrivial compact connected C_s^∞ monoid. If S has a right or left zero, then $L(S)$ has an ideal of codimension one</u>.

<u>PROOF</u>. Since S is compact, S is finite-dimensional by Proposition 1.2. Suppose that e is a right zero for X, i.e. $xe = e$ for all $x \in S$. Then e is a fixed point of each of the maps λ_x, and $d\lambda_x(e): T_eS \to T_eS$ for all $x \in S$. Moreover, the map $\psi: S \to L(T_eS, T_eS)$, defined by $\psi(x) = d\lambda_x(e)$, is a C_s^∞ homomorphism.

Since T_eS is finite-dimensional, we may assume that ψ is a homomorphism into $M_n(\mathbb{R})$ for some $n \geq 1$. Define $\varphi: S \to (\mathbb{R},\cdot)$ by $\varphi(x) = \det(\psi(x))$. Then φ is a C_s^∞ homomorphism. If $f^2 = f \in S\setminus\{1\}$, then $\psi(f)$ and $\varphi(f)$ are idempotents. Now fS

has empty interior by Proposition 2.4 and thus $\psi(f) \neq I$ and $\varphi(f) = 0$.

We claim that $L(\varphi)\colon L(S) \to (\mathbb{R},+) \cong L((\mathbb{R},\cdot))$ is nontrivial. If $L(\varphi) = 0$, then $d\varphi(1) = 0$. Suppose that $d\varphi(1) = 0$. Since φ is a homomorphism, $\varphi\circ\lambda_x = \lambda_{\varphi(x)} \circ \varphi$ and hence $d\varphi(x) \circ d\lambda_x(1) = d\lambda_{\varphi(x)}(1) \circ d\varphi(1)$ for all $x \in S$. If $x \in P(S)$, then $d\lambda_x(1)$ is an isomorphism and hence $d\varphi(x) = 0$. Now the identity component of $H(1)$ is a compact Lie group by Theorem 2.8. Since S has a right zero, S is not a group and therefore $H(1)$ is not open. Also, there is a neighborhood of 1 containing no other idempotents by the local cancellation near 1. Thus by a theorem of Mostert and Shields [M-S, Theorem A], there is a one-parameter submonoid α such that $\alpha(t) \notin H(1)$ for $t > 0$. Since $P(S)$ is an open submonoid, $\alpha(t) \in P(S)\setminus H(1)$ for $t > 0$. It follows that the compact monoid $\alpha(\mathbb{H})^*$ must contain an idempotent $f \neq 1$. Then $\varphi(f) = 0$, but $\varphi(\alpha(t)) = 1$ for all t since $d\varphi(\alpha(t)) = 0$. This is impossible since $\varphi\circ\alpha$ is a C^∞ map. Thus $L(\varphi)$ is nontrivial and $\ker(L(\varphi))$ is an ideal of $L(S)$ of codimension one.

Finally, if e is a left zero for S, then the map $\psi(x) = d\rho_x(e)$ is a C_s^∞ anti-homomorphism. If we let $\psi^*(x) = (d\rho_x(e))^*\colon T_eS^* \to T_eS^*$, where $*$ denotes dual, then ψ^* is a C_s^∞ homomorphism. The result follows by an analogous argument. □

<u>COROLLARY 6.9</u>. <u>Let</u> S <u>be a finite-dimensional</u> C^∞ <u>monoid on a manifold with smooth boundary and suppose that</u> $1 \in \partial S$. <u>Then</u> $H(1)$ <u>is open in</u> ∂S <u>and is a Lie group</u>.

<u>PROOF</u>. We note that C^k differentiability is equivalent to C_s^k differentiability in this case. Let $X \in L(S)$ such that $X_1 \in T_1\partial S$ and let $c\colon (-\varepsilon,\varepsilon) \to S$ be a C_s^1 curve with $c'(0) = X_1$.

Then $(\rho_p \circ c)'(0) = d\rho_p(1)(X_1) = X_p$ for all $p \in S$. In particular, if $p \in \partial S$ then $X_p \in T_p\partial S$ since X_p can be neither inward pointing nor outward pointing. Thus $X|\partial S$ is a C_s^∞ vector field on the submanifold ∂S. Let $\alpha:(-\varepsilon,\varepsilon) \to \partial S$ be an integral curve of $X|\partial S$ with initial condition 1. If $s \in (-\varepsilon,\varepsilon)$, then $\beta(t) = \alpha(s+t)$ and $\gamma(t) = \alpha(s)\alpha(t)$ are integral curves of X with the same initial condition $\beta(0) = \gamma(0) = \alpha(s)$. By uniqueness of integral curves, α is a local one-parameter subgroup which can be extended to a global one-parameter subgroup. Thus if $X \in L(S)$ and $X_1 \in T_1\partial S$, then X and $-X$ are in $W(S)$. Since S is locally compact, $W(S)$ is a half-space and exp is a diffeomorphism on a neighborhood of 0. Let $H = \{X \in L(S): X_1 \in T_1\partial S\}$. Then $\exp H \subset H(1) \subset \partial S$ and since exp is a local diffeomorphism, $\exp H$ must have interior and thus $H(1)$ is open in ∂S. Finally, the identity component of $H(1)$ is a Lie group by Proposition 2.2. Since $H(1)$ is open in ∂S, $H(1)$ is a Lie group. □

NOTE. The set H in the above proof is a subalgebra of $L(S)$ which is isomorphic to $L(H(1))$. We also note that H has codimension 1.

A fairly complete description of topological monoids on a compact connected manifold with smooth connected boundary was obtained by Mostert and Shields [M-S] under the assumption that the boundary is a submonoid. Corollary 6.9 shows that the assumption that ∂S is a submonoid holds when the multiplication is C^∞ differentiable.

II.7. PROBLEMS: In this section we consider problems which appear to us to be worthy of further investigation. Throughout this section S is a connected C_s^∞ monoid (of arbitrary dimension.)

PROBLEM 1: Is S locally embeddable (near 1) in a C^∞ group G with $L(G) \cong L(S)$? Can this result be obtained if the order of differentiability of S is reduced to C_s^2? If S has a smooth boundary, must $H(1)$ be open in ∂S?

PROBLEM 2. Let T be a C_s^∞ monoid and let $f: S \to T$ be a continuous homomorphism. Must f be C_s^∞ differentiable on a neighborhood of 1? In particular, is each one-parameter subsemigroup of S C_s^∞ differentiable?

NOTE. A continuous isomorphism from S to S need not be globally C_s^∞ differentiable. For example, let $S = [0,1]$ under multiplication. Each continuous homomorphism from S to S has the form $f_r(x) = x^r$ where $r \in [0,\infty)$. If $r \in (0,1)$ then f_r is not differentiable at 0. The map f_r is a diffeomorphism if and only if $r = 1$.

PROBLEM 3. Suppose that S is a ray semigroup.

3.1. If S is a subsemigroup of a simply connected Lie group, is S simply connected?

3.2. Does S embed in a C^∞ group G with $L(G) \cong L(S)$?

3.3. If S is finite-dimensional, does S embed in a locally compact C_s^∞ monoid T with S dense in T and $L(T) \cong L(S)$?

3.4. If S is commutative, must $W(S)$ have nonempty interior in $L(S)$ and is exp a C_s^∞ map? If S has dimension n, does S embed in $T^p \times \mathbb{R}^{n-p}$ where T is the circle group?

NOTE. If S is a one-dimensional ray semigroup, then S is the circle group, $(\mathbb{R},+)$, or $(\mathbb{H},+)$.

PROBLEM 4. Let $R(S)$ be the submonoid of S generated by the set of C_s^∞ one-parameter submonoids of S. We note that $1 \in R(S) \subset P(S)$.

4.1. Is $R(S)$ nontrivial when S is locally compact?

4.2. Is $R(S)$ an embedded C_s^∞ submonoid of S?

4.3. Does S embed in a C_s^∞ monoid T with $L(T) \cong L(S)$ in such a way that $S \cap \exp(W(T))$ is an S-neighborhood of 1? (See Section II.6.)

4.4. If S is compact and one-dimensional, then $R(S)$ is open. If S is not the circle group, must S be equivalent to an interval of the form $[a,1]$ under multiplication where $a \in [-1,0]$? In particular, if we define a multiplication m on $S = [-1,1]$ by $m(x,y) = 0$ if $x, y \leq 0$ and $m(x,y)$ = usual product otherwise, can S be given a C_s^k structure so that m is a C_s^k map?

4.5. If S is a compact commutative, and n-dimensional, does S embed in $[-1,1]^p \times T^q \times D^r$ where T is the circle group, D is the unit disc, and $p + q + 2r = n$?

4.6. Let S be compact, noncommutative, and two-dimensional. If $R(S)$ is dense in S, must S be equivalent to the affine triangle or its dual (transpose)? In particular, if S is the affine triangle, can $S/M(S)$ be made into a C_s^∞ monoid?

PROBLEM 5. Recall that the semigroup $P(S)$ of pseudo-units of S is the set of elements $x \in S$ such that λ_x and ρ_x are diffeomorphisms onto admissible subsets of S.

5.1. Are the following statements equivalent for $x \in S$:

(a) $x \in P(S)$.

(b) λ_x and ρ_x are one-to-one.

(c) $d\lambda_x(1)$ and $d\rho_x(1)$ are isomorphisms.

(d) $d\lambda_x(y)$ and $d\rho_x(y)$ are isomorphisms for all $y \in S$.

5.2. If S is finite-dimensional, is rank $d\lambda_x(y)$ independent of y?

5.3. Is P(S) open, dense, and embeddable in a C^∞ group G where $L(G) \cong L(S)$ and P(S) and G have the same number of components?

PROBLEM 6. (Horne [Hr]) If S has a right zero e, is the map $\varphi: S \to L(T_eS, T_eS)$, defined by $\varphi(x) = d\lambda_x(e)$, a one-to-one map? (See Section II.5.)

PROBLEM 7. If S is compact, does S have a faithful matrix representation?

PROBLEM 8. Does S embed in a C_s^∞ monoid T with $L(T) \cong L(S)$ where ∂T is either smooth or empty?

PROBLEM 9. (Control Theory) Let M be a C^∞ n-manifold (without boundary) and let $D \subset V^\infty(M)$ such that: (i) $X \in D$ implies that X is semi-integrable, and (ii) the Lie algebra $L(D)$ generated by D is finite-dimensional. Let $C^\infty(M)$ be the semigroup of all C^∞ maps from M to M (under composition) and let S be the subsemigroup of $C^\infty(M)$ generated by the set of elements of the form π^t where $t \geq 0$ and π is the semi-flow for some $X \in D$. Is S a C_s^∞ monoid with $L(S) \cong L(D)$? (See Section I.6.)

NOTE. If, in addition, X is completely integrable for each $X \in D$ and S includes maps of the form π^t for $t \in \mathbb{R}$, then by a theorem of Palais [Pl], S is a Lie group with $L(S) \cong L(D)$.

REFERENCES

[B] Bartle, R. G., The Elements of Real Analysis, Wiley, New York, 1964.

[D] Dieudonné, J., Foundations of Modern Analysis, Academic Press, New York, 1960.

[E] Ellis, R., Locally compact transformation groups, Duke Math. 24 (1957), 119-126.

[G] Graham, G., Manifolds with Generalized Boundary and Differentiable Semigroups, Dissertation, University of Houston, 1979.

[G_1] Graham, G., Differentiable manifolds with generalized boundary, (to appear).

[G_2] Graham, G., Differentiable transformation semigroups, (to appear).

[Gr] Graves, L. M., Some mapping theorems, Duke Math. J. 17 (1950), 111-114.

[H] Hofmann, K. H., Topological semigroups history, theory, applications, Sonderdruck aus Jber. Deutsch. Math.-Verein. 78, H.1 (1976), 9-59.

[H-L] Hofmann, K. H. and J. D. Lawson, Foundations of Lie semigroups (this volume).

[H-M] Hofmann, K. H. and P. S. Mostert, Elements of Compact Semigroups, C. E. Merrill, Columbus, Ohio, 1966.

[Hl] Holmes, J. P., Rees products in differentiable semigroups, Semigroup Forum 25 (1982), 145-152.

[Hr] Horne, J. G., Sl(2) has no C^1 extension to a half space, Semigroup Forum 7 (1974), 286-329.

[J-S] Jurdjevic, V. and H. Sussmann, Control systems on Lie groups, J. Diff. Eq. 12 (1972), 313-329.

[L] Lang, S., Differential Manifolds, Addison-Wesley, Reading, Mass., 1972.

[Le] Leach, E. B., A note on inverse function theorems, Proc. Amer. Math. Soc. 12 (1961), 694-697.

[M-Z] Montgomery, D. and L. Zippin, Topological Transformation Groups, Interscience, New York, 1955.

[M-S] Mostert, P. S. and A. L. Shields, On the structure of semigroups on a compact manifold with boundary, Ann. Math. 65 (1957), 117-143.

[M-S_1] Mostert, P. S. and A. L. Shields, Semigroups with identity on a manifold, Trans. Amer. Math. Soc. 91 (1959), 380-389.

[N] Nashed, M. Z., Generalized inverse mapping theorems and related applications of generalized inverses in nonlinear analysis, Nonlinear Equations in Abstract Spaces, V. Lakshmikantham, ed., Academic Press, New York, 1978, 217-252.

[Nj] Nijenhuis, A., Strong derivatives and inverse mappings, Amer. Math. Monthly, 81 (1974), 969-981.

[Pl] Palais, R. S., A Global Formulation of the Lie Theory of Transformation Groups, Mem. Amer. Math. Soc. 22 (1957).

[Pn] Peano, G., Sur la définition de la dérivée, Mathesis (2) 2 (1892), 12-14.

[S] Spivak, M., A Comprehensive Introduction to Differential Geometry, Vol. I, Publish or Perish, Berkeley, 1970.

Department of Mathematics
Texas A&M University
College Station, Texax 77843-3368

FOUNDATIONS OF LIE SEMIGROUPS

K.H. Hofmann* and J.D. Lawson*

The purpose of this paper is the formulation of a useful notion of Lie semigroups and the development of their basic theory. We will not now pursue many questions of detail which arise in this context, but concentrate on laying broad and solid foundations on which further more specialized studies can be based.

Essentially three reasons motivate us: Firstly, there are numerous attempts in the literature to complement Lie group theory by a Lie semigroup theory and to illustrate its applications. But in comparison with the highly developed theory of Lie groups, these attempts appear unsystematic and sporadic, so that a comprehensive approach to the definition of Lie semigroups and their basic theory seems overdue. Secondly, there is a variety of results in topological semigroup theory which now appears somewhat dispersed, but which can be unified by a systematic Lie semigroup theory. Thirdly a theory of a Lie semigroups is needed in geometric control theory.

---

* The authors gratefully acknowledge the support of NSF grants MCS-7900241 and MCS-7900295, respectively.

Let us dwell on these points for a moment.

Hille introduced around 1950 the study of topological semigroups on closed cones (and more particularly, on hyperquadrants) of $\mathbb{R}^n$; this theory occupies a whole chapter in the book by Hille and Phillips [H-P] and was pursued further in an article by Langlands [La] in 1960. If sufficiently many conditions of a technical nature are imposed, then an exponential function can be constructed; the cone itself appears as the set of infinitesimal generators and can be identified as a portion of a Lie algebra which is defined on the ambient $\mathbb{R}^n$. These authors then study strongly continuous representations of the semigroup into the semigroup of bounded linear operators of a Banach space and investigate the representation induced on the infinitesimal generators. This clearly generalises the program of studying differentiable one parameter semigroups of operators on a Banach space and their infinitesimal generators. However, from the viewpoint of a Lie semigroup theory of the kind we envisage, the class of semigroups considered here is still very special. For many years, another author, Charles Loewner studied semigroups in transformation Lie group; this work is surveyed in 1962 and published in [Lo]. He was interested in various particular types of semigroups of linear transformations which leave a positive cone invariant and in certain minimal semigroup extensions of specific classical Lie groups. In none of these lines of investigation does there appear to be a concerted effort to formulate a clear-cut definition of a Lie semigroup, nor to consider the infinitesimal generators and what algebraic structure is carried by the set of these generators (although it was observed by all of these authors that they form a convex cone); and finally no one concentrates on the exponential function for Lie semigroups in the same spirit as one does in Lie group theory. But this is exactly what our approach intends to do: Draw from the rich source of Lie group theory in an effort to deal with Lie semigroups and their infinitesimal generation systematically.

In order to illustrate the second point, we remark that a variety of results in topological semigroup theory heavily involve the consideration of semigroups of Lie groups and employ techniques of Lie group theory. Examples are provided by the centralizing result of Hofmann and Mostert for semigroups locally embeddable into a Lie group [H-M, p.92], the work of Mostert and Shields on one-parameter semigroups and semigroups on manifolds [M-S], the paper of Borrego, Cohen, and Devun on uniquely representable semigroups [B-C-D], the characterisation of certain uniquely divisible semigroups by Brown and Friedberg [B-F] and by Keimel [K], and the results of Dobbins [Do] and Poguntke [Po] on subsemigroups of Lie groups whose boundary is a subsemigroup. R. Houston [Ho] considered several problems on subsemigroups of Lie groups, such as e.g. the question when a topological semigroup is embeddable into a Lie group. The work of G. Graham on differentiable semigroups [Gr] again provides criteria for embedding into Lie groups.[1)]

Our third point concerns the need for a Lie semigroup theory in geometric control theory. Let us consider an example. One would like to study matrix differential equations of the form

$$X'(t) = (A + \sum_{i=1}^{m} u_i(t) B_i)X(t), \; X(0) = X_0$$

where $\{A, B_1, \ldots, B_m\}$ are $n \times n$ matrices in $M(n)$, $X(t)$ is a function from the non-negative reals to $M(n)$, and $u(t)$ is a piecewise continuous function from the non-negative reals to the reals, where u_i is chosen out of a set of controls. Besides solving the equation one is interested in the set of "accessible" points, points $Y \in M(n)$ such that $Y = X(t)$ for some solution for some set of controls.

The point of interest is that if $X_0 = I$, the accessible points form a subsemigroup of $Gl(n)$ generated by the exponential of a cone in $M(n)$ [Hi]. These considerations have generalizations to Lie groups and Lie algebras and wider classes of differential equations (see e.g. [J-S] and [Br]). It is therefore of interest to have a general semigroup theoretical method to deal with accessibility. The semigroups arising in this context are what we will call ray semigroups: They are the subsemigroups of Lie groups which are generated by the image of a cone in the Lie algebra under the exponential function. This calls for a comprehensive theory of these semigroups.

---

1) For details we refer to the preceding article by G.E.Graham.

We proceed as follows: In the first section we illustrate our topic by discussing some characteristic examples. In Section 2 we turn to the very subject of ray semigroups and develop some fundamental properties. The core result of this section will imply among other things that a ray semigroup generating a Lie group has dense interior in this group.

In Section 3 we turn to a very basic issue for the treatment of semigroups in Lie groups: We investigate in which way a subsemigroup of a Lie group can be "infinitesimally generated." In the same way as the tangent vectors to a subgroup of a Lie group at the origin form a subalgebra we associate with a subsemigroup tangent vectors accordingly. We first observe that they always form a closed convex cone, or wedge, as we will say to emphasise the fact that the cone need by no means be proper. The difficulty is to characterize exactly those wedges which occur as generating wedges of ray semigroups. In order to accomplish this characterisation we develop the theory of local semigroups (or germs of semigroups) and introduce the notion of local Lie semigroup which in many respects is simpler than the global concept. Some of the technical complications are considerable, and our examples demonstrate that these complications have to be faced squarely.

The algebraic objects which function as the analog of Lie algebras for semigroups are the so called "Lie wedges" which we introduce in 3.14. The crucial role they play in the entire theory begins to emerge in 3.15.

In Section 5 we will show how every analytic semigroup in a Lie group canonically determines a Lie wedge which generates it.

The section concludes with a precise statement of the way in which the Lie wedge of a local semigroup "approximates locally" this semigroup (3.20).

In Section 4 we study the local groups which are canonically associated with a local semigroup, namely, the largest subgroup contained in it and the smallest local subgroup containing it. We prove that on the Lie algebra level, the expected phenomena occur, although for a complete proof of some of them we have to wait for Section 5. Curiously, they require the global theory, which follows in the next section. Here we finally have all the background to define the concept of an analytic semigroup. For closed subsemigroups the terms analytic semigroup and ray semigroup are synonymous. In the absence of closedness, the concept is more sophisticated due to the complication arising from non-closed analytical subgroups of a Lie group. We establish the link to the local (i.e. the algebraic) theory which was previously developed. This section concludes with a discussion of the group H(S) of invertible elements in a Lie semigroup S. It turns out to be a (connected) analytic subgroup; and if one considers on the group G(S) generated by S its intrinsic Lie group topology (which may be finer than the induced topology!), then H(S) is closed in G(S). Finally we show that with each ray semigroup T we can canonically assoicate an analytic semigroup S and a Lie semigroup $\overline{T}$ such that all three have the same Lie wedge and the same interior, and that $T \subseteq S \subseteq \overline{T}$.

The paper begins with the discussion of examples, and so it ends: In Section 6 we discuss in some detail a three dimensional Lie semigroup in a solvable Lie group and a three dimensional Lie semigroup in $Sl(2, \mathbb{R})$ whose wedge is a semialgebra which is canonically constructed from a Cartan subalgebra and a root system.

SECTION 1. DEFINITIONS and EXAMPLES

There is no completely obvious way to define a Lie semigroup. Indeed, off-hand, there are a variety of reasonable candidates. In this paper we focus on one particular definition (together with some closely related concepts) which arises in a number of various situations. Our motivation is two-fold: (i) Lie subsemigroups of Lie groups which are also subgroups should be Lie subgroups, and, more importantly, (ii) a Lie subsemigroup should be recoverable from its set of tangent vectors.

1.1 DEFINITION. Let G be a topological group (Hausdorffness is always assumed). A one-parameter semigroup (group) is a continuous function σ from the set $\mathbb{R}^+$ of non-negative real numbers (resp. all real numbers $\mathbb{R}$) into G such that $\sigma(r+s) = \sigma(r)\cdot\sigma(s)$ for all r,s in the domain. □

Note that a one-parameter semigroup has a unique extension to a one-parameter group by defining $\sigma(-t) = [\sigma(t)]^{-1}$. In a Lie group with exponential function exp: $L(G) \to G$, there is a bijection $X \to \sigma_X : L(G) \to \mathrm{Hom}(\mathbb{R},G)$ between the Lie algebra $L(G)$ and the one-parameter group $\mathrm{Hom}(\mathbb{R},G)$ characterized by the equation $\sigma_X(t) = \exp tX$ for all $t \in \mathbb{R}$.

A connected Lie group is generated (either as a group or a semigroup) by exp L(G), the union of the images of all its one-parameter groups. In fact there exist open neighborhoods U of 0 in L(G) and V of 1 in G respectively such that $\exp|U:U \to V$ is a homeomorphism. Moreover, a connected subgroup of a Lie group is a Lie subgroup if and only if it is generated by the images of the one-parameter subgroups contained in it (this is further discussed in the appendices). This fact motivates the following definition (essentially due to R. Hirshorn [Hi]).

1.2 DEFINITION. Let G be a Lie group. A subsemigroup S is a <u>ray subsemigroup</u> if S is generated (as a semigroup) by the union of the ranges of the one-parameter semigroups which it contains. A <u>Lie subsemigroup</u> of G is the closure of a ray subsemigroup. A semigroup S is a <u>ray semigroup</u> resp. <u>Lie semigroup</u> if it is simultaneously homeomorphic and isomorphic to a ray resp. Lie subsemigroup of some Lie group. □

1.3 REMARKS. (i) If S is a ray semigroup and $\{\sigma_\alpha\}$ is the collection of all one-parameter semigroups into S, then each $x \in S$ can be written as a product

$x = \sigma_{\alpha_1}(t_1)\,\sigma_{\alpha_2}(t_2)\ldots\sigma_{\alpha_n}(t_n)$. Conversely given any collection of one-parameter semigroups into a Lie group G, the set of all finite products of elements of the preceding form is a ray semigroup. (One needs only to make the trivial observation that if <u>any</u> set of one-parameter semigroups generates a semigroup S, then all one-parameter semigroups in S also generate it).

(ii) Clearly every closed ray subsemigroup is a Lie subsemigroup. The converse may fail (see Example 6.2).

1.4 <u>PROPOSITION</u>. Let S be a ray subsemigroup of a Lie group G. Then the subgroup $G(S)$ generated by S is a connected analytic subgroup.

Proof. If each of the one-parameter semigroups in S is extended to a one-parameter group, then it is easily seen that these one-parameter subgroups generate the same subgroup as S does. Since any set of one-parameter groups generates an analytic subgroup [Ya, Go, Bo], it follows that $G(S)$ is analytic. □

It should be recalled that the topology induced on $G(S)$ from G may be strictly coarser than the Lie group topology of $G(S)$. The Lie group topology of $G(S)$ induces on S a possibly finer topology, relative to which it is still a ray semigroup. For most purposes we can therefore assume that a ray subsemigroup generates the Lie group containing it. In most examples and results we assume this is true.

We now give some examples to illustrate the notions.

1.5 EXAMPLE. Let G be the nilpotent Heisenberg group of all unipotent matrices of the form $\begin{bmatrix} 1 & a & c \\ 0 & 1 & b \\ 0 & 0 & 1 \end{bmatrix}$.

Represent this matrix by the triple (a,b,c). Then multiplication is given by the formula $(a,b,c)(x,y,z) = (a+x, b+y, c+z+ay)$. Let $S = \{(a,b,c): 0 \le a,b,;\ 0 \le c \le ab\}$. A straightforward calculation verifies that S is a subsemigroup of G. Graphically it is the region in the first octant bounded by the surface $z = xy$ and the xy-plane.

We claim that S is a ray subsemigroup. Note that if we define $\sigma, \tau : [0,\infty) \to S$ by $\sigma(t) = (t,0,0)$ and $\tau(t) = (0,t,0)$, then σ and τ are one-parameter semigroups. Furthermore they generate S since if $(a,b,c) \in S$ and $b > 0$, then $(a,b,c) = \sigma(\frac{c}{b})\ \tau(b)\ \sigma(a - \frac{c}{b})$. If $b = 0$ then $c = 0$; hence $(a,b,c) = \sigma(a)$. Thus S is a three-dimensional closed ray subsemigroup which is generated by two one-parameter semigroups. □

We turn now to some examples which arise inside of solvable Lie groups.

1.6 EXAMPLE. Let D be the solvable Lie group of all $(n+1) \times (n+1)$ upper triangular matrices with real entries and positive entries along the diagonal. Let G be the subgroup of all matrices of the form:

$$\begin{bmatrix} t & 0 & \cdots & 0 & x_1 \\ 0 & t & \cdots & 0 & x_2 \\ \vdots & \vdots & & \vdots & \vdots \\ 0 & 0 & \cdots & t & x_n \\ 0 & 0 & \cdots & 0 & 1 \end{bmatrix}$$

where $t > 0$ and $x_1, \ldots, x_n$ are arbitrary real numbers. Let S be the subsemigroup satisfying the further restrictions that $0 < t \leq 1$ and $0 \leq x_i \leq 1 - t$ for $i = 1, \ldots, n$. Then S is a closed ray semigroup; indeed in this example every element of S is the member of a unique one-parameter semigroup.

The group G may be viewed alternately as a split extension of Euclidean n-space by the multiplicative reals $\mathbb{R}^*$. We have $(x_1, \ldots, x_n, t)(y_1, \ldots, y_n, s) = (x_1 + ty_1, \ldots, x_n + ty_n, ts)$. The semigroup S is defined by the same inequalities as before. One verifies that $\sigma(r) = (\frac{t^r - 1}{t - 1}(x_1, \ldots, x_n), t^r)$ is a one-parameter semigroup in S such that $\sigma(1) = (x_1, \ldots, x_n, t)$ if the later element is in S.

This example is due to Brown and Friedberg [B-F]. For $n = 1$ it is an old example long known to semigroupers as the "triangle" in the affine plane.
For a more systematic discussion of this example, see [HL, sec. 3].

1.7 <u>EXAMPLE</u>. Let $G = Sl(2, \mathbb{R})$ be the semisimple group of all 2×2 matrices of determinant 1. Let S be the subsemigroup of all elements of G which have all non-negative entries. Then S is a semigroup with a group of units consisting of all the diagonal matrices which are in S.

Define a one-parameter group in S by

$$\sigma(t) = \begin{bmatrix} a^t & 0 \\ 0 & a^{-t} \end{bmatrix}$$

where $a > 1$, and two one-parameter semigroups ρ, λ by

$$\rho(t) = \begin{bmatrix} 1 & t \\ 0 & 1 \end{bmatrix} \text{ and } \lambda(t) = \begin{bmatrix} 1 & 0 \\ t & 1 \end{bmatrix}. \text{ If } \begin{bmatrix} a & b \\ c & d \end{bmatrix} \in S,$$

then $a \neq 0$ (for otherwise b or c must be negative). We can then write

$$\begin{bmatrix} a & b \\ c & d \end{bmatrix} = \begin{bmatrix} 1 & 0 \\ c/a & 1 \end{bmatrix} \begin{bmatrix} a & 0 \\ 0 & 1/a \end{bmatrix} \begin{bmatrix} 1 & b/a \\ 0 & 1 \end{bmatrix}$$

(since $(1 + cb)/a = d$), and the matrices on the right are in the images of λ, σ, and ρ resp. □

These examples provide a glimpse of some of the diversity exhibited by ray semigroups. We shall consider other examples as we proceed.

SECTION 2. Ray Semigroups

We consider first some elementary properties of the interior of a semigroup.

2.1 PROPOSITION. Let G be a topological group, S a subsemigroup with non-empty interior.

(i) $\text{int}(S)$ is an ideal of S;

(ii) If 1 is in the closure of $\operatorname{int}(S)$, then $\operatorname{int}(S) = \operatorname{int}(\overline{S})$ and $\operatorname{int}(S)$ is dense in S.

(iii) If G is compact and connected, then $S = G$.

Proof. (i) Let $s \in S$. Then $s \cdot \operatorname{int}(S) \subseteq S$ and is open. Hence $s \cdot \operatorname{int}(S) \subseteq \operatorname{int}(S)$.

(ii) Obviously $\operatorname{int}(S) \subseteq \operatorname{int}(\overline{S})$. Let $s \in U = \operatorname{int}(\overline{S})$. There exists V open containing 1 with $sV^{-1} \subseteq U$. Let $W = V \cap \operatorname{int}(S)$ (by hypothesis $W \neq \emptyset$). Then $sW^{-1} \subseteq U \subseteq \overline{S}$; since sW^{-1} is open, there exists $t \in S$, $w \in W$ such that $sw^{-1} = t$. Then $s = tw \in tW \subseteq \operatorname{int}(S)$.

Let $s \in S$; then $s = s \cdot 1 \in s \cdot (\operatorname{int} S)^{-} = (s \cdot \operatorname{int} S)^{-} \subseteq (\operatorname{int} S)^{-}$. Thus $\operatorname{int}(S)$ is dense in S.

(iii) $\overline{S}$ is a compact subsemigroup of G, hence a group. As $\overline{S}$ has interior, it is open and closed; thus $G = \overline{S}$. Then $G = \operatorname{int}(\overline{S}) \subseteq S$ by (ii). □

2.2 PROPOSITION. Let the ray semigroup S generate the connected Lie group G. Let Σ be a family of one-parameter semigroups generating S. Then some finite subset $\Gamma \subseteq \Sigma$ (group-) generates G.

Proof. Let the dimension of G be n. Let Δ be any non-empty finite subset of Σ. Let H be the subgroup generated by Δ. By Proposition 1.4 H is a Lie subgroup determined by a Lie subalgebra $L(H)$ of $L(G)$. We write $n(\Delta) = \dim L(H)$. Thus the function $\Delta \to n(\Delta)$ is defined on the set of finite subsets of Σ and takes values in $\{1,2,\ldots,\dim G\}$. This function attains its maximum at Δ, say. Suppose that $n(\Delta) < \dim G$.

Then H is proper in G, since $\dim L(H) < \dim L(G)$. Therefore $\Delta \neq \Sigma$, otherwise $S \subseteq H$, whence $H = G$. Now pick $\sigma \in \Sigma\backslash\Delta$ such that $\sigma(\mathbb{R}^+) \not\subseteq H$. Let K be the subgroup generated by $H \cup \sigma(\mathbb{R}^+)$. Again by Proposition 1.4, K is a connected Lie group and $L(K) \supseteq \operatorname{span} L(H) + \mathbb{R} \cdot X$ with $\sigma(t) = \exp t X$, $t \in \mathbb{R}$. Thus $n(\Delta \cup \{\sigma\}) = \dim L(K) > \dim L(H) = n(\Delta)$, contradicting maximality of $n(\Delta)$. □

We turn now to the important notion of reachability.

2.3 <u>DEFINITION</u>. Let S be a ray semigroup and let $\Sigma = \{\sigma_i : i \in J\}$ be a family of one-parameter semigroups which generate S. A point $s \in S$ is said to be *reachable at time* T *with respect to* Σ if there exist positive real numbers $t_1,\ldots,t_n$ and $\sigma_{i(1)},\ldots,\sigma_{i(n)} \in \Sigma$ such that $T = \Sigma_{i=1}^n t_i$ and $s = \sigma_{i(1)}(t_1)\ldots\sigma_{i(n)}(t_n)$. If Σ is understood, we frequently omit the phrase "with respect to Σ" and say that the point s is "T-reachable." Note that a point may be reachable for many different times, but that every point in S is reachable for at least one time.

Let the points reachable at time T or less be denoted by

$S(\Sigma,T) = \{s \in S : s$ is reachable at time t with respect to Σ for some t, $0 \leq t \leq T\}$. □

We come now to a very important theorem concerning the reachable points. The theorem is basically the translation of results out of differential geometry into the language of ray semigroups. To avoid interrupting the flow of material, we defer the discussion of the necessary differential geometry and the translation process to Appendix I. The final results themselves betray little evidence of their origins; the reader may wish simply to assume them.

We now have the following theorem:

2.4 THEOREM. Let G be a Lie group which as a group is generated by a ray subsemigroup S. Let Σ be a collection of one-parameter semigroups which generate the semigroup S.

Then the following conclusions hold:

(i) For each $T > 0$, the set $S(\Sigma, T)$ of points reachable in time T or less has non-empty interior. Furthermore, the interior is dense in $S(\Sigma, T)$.

(ii) If Σ is finite, then for any open set U with $1 \in U$, there exists an $\epsilon > 0$ such that if $T < \epsilon$, then $S(\Sigma, T) \subseteq U$. $\sqcap$

Part (i) is a restatement of Theorem 3.1 of [S-J] (See Proposition 5 of [Hi] for a translation into the semigroup setting). Part (ii) is a translation of some basic facts concerning integral curves in differential geometry (cf. Appendix A).

2.5 COROLLARY. Let G be a Lie group which is generated by a ray subsemigroup S. Then the interior of S is dense in S.

Proof. By Proposition 2.2 there exists a finite set $\Gamma \subseteq \Sigma$ such that G is the smallest subgroup containing the range of every member of Γ. Let $s \in S$. Then there exist $\sigma_1, \ldots, \sigma_n \in \Sigma$ and $t_1, \ldots, t_n$ such that $s = \sigma_1(t_1) \ldots \sigma_n(t_n)$. Let $\Delta = \Gamma \cup \{\sigma_i : 1 \leq i \leq n\}$. Let S' be the ray semigroup generated by Δ. Then G is generated by S' and $s \in S'$.

Let U be open, $1 \in U$. By Theorem 2.4 (ii), there exists $\epsilon > 0$ such that $T < \epsilon$ implies $S'(\Delta, T) \subseteq U$. By 2.4 (i) $S'(\Delta, T)$ has interior in G. Thus since $\mathrm{int}(S'(\Delta, T)) \subseteq \mathrm{int}(S') \subseteq \mathrm{int}(S)$, we have $U \cap \mathrm{int}(S) \neq \emptyset$, i.e. 1 is in the closure of $\mathrm{int}(S)$. By Proposition 2.1, $\mathrm{int}(S)$ is dense in S.

Section 3. Local Semigroups and Lie Algebras

The Lie algebra $L(G)$ of a Lie group G is a most important tool in the study of a Lie group since it allows many statements about the structure of G to be translated into algebraic language in $L(G)$. Our goal is to associate with a ray semigroup a suitable algebraic object within the Lie algebra. In this section, however, we restrict our attention to the Lie algebras themselves and develop the foundations of an "algebraic" machinery suitable to deal with semigroups. (The term "algebraic" means simply that the theory is formulated entirely in terms of Lie algebras). In the next sections we connect this material with earlier work.

We shall use freely many of the basic properties of finite-dimensional Lie algebras. The reader can find a brief review of these in Appendix B.

In this entire section we work within the following setting. Let L be a finite-dimensional real Lie algebra with a norm $\|\cdot\|$ such that $\|[x,y]\| \leq \|x\| \, \|y\|$ for all $x,y \in L$. (For any given finite-dimensional Lie algebra, there is at least one and usually many such norms). Let B be an open ε-ball around 0 with respect to this norm such that (i) $(x,y) \to x * y = x + y + \frac{1}{2}[x,y] + \ldots$ is defined and continuous on $B \times B$ where $*$ is the multiplication given by the absolutely convergent Campbell-Hausdorff series, and (ii) all triple products are defined and associative, i.e. $(x * y) * z = x * (y * z)$ for all $x,y,z \in B$. Such neighborhoods always exist (indeed one may choose $\varepsilon = \frac{1}{3} \ln (\frac{3}{2})$). We assume B is fixed in the remainder of this section.

3.1 <u>DEFINITION</u>. We say $S \subseteq B$ is a <u>local semigroup with respect to</u> B if $0 \in S$ and $(S*S) \cap B \subseteq S$; further $G \subseteq B$ is a <u>local group with respect to</u> B if G is a local semigroup with respect to B and $G = -G$. Often we consider only semigroups and groups with respect to B and hence omit the modifier "with respect to B." □

3.2 REMARK. $H(S) = -S \cap S$ is the maximal local subgroup of a local semigroup S.

Proof. Suppose $x, y \in H(S)$. Then $x * y \in B$ implies $x * y \in S$, since S is a local semigroup. Also $-(x*y) = (-y) * (-x)$ is in B (since $B = -B$) and thus in S since $-y, -x \in S$. Thus $x * y \in H(S)$, and hence $H(S)$ is a local semigroup.

Clearly $H(S) = -H(S)$; thus $H(S)$ is a local group. Finally, if a local group G is contained in S, then $G = -G \cap G \subseteq -S \cap S = H(S)$. □

3.3 REMARK. If S is a local semigroup, then $S' = \overline{S} \cap B$ is also a local semigroup. (We henceforth denote $\overline{S} \cap B$ by S').

Proof. Suppose $x, y \in S'$. Then $x = \lim x_n$, $y = \lim y_n$ where $x_n, y_n \in S$ for all n. Then $x * y = \lim x_n * y_n$. If $x * y \in B$, then eventually $x_n * y_n \in B$ and hence $x_n * y_n \in S$. Thus $x * y \in S'$. Thus S' is a local semigroup. □

3.4 DEFINITION. Let $X \subseteq L$. An element $b \in B$ is <u>X-generated</u> (<u>with respect to</u> B) if $b \in \bigcup_{n \in \omega} X^n$ where X^n is defined inductively by $X^1 = (B \cap X) \cup \{0\}$, $X^n = (X^{n-1} * X^{n-1}) \cap B$. Let $\langle X \rangle_B$ (or $\langle X \rangle$ if B is understood) denote all elements of B that are X-generated with respect to B. □

3.5 PROPOSITION. The set $\langle X\rangle$ of all X-generated elements is the smallest local semigroup in B containing $X \cap B$.

Proof. Suppose $a \in X^m$, $b \in X^n$, where $m \leq n$. Since $0 \in X^k$ for all k, $a \in X^n$. If $a * b \in B$, then $a * b \in X^{n+1} \subseteq \langle X\rangle$. Thus $\langle X\rangle$ is a local semigroup.

If S is a local semigroup containing $X \cap B$, then an easy induction gives $X^n \subseteq S$ for all n. Thus $\langle X\rangle$ is the smallest local semigroup containing $X \cap B$. □

3.6 COROLLARY. If S is a local semigroup, then $G(S) = \langle S \cup -S\rangle$ is the smallest local group containing S.

Proof. Clearly any local group containing S contains $S \cup -S$ and hence $\langle S \cup -S\rangle$ by Proposition 3.5. Since $\langle S \cup -S\rangle$ is a local semigroup and since $B = -B$, an easy induction shows $(S \cup -S)^n$ is closed with respect to $-$ for all n. Hence $\langle S \cup -S\rangle$ is a local group. □

The following lemma is crucial.

3.7 LEMMA. Let S be a local semigroup. Then $\{x \in L: x = \lim nx_n, x_n \in S\} = \{x \in L: \mathbb{R}^+ x \cap \overline{B} \subseteq \overline{S}\} = \{x \in L: \mathbb{R}^+ x \cap B \subseteq S'\}$. Furthermore, if $x = \lim nx_n$, then $\lim x_n = 0$.

Proof. Suppose $x = \lim nx_n$, $x_n \in S$. Let $r > 0$ such that $rx \in B$. Let $m_n = [rn]$, the greatest integer less than or equal to rn. Then $rx = \lim rnx_n = \lim([rn]x_n + \epsilon_n x_n)$ where $0 \leq \epsilon_n < 1$. Now $x = \lim nx_n$ implies

$\|x\| = \lim n\|x_n\|$, which implies $\lim \|x_n\| = 0$, i.e., $\lim x_n = 0$. Thus $\lim \varepsilon_n x_n = 0$. We conclude $rx = \lim\,[rn]x_n$. Since $rx \in B$, eventually $[rn]x_n \in B$, and hence $x_n, 2x_n, \ldots, [rn]x_n \in B$ since B is an ε-ball. Since S is a local semigroup and $x_n \in S$, $[rn]x_n \in S$. Thus $rx \in \overline{S} \cap B = S'$, and hence the first set is contained in the third.

Suppose x is in the third set and $rx \in \overline{B}$. Then $rx = \lim\,(r-\frac{1}{n})x$ and each $(r-\frac{1}{n})x \in B$ for $\frac{1}{n} < r$. Thus eventually $(r-\frac{1}{n})x \in \overline{S}$, and hence $rx \in \overline{S}$. This shows the third set is contained in the second.

Suppose finally that $\mathbb{R}^+ x \cap \overline{B} \subseteq \overline{S}$. For each n large enough, pick $x_n \in S$ such that $d(\frac{1}{n}x, x_n) < \frac{1}{n^2}$ (let $x_n = 0$ for smaller n). Then $d(x, nx_n) = nd(\frac{1}{n}x, x_n)$ (where d is the metric derived from the norm).
But $nd(\frac{1}{n}x, x_n) < n \cdot \frac{1}{n^2} = \frac{1}{n}$. Thus $x = \lim nx_n$, and hence the second set is contained in the first. □

We are now in a position to associate with a local semigroup its set of tangent vectors at 0.

3.8 <u>DEFINITION</u>. Let S be a local semigroup.
We set
$$L(S) = \{x \in L : x = \lim nx_n,\ x_n \in S,\ \lim x_n = 0\}$$
$$= \{x \in L : \mathbb{R}^+ x \cap B \subseteq S'\}.$$
(Note that the two sets are equal by Lemma 3.7). □

3.9 <u>REMARK</u>. $L(S) \cap B \subseteq S'$ since for $x \in L(S) \cap B$, $x = 1 \cdot x \in \mathbb{R}^+ x \cap B \subseteq S'$. Consequently $\langle L(S) \cap B \rangle \subseteq S'$ by Remark 3.3 and Proposition 3.5. □

A non-empty subset K of a vector space V which is topologically closed and closed under vector addition and multiplication by non-negative scalars is called a <u>wedge</u>. The wedge K is a <u>cone</u> if $K \cap -K = \{0\}$. The wedge K <u>generates</u> V or <u>is generating</u> if $V = K - K$, i.e., V is the smallest subspace containing K.

3.10 <u>PROPOSITION</u>. Let S be a local semigroup.

(i) $L(S)$ is a wedge.

(ii) $L(S) = L(S')$.

(iii) $L(H(S)) \subseteq L(S) \cap (-L(S))$.

(iv) If $L(S) \cap B \subseteq S$ (which is the case for $S = S'$) then $L(H(S)) = L(S) \cap (-L(S)) = L(S') \cap (-L(S'))$ $= L(H(S'))$.

Proof. (i) Suppose $a, b \in L(S)$. Then $a = \lim nx_n$, $b = \lim ny_n$ where $\lim x_n = 0 = \lim y_n$ and $x_n, y_n \in S$ for each n. Then by basic properties of Lie algebras we have $a + b = \lim n(x_n * y_n)$, and eventually $x_n * y_n \in B$ and hence in S. Thus $a + b \in L(S)$. The second characterization of $L(S)$ shows that $L(S)$ is closed under scalar multiplication by $\mathbb{R}^+$. A straightforward argument from the same characterization gives that $L(S)$ is closed.

(ii) Clearly $L(S) \subseteq L(S')$. Suppose $x = \lim nx_n$ where $x_n \in S'$. Then $x = \lim ny_n$, $y_n \in S$, by picking y_n such that $d(x_n, y_n) < 1/n^2$ (see the last part of the proof of Lemma 3.7). Thus $L(S') \subset L(S)$.

(iii) Note that $-L(S) = L(-S)$. Since $H(S) = S \cap -S$, $L(H(S)) \subseteq L(S) \cap L(-S) = L(S) \cap (-L(S))$.

(iv) The middle equality follows from (ii). Suppose $x \in L(S) \cap (-L(S))$. Then $\mathbb{R}^+ x \cap B \subseteq S'$ and $R^+(-x) \cap B \subseteq S'$. If $rx \in B$, then $rx \in L(S) \cap B \subseteq S$ by hypothesis. Also $rx \in B$ implies $r(-x) \in B$; thus $r(-x) \in B \cap L(S) \subseteq S$. Since $-rx = r(-x)$, we have $rx \in H(S)$ if $rx \in B$. Thus $R^+ x \cap B \subseteq H(S)$, i.e., $x \in L(H(S))$. Thus $L(H(S)) = L(S) \cap (-L(S))$ follows from (iii). This equality applies to S' in place of S, which proves the rest.

If $S = S'$, then $L(S) \cap B \subseteq S$ by Remark 3.9. Hence the equalities of (iv) follow. □

<u>NOTATION</u>. For $Y \subseteq L$, let $\langle\langle Y \rangle\rangle$ be the Lie algebra generated by Y.

The following is, of course, well-known:

3.11 <u>LEMMA</u>. Let M be a subalgebra of L. Then $M \cap B$ is a local group.

Proof. Since M is a subspace, M is closed in L. By the Campbell-Hausdorff formula, M is closed with respect to the $*$-operation since it is a subalgebra, subspace, and closed. Thus $M \cap B$ is a local group since

$-(M \cap B) = M \cap -B = M \cap B.$ □

3.12 PROPOSITION. Let S be a local semigroup such that $S \subseteq \langle\langle L(S) \rangle\rangle$. Then $L(G(S)) = L(\langle S \cup -S \rangle) = \langle\langle L(S) \rangle\rangle$. In particular if H is a local group, $H \subseteq L(H)$, then $L(H)$ is a subalgebra.

Proof. We first show that $L(G(S))$ is a subalgebra. Since $G(S)$ is a local semigroup, by Proposition 3.10 $L(G(S))$ is a closed wedge. Since $G(S) = -G(S)$, $L(G(S)) = -L(G(S))$. Thus $L(G(S))$ is a vector space.

Suppose $x, y \in L(G(S))$. Then $x = \lim nx_n$, $y = \lim ny_n$, where $x_n, y_n \in G(S)$. Then also $-x_n, -y_n \in G(S)$. By Lie algebra properties $[x,y] = \lim n^2(-x_n * -y_n * x_n * y_n)$ $= \lim n(n(-x_n * -y_n * x_n * y_n)$. Thus it must be the case that $\lim n(-x_n * -y_n * x_n * y_n) = 0$; hence eventually the element $k(-x_n * -y_n * x_n * y_n) \in B$ for all k, $1 \leq k \leq n$, and thus is a member of $G(S)$. We conclude $[x,y] \in L(G(S))$.

Now $L(S) \subseteq L(G(S))$ and the fact that $L(G(S))$ is a subalgebra imply $\langle\langle L(S) \rangle\rangle \subseteq L(G(S))$.

Conversely by Lemma 3.11 $\langle\langle L(S) \rangle\rangle \cap B$ is a local group. Thus $S \subseteq \langle\langle L(S) \rangle\rangle \cap B$ implies $G(S) \subseteq \langle\langle L(S) \rangle\rangle \cap B$, and hence $L(G(S)) \subset \langle\langle L(S) \rangle\rangle$. □

3.13 PROPOSITION. Let S be a local semigroup. Then $\overline{\langle L(S) \cap B \rangle} \subseteq \langle\langle L(S) \rangle\rangle$. Hence if $S \subseteq \overline{\langle L(S) \cap B \rangle}$, then $S \subseteq \langle\langle L(S) \rangle\rangle$.

Proof. Finite products of elements in $L(S) \cap B$ which are in B are also in $\langle\langle L(S) \rangle\rangle$ since $\langle\langle L(S) \rangle\rangle \cap B$ is a local group (Lemma 3.11). Thus $\overline{\langle L(S) \cap B \rangle} \subseteq \langle\langle L(S) \rangle\rangle$ since $\langle\langle L(S) \rangle\rangle$ is a subspace (hence closed). □

3.14 DEFINITION. Let L be a Lie algebra. A subset $K \subseteq L$ is called a Lie wedge if there exists a local semigroup S with respect to some open ϵ-ball B such that $K = L(S)$. By Proposition 3.10 K is closed under vector addition and scalar multiplication by non-negative scalars and is closed topologically (i.e. K is a wedge).

A Lie wedge K is called a Lie semialgebra (or sometimes simply a semialgebra) if $K \cap B$ is a local semigroup with respect to some (hence any sufficiently small) open ϵ-ball B.

A local semigroup S with respect to B is called a local analytic semigroup if $S = \langle L(S) \cap B \rangle$. If, in addition, S is closed in B, then S is called a local Lie semigroup (with respect to B).

3.15 PROPOSITION.

1) Let W be a Lie wedge in L. Then there is a local semigroup S in B for a suitable open ball B around 0 such that $W = L(S)$. Let C

be any open ball around 0 in B and set $T = \langle W \cap C \rangle_C$. Then T is a local analytic semigroup in C and $L(T) = W$.

2) For any C as in 1) above we have $L(S \cap C) = W$.

3) If S is a local analytic semigroup in B with $L(S) = W$ then for any C as in 1) the set $S \cap C$ is a local subsemigroup of C with $L(S \cap C) = W$. For the local analytic semigroup $T = \langle W \cap C \rangle_C$ of C we have (i) $L(T) = W$, (ii) $T \subset S \cap C$ and (iii) $\langle T \rangle_B = S$.

Proof.

1) By definition $W \cap C \subseteq L(T)$, since W is a wedge and $W \cap C \subseteq T$. Thus $W \subseteq L(T)$ is also a wedge. By Remark 3.9 we have $W \cap B \subseteq S'$, whence $T = \langle W \cap C \rangle \subseteq \langle W \cap C \rangle_B \subseteq S'$, since S' is a local subsemigroup of B. Thus $L(T) \subseteq L(S') = L(S)$ by Proposition 3.10. We conclude $L(T) = W$. Hence $T = \langle L(T) \cap C \rangle_C$ and thus T is a local analytic subsemigroup of C by definition.

2) We have $W \cap C = L(S) \cap C \subseteq S' \cap C = (S \cap C)'$. From Remark 3.9 and 3.10 (ii), we deduce $W = L(\langle W \cap C \rangle_C) \subseteq L((S \cap C)') = L(S \cap C) \subseteq L(S) = W$.

3) (i) From 2) we have $L(S \cap C) = W$. We must show that $S \cap C \supseteq \langle W \cap C \rangle_C$. By hypothesis on S we know $S = \langle W \cap B \rangle_B$. Thus $\langle W \cap C \rangle_C \subset \langle W \cap B \rangle_B = S$, whence $\langle W \cap C \rangle_C \subseteq S \cap C$.

(ii) Since $T = \langle W \cap C\rangle_C$ we have $T \subseteq \langle W \cap B\rangle_B = S$.

(iii) $\langle T\rangle_B \subseteq \langle S \cap C\rangle_B$ (by (ii)) $\subseteq S = \langle W \cap B\rangle_B$. Since W is a wedge, $\langle W \cap C\rangle_B$ contains $W \cap B$, whence $\langle W \cap B\rangle_B = \langle W \cap C\rangle_B \subseteq \langle T\rangle_B$. □

REMARK. The information provided in Proposition 3.15 is adequate with regard to 1) and 2). Every Lie wedge is a wedge associated to a local analytic semigroup of an open ball C which, in addition, we may assume to be as small as we like. The Lie wedge is, therefore, defined independently of the choice of the ball B of reference. In this case, it is defined in terms of L alone and is, therefore, an "algebraic" concept. Indeed in [H-L, 82] we introduce the following concept which is defined without reference to local semigroups:

<u>DEFINITION</u>. A wedge W in a Lie algebra L is called <u>compatible</u>, provided it satisfies the following condition:

For each $X \in W \cap -W$ we have $e^{\operatorname{ad} X} W = W$. □

We show in [H-L, 82] that every Lie wedge is compatible, and we conjecture that the converse is true, too, but we are able to verify this conjecture only if the edge $W \cap -W$ of the wedge is a semisimple algebra (or $\{0\}$).

The information we give in 3.15 on local analytic semigroups S in a ball of reference B is not entirely satisfactory. One would hope that the property of being a local analytic semigroup is truly "local," i.e., that

$S \cap C$ is a local analytic semigroup in C for any open ball of reference within B. We do not know whether this is the case. The information given in 3.15.3) is the best we have.

3.16 <u>DEFINITION</u>. Let S be a local semigroup of B. We say that S is divisible iff for each $s \in S$ and each natural number n the element $\frac{1}{n}s$ is also in S.

3.17 <u>PROPOSITION</u>. Let S be a divisible local semigroup of B. Then a) $S' = L(S) \cap B$;

b) $L(S)$ is a Lie semialgebra;

c) $S' \cap C$ is a divisible local Lie semigroup for all open balls C in B around 0.

Proof. a) By Remark 3.9 we have $L(S) \cap B \subset S'$. Now let $s \in S$. If r is any rational number which satisfies $0 \leq r \leq 1$, then $r = m/n$ with natural numbers $0 \leq m \leq n \neq 0$. Since S is divisible, $\frac{1}{n}s \in S$, and since S is a local semigroup, by induction we find $rs = \frac{m}{n}s \in S \subseteq S'$. Since S' is closed in B, we have $rs \in S'$ for all $r \in [0,1]$. Thus $s \in L(S)$ by 3.7 and 3.8. Thus $S \subseteq L(S)$, and since $L(S)$ is closed, we have $S' \subseteq L(S)$, and hence a) is proved since $S' \subseteq B$. The claim b) is immediate from the definition of a semialgebra in 3.14. In order to prove c), we take a ball C around 0 in B; then $S' \cap C$ is clearly divisible, and since $S' \cap C = L(S) \cap C$ by a) and

3.10 (ii) we notice $L(S'\cap C) = L(S)$. Since $S'\cap C$ is a local semigroup, we have $S'\cap C = \langle L(S)\cap C\rangle_C = \langle L(S'\cap C)\cap C\rangle_C$ whence $S'\cap C$ is a local Lie semigroup. □

Thus divisible closed local semigroups in B are local Lie semigroups for which the information lacking in 3.15 is indeed available. Of course, whenever K is a Lie semialgebra, then $K\cap B$ is divisible whenever $K\cap B$ is a local semigroup. What we do not know is whether for any open ball B' on which multiplication $*$ is defined, the set $K\cap B'$ is a local subsemigroup of B' whenever $K\cap B$ is a local semigroup for some open ball B (which may be smaller than B'). In any case we have the following corollary:

3.18 <u>COROLLARY</u>. Let S be a local subsemigroup of B. Then the following statements are equivalent:

(1) S is divisible and closed in B.

(2) S is a divisible analytic subsemigroup of B.

(3) L(S) is a Lie semialgebra with $S = L(S)\cap B$.

Proof. By 3.17 we have (1) => (3) => (2). Suppose (2), then by 3.17, L(S) is a semialgebra with $S' = L(S)\cap B$. But since S is an analytic semigroup, we have $S' = L(S)\cap B \subseteq S$, whence S is closed in B. This shows (1). □

3.19 <u>PROPOSITION</u>. Let K be a Lie semialgebra in L. Then $\langle\langle K\rangle\rangle = K - K$.

Proof. Clearly $K-K$ is a vector space which is contained in $\langle\langle K\rangle\rangle$. In order to show the reverse containment, it suffices to show that $K-K$ is a Lie algebra, and for this it suffices to show that for $a,b \in K$ we have $[a,b] \in K-K$. We consider the element $c_n = (\frac{1}{n}a * \frac{1}{n}b) - (\frac{1}{n}b * \frac{1}{n}a)$ for large enough n. Since K is a semialgebra, $c_n \in K-K$ for large enough n. It follows that the element $n^2 c_n$ is also in the vector space $K-K$. But one calculates that $n^2 c_n = [a,b] + \frac{1}{n}0(n)$, where $0(n)$ is a vector which remains bounded as n grows. Thus $[a,b] = \lim n^2 c_n \in K-K$. □

We recall the example of the three-dimensional Heisenberg algebra L of all upper triangular 3 by 3 matrices with zero diagonal. The set of all matrices

$$\left\{ \begin{bmatrix} 0 & a & 0 \\ 0 & 0 & b \\ 0 & 0 & 0 \end{bmatrix}, \quad a,b \in \mathbb{R}, \quad a,b \geq 0 \right\}$$

is a Lie wedge W: Indeed if $\exp\colon L \to G$ is the exponential function of the Heisenberg group G in 1.5, then $T = \exp^{-1}(S)$ is a subsemigroup of $(L,*)$, where we note that $*$ is globally defined since L is nilpotent. Now we observe $W = L(T)$, which shows that W is a Lie wedge. But the two-dimensional subvector space $W-W$ is not a subalgebra. Thus $\langle\langle W\rangle\rangle = L \neq W-W$. Hence 3.19 does not hold in general for Lie wedges.

The Lie wedges associated with the examples 1.6 and

1.7 are semialgebras.

The local structure of a local subsemigroup S in B remains somewhat obscure even if S is a local analytic semigroup and $L(S)$ can be calculated; this phenomenon is illustrated by Example 1.5, and we will see more of it in 5.2 below. The only general situation in which the local structure of S near 0 is completely given in terms of $L(S)$ is the case that S is divisible (see 3.17). Nevertheless we can say something about the way in which $L(S)$ locally "approximates " S near 0. First we define the necessary tool for our concept of approximation.

3.20 DEFINITION. Let W be a closed wedge in finite dimensional real vector space V. We say that a closed wedge W' surrounds W iff $W \subseteq W'$ and $W \setminus (W \cap -W)$ is contained in the interior of W'.

Except for the edge of the wedge W a surrounding wedge W' is a neighborhood of W. Notice that a vector subspace is surrounded by itself.

Now we have the following local structure theorem for local semigroups.

3.21 THEOREM. Let S be a local semigroup in B. Suppose that the closed wedge W in L surrounds $L(S)$. Then there is an open ball C around 0 in B such that $C \cap S \subseteq W$.

Proof. Assume that such a C does not exist for some wedge W which surrounds $L(S)$. Then there is a sequence of elements $x_n \in S \setminus W$ with $0 = \lim x_n$. By picking a subsequence, if necessary, we may assume that $x = \lim x_n / \| x_n \|$ exists. We denote the edge $L(S) \cap -L(S)$ of $L(S)$ by E. We select a vector space complement V for E so that $L = E \oplus V$. We then write $x_n = e_n + v_n$ according to the direct sum decomposition. Since $e_n \in E \subseteq L(S) \subseteq W$ and $x_n \notin W$, we conclude $v_n \notin W$. We set $\|v_n\| = r_n$. Since $L(S) \cap B \subseteq S'$ by 3.9 and since eventually $e_n \in L(S) \cap B$, we find for each n (sufficiently large) an element $s_n \in S$ such that $\|s_n * x_n - (-e_n) * x_n\| < r_n^2$. We note that $\|(-e_n) * x_n - v_n\| = \|(-e_n) * (e_n + v_n) - v_n\|$. An inspection of the Campbell - Hausdorff formula shows that the power-series $(-X) * (X+Y) - Y$ is of the form $P(\text{ad}\, X, \text{ad}\, Y)([X,Y])$ with a suitable power series $P(s,t)$ (with constant term 1/2). From this we conclude $\|(-e_n) * (e_n + v_n) - v_n\| \leq C\|e_n\| \, \|v_n\|$ for a suitable constant C and for large enough n. Then we observe that

$$\|r_n^{-1}(s_n * x_n) - r_n^{-1} v_n\| \leq r_n^{-1} \|(s_n * x_n) - (-e_n) * x_n\|$$
$$+ r_n^{-1}\|(-e_n) * (e_n + v_n) - v_n\| \leq r_n + C\|e_n\| \qquad \to 0 .$$

We may assume (upon picking a suitable subsequence and renaming) that $v = \lim v_n / r_n \in V$ exists. Note $\|v\| = 1$. By the preceding we have $v = \lim r_n^{-1}(s_n * x_n) =$ $\lim\, [r_n^{-1}](s_n * x_n) \in L(S)$ in view of the definition of $L(S)$ (see 3.8). Thus $v \in (L(S) \cap V) \setminus \{0\} \subseteq L(S) \setminus E \subseteq \text{int}\, W$

since W surrounds $L(S)$. But since $v_n \notin W$ and thus $v_n/r_n \notin W$, this is a contradiction. □

As a special case of this result we retrieve the fact we know from local group theory: A local group S in B is locally contained in $L(S)$.

Section 4. Local groups associated with local semigroups

In semigroup theory it is important to study the groups which are associated with a given semigroup; in particular, if S is a submonoid of a group G there is always the subgroup $G(S)$ generated by S in G and the largest subgroup $H(S) = S \cap S^{-1}$ contained in S.

This is a guiding idea in the following discussion. We consider, however, a local semigroup S with respect to suitable ball B in a Lie algebra L and study the associated local groups. We begin with the particular definition of local Lie group which we wish to use:

4.1 DEFINITION. Let B be a ball around 0 in a Lie algebra such that $B * B$ is defined. A local Lie group H in B is a local group of the form $B \cap M$ where M is some some subalgebra of L.

4.2 PROPOSITION. Let G be a local group. The following statements are equivalent:

(1) G is a local analytic semigroup;

(2) G is a local Lie semigroup;

(3) G is a local Lie group;

(4) $G = L(G) \cap B$.

In any of these cases, G is connected.

Proof. (1) implies (4): We have $G = \langle L(G) \cap B \rangle \supseteq L(G) \cap B$ by definition. Conversely by Proposition 3.12 and 3.13 $G = \langle L(G) \cap B \rangle \subseteq \langle\langle L(G) \rangle\rangle \cap B = L(G) \cap B$. Thus $G = L(G) \cap B$.

(4) implies (3): Immediate from Proposition 3.12.

(3) implies (2): Let M be a subalgebra of L such that $G = M \cap B$. By Lemma 3.11 $\langle M \cap B \rangle = M \cap B$.

(2) implies (1): Trivial.

Also we have $M = \langle\langle B \cap M \rangle\rangle$ (indeed M is in the linear span of $B \cap M$). Thus $M = \langle\langle B \cap M \rangle\rangle = \langle\langle G \rangle\rangle = L(G)$ by Proposition 3.12. Hence $G = \langle L(G) \cap B \rangle$. Since $L(G) \cap B$ is convex, G is connected. □

At this point we would like to be able to show that the local group generated by a local analytic semigroup is a local Lie group. This indeed turns out to be the case, but it is not obvious how to derive this result with our present tools. Hence we defer the derivation of this result until Section 5 (see Proposition 5.11) and give here a weakened version.

4.3 <u>PROPOSITION</u>. Let S be a local analytic semigroup, and let $G(S) = \langle S \cup -S \rangle$. If $G(S)' = \overline{G(S)} \cap B$, then $G(S)'$ is a local Lie group and $L(G(S)') = L(G(S)) = \langle\langle L(S) \rangle\rangle$.

Proof. Since $S = \langle L(S) \cap B \rangle$ by Definition 3.14, then by 3.13 $S \subseteq \langle\langle L(S) \rangle\rangle$. Thus $L(G(S)) = \langle\langle L(S) \rangle\rangle$ by Proposition 3.12.

Again from $S = \langle L(S) \cap B \rangle$, we deduce $S \cup -S \subseteq \langle L(S) \cap B \rangle \cup \langle L(-S) \cap B \rangle \subseteq \langle L(G(S)) \cap B \rangle$. Since $\langle L(G(S)) \cap B \rangle = \langle\langle L(S) \rangle\rangle \cap B$ is a local group by Lemma 3.11 which clearly is closed in B, we conclude $G(S)' \subset \langle\langle L(S) \rangle\rangle \cap B = \langle L(G(S)) \cap B \rangle = \langle L(G(S))' \cap B \rangle$ by Proposition 3.10 (ii).

On the other hand $\langle L(G(S))' \cap B \rangle \subseteq G(S)'$ by Remark 3.9. Thus $G(S)'$ is a local analytic semigroup, and hence a local Lie group by Proposition 4.2. □

We turn now to a detailed investigation of the group of units of a local Lie semigroup. We have the following elementary lemma.

4.4 <u>LEMMA</u>. If S is a local semigroup, $x, y, x * y \in S$ and $x * y$ is a unit, then x and y are units. In general if a finite product of members of S is a unit of S, then so is each factor.

Proof. Since $x * y$ is a unit, we have $-(x * y) \in S$. Thus since $-x \in B$ and $-x = y * ((-y) * (-x)) = y * (-(x * y))$, we have $-x \in S$. Thus x is a unit in S. Similarly y is a unit in S. The last statement follows from the first by a straightforward induction. □

We come now to a key proposition.

4.5 PROPOSITION. Let S be a local analytic semigroup. Then $H(S)$ is a local Lie group (in particular connected). Furthermore we have

(i) $L(H(S)) = L(S) \cap -L(S) = \langle\langle H(S)\rangle\rangle$;

(ii) $H(S) = S \cap -S = L(H(S)) \cap B$.

Proof. Suppose (ii) holds. Then by Proposition 4.2 $H(S)$ is a local Lie group. By the definition of a local analytic semigroup (3.14) and Proposition 3.10 (iv), $L(H(S)) = L(S) \cap -L(S)$. By Proposition 4.3 (applied to $S = H(S)$), $L(H(S)) = \langle\langle L(H(S))\rangle\rangle$. But $\langle\langle L(H(S))\rangle\rangle = \langle\langle L(H(S)) \cap B\rangle\rangle = \langle\langle H(S)\rangle\rangle$ (the last equality by (ii)). Thus if (ii) is established, the rest of the proposition follows.

By Remark 3.2 $H(S) = S \cap -S$. As usual $S' = \overline{S} \cap B$. Then S' is also a local semigroup which is closed in B, and hence $H(S') = S' \cap -S'$ is a local subgroup which is closed in B. But if Γ is a closed local subgroup of B, then $L(\Gamma) \cap B$ is an open-closed subgroup of Γ (see Appendix B). This result yields an open ball B_1 around 0 contained in B such that $K = H(S') \cap B_1$ is a local Lie group with respect to B_1 such that $K = L(K) \cap B_1$. Since B_1 is open and contains 0, it follows easily from Definition 3.8 that $L(K) = L(H(S'))$. Also by Proposition 3.12 $L(K) = \langle\langle L(K)\rangle\rangle$, i.e. $L(K)$ is a subalgebra.

Let $x \in L(S) \cap H(S)$. Then there exists a positive integer n such that $(1/n)x \in B_1$. Since $L(S)$ is a wedge,

$(1/n)x \in L(S) \cap B \subseteq S$. Thus by Lemma 4.4 $(1/n)x \in H(S) \cap B_1 \subset H(S') \cap B_1 = K \subseteq L(K)$. Since $L(K)$ is a wedge, $x \in L(K)$. Thus $L(S) \cap H(S) \subseteq L(K) = L(H(S'))$.

Now let $y \in H(S)$. Then since $S = \langle L(S) \cap B \rangle$, y is a finite product of members of $L(S) \cap B$. Again by Lemma 4.4 each of these factors is in $H(S)$. By the preceding paragraph, each factor is a member of $L(K)$. Since by Lemma 3.11 $L(K) \cap B$ is a local group, we conclude $y \in L(K) \cap B$. But $L(K) \cap B = L(H(S')) \cap B = L(H(S)) \cap B$ (by Proposition 3.10 (iv)). We conclude $H(S) \subseteq L(H(S)) \cap B$.

Conversely $L(H(S)) \cap B = L(S) \cap -L(S) \cap B$ (again by Proposition 3.10 (iv)) $= (L(S) \cap B) \cap -(L(S) \cap B) \subseteq S \cap -S$ (since S is a local analytic semigroup) $= H(S)$.

Thus $H(S) = L(H(S)) \cap B$ and the proof is concluded. □

4.6 <u>COROLLARY</u>. If W is a Lie wedge in L, then $W \cap -W$ is a subalgebra of L.

Proof. The proof is immediate form Propositions 3.15 and 4.5. □

Section 5. Analytic Semigroups

In the last section we presented the definition of a local analytic semigroup and derived some of their basic properties. In this section we turn to the global setting. We assume throughout this section that G is a Lie group with Lie algebra $L(G)$. In order to employ the machinery of the last sections, we depend heavily on the exponential functions $\exp : L(G) \to G$. The basic properties of the exponential function may be found in Appendix B.

In Definition 3.14 we defined a local analytic semigroup (with respect to B) to be one which was generated by a Lie wedge intersected with B (see also Proposition 3.15). The global definition requires more care.

5.1 <u>DEFINITION</u>. Let G be a Lie group with Lie algebra $L(G)$. A subsemigroup S of G is a <u>wedge subsemigroup</u> if there exists a Lie wedge W in $L(G)$ (see Definition 3.14) such that S is the smallest subsemigroup of G containing $\exp(W)$. A topological semigroup is a <u>wedge semigroup</u> if it is simultaneously homeomorphic and isomorphic to a wedge subsemigroup of a Lie group. □

Clearly wedge semigroups are ray semigroups; the interior of the first quadrant with 0 adjoined in the additive plane shows the converse may fail.

5.2 <u>PROPOSITION</u>. A subsemigroup S of G is a wedge subsemigroup if and only if there exists a local analytic subsemigroup T of $L(G)$ such that $S = \langle \exp T \rangle$, the semigroup generated by $\exp T$.

Proof. Suppose S is a wedge subsemigroup of G, i.e., $S = \langle \exp (W) \rangle$ for some Lie wedge $W \subseteq L(G)$. By Proposition 3.15 there exists an open ball B such that if T is the local semigroup with respect to B generated by $W \cap B$,

then T is a local analytic semigroup and $L(T) = W$. By 3.15 we may assume that B is small enough so that exp is one-to-one on B and satisfies
$\exp(a)\exp(b) = \exp(a*b)$ for all $a,b \in B$ (where $*$ is the multiplication given by the Campbell-Hausdorff formula). Then $W \cap B \subseteq \exp^{-1}(S) \cap B$, and the latter is a local semigroup with respect to B. Thus $T \subseteq \exp^{-1}(S) \cap B$. Thus we have $S = \langle \exp W \rangle = \langle \exp(W \cap B) \rangle$ (since W is a wedge) $\subseteq \langle \exp T \rangle \subseteq \langle \exp(\exp^{-1}(S)) \rangle \subseteq \langle S \rangle = S$. Thus $S = \langle \exp T \rangle$.

Conversely suppose $S = \langle \exp T \rangle$ where T is a local analytic semigroup. Then T is locally generated by $L(T)$. It thus follows that $S = \langle \exp L(T) \rangle$. □

The preceding discussion provides a method for obtaining a semigroup in G from a Lie wedge in $L(G)$, namely take the semigroup generated by the exponential of the wedge. We need also a method for going in the reverse direction. The technique of the last section of obtaining a Lie wedge from a local subsemigroup works here with a slight modification. The example of a one-parameter semigroup S winding densely in the torus shows that some care is necessary in our definitions to assume that the wedge associated with S is indeed a half line and not the entire Lie algebra of the torus.

5.3 <u>DEFINITION</u>. Let S be a subsemigroup of a Lie group G. We define S_R to be the ray semigroup generated by the union of the images of all one-parameter semigroups lying entirely in S, and $G(S_R)$ to be the subgroup generated by S_R. (Note by Proposition 1.4 that $G(S_R)$ is a analytic subgroup). Let S^* be the closure of S_R in the unique topology of $G(S_R)$ making it a Lie group. Then we define

$$L(S) = \{X \in L(G) : \exp(tX) \in S^* \text{ for all } t \geq 0\}$$
$$= \bigcap_{r>0} r \exp^{-1} S^*.$$

We call $L(S)$ the <u>wedge associated to</u> S. □

Note that for a closed subsemigroup S of G we have

$$L(S) = \bigcap_{r>0} r \exp^{-1} S,$$

and that this set is the set of tangent vectors of $\exp^{-1} S$ at 0 in the sense of 3.8.

5.4 <u>PROPOSITION</u>. Let S be a subsemigroup of a Lie group G. Then $L(S)$ is a Lie wedge.

Proof. Since $L(S) = L(S_R)$, we may assume without loss of generality that $S = S_R$, i.e. that S is a ray semigroup.

Let B be a small open ball around 0 in $L(G)$ such that the multiplication $*$ is defined on B and exponential mapping restricted to B is a homeomorphism. Since $G(S)$ is an analytic subgroup, there is a Lie subalgebra $L(G(S))$ of $L(G)$ such that the restriction of the exponential function to $L(G(S))$ yields the exponential function of $G(S)$ with its Lie group structure. Now S^* is closed in the Lie group topology of $G(S)$. Hence $\exp_{G(S)}^{-1}(S^*)$ is closed in $L(G(S))$, and thus $T = B \cap \exp_{G(S)}^{-1}(S^*)$ is closed in B. Also T is a local semigroup with respect to B since S^* is a semigroup in G.

Now $X \in L(T)$ iff $\mathbb{R}^+ X \cap B \subseteq T$ (from Definition 3.8 and the fact T is closed in B). But $tX \in T$ for small positive t iff $\exp(tX) \in S^*$ for small positive t iff $\exp(tX) \in S^*$ for all non-negative t (since S is a subsemigroup). The last statement is equivalent to $X \in L(S)$. Thus $L(T) = L(S)$. By definition 3.14, $L(T)$ is a Lie wedge, and hence $L(S)$ is. □

5.5 <u>DEFINITION</u>. Let S be a subsemigroup of Lie group G. Then S is an <u>analytic subsemigroup</u> of G if S is the semigroup generated by $\exp L(S)$. <u>Analytic semigroups</u> are ones simultaneously homeomorphic and isomorphic to some analytic subsemigroup of a Lie group. □

Note that not all Lie semigroups are analytic and vice-versa (See Example 6.2). This is in contrast with the case of groups.

The next two propositions summarize some of the basic properties developed thus far.

5.6 <u>PROPOSITION</u>. Let G be a Lie group and S a subsemigroup.

(i) If S is an analytic subsemigroup, then S is a wedge (and hence ray) subsemigroup.

(ii) If S is closed in G, then the following statements are equivalent:

(1) S is a ray semigroup,

(2) S is a wedge semigroup.

(3) S is analytic;

and these conditions imply

(4) S is a Lie semigroup.

(iii) If S is an analytic subsemigroup, then the subgroup generated by S is an analytic subgroup. Hence if S is a subgroup, it is an analytic subgroup.

Proof. (i) By Proposition 5.4 $L(S)$ is a Lie wedge. Hence $S = \langle \exp L(S) \rangle$ implies S is a wedge semigroup.

(ii) S is a closed ray semigroup. Then $S = S_R = S^*$. Hence $\exp(L(S))$ is just the union of the image of all one-parameter semigroups contained in S. Since S is a ray semigroup, this set generates S, i.e. S is an analytic subsemigroup. This shows (1) implies (3). We saw in part (i) that (3) implies (2). The remainder is clear.

(iii) Follows from Proposition 1.4. □

From 2.5 and 5.6 (i) we have immediately

5.7 PROPOSITION. Let G be a Lie group which (as a group) is generated by an analytic subsemigroup S. Then the interior of S is non-empty and dense in S. □

From 2.1 (iii) and 5.7 we record

5.8 COROLLARY. If a compact Lie group G is generated by a Lie subsemigroup S, then $G = S$. □

We come now to a global analog of Proposition 4.5.

5.9 THEOREM. Let S be an analytic subsemigroup of a Lie group G. Then $H(S) = S \cap S^{-1}$ is the group of units of S, $H(S)$ is a connected analytic subgroup of G, and $L(H(S)) = L(S) \cap -L(S)$.

Proof. One verifies easily that $H(S) = S \cap S^{-1}$ (where S^{-1} is the set of inverses of elements of S) is the group of units of S.

Since $H(S) \subseteq S$, we have $L(H(S)) \subseteq L(S)$. Similarly $L(H(S)) \subseteq L(S^{-1}) = -L(S)$. Thus $L(H(S)) \subseteq L(S) \cap -L(S)$.

Conversely suppose $X \in L(S) \cap -L(S)$. Then $tX \in L(S)$ for all $t \in \mathbb{R}$. Thus $\exp(tX) \in S$ for all t, which implies $\exp(tX) \in H(S)$ for all t. Hence $X \in L(H(S))$. Thus $L(H(S)) = L(S) \cap -L(S)$.

Since $\exp(L(H(S))) \subseteq \exp L(S) \subseteq S$, we have that the subgroup generated by $\exp(L(H(S)))$ is contained in S (note the subgroup and subsemigroup generated by $\exp L(H(S)))$ coincide). Hence $\exp(L(H(S)))$ is contained in $H(S)$. To finish the proof we show that $H(S)$ is generated by $\exp(L(H(S)))$.

Let $g \in H(S)$. Then $g = \exp(X_1) \ldots \exp(X_n)$ for some $X_1, \ldots, X_n \in L(S)$. Since the complement of $H(S)$ in S is a (semigroup) ideal in S, we have $\exp(X_1)$, ..., $\exp(X_n) \in H(S)$. It follows that $\exp(tX_i) \in H(S)$ for all positive $t < 1$ (since the complement is an ideal), hence for all positive t (since $H(S)$ is a subsemigroup). Thus $X_1, \ldots, X_n \in L(H(S))$, and g is in the semigroup generated by $\exp(L(H(S)))$. □

Notice that it is not at all clear a priori even that $H(S)$ is merely connected.

In the proof of Proposition 5.6 we passed to the Lie algebra and used some of the machinery developed in the previous sections on local semigroups. Occasionally it is useful to go in the other direction, i.e. to deduce properties of local analytic semigroups from known properties of the Lie group via the exponential function.

5.10 PROPOSITION. Let S be a local analytic semigroup with respect to an open ball B in the Lie algebra L, and suppose L is the smallest Lie subalgebra containing $L(S)$. Then S has dense interior. In particular, $S-S$ and $S * (-S)$ are neighborhoods of 0.

Proof. It suffices to show that any neighborhood of 0 meets the interior of S since on B the translation of an open set is open. Let U be an open set containing 0, and small enough that the exponential mapping is a homeomorphism on U, where $\exp: L \to G$ is the exponential mapping into some connected Lie group G.

By properties of the exponential function for all $X, Y \in U$, $\exp(X)\exp(Y) = \exp(X * Y)$, where $*$ is the operation given by the Campbell-Hausdorff formula. Since $L(S)$ generates L, there exists a finite set $\{X_1, \ldots, X_n\} \subseteq U \cap L(S)$ which generates L. Let $\Sigma = \{\sigma_1, \ldots, \sigma_n\}$ be the set of one-parameter semigroups defined by $\sigma_j(t) = \exp(tX_j)$ for all $t \geq 0$. By Theorem 2.4 there exists $\epsilon > 0$ such that the set $S(\Sigma, T)$ of points reachable with respect to Σ in time T or less is contained in $\exp(U)$ for all $T < \epsilon$.

Since the set $\{X_1, \ldots, X_n\}$ generates $L = L(G)$, we conclude that $\{\sigma_i(t) : 1 \leq i \leq n, t \geq 0\}$ generates (as a group) G, since the group generated by all $\sigma_i(t)$ is an analytic subgroup of G whose Lie algebra contains $X_1, \ldots, X_n$, hence agrees with L. Fix T such that $0 < T < \epsilon$. Then by Theorem 2.4 again $S(\Sigma, T)$ has dense interior. The fact that $S(\Sigma, T) \subseteq \exp(U)$ and $\exp(x) \cdot \exp(y) = \exp(x * y)$ implies $\exp^{-1}(S(\Sigma, T)) \cap U \subseteq S$. Since $\exp$ restricted to U is a homeomorphism, $S \cap U$ has interior. Since U was an arbitrarily small neighborhood, we conclude that 0 is in the closure of the interior of S. □

Proposition 5.10 allows us to give an improved version of Proposition 4.3 of the last section.

5.11 <u>THEOREM</u>. Let S be a local analytic semigroup, and let $G(S) = \langle S \cup -S \rangle$. Then $G(S)$ is a local Lie group, $L(G(S)) = \langle\langle L(S) \rangle\rangle$, and $G(S) = \langle\langle L(S) \rangle\rangle \cap B$.

Proof. By Proposition 5.10 S has non-empty interior in $\langle\langle L(S) \rangle\rangle$; thus $G(S)$ also does. Let U be an open set in $G(S)$, $g \in U$. We may assume that U is small enough so that $(-g) * U \subseteq B$, the open ϵ-ball in which we are working. Since $V = (-g) * U$ is an open set containing 0, we have 0 in the interior of $G(S)$.

Let $y \in G(S)' = \overline{G(S)} \cap B$. There exists an open set W containing 0 and contained in $G(S)$ such that $y * W \subseteq B$. Since $y \in y * W$ and $y * W$ is open, there exists $x \in (y * W) \cap G(S)$. Then $x = y * w$ for some $w \in W$. Thus $y = x * (-w)$, which implies $y \in G(S)$.

Since y was arbitrary, we conclude $G(S) = G(S)'$. The proposition now follows from Proposition 4.3. □

As a consequence of the results of this section we have the following global statement:

5.12 <u>THEOREM</u>. Let S be an analytic semigroup in a Lie group G. Then we have the following conclusions concerning the minimal group containing S and the maximal group contained in S:

(1) Let $G(S)$ be the semigroup generated by

$S \cup S^{-1}$.

Then $G(S)$ is the analytic subgroup of G with

$L(G(S)) = \langle\langle L(S) \rangle\rangle$.

(2) Let $H(S) = S \cap S^{-1}$ be the group of invertible elements in S. Then $H(S)$ is closed in $G(S)$ in the Lie group topology of $G(S)$, and $L(H(S)) = L(S) \cap -L(S)$.

Proof. From previous discussions we know that $G(S)$ is an analytic subgroup. For the purposes of the proof of (1) and (2) it is no loss of generality to suppose that $G(S) = G$.

(1) The only thing which remains to be shown for (1) is that $L(G) = L(G(S)) \subset \langle\langle L(S)\rangle\rangle$. By 5.2 we know that $S = \langle \exp T\rangle$ for a local analytic semigroup T in B, where B is a suitable open ball in $L(G)$. Let $G(T)$ be the smallest local group in B containing T, i.e. $G(T) = \langle T \cup -T\rangle_B$. Then $\exp G(T)$ generates $\langle S \cup S^{-1}\rangle = G$, since $\exp T$ generates S. By 5.11 we know that $L(G(T)) = \langle\langle L(T)\rangle\rangle = \langle\langle L(S)\rangle\rangle$ and $G(T) = \langle\langle L(S)\rangle\rangle \cap B$. The group generated by $\exp G(T)$, therefore, is the analytic group determined by the subalgebra $\langle\langle L(S)\rangle\rangle$ of $L(G)$. On the other hand, as we just saw, this group is all of G. Hence $L(G) = \langle\langle L(S)\rangle\rangle$.

(2) Since we assume that $G = G(S)$ we have to show that $H(S)$ is closed. By 5.9 we know that $H(S)$ is an analytic subgroup with $L(H(S)) = L(S) \cap -L(S)$. Then $H(S)^-$ is a closed connected Lie subgroup. Every element X of $L(H(S)^-)$ has the property that $\exp tX \in H(S)^- \subseteq S^-$ for all $t \in \mathbb{R}$. By 5.3 this implies that $X \in L(S) \cap -L(S)$.

Hence $L(H(S)^-) \subseteq L(S) \cap -L(S) = L(H(S)) \subseteq L(H(S)^-)$. Thus

$L(H(S)) = L(\overline{H(S)})$ which implies $H(S) = \overline{H(S)}$ as asserted. □

In the local theory we saw in 3.21 how the Lie wedge $L(S)$ of a local analytic semigroup S approximated S locally near the identity. We now prove a global theorem which says in effect that information on $L(S)$ gives us an approximation of a Lie semigroup S in a Lie group G in a whole neighborhood of the group $H(S)$ of units of S.

5.13 PROPOSITION. Let S be an analytic semigroup in the group G, and assume that S generates G. Suppose that a closed wedge W surrounds $L(S)$ (see 3.20) and that $L(G)$ is decomposed into a vector space direct sum of $E = L(S) \cap -L(S) = L(H(S))$ and a complementary subvector space F. We take a standard open ball B around 0 in $L(G)$ such that $\exp|B$ is a homeomorphism from B onto an open neighborhood U of 1 in G, and that $\exp(X*Y) = \exp X \exp Y$ for $X, Y \in B$.

Then we can find an open ball C around 0 in B with the following properties:

> If $V_F = \exp(C \cap F)$, $K_F = \exp(C \cap F \cap W)$ then $(h, v) \to hv : H(S) \times V_F \to V = H(S)V_F$ is an $H(G)$-equivariant homeomorphism onto an open neighborhood of $H(S)$, and
>
> $S \cap V = H(S)(S \cap V_F) \subseteq H(S)K_F$.

Proof. There is an open neighborhood C of 0 in B such that $(\exp|C)^{-1}(\exp C \cap S) \subseteq W$ (see 3.21). There is no loss of generality in assuming that C is of the form $C_E * C_F$ with an open neighborhood C_E of 0 in E and an open neighborhood C_F of 0 in F; indeed the function $(X,Y) \to X*Y : (E \cap B) \times (F \cap B) \to L(G)$ has differential 1 at 0 and is therefore a local homeomorphism at 0.

We first note that $H(S) \exp C = H(S) \exp C_E \exp C_F = H(S) \exp C_F$ and that $H(S) \exp C$ is an open neighborhood V of $H(S)$. Also note that $C_F = (C_E * C_F) \cap F = C \cap F$. Since $C = C_E * C_F$, the set $V_F = \exp C_F$ is a local cross section for the local orbits of the action of $H(S)$ on S by left multiplication. Since $H(S)$ is closed by 5.12, we may assume (after making C smaller, if necessary) that V_F is in fact a local corss section to the global orbits of $H(S)$. This means in effect that the function $(h,v) \to hv : H(S) \times V_F \to H(S)V_F$ is a homeomorphism. Now let $s \in S \cap V$. Then $s = hv$ with some $h \in H(S)$ and $v \in V_F$. Then $v = h^{-1}s \in S \cap V$, and so $S \cap V \subseteq H(S)(S \cap V_F)$; the reverse containment is trivial. If we further consider the element $s = hv$ as before, we observe $v = \exp X$ with $X \in (\exp|C)^{-1}(\exp C \cap S)$; thus $X \in W$ by the choice of C and thus $v \in K_F$. Thus $s \in H(S)K_F$ and the above is proved completely. □

We close this section by showing further important relationships between ray semigroups and Lie semigroups.

5.14 THEOREM. Let T be a ray semigroup in a Lie group G and suppose that T generates G. Then there exists a unique analytic semigroup S such that the ray semigroup T, the Lie semigroup $\overline{T}$, and S are related as follows:

(1) $T \subseteq S \subseteq \overline{T}$.

(2) $L(T) = L(S) = L(\overline{T})$.

(3) interior T = interior S = interior $\overline{T}$.

Proof. Let S be the semigroup generated by $\exp(L(T))$ (equivalently the semigroup generated by all one-parameter semigroups lying entirely in $\overline{T}$). Since T is a ray semigroup, $T \subseteq S$, and from definition $S \subseteq \overline{T}$. Thus $L(T) \subseteq L(S) \subseteq L(\overline{T})$; but by definition $L(T) = L(\overline{T})$ (since T is a ray semigroup); hence S is generated by $\exp L(S)$, i.e. S is a Lie semigroup. The last assertion follows from Proposition 2.1 and Corollary 2.5. □

Theorem 5.14 says that a ray semigroup generating a Lie group deviates from an analytic and from a Lie semigroup at most in boundary points, and that its infinitesimal generators are those of an analytic and a Lie semigroup. If T is an arbitrary ray semigroup in some Lie group, then T generates the analytic subgroup $G(T)$. Then Theorem 5.14 applies to the pair $(T, G(T))$ in the intrinsic Lie group topology of $G(T)$. In this sense, 5.14 is quite general.

In the following we see that indeed with <u>any</u> closed subsemigroup of a Lie group there is naturally associated an analytic subsemigroup:

5.15. <u>PROPOSITION</u>. Let G be a Lie group and let S be a closed subsemigroup which contains the identity. Then the semigroup S_R generated by all one-parameter semigroups in S is an analytic subsemigroup and $L(S) = L(S_R)$.

Proof. By definition of $L(S)$ we have $L(S) = L(S_R)$. Since S is closed, we have $S^* \subseteq (G(S_R) \cap S)^- \cap G(S_R) = S \cap G(S_R)$ (see Definition 5.3). Hence, if a one-parameter subsemigroup lies entirely in S^*, then it lies in S, and hence in S_R by the definition of S_R. Thus S_R is generated by $\exp L(S) = \exp L(S_R)$, i.e., S_R is an analytic subsemigroup. ⊓

Section 6. Further Examples and Computations

In general it is difficult to determine whether a wedge in a Lie algebra is a Lie wedge; some contributions to this question were made in [H-L,82]. Another often difficult problem is to determine the semigroups generated by the exponential image of a Lie wedge. In this section, we carry out these computations for some specific examples.

The first example shows that an analytic semigroup need not necessarily be closed in the Lie group it generates.

6.1 EXAMPLE. Let G be a semidirect product of the com plex number plane with addition and the real number line with addition such that multiplication is defined on $G = \mathbb{C} \times \mathbb{R}$ by $(c,r)(d,s) = (c + e^{ir}d, r+s)$. Set $K = \{(c,r) \in G : |c| \leq r\}$. Then K is easily seen to be a closed subsemigroup of G. By Proposition 5.4 $L(K)$ is a closed convex wedge.

An alternate approach to this example is to consider the Lie algebra (with respect to the usual bracket product) of all matrices of the form

$$\begin{bmatrix} is & z & 0 \\ 0 & 0 & 0 \\ 0 & 0 & s \end{bmatrix}$$

where $s \in \mathbb{R}$, $z \in \mathbb{C}$. Calculating the exponential of such a matrix by the usual power series, one obtains

$$\begin{bmatrix} e^{is} & z(e^{is}-1)(is)^{-1} & 0 \\ 0 & 1 & 0 \\ 0 & 0 & e^{s} \end{bmatrix}$$

If one lets $c = z(e^{is}-1)(is)^{-1}$, then the function which assigns the value (c,s) to the above matrix is a topological ismorphism from all such matrices onto G.

The preceding identification allows one to compute the one-parameter groups in G by use of the exponential function. These turn out to be of the form $t \to (z(e^{it}-1),t)$ (and reparameterized versions thereof). In order for such a one-parameter semigroup to lie in K we must have $|z(e^{it}-1)| \leq t$ for all $t \geq 0$. We have

$$|z(e^{it}-1)| \leq t \text{ iff } 2|z|^2(1 - \cos t) \leq t^2$$
$$\text{iff } 4|z|^2 \sin^2(\tfrac{1}{2} t) \leq t^2 \text{ iff } |z|\,|\sin(\tfrac{1}{2}t)| \leq \tfrac{1}{2} t.$$

Thus the one-parameter semigroup lies entirely in K iff $\frac{|\sin x|}{x} \leq \frac{1}{|z|}$ for all $x > 0$. This happens iff $|z| \leq 1$. From this it follows that $(c,t) \in K$ lies on a one-parameter semigroup which lies entirely in K iff $|c| \leq 2|\sin(\tfrac{1}{2} t)|$. Note that $2|\sin(\tfrac{1}{2} t)| < t$ for all $t > 0$.

We claim that the semigroup generated by the images of all one-parameter semigroups lying entirely in K is precisely the set $S = \{(c,t) : |c| < t\} \cup \{(0,0)\}$. This set S is easily verified to be a semigroup, so from the preceding paragraph it contains the subsemigroup generated by the one-parameter semigroups in K. Conversely let $(c,t) \in S$. Using the fact that rotations around the $\mathbb{R}$-axis are topological automorphisms, we may assume $c = |c|$, i.e., c is real. For $n \in \mathbb{N}$, set $c_k = d\, e^{ia_k}(e^{i(t/n)}-1)$ where $|d| \leq 1$, $a_k \in \mathbb{R}$. Then

$(c_1,t/n)\dots(c_n,t/n) = (d(e^{i(t/n)}-1)\Sigma_{k=1}^{n} e^{i(a_k+(k-1)/n)}, t)$.

If we set $a_k = -(k-1)/n$, then $\Sigma_{k=1}^{n} e^{i(a_k+(k-1)/n)} = n$.

Choose n such that $c/t \leq |(e^{i(t/n)}-1)/(t/n)| < 1$. Then find $|d| \leq 1$ such that $c = d(e^{i(t/n)}-1)n$. If $|c|$ is close to t, then a large n will be required. This proves the claim.

By Proposition 5.15 we have that S is an analytic semigroup and that $L(K) = L(S)$.

Finally we wish to compute $L(S)$. The Lie algebra of G can be identified with $L(G) = \mathbb{C} \times \mathbb{R}$ with Lie product $[(z_1,r_1),(z_2,r_2)] = (r_1z_2 - r_2z_1, 0)$. Since the one-parameter groups are all reparameterized versions of $t \to (z(e^{it}-1),t)$, we have $\exp(tz,t) = (z(e^{it}-1),t)$. We saw that a one-parameter semigroup was in K (and hence S) iff $|z| \leq 1$. Thus
$L(S) = \{(tz,t) \in \mathbb{C} \times \mathbb{R} : |z| \leq 1,\ t \geq 0\}$. Strangely enough, in this example K and $L(K)$ turn out to be the same subsets of $\mathbb{C} \times \mathbb{R}$. □

In order to enrich our understanding of examples in solvable Lie groups we return to the example in 1.6.

6.2 <u>EXAMPLE.</u> Let V be a finite dimensional real vector space and let $\mathbb{R}^+$ be the semiring of non-negative real numbers. We consider the semidirect product $V \times \mathbb{R}^+$

under the operation $(u,r)(v,s) = (u+rv, rs)$. We obtain in this fashion an affine semigroup. Let C be any convex set in V. We observe that the set $C \times \{0\}$ is a subsemigroup with left zero multiplication $xy = x$. The closed convex hull of $C \times \{0\} \cup \{(0,1)\}$ is a locally compact affine semigroup S_0. Let $S = \{(v,t) \in S_0 | 0 < t\}$. Then S is a divisible closed subsemigroup of the Lie group $V \times P$, where P is the multiplicative group of positive numbers. Every element lies on precisely one one-parameter semigroup: If $c \in C$, then $t \to ((1-e^{-t})c, e^{-t})$ is a one-parameter group on the straight line segment connecting the identity $(0,1)$ with the point $(c,0)$. The semigroup S is a Lie subsemigroup, and its associated wedge $L(S)$ is a semialgebra. The Lie algebra of the group $V \times P$ is the semidirect product $V \times \mathbb{R}$ with componentwise addition and bracket $[(u,r), (v,s)] = (r \cdot v - s \cdot u, 0)$. We have shown in [H-L,82] that every wedge in the Lie algebra $V \times \mathbb{R}$ is a semialgebra, and the Lie algebras of the type $V \times \mathbb{R}$ are the only non-abelian in which this phenomenon occurs. □

We will show in [H-L, 82] that a nilpotent Lie algebra can contain a generating proper semialgebra (i.e. one not containing a non-degenerate subvector space) only if it is abelian.

Semialgebras, however, do occur in simple Lie algebras, as the Example 1.7 shows.

6.3 EXAMPLE. We consider the algebra $L = sl(2, \mathbb{R})$ of 2 by 2 matrices with trace 0. We set

$$H = \begin{bmatrix} 1 & 0 \\ 0 & -1 \end{bmatrix}, \quad X_+ = \begin{bmatrix} 0 & 1 \\ 0 & 0 \end{bmatrix}, \quad X_- = \begin{bmatrix} 0 & 0 \\ 1 & 0 \end{bmatrix}.$$

Then H spans a Cartan algebra of L and we have $[X_+, X_-] = H$, $[H, X_+] = 2X_+$, $[H, X_-] = -2X_-$. In the notation of 1.7 we have $\exp tH = \sigma(t)$ (with $a = e$), $\exp tX_+ = \rho(t)$ and $\exp tX_- = \lambda(t)$.

The closed cone A_H spanned by H, X_+ and X_- is a proper Lie cone which is not a semialgebra. It is the tangent cone of the semigroup of all matrices of $Sl(2,\mathbb{R})$ with non-negative entries and with left upper entry $a \geq 1$.

The closed wedge W, spanned by the full Cartan algebra $\mathbb{R} \cdot H$ and the vectors X_+ and X_- is a semialgebra containing the Cartan algebra as largest subvectorspace. The semigroup S^+ of all matrices in $Sl(2, \mathbb{R})$ with non-negative entries, is a Lie semigroup with $L(S^+) = W$, and $\exp|W : W \to S^+$ is an analytic homeomorphism.

The closed half space E of L consisting of all vectors of the form $h \cdot H + x \cdot X_+ + y \cdot X_-$ with $h, x \in \mathbb{R}$ is a semialgebra. Its edge $\mathbb{R} \cdot H + \mathbb{R} \cdot X_+$ is a Borel subalgebra. The image $\exp^\sim E$ of the half space in $Sl(2, \mathbb{R})^\sim$ in the universal covering group of $Sl(2, \mathbb{R})$ is a Lie semigroup such that $\exp^\sim|E : E \to \exp^\sim E$ is an analytic homeomorphism. This is one of the three possible reduced half space Lie semigroups which were investigated by Dobbins [Do] and Poguntke [Po]. Poguntke elucidated the semigroup structure of $\exp E$ [Po]. The image of E in $Sl(2, \mathbb{R})$ itself under the exponential function of this group is the entire group. If B is a small enough open ball around 0 in L, then $T = E \cap B$ is a local semigroup and $\exp T$ generates $G = Sl(2, \mathbb{R})$ as does $\exp B$ itself. Thus the local semigroup T of 5.2 is not uniquely determined. The discrepancy between the local and the global theory is greater in Lie semigroup theory than in Lie group theory.

We conclude the discussion of this example by noting that in $Sl(2, \mathbb{R})$ we observed an ascending sequence $A \subseteq W \subseteq E$ of generating semialgebras in which the first one is proper. Of course, A itself contains a non-generating two dimensional proper semialgebra, namely $\mathbb{R}^+ \cdot H + \mathbb{R}^+ \cdot X_+$ which in fact is one of the semialgebras we described in Example 7.4 above.

APPENDIX A. Differential Geometry and Lie Groups

The purpose of this section is to survey some of the aspects of differential geometry which are important for this paper. Some familiarity with the fundamental notions of differential geometry will be necessary to read this material with comprehension. The concepts needed appear in any standard book on the subject.

Let M denote a real, analytic manifold (without boundary) of dimension n (all manifolds are assumed to be paracompact Housdorff). Let TM denote the tangent bundle of M and let M_x denote the tangent space to the manifold M at the point x. Let $V(M)$ denote the set of all analytic vector fields on the manifold M. For any $X, Y \in V(M)$, let $[X,Y]$ denote the Lie product of X and Y (i.e., $[X,Y] = XY - YX$). Equipped with this product $V(M)$ may be viewed as a Lie algebra.

A subset D of $V(M)$ is <u>involutive</u> if it is closed with respect to Lie products. A <u>subalgebra</u> is an involutive subspace. An integral manifold of $D \subseteq V(M)$ is a connected submanifold N of M such that $N_x = \langle D(x) \rangle$ for every $x \in N$, where $D(x) = \{X(x) : X \in D\}$, and where $\langle D(x) \rangle$ is the subspace of M_x spanned by $D(x)$. If D is an involutive subset of $B(M)$ and $x \in M$, then there is a unique maximal integral manifold of D which contains x (this result is the classical Frobenius theorem if the dimension of $\langle D(x) \rangle$ is the same for each $x \in M$; for a very nice treatment of the result in this generality, see [Su]). If $D \subset V(M)$, we denote the smallest subalgebra of $V(M)$ which contains D by $\mathcal{L}(D)$, and the maximal integral manifold of $\mathcal{L}(D)$ through x by $I(D,x)$.

If X is a vector field on M, then α is an <u>integral curve</u> for X if α is a smooth mapping from a closed interval I contained in the real numbers into M such that

$$\frac{d\alpha(t)}{dt} = X(\alpha(t)) \quad \text{for all} \quad t \in I.$$

If $D \subset V(M)$, then an <u>integral curve</u> of D is a continuous mapping α from a real interval $[s,t]$ into M such that there exist $s = t_0 < t_1 < \ldots < t_k = t$, and elements $X_1, \ldots, X_k$ of D with the property that the restriction of α to $[t_{i-1}, t_i]$ is an integral curve of X_i for each $i = 1, \ldots, k$. For $x \in M$, a point Y <u>is</u> D-<u>reachable from</u> x <u>at time</u> T if there exists an integral curve α of D defined on $[0,T]$ such that $\alpha(0) = x$ and $\alpha(T) = y$. Denote the set of all points D-reachable from x at time T by $L_0(D,x,T)$, the union of all $L_0(D,x,t)$ for all $t \in [0,T]$ by $L(D,x,T)$, and the union of all $L(D,x,t)$ for all $t \geq 0$ by $L(D,x)$.

We have the following facts:

(1) If $\alpha : [t_0, t_1] \to M$ is an integral curve of D and $\alpha(t) = x$ some $t, t_0 \leq t \leq t_1$, then $\alpha(s) \in I(D,x)$ for all $s \in [t_0, t_1]$. Hence $L(D,x) \subset I(D,x)$.

(2) (Chow's Theorem) If D is symmetric (i.e., $X \in D$ implies that $-X \in D$) then for every $y \in I(D,x)$ there exists an integral curve $\alpha : [0,T] \to M$ of D with $T \geq 0$ such that $\alpha(0) = x$ and $\alpha(T) = y$. Thus in this case $L(D,x) = I(D,x)$.

We now specialize to the case that the analytic manifold is a connected Lie group G and the vector fields are members of the Lie algebra $L(G)$, i.e. right invariant vector fields. (A vector field X is right-invariant if for each right translation $\rho_x : G \to G$, we have $(\rho_x)_*(X(y)) = X(yx)$ where $(\rho_x)_* : T(G) \to T(G)$ is the mapping induced on the tangent bundles). It is known that every $X \in L(G)$ is analytic, and that $L(G)$ is a Lie algebra with respect to the usual vector operations and standard Lie product of vector fields, $[X,Y] = XY - YX$. The mapping $X \to X(1)$ from $L(G)$ to G_1, the tangent vectors at 1, is a linear isomorphism. The two are frequently identified via this isomorphism, where G_1 is endowed with the Lie product induced by the isomorphism.

For a Lie group G a continuous mapping $\alpha : \mathbb{R} \to G$ with $\alpha(0) = 1$ is an integral curve for a right invariant vector field X if and only if α is a one-parameter group $(\alpha(s+t) = \alpha(s)\cdot\alpha(t))$ and $\frac{d\alpha}{dt}\big|_0 = X(1)$. Hence all integral curves are of the form $\rho_x \circ \alpha$ where α is a one-parameter group and ρ_x is right translation by $x \in G$. For notational convenience we denote $\rho_x \circ \alpha(t)$ by $X_t(x)$ where α is the one-parameter group which is an integral curve for X. Note that the family $\{X_t : t \in \mathbb{R}\}$ is a one-parameter group of analytic homeomorphisms on G for each $X \in L(G)$. Since $X_t(x)$ is defined for all

t and x, G is <u>complete</u> with respect to $L(G)$. Also note that if α is a one-parameter group which is an integral curve for X, then $\alpha(t) = X_t(e)$. Hence $X_t(x) = (\rho_x \circ \alpha)t = \alpha(t) \cdot x = X_t(e) \cdot x$. Thus $X_t Y_s(x) = X_t(e) \cdot Y_s(e) \cdot x$; a similar statement holds for finitely many compositions.

If D is a set of right-invariant vector fields on G, then y <u>is</u> D-<u>reachable</u> <u>from</u> x <u>at</u> <u>time</u> T if and only if there exist $X_1, \ldots, X_n \in D$ and $0 = t_0 < t_1 < t_2 < \ldots < t_2 < \ldots < t_n = T$ for some n such that $y = (X_n)_{\Delta t_n} (X_{n-1})_{\Delta t_{n-1}} \ldots (X_1)_{\Delta t_1}(x)$ where $\Delta t_i = t_i - t_{i-1}$.

Note that integral curves of D consist of right translates of finite sections of one-parameter groups for members of D pieced together. Hence the right translate of an integral curve is again an integral curve.

A.1 <u>PROPOSITION</u>. Let G be a Lie group and let $D \subseteq L(G)$, the Lie algebra of G. If x is D-reachable at time t and y is D-reachable at time s, then xy is reachable at time $t+s$.

Proof. Let $\alpha : [0,t] \to G$ be an integral curve of D such that $x(t) = x$, $\alpha(0) = e$, and let $\beta : [0,t] \to G$ be an integral curve of D such that $\beta(s) = y$, $\beta(0) = e$. Define $\gamma : [0,s+t] \to G$ by $\gamma(u) = \alpha(u)$ if $0 \leq u \leq t$ and

$\gamma(u)\beta(u-t)\cdot x$ if $t \leq u \leq s+t$. Then γ is an integral curve of D. Hence sy is reachable at time $t+s$. □

(Note that this proposition does not say that xy isn't also D-reachable at an earlier time).

A.2 <u>COROLLARY.</u> If $D \subseteq L(G)$, then the set $L(D,e)$ of points which are D-reachable from the identity e form a subsemigroup of G.

Proof. If $x, y \in L(D,e)$, then x is reachable at some time t and y is reachable at some time s. Hence xy is reachable at time $t+s$ by Proposition A.1, and thus $xy \in L(D,e)$.

A.3 <u>PROPOSITION</u>. Let G be a Lie group with Lie algebra $L(G)$. Let $S \subseteq G$. The following statements are equivalent:

(1) There exists $D \subseteq L(G)$ such that $S = L(D,e)$.

(2) There exists a family of one-parameter semigroups Δ such that S is the semigroup generated by the union of the images of this family.

(3) There exists a family of one-parameter semigroups Δ such that $S = \{\sigma_1(t_1)\sigma_2(t_2)\ldots\sigma_n(t_n) : \sigma_i \in \Delta, 0 \leq t_i \text{ for } i = 1,\ldots,n\}$.

(4) There exists $D \subseteq L(G)$ such that $S = \{(X_1)_{t_1}(e)\ldots(X_n)_{t_n}(e) : X_i \in D, 0 \leq t_i \text{ for } i = 1,\ldots,n\}$.

Proof. (2) ⇔ (3). Straightforward.

(1) ⇒ (2). Let $\Delta = \{\sigma$: there exists $X \in D$ such that $\sigma(t) = X_t(e)$ for all $t \in R^+\}$. Then each σ in Δ is a one-parameter semigroup and the image of σ is contained in $L(D,e)$. Hence the subsemigroup T generated by the union of the images is contained in $L(D,e)$ since the latter is a subsemigroup.

Conversely if $x \in L(D,e)$, then

$$x = (X_n)_{t_n}(X_{n-1})_{t_{n-1}}\cdots(X_1)_{t_1}(e) \quad \text{for some } X_1,\ldots,X_n \in D$$

and $t_1,\ldots,t_n \in \mathbb{R}^+$. Now

$$(X_n)_{t_n}(X_{n-1})_{t_{n-1}}\cdots(X_1)_{t_1}(e) = (X_n)_{t_n}(e)\cdot(X_{n-1})_{t_{n-1}}\cdots$$

$$\cdot(X_1)_{t_1}(e) \quad \text{which is in } T. \quad \text{Thus } L(D,e) = T.$$

(2) ⇒ (1). Let $D = \{X \in L(G)$: there exists $\sigma \in \Delta$ such that $X_t(e) = \sigma(t)$ for all $t \in \mathbb{R}^+\}$. (By our discussion of Section 2 each σ in Δ corresponds to a unique $X \in L(G)$ such that $\sigma(t) = X_t(e)$.) If

$x = \sigma_1(t_1)\ldots\sigma_n(t_n)$, then

$$x = (X_1)_{t_1}(e)\ldots(X_n)_{t_n}(e) = (X_1)_{t_1}(X_2)_{t_2}\cdots(X_n)_{t_n}(e) \in L(D,e).$$

The argument reverses to show $L(D,e) \subseteq S$. Thus $L(D,e) = S$.

(3) ⇔ (4). The proof is similar to the equivalence of (1) and (2). □

Let G be a Lie group and let S be a subsemigroup of G. If S satisfies any of the equivalent conditions of Proposition 3.3, S is called a ray semigroup (see Section 1).

A.5 PROPOSITION. Let G be a Lie group, $D \subseteq L(G)$. Then the subgroup H generated by $L(D,e)$ is equal to $L(D \cup -D,e)$ and is the integral manifold for the subalgebra M generated by D. Hence H is a Lie subgroup with Lie algebra M.

Proof. By Chow's Theorem $L(D \cup -D,e)$ is the integral manifold for M. By Proposition A.3 $L(D,e)$ is generated by one-parameter semigroups $\{t \to X_t(e) : X \in D\}$ and $L(D \cup -D,e)$ will hence be generated by the corresponding one-parameter groups. Since the semigroup generated by a union of groups is a subgroup, $L(D \cup -D,e)$ is a group. Since the group generaged by $L(D,e)$ will contain all the one-parameter groups generating $L(D \cup -D,e)$, we conclude the two groups are equal. Since $L(D \cup -D,e)$ is the integral manifold is M containing e and is a group, it is a Lie group and its Lie algebra is M. □

We now quote an important result for our development [S-J, Theorem 3.1].

A.6 THEOREM. Let M be an n-dimensional analytic manifold, and let $D \subseteq V(M)$ be a family of complete vector fields. A necessary and sufficient condition for $L(D,x)$

to have non-empty interior in M is that $\dim\langle\langle D\rangle\rangle_x = n$. Moreover, if this condition is satisfied, then for each $T > 0$, the interior of $L(D,x,T)$ is non-empty and dense in $L(D,x,T)$. □

A.7 COROLLARY. Let G be a Lie group and let $D \subseteq L(G)$ such that $\langle\langle D\rangle\rangle = L(G)$. Then $L(D,e)$ (resp. $L(D,e,T)$) has interior dense in $L(D,e)$ (resp. $L(D,e,T)$). □

We now introduce some more facts about Lie groups (see the latter part of Kaplansky [Ka]). Let G be a locally compact Lie group. A canonical neighborhood in G is a compact symmetric neighborhood of 1 which contains no subgroup $\neq 1$ and has the property that squaring is one-to-one on it. G has a basis of canonical neighborhoods at 1.

A.8 THEOREM. Let U be a canonical neighborhood in a Lie group G. Define Q_i to be the set of all x satisfying $x, x^2, \ldots, x^i \in U$. Let n_i be the smallest positive integer such that $(Q_i)^{n_i} \not\subseteq U$. Then i/n_i is bounded. □

Theorem A.8 allows us to derive the following important result.

A.9 THEOREM. Let G be a Lie group, U a canonical neighborhood, and $D \subseteq L(G)$. If $X_t(e) \in U$ for all $X \in D$ and all $t \in [0,1]$, or if D is finite, then for any open set V containing 1, there exists $T > 0$ (depending on

V) such that $L(D,e,T) \subseteq V$. $\square$

The proof of Theorem A.9 is a straightforward, but tedious, application of Theorem A.8; hence we omit the proof.

There is an alternate approach to Theorem A.9 via differential geometry. The idea is that if one passes via a chart around e to $\mathbb{R}^n$, then the set D gives rise to a set of vector fields on an open ball in $\mathbb{R}^n$, and the vectors in the vector field are uniformly bounded. Hence the speed at which one can travel along integral curves for those vector fields is uniformly bounded. Hence for small times $T > 0$, only points within a fixed distance ε from the origin can be reached.

APPENDIX B. Lie Algebras

In order to make our discussion as self-contained as possible we provide some background material from Lie group theory which we use in the text. The standard source of reference is Bourbaki, Theorie des groupes et algebres de Lie [Bo] although this is not strictly an introductory reference.

A Lie algebra is an algebra over a field with a bilinear multiplication $(x,y) \to [x,y]$ satisfying the conditions $[x,x] = 0$ and $[x,[y,z]] + [y,[z,x]] + [z,[x,y]] = 0$ for all x,y,z. The first condition implies $[x,y] = -[y,x]$ for all x,y; if the characteristic is different from 2 then both conditions are equivalent. We are only interested here in the case of characteristic 0, in fact only in the case that the ground field is $\mathbb{R}$. Any finite dimensional real Lie algebra possesses a (not uniquely determined) norm for which

(B) $\|[x,y]\| \leq \|x\| \, \|y\|$ for all x,y.

In general we say that L is a Banach Lie algebra (over $\mathbb{R}$) if it is a Banach space together with a norm satisfying (B). Thus every finite dimensional Lie algebra may automatically be considered as a Banach Lie algebra once a suitable norm is chosen.

The fundamental fact on Banach Lie algebras is the following: For each natural number $n = 1,2,\ldots$ there are Lie polynomials $H_1(x,y) = x+y$, $H_2(x,y) = \frac{1}{2}[x,y]$, $H_3(x,y)$ which, in principle, are completely known. The following result is now true in any Banach Lie algebra:

B.1. For $\|x\|, \|y\| < \frac{1}{2} \log 2$, the series $\Sigma_{n=1}^{\infty} H_n(x,y)$ is absolutely convergent and yields an element $x * y \in L$.

If $B = \{x \in L \mid \|x\| < \frac{1}{2} \log 2\}$, then there is a continuous partial multiplication $(x,y) \to x * y : B \times B \to L$ such that B is made into a local topological group in the following sense:

(i) For $x,y,x \in B$ with $x * y, y * z \in B$, then $(x * y) * z = x * (y * z)$.

(ii) $0 * x = x * 0 = x$ for all $x \in B$.

(iii) $x * (-x) = (-x) * x = 0$ for all $x \in B$. □

For all practical purposes it suffices to know that $x * y = x + y + \frac{1}{2}[x,y] + R(x,y)$ for small enough x, y with $R(tx,ty) = t^3 S(x,y,t)$, $t \in \mathbb{R}$ where S is bounded for $t \to 0$. Thus, irrespective of its particular structure, each Banach Lie algebra carries the additional structure of a local topological group on an open neighborhood of 0 which is given by a formula which does not depend on the particular structure of L.

A vector space endomorphism D of a Lie algebra L is called a derivation iff it satisfies $D[x,y] = [Dx,y] + [x,Dy]$. For each $x \in L$ we obtain a derivation $\operatorname{ad} x$ given by $(\operatorname{ad} x)(y) = [x,y]$; these derivations are called inner.

B.2. Let D be a continuous derivation on a Banach Lie algebra L. Then $e^D = 1 + D + \frac{1}{2!} D^2 + \ldots$ is absolutely convergent in $\operatorname{End} L$ and yields an element of the automorphism group $\operatorname{Aut} L$ of L.

If $x \in L$ then $e^{\operatorname{ad} x}$ is an automorphism of L for which we have formula

(I) $x * y * (-x) = e^{\operatorname{ad} x} y$ whenever the left hand side makes sense. □

The formula (I) relates inner derivations and inner automorphisms of the local group on L. A closed subalgebra A of L is invariant under $e^{\text{ad } x}$ iff $[x,A] \subseteq A$. In particular, A is invariant under all $e^{\text{ad } x}$ iff A is an ideal of L, i.e. satisfies $[L,A] \subseteq A$.

We have seen in Section 3, that local group and even semigroup theory is possible in B. In particular, a subset $G \subseteq B$ is a local subgroup of B iff (i) $G * G \cap B \subseteq G$ and (ii) $-G = G$. We say that G is a local Lie group iff there is a closed subalgebra A of L such that $G = A \cap B$.

B.3. The following conditions are sufficient for a local subgroup G of a Banach Lie algebra L to be a local Lie subgroup:

(1) G is locally compact. (This is the case if L is finite dimensional and G is closed in B).

(2) L is finite dimensional and G is arcwise connected. □

We note that neither of these assertions is trivial, and that the proof on the basis of condition (2) is in fact hard, technically.

There are several possibilities to start a global theory. The following definition is suitable for our purposes, and it is indeed equivalent to other definitions.

B.4. A Lie group is a topological group G for which

there is a Banach Lie algebra L and a function $\exp: L \to G$ called exponential function which satisfies the following conditions:

(1) There is an open neighborhood B of 0 in L and an open neighborhood U of 1 in G such that $*$ is defined on $B \times B$ and that

(i) $\exp | B : B \to U$ is homeomorphism

(ii) $\exp (x * y) = \exp x \exp y$ for $x, y \in B$.

(2) For each $x \in L$ and all $r, s \in \mathbb{R}$ one has $\exp (r+s)x = \exp rx \exp sx$.

Except for the norm on L, these conditions determine L uniquely (up to natural isomorphism compatible with exp); one may therefore write $L(G)$. Every continuous group homomorphism $f : G \to H$ between Lie groups automatically induces a bounded Lie algebra map $L(f) : L(G) \to L(H)$ such that the following diagram commutes:

$$\begin{array}{ccc} L(G) & \xrightarrow{L(f)} & L(H) \\ {\scriptstyle \exp_G}\downarrow & & \downarrow{\scriptstyle \exp_H} \\ G & \xrightarrow[f]{} & H \end{array}$$

It is, under these circumstances, reasonable to think of a Lie group as being given by the data $\exp_G : L(G) \to G$. Much of the basic theory is concerned with the behavior of the functor L from the category of Lie groups to the category of completely normable Lie algebras.

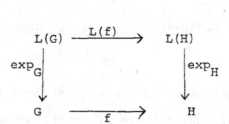

A prime example for the concept of a Lie group is furnished by what we call linear Lie groups: Let E be an associative Banach algebra with identity over the reals, e.g. the algebra End V of all bounded operators on a Banach space V endowed with the operator norm. Then the group E^{-1} of all invertible elements of E is a Lie group where $L(E)$ is given by the Banach space E together with the Lie algebra product $[a,b] = ab - ba$, and where the exponential function is given by the exponential series $\exp x = 1 + x + \frac{1}{2!} x^2 + \ldots$ which is everywhere convergent.

A subgroup of E^{-1} is then called a linear Lie group if there is a closed Lie subalgebra A of $L(E)$ such that the exponential function maps some open ball around 0 onto a neighborhood of 1 in G. Thus applies in particular to $E = \text{End } V$ in which case we write $Gl(V)$ for E^{-1}. The classical linear groups fall under our definition.

One aspect of great importance is the correspondence between connected Lie subgroup or, as they are alternatively called, analytic subgroups of a Lie group G. The simplest illustration of the phenomenon which causes problems in this context is the two torus $G = \mathbb{R}^2/Z^2$ with the exponential function $\exp : \mathbb{R}^2 \to \mathbb{R}^2/Z^2$, $\exp r = r + Z^2$.

The subalgebras of the Lie algebra $L(G) = \mathbb{R}^2$ are (except for $L(G)$ and (0)) the one dimensional subvector spaces. Let $A = \{(ta, tb) : t \in \mathbb{R}\}$ be one of them. The subgroup generated by A in this case is simply $\exp A = (A + \mathbb{Z}^2)/\mathbb{Z}^2$. If a and b are linearly dependent over the field of rationals, then A is a circle group, if not, then $\exp A$ is a dense subgroup of G and $t \to \exp(ta, tb)$ is a continuous bijective map from $\mathbb{R}$ onto the subgroup $\exp A$ which is not a homeomorphism. The induced topology on $\exp A$ is not complete and not locally compact and not locally path connected. Indeed A is locally isomorphic to the dense subgroup $A + \mathbb{Z}^2$ of $\mathbb{R}^2$ which, in turn is isomorphic to the direct product of $\mathbb{R}$ and some countable dense subgroup of $\mathbb{R}$. In general, a closed connected subgroup H of a Lie group G is a Lie subgroup and uniquely determines a closed subalgebra $L(H)$ of $L(G)$ such that $\exp L(H)$ generates H and is, in fact, a neighborhood of 1 in H. The example shows that the converse is not true. What is true is summarized as follows:

B.5. Let G be a finite dimensional Lie group. For each subalgebra A of $L(G)$ there exists a unique subgroup H of G and a unique Lie group structure $\exp_G | A : A \to (H, \tau)$ where τ is a Lie group topology which is possibly finer than the induced topology. Any subgroup H arising in this fashion is called an analytic subgroup. The subalgebra A is denoted $L(H)$. The analytic subgroup corresponding to a subalgebra A of $H(G)$ is the subgroup algebraically generated by $\exp A$. If H is an analytic subgroup, then $L(H) = \{x \in L(G) \mid \exp tx \in H$ for all $t \in \mathbb{R}\}$. □

We say A generates H.

It is most noteworthy, that there is a purely topological characterization of analytic subgroups of a finite dimensional Lie group. This is the global version of the second part of B.3.

B.6. A subgroup H of a Lie group G is an analytic subgroup (therefore uniquely determines a subalgebra $L(H)$ of G) if and only if H is arcwise connected. □

One remark in passing is that there are dense connected subgroups in $\mathbb{R}^2$ which do not contain any non-degenerate arc. Thus it is essential in B.6 to speak of arcwise connectedness. Theorem B.4 is nontrivial; it was first formulated by Yamabe with an extremely terse proof. Complete proofs were given by Gotô and Bourbaki.

REFERENCES

[B-C-D] J.T. Borrego, H. Cohen, and E. Devun, Uniquely representable semigroups II, Pac. J. Math. 39 (1971), 573-579.

[Bo] N. Bourbaki, Groupes et algebres de Lie, Chapitres 2 et 3, Hermann, Paris, 1972.

[Br] R. Brockett, Lie algebras and Lie groups in control theory, Geometric Methods in System Theory, Proc. of 1973 London NATO Advanced Study Institute, D. Reidel, Hingham, Mass., 1973, pp 43-82.

[B-F] D.R. Brown and M. Friedberg, Linear representations of certain compact semigroups, Trans. Am. Math. Soc. 160 (1971), pp 453-465.

[Do] J.G. Dobbins, Well-bounded semigroups in locally compact groups, Math. Z. 148 (1976), pp 155-167.

[Go] M. Goto, On an arcwise connected subgroup of a Lie group, Proc. Amer. Math. Soc. 20 (1969), pp 157-162.

[Gr] G.E. Graham, Differentiable semigroups, these proceedings, pp 57-127.

[He] H. Heyer, Probability Measures on Locally Compact Groups, Springer-Verlag, New York, 1977.

[Hi] R. Hirschorn , Topological semigroups, sets of generators and controllability, Duke J. Math. 40 (1973), pp 937-949.

[H-L] K.H. Hofmann and J.D. Lawson, The local theory of semigroups in nilpotent Lie groups, Semigroup Forum 23 (1981), pp 343-357.

[H-M] K.H. Hofmann and P. Mostert, Elements of Compact Semigroups, Merrill, Columbus, Ohio, 1966.

[Ho] R. Houston, Cancellative Semigroups on Manifolds, Dissertation, University of Houston, 1973.

[H-P] E. Hille and R.S. Phillips, Functional Analysis and Semi-groups, Am. Math. Soc., Providence, 1957.

[J-S] V. Jurdjevic and H.J. Sussman, Control systems on Lie groups, J. Differential Equations 12 (1972), pp 313-329.

[K] K. Keimel, Eine Exponential Funktion für kompakte abelsche Halbgruppen, Math. Z. 96 (1967), pp 7-25.

[La] R.P. Langlands, On Lie semi-groups, Canad. J. Math. 12 (1960), pp 686-693.

[Lo] C. Loewner, On semigroups and geometry, Bull. Am. Math. Soc. 70 (1964), pp 1-15.

[M-S] P.S. Mostert and A.S. Shields, On the structure of semigroups on a compact manifold with boundary, Ann. Math 65 (1957), pp 117-143.

[Po] D. Poguntke, Well-bounded semigroups in connected groups, Semigroup Forum 15 (1977), pp 159-167.

[Ra] M. Raghunathan, Discrete Subgroups of Lie Groups, Springer-Verlag, Heidelberg, 1972.

[Su] H. Sussman, Orbits of families of vector fields and integrability of distributions, Trans. Amer. Math. Soc. 180 (1973), pp 171-188.

[S-J] H. Sussman and V. Jurdevic, Controllability of nonlinear systems, J. Differential Equations 12 (1972), pp 95-116.

[Ya] H. Yamabe, On an arcwise connected subgroup of a Lie group, Osaka Math. J. 2 (1950), pp 13-14.

Fachbereich Mathematik
Technische Hochschule Darmstadt
Schloßgartenstraße 7

D-6100 Darmstadt
Germany

and

Department of Mathematics
Louisiana State University

Baton Rouge, La. 70803

MEASURE ALGEBRAS OF LOCALLY COMPACT SEMILATTICES

by

John R. Liukkonen and Michael Mislove

0. Introduction. If G is a locally compact abelian group, then it is well-known that the algebra $M(G)$ of all finite regular Borel measures on G is symmetric if and only if the group G is discrete. In fact, if G is not discrete, then there is some compact independent Cantor set X in G such that every continuous linear functional on $M_c(X)$ -- the space of continuous measures supported on X -- of norm at most 1 extends to a complex homomorphism of $M(G)$ (this is the so-called Hewitt-Kakutani phenomenon; for details, see [R]). Our goal in this paper is to establish a similar dichotomy for the class of locally compact Lawson semilattices. These are the locally compact semilattices which have a neighborhood basis of subsemilattices at each point. Our results show that such a dichotomy does exist for these semilattices. Indeed, we show that either every complex homomorphism of $M(S)$ is given by integration against some Borel semicharacter of S, or else $M(S)$ is asymmetric. In fact, we show further that the dichotomy is detectable in terms of the semilattice S itself: Either, for every compact subset X of S, there is a finite subset F of X with $\inf X = \inf F$, or else S contains a compact independent set Y such that every continuous linear functional on $M_c(Y)$ of norm at most 1 extends to a complex homomorphism of $M(S)$, and so $M(S)$ is asymmetric in this case.

This work extends the results of [LLM], where it was shown that every locally compact semilattice of finite breadth (which is automatically Lawson) has a symmetric measure algebra, and, in fact, that every complex homomorphism of the measure algebra of such a semilattice is given by integration against some Borel semicharacter of S. That work was, in turn, inspired by the (independent) work of Baartz [B] and Newman [N], who showed that $M(S)$ is symmetric for any locally compact semilattice which is embeddable in a finite product of chains; they also showed that $M(2^{\mathbb{N}})$ is asymmetric, where 2 denotes the two point lattice.

Much of the research for this paper was carried out while the second author was a fellow of the Alexander von Humboldt Stiftung and a guest of the Fachbereich Mathematik der Technische Hochschule Darmstadt. Thanks are due to both for their kind hospitality and generous support.

I. Conditions for a copy of $2^{\mathbb{N}}$. In this section we characterize when a locally compact Lawson semilattice S contains a copy of the compact semilattice $2^{\mathbb{N}}$ as a subsemilattice. This is crucial for our work, since we show that S contains a copy of $2^{\mathbb{N}}$ precisely when $M(S)$ is asymmetric.

Recall that a locally compact semilattice is a locally compact space S together with a continuous idempotent multiplication which is associative and commutative. Such a semilattice is called Lawson if each point has a neighborhood basis of subsemilattices. Since we are assuming that the semilattice itself is locally compact, we may also assume that this basis of subsemilattices consists of compact subsemilattices. If X is a locally compact space, then by $Q(X)$ we denote the family of all non-empty compact subsets of X. Equipped with the Vietoris topology, which has for a subbasis all sets of the form $W(U,V) = \{A \in Q(X) : A \subset U \text{ and } A \cap V \neq \emptyset\}$, where U and V range over the open subsets of X, the space $Q(X)$ is a locally compact Lawson semilattice under the operation $(A,B) \mapsto A \cup B$: Indeed, if $A \in Q(X)$, and if $B \subset X$ is a compact neighborhood of A, then $A \in W(B^\circ,B^\circ) \subset Q(B)$, and $Q(B)$ is compact. In particular, if X is compact, then $Q(X)$ is a compact Lawson semilattice. The following proposition contains several related results we shall need:

1.1. Proposition. Let S be a locally compact semilattice. Then:

(a) S is Lawson iff the map $A \mapsto \inf A : Q(S) \to S$ is a continuous surmorphism.

(b) If S is Lawson and $X \subset S$ is compact, then X has a compact semilattice neighborhood in S.

(c) If S is Lawson and $X \subset S$ is compact, then there is a minimal compact subset $Y \subset X$ with $\inf X = \inf Y$.

Proof. We prove (b) first: Let $A \in Q(S)$. Then since S is Lawson, for each $a \in A$, there is some compact semilattice neighborhood T_a of a in S, and since A is compact, finitely many of these, say $T_1, \ldots, T_n$ cover A. Thus we have compact semilattices $T_1, \ldots, T_n$ with $A \subset (T_1 \cup \ldots \cup T_n)^\circ \subset T_1 \cup \ldots \cup T_n = T$. It is then routine to show that $T' = T \cup T^2 \cup \ldots \cup T^n$ is a subsemilattice of S, and T' is clearly compact and contains A in its interior.

(a): Suppose that $A \mapsto \inf A : Q(S) \to S$ is continuous, and let $s \in S$ and $X \subset S$ be a compact neighborhood of s. If $\mathcal{A} = \{A \in Q(S) : s \in A^\circ\}$, then $\{s\} = \lim_{A \in \mathcal{A}} Q(A)$, so there is some $A \in \mathcal{A}$ with $\inf Q(A) \subseteq X$. Now, $s \in A^\circ$ implies $Q(A)$ is a compact semilattice neighborhood of $\{s\}$ in $Q(S)$, and so $\inf Q(A)$ is a compact subsemilattice of S. Moreover, for each $a \in A$, $\{a\} \in Q(A)$, so $a \in \inf Q(A)$; i.e. $A \subseteq \inf Q(A)$. Hence $s \in A^\circ \subseteq \inf Q(A) \subseteq X$, so X contains the compact subsemilattice of S which A generates. Hence, since X is an arbitrary compact neighborhood of s, we conclude that S is Lawson.

Conversely, suppose that S is a Lawson semilattice. We show that the map $A \mapsto \inf A : Q(S) \to S$ is continuous. From (b) we conclude that the map $A \mapsto \inf A$ is well-defined, and it is clearly a surjection preserving finite infima. Hence, we only need show that this map is continuous. Now if $A \in Q(S)$, then (b) implies there is a compact semilattice $T \subset S$ with $A \subset T^\circ$, and our comments preceding this proposition show that $Q(T)$ is a neighborhood of A in $Q(S)$. Moreover, since T is a compact Lawson semilattice, the map $B \mapsto \inf B$ from $Q(T)$ to T is continuous (see [G, Proposition 3.9, p. 285]). It then follows that $A \mapsto \inf A : Q(S) \to S$ is also continuous.

Finally, we show (c): Suppose that $X \subset S$ is a compact subset, and let $s = \inf X$. Then $Q(X)$ is a compact semilattice, and the map $A \mapsto \inf A : Q(X) \to S$ is continuous by (a). Since $Q(X)$ is compact, there are minimal elements Y in $Q(X)$ satisfying $\inf Y = s$, and any these satisfies (c). □

We now begin to establish our dichotomy for locally compact Lawson semilattices in terms of the topological semilattice structure itself.

1.2. Proposition: Let S be a locally compact Lawson semilattice, and suppose there is a nonempty compact subset $X \subset S$ such that

$$(*) \qquad \inf X < \inf F \quad \text{for every finite subset } F \subset X.$$

Then S contains a copy of $2^{\mathbb{N}}$.

Proof. By 1.1 (c), X contains a minimal compact subset Y with $\inf Y = \inf X$. By (*), Y must be infinite. Since Y is an infinite compact Hausdorff space, we may define inductively a sequence of nonempty, relatively open (in Y), pairwise disjoint subsets $\{U_k\}$ of Y such that for each n, $Y \setminus \bigcup\{U_k \mid k \leq n\}$ is infinite with nonvoid interior (in Y).

Now define $f : 2^{\mathbb{N}} \to Q(Y)$ by $f(M) = Y \setminus \bigcup\{U_k : k \notin M\}$. Notice that $f(\phi) \neq \emptyset$ and hence $f(\phi)$ is a well-defined element of $Q(Y)$; in fact, since Y is compact, $\bigcup U_k \neq Y$, and so $f(\phi) = Y \setminus \bigcup U_k \neq \emptyset$. Clearly f preserves unions and intersections; i.e., by [G, Theorem 3.4, p. 282] is a morphism of compact Lawson semilattices (in particular is continuous). Now $\inf : Q(Y) \to S$ is continuous by 1.1(a), so $\phi : 2^{\mathbb{N}} \to S$ defined by $\phi(M) = \inf f(M)$ is also continuous.

It remains to check that ϕ is injective. If we had distinct $M_1, M_2 \subset \mathbb{N}$ with $\phi(M_1) = \phi(M_2)$, say $k \in M_1 \setminus M_2$, we would have

$$\begin{aligned}\inf(Y\setminus U_k) &= \phi(\mathbb{N}\setminus\{k\}) = \phi(\mathbb{N}\setminus\{k\})\phi(M_2)\\ &= \phi(\mathbb{N}\setminus\{k\})\phi(M_1) = \phi(\mathbb{N}) = \inf Y,\end{aligned}$$

contradicting the minimality of Y. This shows ϕ is injective, and so $2^{\mathbb{N}} \subset S$. □

We now present two further results, before establishing our structural dichotomy for locally compact Lawson semilattices. For the following, we recall that a finite set $F \subset S$ is called meet irredundant if for any proper subset $G \subset F$, $\inf F < \inf G$, and that an arbitrary subset $Y \subset S$ is called meet irredundant if F is meet irredundant for any finite subset $F \subset Y$.

1.3. Lemma. Let S be a locally compact Lawson semilattice, and $Y = \{y_1,\ldots,y_n\} \subset S$ a finite meet irredundant subset. Then there are compact neighborhoods $W_1, \ldots, W_n$ of $y_1, \ldots, y_n$, respectively, such that whenever K_1 and K_2 are subsets of $\bigcup\{W_i : i = 1,\ldots,n\}$ such that $\{j : K_1 \cap W_j \neq \emptyset\} \neq \{j : K_2 \cap W_j \neq \emptyset\}$, we have $\inf K_1 \neq \inf K_2$.

Proof. Since Y is meet irredundant, the map $\inf : Q(S) \to S$ sends distinct subsets of Y to distinct points in S. So we can find a family $\{V_G : G \subset Y\}$ of disjoint neighborhoods in S such that for each $G \subseteq Y$, V_G is a neighborhood of $\inf G$. Now by 1.1(a), inf is continuous, and so for each $G \subset Y$, there is a neighborhood U_G of G in $Q(S)$ such that $\inf(U_G) \subset V_G$. We now argue that these U_G may be taken of a particular form. In fact, assuming $G = \{g_1,\ldots,g_k\}$, a basic neighborhood U_G of G can be taken of the form

$$U_G = \{A \subset S : A \subset V'_1 \cup \ldots \cup V'_k \text{ , } A \cap V'_i \neq \emptyset \text{ for } i = 1,\ldots,k\},$$

where $V'_1, \ldots, V'_k$ are neighborhoods of $g_1,\ldots,g_k$ respectively. Since each y_j belongs to several subsets G of Y, we obtain in this fashion several neighborhoods of each y_j. Intersecting these neighborhoods, we can get for each y_j a compact neighborhood W_j, and arrive at the following definition of U_G for each $G \subset Y$:

$$U_G = \{A \subseteq S : A \subset \bigcup\{W_j : G \cap W_j \neq \emptyset\} \text{ and }$$
$$A \cap W_j \neq \emptyset \text{ whenever } G \cap W_j \neq \emptyset\}.$$

To say that $\inf(U_G) \subset V_G$ for each $G \subset Y$ is to say in particular that $\inf K_1 \neq \inf K_2$ whenever $K_1 \in U_{G_1}$ and $K_2 \in U_{G_2}$ for $G_1, G_2 \subset Y$, $G_1 \neq G_2$. But this proves the lemma. □

1.4. Proposition. Let S be a locally compact Lawson semilattice, and $K \subset S$ be a compact subset. Suppose that for each $n \geq 2$ and each family $\{U_0,\ldots,U_n\}$ of nonempty relatively open subsets of K, we have

$$U_0 \times U_1 \times \ldots \times U_n \not\subset \{(x_0,\ldots,x_n) \in S^{n+1} : x_1 \ldots x_n < x_0\}$$

Then K contains a compact subset X such that $\inf X < \inf F$ for every finite subset $F \subset X$.

Proof. We first rewrite the condition on K as:

(§) For each $n \geq 2$ and every family $\{U_1, \ldots, U_n\}$ of nonempty relatively open subsets of K, we have

$$U_1 \times \ldots \times U_n \not\subset \bigcup_{j=1}^{n} \{(x_1,\ldots,x_n) \in S^n : x_1 \ldots \hat{x}_j \ldots x_n < x_j\}.$$

Clearly (§) implies the condition on K in the statement of 1.4, and it is easy to check that the converse holds as well.

We now construct inductively a sequence $\mathcal{C}_n = \{C(n,1),\ldots,C(n,2^n)\}$ of subsets of K satisfying for each n:

(i) $C(n,i)$ is compact with the relative interior of $C(n,i)$ in K nonempty for each i;

(ii) $C(n,i) \neq C(n,j)$ whenever $i \neq j$;

(iii) $C(n+1,2i-1)$, $C(n+1,2i) \subset C(n,i)$ for each i;

(iv) Whenever L_1 and L_2 are compact subsets of $\bigcup\{C(n,i) : i = 1,\ldots,2^n\}$ such that $\{j : L_1 \cap C(n,j) \neq \emptyset\} \neq \{j : L_2 \cap C(n,j) \neq \emptyset\}$, we have $\inf L_1 \neq \inf L_2$.

To construct $\mathcal{C}_1$, we first choose $x_1, x_2 \in K$ which are incomparable (this is possible by (§)), and then apply Lemma 1.3 to choose $C(1,1)$, $C(1,2)$ satisfying (i), (ii), (iv). We regard (iii) as vacuously satisfied. Now suppose given $\mathcal{C}_n$ satisfying (i) - (iv); we construct $\mathcal{C}_{n+1}$. Since (§) holds, we know

$$C(n,1) \times C(n,1) \times C(n,2) \times C(n,2) \times \ldots \times C(n,2^n) \times C(n,2^n)$$

is not a subset of $\bigcup_{j=1}^{2^{n+1}} \{(x_1,\ldots,x_{2^{n+1}}): x_1 \ldots \hat{x}_j \ldots x_{2^{n+1}} < x_j\}$.
So, let $x_{n+1,1}, x_{n+1,2}, \ldots, x_{n+1,2^{n+1}}$ be the coordinates of a point in the first set which is not in the second. Then $\{x_{n+1,1}, \ldots, x_{n+1,2^{n+1}}\}$ is meet irredundant, and we can use Lemma 1.3 to find $\mathcal{C}_{n+1} = \{C(n+1,1),\ldots,C(n+1,2^{n+1})\}$ satisfying (i)-(iv). This constructs inductively a sequence of families $\mathcal{C}_n = \{C(n,1),\ldots,C(n,2^n)\}$ satisfying (i)-(iv).

Now we set $X = \bigcap_n (\bigcup \mathcal{C}_n)$, and we note that $X \subset K$ is compact. we check that $\inf X < \inf F$ for any finite subset $F \subset X$. Indeed, suppose that $F = \{x_1,\ldots,x_n\}$ is a finite subset of X. Then in particular, $F \subseteq X \subset \bigcup \mathcal{C}_n = \bigcup\{C(n,i) : i = 1,\ldots,2^n\}$. Since $F = \{x_1,\ldots,x_n\}$, $F \cap C(n,i) \neq \emptyset$ for at most n indices i and

$X \cap C(n,i) \neq \emptyset$ for every $i = 1,\dots,2^n$. So by (iv), $\inf X \neq \inf F$. Since $X \supset F$, $\inf X < \inf F$. □

Now we collect our structural results on locally compact Lawson semilattices in a theorem.

1.5. Theorem. Let S be a locally compact Lawson semilattice. The following conditions are equivalent:

a) S contains no copy of $2^{\mathbb{N}}$

b) For every nonempty compact subset $X \subset S$, there is a nonempty finite subset $F \subset X$ with $\inf F = \inf X$.

c) S contains no meet irredundant Cantor set.

d) For every nonempty compact subset $X \subset S$, there are nonempty relatively open subsets $U_0, U_1, \dots, U_n$ of X such that $U_0 \times U_1 \times \dots \times U_n \subset \{(x_0,x_1,\dots,x_n) \in S^{n+1}: x_1 \dots x_n \leq x_0\}$.

Proof. The implication (a) ===> (b) is the content of 1.2. That (b) ===> (c) is obvious. The implication (b) ===> (d) is the content of 1.4. To prove (d) ===> (a), we merely observe that if $2^{\mathbb{N}} \subset S$, we could set $X = 2^{\mathbb{N}}$ and (d) would fail. So if (a) fails, (d) fails, and we conclude (d) ===> (a). So we know (a), (b) and (d) are equivalent, and that they imply (c).

To see (c) ===> (a), it is enough to establish the existence of a meet irredundant Cantor set in $2^{\mathbb{N}}$. First, we observe that $2^{\mathbb{N}}$ has the property (§) of 1.4: If $U_1,\dots,U_n \subset 2^{\mathbb{N}}$ are open subsets, then $U_1 \times \dots \times U_n \not\subset \bigcup_{j=1}^{n} \{(x_1,\dots,x_n) \in S^n : x_1 \dots \hat{x}_j \dots x_n \leq x_j\}$. Proceeding with the standard Cantor construction as in 1.4, but assuming in addition that (under some metric on $2^{\mathbb{N}}$) $\max\{\text{diameter } C(n,i): i = 1,\dots,2^n\} \to 0$ as $n \to \infty$, we obtain $X = \bigcap(\bigcup C_n)$ is a meet irredundant Cantor set. The only difference between the X actually developed in 1.4 and the X here is that the X in 1.4 maps onto a Cantor set, whereas the X here is homeomorphic to a Cantor set. □

Definition: In the sequel, we will call locally compact Lawson semilattices tame if they satisfy (a) - (d) of 1.5. Otherwise they will be called wild.

II. Dichotomy for Measure Algebras.

In this section, we show that for tame locally compact Lawson semilattices S, every complex homomorphism of $M(S)$ is given by integration against a semicharacter, while for wild locally compact Lawson semilattices, we have phenomena along the lines of Hewitt-Kakutani [HK]. We recall some terminology from semilattice theory. Fix a locally compact Lawson semilattice S. A semicharacter χ is a map $\chi : S \to \{0,1\}$ (not assumed continuous) such that $\chi(x_1,x_2) = \chi(x_1)\chi(x_2)$ for all x_1, x_2 in S. The filter associated with χ is the set $F_\chi = \{x \in S : \chi(x) = 1\}$ and the ideal associated with χ is the set $I_\chi = \{x \in S : \chi(x) = 0\}$. In general, an ideal in S is a set I such that $SI \subset I$, and a filter in S is a set F such that F is a subsemilattice such that $S \setminus F$ is an ideal. Equivalently, F is a subsemilattice such that $y \geq x \in F \implies y \in F$. As noted above, every semicharacter defines a filter and complementary ideal; it is easily checked that every filter arises in this way. For more details, see [LLM] and the references cited there.

For the first part of this section, we will assume that S is a tame locally compact Lawson semilattice.

2.2. <u>Theorem.</u> Let S be a tame locally compact Lawson semilattice and let $h \in \Delta M(S)$ be a complex homomorphism of $M(S)$, the Banach algebra of all bounded Borel measures on S under convolution. Let $F = \{x \in S : h(\delta_x) = 1\}$ and let χ_F be the semicharacter associated with the filter F.

a) F and χ_F are μ measurable for all $\mu \in M(S)$.

b) $h(\mu) = \mu(F) = \int \chi_F \, d\mu$ for all $\mu \in M(S)$.

<u>Proof.</u> a) We may reduce quickly to the case where μ is positive with compact support $X \subset \overline{F}$. Let $\mathcal{F} = \{(\uparrow x) \cap X : x \in F\}$. We want to write $\mu = \Sigma \mu_n$, where each μ_n is concentrated on a member of .

According to Lemma 2.1 of [LLM], we can write $\mu = \mu_0 + \sum_{n=1}^{\infty} \mu_n$, where each μ_n $(n \geq 1)$ is concentrated on $(\uparrow x_n) \cap X$ for some $x_n \in F$, and where $\mu_0((\uparrow x) \cap X) = 0$ for all $x \in F$. The sum is convergent in the total variation norm. We claim $\mu_0 = 0$. In fact, suppose $\mu_0 \neq 0$, and let $Y = \operatorname{supp} \mu_0$. Then since S is tame, we have by 1.5 (d) nonempty relatively open subsets $U_0, U_1, \ldots, U_n \subset Y$ such that $x_1 \ldots x_n \leq x_0$ whenever $(x_0, x_1, \ldots, x_n) \in U_0 \times U_1 \times \ldots \times U_n$. This means we can choose $x_1, \ldots, x_n \in F$ and have a nonempty relatively open (in Y) subset $U_0 \subset (\uparrow x_1 \ldots x_n) \cap Y \subset (\uparrow x_1 \ldots x_n) \cap X \in \mathcal{F}$. Since U_0 is a relatively open subset of $Y = \operatorname{supp} \mu_0$, we have

$\mu_0((\uparrow x_1 \ldots x_n) \cap X) \geq \mu_0(U_0) > 0$, a contradiction. So $\mu_0 = 0$, and $\mu = \sum_{n=1}^{\infty} \mu_n$, where each μ_n is concentrated on $\uparrow x_n$ for some $x_n \in F$. Since $\uparrow x_n \subseteq F$ is closed, F is μ_n measurable for each n. It follows that F and χ_F are μ measurable as well.

b) Let $\mu \in M(S)$. We want to show $h(\mu) = \mu(F)$. The first step is to consider the restriction $\nu = \mu|I$, and to show $h(\nu) = 0$ where $I = S\backslash F$. Since ν is inner regular (and by a) I is ν measurable), we may reduce to the case where supp ν is a compact subset of I. Define the measure $\tau \in M(S)$ by $\tau(E) = h(\nu|E)$. We want to show τ is trivial. Suppose τ is not trivial, and let $X = \text{supp } \tau$. By 1.5 (d), there are nonempty relatively open subsets $U_0, \ldots, U_n$ of X such that $x_1 \ldots x_n \leq x_0$ for all $(x_0, x_1, \ldots, x_n) \in U_0 \times U_1 \times \ldots \times U_n$. It follows that for every choice of Borel sets $E_i \subset U_i$, $i = 1,\ldots,n$, we have $\delta_{x_0} * (\nu|E_1) * \ldots * (\nu|E_n) = (\nu|E_1) * \ldots * (\nu|E_n)$. Applying h to this equation and keeping in mind $h(\delta_{x_0}) = 0$, we get $\tau(E_1) \ldots \tau(E_n) = 0$ for all choices $E_i \subset U_i$, $i = 1,\ldots,n$. It follows that for at least one i , $\tau(E_i) = 0$ for all Borel $E_i \subset U_i$, and so $U_i \cap \text{supp } \tau = \emptyset$, a contradiction. So τ is trivial, and we have shown $h(\nu) = 0$, when supp $\nu \subset I$ is compact. It is easy to extend (by approximation) this result to the general case where ν is supported on I.

So to prove (b), namely that $h(\mu) = \mu(F)$, we may now assume μ is supported on F. Again, we reduce to the case that μ is positive and supported on a compact subset of F. Then, the proof of (a) shows in particular that $\mu = \sum_{k=1}^{\infty} \mu_k$, where the μ_k are positive, pairwise mutually singular, and each μ_k is supported on $\uparrow x_k$ for some $x_k \in F$. Now

$$\begin{aligned} h(\mu_k) = h(\delta_{x_k})h(\mu_k) &= h(\delta_{x_k} * \mu_k) = h(\mu_k(\uparrow x_k)\delta_{x_k}) \\ &= \mu_k(\uparrow x_k) = \mu_k(F) , \quad \text{for each } k. \end{aligned}$$

So $h(\mu) = \mu(F) = \int \chi_F \, d\mu$ for each positive μ compactly supported in F , and that suffices to establish (b). □

The thrust of Theorem 2.2 is to concretely identify the maximal ideal space of the measure algebra $M(S)$ for a tame locally compact

Lawson semilattice S. Since S is Lawson, Theorem 3.2 of [LLM] shows that M(S) is semisimple. Moreover, the discussion of pages 134-136 of [LLM] following the proof of Proposition 2.5 applies verbatim to our present situation, and so the same conclusions are valid. Namely, for a tame locally compact Lawson semilattice S, the maximal ideal space $\Delta M(S)$ of M(S) is precisely the space $\hat{S}_d$ of all semilattice semicharacters of S_d, the discrete semilattice S, this space endowed with the compact-open topology. Further, the structure semigroup of M(S) is then the space $(\hat{S}_d)^{\wedge}_d$ of all semilattice semicharacters of the discrete semilattice $(\hat{S}_d)_d$. Finally, we can identify the idempotents of M(S) and the invertible measures on S: Namely, the idempotents are precisely the measures of the form $\mu = \sum_{k=1}^{n} a_k \delta_{s_k}$, where $s_1, \ldots s_n \in S$ and, for every $s \in S$, we have $\sum_{s \leq s_k} a_k = \begin{cases} 0 \\ 1 \end{cases}$.

Moreover, the space $\Delta M(S)$ is totally disconnected, and so its first cohomology group is trivial. This means, by the Arens-Royden Theorem, that the invertible measures in M(S) are precisely those which are exponential measures.

We conclude the paper by pointing out some of the pathologies in M(S) in case S is a wild locally compact Lawson semilattice. In that case, by 1.5 (c), S must contain a meet irredundant set X. We state the analogue here of the famous result of Hewitt-Kakutani [HK], and sketch out its proof in a sequence of propositions. The virtue we find in these propositions is that they get at the central issues more clearly than the usual proofs (cf [R]); we will not burden the reader with detailed proofs. This viewpoint is not new; it was given to us in various conversations with J. L. Taylor. Responsibility for the particular assertions, however, rests with us.

2.3. <u>Theorem</u>. Let S be a wild locally compact Lawson semilattice and let $X \subset S$ be a meet irredundant Cantor set. Let F be a bounded linear functional of norm ≤ 1 on the space $M_c(X)$ of continuous measures on X. Then F extends to be a multiplicative linear functional on M(S).

<u>Remark</u>. Thus on M(S) the multiplicative linear functionals are anything but highly determined. We give now a few propositions setting out the proof of 2.3. The sketch of the proof of 2.3

follows 2.7.

Notation. We establish here some notation for use in the rest of the paper. We will use $Ⓢ^n$ to denote the symmetrized n-fold tensor power in various categories. For a compact space X, $Ⓢ^n X$ will denote the quotient space of the n-fold Cartesian product X^n modulo the relations imposed by the coordinate permutations. Similarly, for for an L-space L, $Ⓢ^n L$ will denote the n-fold projective tensor product $\hat{\otimes}^n L$ modulo the closed subspace generated by the relations imposed by permutations of elementary n-fold tensors. For a compact space X, $\Delta_n(X) \subset Ⓢ^n X$ will denote the closed subspace of symmetrized n-tuples in which two or more elements are the same. For a compact subset K of a semilattice S, $K^{(n)} = \{x_1 \ldots x_n : x_i \in K \text{ for } i = 1,\ldots,n\}$, and for an L-subspace of a convolution measure algebra, $L^{(n)}$ will denote the closed L-subspace generated by $\{\mu_1 * \ldots * \mu_n : \mu_i \in L \text{ for } i = 1,\ldots,n\}$.

In the rest of this section, S will be a wild locally compact Lawson semilattice, and $X \subset S$ will be a meet irredundant Cantor subset. We will let $L = M_c(X)$, the space of continuous measures on X.

2.4. Proposition. The multiplication map $m : X^n \longrightarrow X^{(n)}$ induces a homeomorphism $\bar{m} : Ⓢ^n X/\Delta_n(X) \longrightarrow X^{(n)} / \bar{m}(\Delta_n(X))$.

Proof. Straightforward, using the fact that X is meet irredundant.

□

2.5. Proposition. The convolution map $m : \hat{\otimes}^n L \longrightarrow L^{(n)}$ induces an (onto) isomorphism $\bar{m} : Ⓢ^n L \longrightarrow L^{(n)}$ of L spaces.

Proof. We provide only the barest outline. The details are tedious, but not deep. We may identify $\hat{\otimes}^n L$ with the space of all measures on X^n absolutely continuous with respect to n-fold products of continuous measures (see [T, Proposition 2.5.2].) Further, $Ⓢ^n L$ is identified with the L space on $Ⓢ^n X$ generated by the images on $Ⓢ^n X$ of all n-fold products of continuous measures. Now, a standard calculation with the Fubini theorem shows that $\Delta_n(X)$ is a null set for all such images, and hence for the entire L-space. So $Ⓢ^n L$ is identified with an L-space of measures on $Ⓢ^n X/\Delta_n(X)$. Using the homeomorphism of 2.4, we transfer this L-space to $X^{(n)} / \bar{m}(\Delta_n(X))$. On the other hand, using another standard calculation and the meet irredundancy of X, we can see that $\bar{m}(\Delta_n(X))$ is a null set for every $\mu_1 * \ldots * \mu_n$ such that each $\mu_i \in L$. Consequently $\bar{m}(\Delta_n(X))$ is a null set for all of $L^{(n)}$, and it is now clear that $\circledS^n L \cong L^{(n)}$, as claimed. □

2.6. <u>Proposition.</u> If $m \neq n$, $L^{(m)} \perp L^{(n)}$.

<u>Proof.</u> Standard. □

Proposition 2.5 and 2.6 lead immediately to the following theorem.

2.7. <u>Theorem.</u> If A is the L subalgebra with 1 of $M(S)$ generated by L, then

$$A \simeq C \cdot \delta_0 \oplus L \oplus \circledS^2 L \oplus \circledS^3 L \oplus \ldots .$$

where the norm on the right hand side is given by $\|c\delta_0 + \mu_1 + \mu_2 + \ldots\| = |c| + \|\mu_1\| + \|\mu_2\| + \ldots$, for each $c \in C$, $\mu_n \in \circledS^n L$.

<u>Proof.</u> Clear. □

<u>Proof of 2.3.</u> Again we provide only an outline. We first argue that the maximal ideal space of A (A as in 2.7), can be identified by restriction with the entire unit ball of L^*. The point is that for F in the unit ball of L^*, we must define F on elementary tensors $\mu_1 \otimes \ldots \otimes \mu_n$ as $F(\mu_1) \ldots F(\mu_n)$; but this serves to define F on $\hat{\otimes}^n L$ by the universal properties of $\hat{\otimes}$, and then on $\circledS^n L$ by the definition of $\circledS^n L$. So every F in the unit ball of L^* extends to a multiplicative linear functional on A.

The remaining step is to check that the entire unit ball of L^* is the Shilov boundary for A (and so by a standard result in Banach algebras, the entire unit ball of L^* extends to be multiplicative linear functionals on $M(S)$). This is carried out in steps, as follows: First, for any probability measure μ in L, $\delta_0 + \mu$ assumes its maximum modulus over ΔA at the constant function 1, and nowhere else. Next (using the semigroup structure of ΔA and its action on A, cf [T]), if $u \in \Delta A$ is a unit, then $u^* \cdot (\delta_0 + \mu)$ assumes its maximum modulus at u, and nowhere else. So the units of ΔA are contained in the Shilov boundary. But since every measure in L is continuous, one can argue that the units in the unit ball of

L (here in ΔA) are weak * dense in the whole unit ball (i.e., all of ΔA), so ΔA is all Shilov boundary, and 2.3 is proved. □

References

[B] Baartz, A., The measure algebra of a locally compact semigroup, Pac. J. Math. 21 (1967), 199-214.

[G] Gierz, G., et. al., A Compendium of Continuous Lattices, Springer-Verlag, Heidelberg, New York (1980), 371 pp.

[HK] Hewitt, E. and S. Kakutani, A class of multiplicative linear functionals on the measure algebras of a locally compact abelian group, Illinois J. Math. 4 (1960), 553-574.

[LLM] Lawson, J. D., J. Liukkonen, and M. Mislove, Measure algebras of semilattices of finite breadth, Pac. J. Math. 69 (1977), 125-139.

[N] Newman, S. E., Measure algebras on idempotent semigroups, Pac. J. Math. 31 (1969), 161-169.

[R] Rudin, W., Fourier Analysis on Groups, John Wiley & Sons, New York (1960), 285 pp.

[T] Taylor, J., Measure algebras, CBMS Regional Conference Series in Mathematics 16 (1972), 108 pp.

Department of Mathematics
Tulane University
New Orleans, Louisiana 70118

ON STRUCTURAL METHODS AND RESULTS IN THE THEORY OF COMPACT SEMITOPOLOGICAL SEMIGROUPS

Wolfgang Ruppert

Since its very beginning, the theory of compact semitopological semigroups has always relied heavily on the use of functional analytic methods; many important results have been found and proved solely (or at least almost solely) with the help of auxiliary theorems from functional analysis This fact is not surprizing; after all, the main applications of semitopological semigroups lie in the theory of certain function spaces; functional analysis is a long-established and well-developed theory which provides powerful tools for many branches of topological algebra. However it seems to be desirable also to have a " structural approach " to the theory of compact semitopological semigroups, an approach which emphasizes algebraico-topological methods and general results about the algebraic and/or topological structure of compact semitopological semigroups - without immediate reference to functional analysis. Such methods and results could in turn be most useful for applications in functional analysis. Furthermore, it is to be hoped that in this way some of the traditional proofs can be simplified or extended and that new properties of semigroups or topological groups can be exhibited.

It is the aim of the present contribution to give a survey over various topics in the theory of compact semitopological semigroups which can be treated by this structural approach.

1. Joint continuity and Ryll-Nardzewsky's Theorem

Perhaps the earliest structural result in the theory of semitopological semigroups was Ellis' discovery that a locally compact semitopological semigroup S is a topological group iff it is algebraically a group ([6], 1958); this amounts to the proof that the multiplication of such a semigroup is jointly continuous at (1,1). By adding an ingenious idea to the method of Ellis' proof, J.D. Lawson ([11], 1973) was able to show that the multiplication of any compact semitopological semigroup S with identity 1 is jointly continuous at every point (1,s) or (s,1), $s \in S$. (It is easy to see that this result implies the earlier one of Ellis; note that every locally compact semitopological group can be imbedded into a compact semitopological semigroup by the adjunction of a zero element.)

A result apparently independent of the above was the Theorem of Ryll-Nardzewsky ([21], 1964) that the minimal ideal of a compact semitopological semigroup S is a (necessarily compact topological) group if the group of units is dense in S. Its original proof was based on a fixed-point theorem, which established the existence of an invariant mean on such semigroups. For many years no attempts were undertaken to devise a direct, "structural" proof; the first fixed-point free treatment of this subject was given by Veech ([24], 1973).

However, it can be shown that - with only minor modifications - the method of Ellis and Lawson can be used also for the Theorem of Ryll-Nardzewsky; thereby allowing a unified treatment of these two theorems, which are central to the theory of semitopological semigroups. In this section we give a brief sketch of the necessary alterations; details will be published later. We also confer the reader to the recent work of Troaillic ([25], 1981), which, though not connected directly with the method of Ellis and Lawson, provides an algebraico-topological proof which essentially goes in the same direction.

Actually, the method of Ellis - Lawson yields a result more general than that of Ryll-Nardzewsky. Recall that the well-known structure theorem about the minimal ideal M(S) of a compact semitopological semigroup S implies that M(S) is a group if and only if for any pair of idempotents e,f in M(S) the identities

$$ef = f \quad \text{and} \quad fe = e$$

or

$$ef = e \quad \text{and} \quad fe = f$$

always imply e = f. (cf. Berglund - Hofmann [1] p. 57 . Thus the result of Ryll - Nardzewsky is contained in the following theorem.

Theorem 1. Let S be a compact semitopological monoid and suppose that the group of units H(1) is dense in S. Then the relations $\leqq_R$ and $\leqq_L$, defined on E(S) by

$$e \leqq_R f \quad \text{iff} \quad fe = e$$

$$e \leqq_L f \quad \text{iff} \quad ef = f,$$

coincide (and thus form a partial order; note that in general $\leqq_R$ and $\leqq_L$ are only pre-orders).

This theorem is in turn strongly connected with joint continuity properties. (For the formulation of the next proposition recall that a right topological semigroup S is a semigroup which is defined on a topological space such that the multiplication of S is continuous in the right variable; that is, the map $x \to sx$ is continuous for every $s \in S$.)

Proposition 1. Let S be a compact right topological semigroup and let e be an idempotent in S. If the map

$$\mu_e\colon eS \times S \to eS, \ (x,y) \to xy,$$

is jointly continuous at every point (e,y) with $y \in S$ then for any idempotent f in S the relation $e \leqq_R f$ always implies $e \leqq_L f$.

Proof. Let $\langle g_n \ n \in D\rangle$ be a net in H(1) which converges to e. Since S is compact, we may assume w.l.o.g. that $y = \lim g_n^{-1}$ exists. By supposition , μ_e is jointly continuous at (e,y), so

$$ey = e^2y = \lim (eg_n)(g_n^{-1}) = e$$

and therefore $fe = e$ implies

$$e = efe = efey = \lim (efg_n)(g_n^{-1}) = ef.$$

This proposition shows that it is sufficient to prove the following Theorem.

<u>Theorem 2</u>. <u>Let</u> S <u>be as in Theorem</u> 1. <u>Then for every</u> $s \in S$ <u>the map</u>

$$\mu_s : sS \times S \longrightarrow sS, \ (x,y) \longrightarrow xy,$$

<u>is jointly continuous at any point</u> (s,y), $y \in S$. (Obviously, a similar statement holds for the map $S \times Ss \longrightarrow Ss$, $(x,y) \longrightarrow xy$.)

We shall see from an example at the end of this section that the assumption " H(1) is dense in S " cannot be dropped in this theorem. The author is indebted to J.D. Lawson who in conversation pointed out that the statement of the theorem follows from Namioka's proof of the joint continuity theorem and how it could be used conveniently to derive Ryll-Nardzewsky's result (cf. [16] and [12]). In the authors original version the above statement was formulated only for elements s in the minimal ideal M(S) of S.

Let us now turn to the method of Ellis and Lawwon. The basic idea of this method is to show that

(1) if S is metrizable then points of joint continuity are abundant - that is, they form a residual subset of $S \times S$;

(2) joint continuity can be "transported" to the prescribed points (1,s) or (s,1) whenever the points of joint continuity are abundant;

(3) the general case can be reduced to the metrizable case.

Since it is in general not possible to represent a compact semitopological monoid (in an obvious way) as the projective limit of metrizable ones, the arguments for (3) are somewhat subtle; they require an apparently more general formulation of (1) and (2), as assertions about separately continuous actions of a semitopological semigroup on a compact space. For the sake of brevity we therefore omit the discussion of point (3) and concentrate on (1) and (2). (Sure enough, if the reader is familiar with Lawson's original paper [11] then he will find no difficulties in devising the necessary adaptions in point (3) himself.)

Assertion (1) is the immediate consequence of a Lemma which seems to be due to Bourbaki (cf [2] p.255, ex. 23); it states that if $f: X \times Y \longrightarrow Z$ is a separately continuous map, where X is a Baire space and Y,Z are metric spaces, then for every $y \in Y$ there exists a residual subset X_y of X such that f is jointly continuous at every point (x,y) with $x \in X_y$. We replace it by the following slightly different version.

Lemma 1. *Let X be a Baire space, Y a compact metric and Z an arbitrary metric space. Suppose that $f: X \times Y \longrightarrow Z$ is a map with the following properties:*

(a) *for every $x \in X$ the map $Y \longrightarrow Z$, $y \longrightarrow f(x,y)$, is continuous;*

(b) *for every element y in a dense subset D of Y the map $X \longrightarrow Z$, $x \longrightarrow f(x,y)$ is continuous.*

Then there is a residual subset R of X such that for each $x_o \in R$

(i) *the maps $X \longrightarrow Z$, $x \longrightarrow f(x,y)$, $y \in Y$, are equicontinuous at x_o (note that in general these maps need not even be continuous at every point of X);*

(ii) *f is jointly continuous at every point (x_o, y_o), where $y_o \in Y$.*

Proof. We use the same letter d for the distance functions of both Y and Z. Define for any $x \in X$ and any $\varepsilon > 0$

$$g(x;\varepsilon) = \sup\{\delta \mid \text{if } y,y' \in Y \text{ and } d(y,y') < \delta \text{ then } d(f(x,y),f(x,y')) \leqq \varepsilon\}.$$

Then for fixed $\varepsilon > 0$ the function $X \longrightarrow \mathbb{R}$, $x \longrightarrow g(x;\varepsilon)$, is upper semicontinuous. To see this, let $\langle x_n \rangle$ be a net in X with $\lim x_n = x$, $g(x_n;\varepsilon) \geqq \delta$ for all n. If $y,y' \in D$ and $d(y,y') < \delta$ then $d(f(x,y),f(x,y')) \leqq \varepsilon$ by continuity. For an arbitrary pair (y,y') with $d(y,y') < \delta$ choose sequences $\langle y_m \rangle$ and $\langle y'_m \rangle$

in D with $y = \lim y_m$, $y' = \lim y'_m$ and

$$d(y_m, y) + d(y'_m, y') + d(y, y') < \delta.$$

Then $d(y_m, y'_m) < \delta$ for all m, and therefore

$$d(f(x,y), f(x,y')) = \lim d(f(x,y_m), f(x,y'_m)) \leq \varepsilon.$$

Thus $g(x;\varepsilon) \geq \delta$ and $g(\ ;\varepsilon)$ is upper semicontinuous.

Since by (a) the map $y \longrightarrow f(x,y)$ is uniformly continuous on the compact space Y, we have $g(x;\varepsilon) > 0$ for all $x \in X$ and all $\varepsilon > 0$. Therefore, using the uniform boundedness principle, to every $\varepsilon > 0$ we can find a dense open subset O_ε of X such that every x_o in O_ε has a neighborhood $U(x_o, \varepsilon)$ where the function $x \longrightarrow 1/g(x;\varepsilon)$ is bounded. This means that there exists a positive number $\delta(x_o;\varepsilon)$ such that $d(f(x,y), f(x,y')) \leq \varepsilon$ whenever $x \in U(x_o, \varepsilon)$ and $d(y,y') < \delta(x_o;\varepsilon)$. Let R be the intersection of the sets $O_{1/n}$, $n = 1,2,\ldots$; obviously, R is a residual subset of X.

Take an element $x_o \in R$ and choose an $\varepsilon > 0$. Then by the definition of R there is a neighborhood U of x_o in X and a positive real number δ such that $d(f(x,y), f(x,y')) \leq \varepsilon$ whenever $d(y,y') < \delta$. Since D is dense in the compact space Y, there is a finite subset A of D with $d(y,A) < \delta$ for all $y \in Y$. Choose a neighborhood V of x_o with $V \subset U$ and $d(f(x_o,a), f(x,a)) < \varepsilon$ for all $a \in A$ and all $x \in V$. Let y be an arbitrary element of Y. Then for a suitable $a^* \in A$ we have $d(y,a^*) < \delta$ and therefore

$$\begin{aligned} d(f(x_o,y), f(x,y)) &\leq \\ &\leq d(f(x_o,y), f(x_o,a^*)) + d(f(x_o,a^*), f(x,a^*)) + d(f(x,a^*), f(x,y)) \\ &\leq \varepsilon + \varepsilon + \varepsilon = 3\varepsilon \end{aligned}$$

for all $x \in V$. This establishes assertion (i). The proof of (ii) is left to the reader.

In the proof of the joint continuity theorem , the " transport " arguments in point (2) were based on the observation that the multiplication μ of S is jointly continuous at (x_o,y_o) if there is a unit $g\in H(1)$ such that μ is jointly continuous at (gx_o,y_o) (Ellis [6]); or, more generally, if the left translations by elements of the set

$$D(x_o,y_o) = \{s\in S \mid \mu \text{ is jointly continuous at } (sx_o,y_o)\}$$

separate x_oy_o from any point $z_o \neq x_oy_o$ (that is, $sx_oy_o = sz_o$ for all $s\in D(x_o,y_o)$ implies $x_oy_o = z_o$) (Lawson [11]). For our purposes we need an analogous statement dealing with equicontinuity properties.

Proposition 2. _Let_ S _be a compact right topological semigroup with identity_ 1. _Let_ x_o _be an element of_ S _such that the maps_

$$\rho_s : x_oS \to S,\ x \to xs,\ s\in S,$$

are equicontinuous at some point $y_o\in x_oS$. (Note that in general the maps ρ_s need not even be continuous.)

Then the maps ρ_s _are equicontinuous at_ x_o _if for every neighborhood_ U _of_ y_o _there is a unit_ $g\in H(1)$ _with_ $x_og\in U$ _and such that_ ρ_g _is continuous_.

Proof. Let W be a symmetric entourage in the uniformity of S and choose a neighborhood U of y_o such that $(y_os,ys)\in W$ for all $s\in S$, $y\in U$. Let g be a unit with $x_og\in U$ and such that ρ_g is continuous. Then there is a neighborhood V of x_o with $Vg\subset U$. For every $v\in V$ and every $s\in S$ we have

$$(x_os, y_og^{-1}s) = ((x_og)g^{-1}s,\ y_og^{-1}s)\in W,$$
$$(vs\ ,\ y_og^{-1}s) = ((vg)g^{-1}s\ ,\ y_og^{-1}s)\in W,$$

since $\{x_og,vg\}\subset U$. Thus $(x_os,vs)\in W\circ W$ and the assertion follows.

We are now ready for the following generalized version of Ryll-Nardzewsky's theorem . (The reader is warned however: whereas the metrizability condition in this theorem can be dropped if S is semitopological, it cannot be dropped in the right topological case.)

Theorem 3. *Let S be a compact metrizable right topological semigroup with identity 1. Assume that the group*

$$G = \{g\in H(1)\,|\,\text{the map } S \to S, x \to xg, \text{ is continuous}\}$$

is dense in S. Then the following statements hold:

(i) *For any element $x_o \in S$ the map*

$$x_oS \times S \longrightarrow S,\ (x,y) \longrightarrow xy,$$

is jointly continuous at every point (x_o,s), $s\in S$.

(ii) *If e,f are idempotents in S with ef = f and fe = e then e = f.*

(iii) *S has only one minimal left ideal and every minimal right ideal is a compact topological group.*

Proof. (i) By Lemma 1 there is a residual subset R of x_oS such that the maps $x_oS \to S$, $x \to xs$, $s\in S$, are equicontinuous at all points $x\in R$. Since G is dense in S, we may apply Proposition 2 to conclude that these maps are equicontinuous at x_o; by the usual arguments, this establishes the assertion.

Assertions (ii) and (iii) follow now easily from (i), Proposition 1 and the structure theorem about the minimal ideal of a compact right topological semigroup (cf. [15]).

Obviously, Theorem 3 implies Theorem 1 and Theorem 2 (and hence the Theorem of Ryll-Nardzewsky) for metrizable compact semitopological semigroups. To prove that the statements of the theorem hold for arbitrary compact semitopological monoids with dense group of units one can use the reduction arguments of Ellis - Lawson (point (3) above) in a suitably modified form. But these arguments rely heavily on the fact that in a semitopological monoid a net $\langle s_n \rangle$ converges to a point s if and only if the associated net of left translations $\langle x \to s_n x \rangle$ converges to the left translation $x \to sx$ and similarly for right translations. Thus the reduction step of our program cannot be carried out in the right topological case; in fact, Theorem 3 is false for general, non-metrizable and non-semitopological, right topological monoids (cf. Chou [4]).

We finally show by an example that the assumption of Theorem 2 that "the group of units is dense in S" cannot be dropped.

Example. Let A,B be compact spaces and assume that $f: A \times B \longrightarrow G$, $(a,b) \longrightarrow [a,b]$, is a separately continuous map into a compact topological group G. Furthermore, suppose that $\langle a_n \rangle$, $\langle b_n \rangle$ are convergent nets in A,B respectively, such that

$$\lim a_n = a, \quad \lim b_n = b$$

but

$$\lim [a_n, b_n] = z \neq [a,b].$$

Consider the "sandwich product" $S = B \times G \times A$, with the multiplication

$$(b,g,a)(b',g',a') = (b, g[a,b']g', a').$$

Then S is a compact semitopological (but not topological) semigroup with M(S) = S. (Semigroups of this type have been constructed e.g. in Berglund-Hofmann [1] p.156.) The element $e = (b, [a,b]^{-1}, a)$ is an idempotent and

$$\lim (b, [a_n,b]^{-1}, a_n) = \lim e(b, [a_n,b]^{-1}, a_n) = e,$$

$$\lim (b_n, [a,b_n]^{-1}, a) = e.$$

However,

$$\lim (b, [a_n,b]^{-1}, a_n)(b_n, [a,b_n]^{-1}, a) = (b, [a,b]^{-1} z [a,b]^{-1}, a)$$

is different from e^2; so the map

$$\mu_e : eS \quad S \longrightarrow eS, \ (x,y) \longrightarrow xy,$$

is not jointly continuous at (e,e).

2. Centric subsemigroups

The result of Ryll-Nardzewsky can be expressed also in the following form:

If S is a compact semitopological monoid with dense group of units H(1) then every idempotent in the minimal ideal M(S) of S is central (i.e. commutes with every element of S).

If H(1) is the continuous homomorphic image of a connected locally compact topological group then much more is true, as can be seen from the next theorem (which has been announced already in [19]; the particulars of its proof will appear in [20]).

Theorem 4. Let S be a compact semitopological semigroup and $\phi: G \longrightarrow S$ a continuous homomorphism which maps a connected locally compact topological group G onto a dense subset of S. Then every idempotent in S is central in S. Moreover, if in addition G is solvable then $\phi(G)$ is centric in S; that is,

$$s\phi(G) = \phi(G)s$$

for every $s \in S$. (Clearly, $s\phi(G) = \phi(G)s$ implies $sS = Ss$.)

It might seem natural to conjecture that the idempotents of S are always central if only the group of units is dense (without assuming H(1) to be the continuous homomorphic image of a connected locally compact group). But this conjecture is definitely false for a large class of weak almost periodic compactifications of discrete groups.

Example. Let G be a discrete group which contains an infinite subgroup H such that $H \cap g^{-1}Hg$ is finite for some $g \in G$. Let S be the weak almost periodic compactification of G and $\phi: G \rightarrow S$ the compactification map. Then the minimal idempotent e in the closure of $\phi(H)$ is not the identity (H being infinite) and therefore $e \neq g^{-1}eg$. (Note that the closure of $\phi(H)$ in S is open in S; cf. Milnes [13].)

However it is possible to generalize the second part of Theorem 4; the assumption "G is solvable" can be dropped.

Theorem 5. Let G be a connected locally compact topological group and let $\phi:G \to S$ be a continuous homomorphism mapping G onto a dense subset of a compact semitopological semigroup S. Then $\phi(G)$ is centric in S.

In this section we sketch a proof of Theorem 5; more details (and a still more general formulation of the theorem) will be given in a subsequent publication. As is to be expected, we make extensive use of auxiliary results taken from [20].

We first note that we may suppose that G does not contain any non-trivial compact normal subgroup (and thus is a Lie group). In fact, if N is the maximal compact normal subgroup of G then we may consider the induced homomorphism

$$\overline{\phi}:G/N \to S/\sim ,$$

where $\sim$ is the closed congruence, defined by

$$s \sim t \quad \text{iff } s\phi(N) = t\phi(N)$$

(it follows from the joint continuity theorem and the denseness of $\phi(G)$ that $\sim$ is a closed congruence). Clearly, $\phi(G)$ is centric in S if and only if $\phi(G/N)$ is centric in $S/\sim$.

Next we pick an element $s \in S$ and assert that we may assume that the right isotropy group

$$F_R(s) = \{g \in G | s\phi(g) = s\}$$

of s with respect to the right action $S \times G \to S$, $(s,g) \to s\phi(g)$, of G on S does not contain any non-trivial normal subgroup of G. Suppose that N is a normal subgroup of G with $N \subset F_R(s)$ and let e be the minimal idempotent in the closure $\overline{\phi(N)}$ of $\phi(N)$ in S. Then $eS = Se$ (since N is normal in S we have $\phi(g^{-1})e\phi(g) = e$ for all $g \in G$, hence $es = se$ by continuity) and $e\overline{\phi(N)}$ is a compact normal subgroup of H(e). Since $s \in eS$ we have $s\phi(G) = \phi(G)s$ if and only if $se\phi(G) = e\phi(G)s$, so we may replace $\phi:G \to S$ by the induced map ϕ_e: $G/_N \longrightarrow eS/\overline{e\phi(N)}$ which amounts to the required reduction.

By the above discussion it is sufficient to prove theorem 5 under the following general assumption, which we adopt for the rest of this section.

General assumption. (i) G is a connected Lie group which does not contain any non-trivial compact normal subgroup.
(ii) $\phi: G \longrightarrow S$ is a continuous homomorphism mapping G onto a dense subset of a compact semitopological semigroup.
(iii) s is an element of S such that $F_R(s)$ does not contain any non-trivial normal subgroup of G (compact or not).

The salient feature of our proof is the use of Lie machinery. Let us introduce some handy conventions. If G,H, ... are Lie groups then we denote the associated Lie groups with the corresponding gothic letters $\mathfrak{g}, \mathfrak{h}, \ldots$. (Thus we write $\mathfrak{F}_R(s)$ for the Lie algebra corresponding to the right isotropy group $\mathfrak{F}_R(s)$.)

We tacitely assume that all Lie algebras are equipped with a Euclidean norm $\| \; \|$ (compatible with the norm of any subalgebra). The adjoint action of a Lie group G on its Lie algebra $\mathfrak{g}$ is written

$$G \times \mathfrak{g} \longrightarrow \mathfrak{g}, \quad (g,x) \longrightarrow g.x,$$

so that $\exp(g.x) = g(\exp x)g^{-1}$ for all $x \in \mathfrak{g}$, $g \in G$.

The following Lemma is the key element of our investigations. (The proof can be found in [20]).

Lemma 2. Under our general assumptions, let $\langle g_n | n \in D \rangle$ be a net in G with $s = \lim \phi(g_n)$ and let $\mathfrak{h}$ be a subalgebra of $\mathfrak{g}$ with $g_n \cdot \mathfrak{h} = \mathfrak{h}$ for all $n \in D$. Then

(i) $\mathfrak{h} \cap \mathfrak{F}_R(s) \neq 0$ if there is a non-zero element x in $\mathfrak{h}$ with either $\lim g_n^{-1} \cdot x = 0$ or $\lim \| g_n^{-1} \cdot x \| = \infty$;

(ii) if the net $\langle g_n \rangle$ has no subnet $\langle g_m \rangle$ with $\lim g_n^{-1} \cdot x = 0$ or $\lim \| g_n^{-1} \cdot x \| = \infty$ for some non-zero $x \in \mathfrak{h}$ then there is a Lie automorphism $\rho: \mathfrak{h} \longrightarrow \mathfrak{h}$ satisfying

$$(\phi(\exp x))s = s\phi(\exp \rho(x)).$$

In particular, the above Lemma shows that if $\mathfrak{F}_R(s) = 0$ then the assertion of the theorem is true. The proof of the next Lemma is found easily by simply rephrasing the proof of Lemma 2.13 in [20].

<u>Lemma 3</u>. <u>If $\mathfrak{h}$ is a solvable ideal of $\mathfrak{g}$ then</u> $\mathfrak{F}_R(s) \cap \mathfrak{h} = 0$.

We now show that the statement of this Lemma also holds for semisimple ideals of $\mathfrak{g}$.

<u>Lemma 4</u>. <u>If $\mathfrak{h}$ is a semisimple ideal of $\mathfrak{g}$ then</u> $\mathfrak{F}_R(s) \cap \mathfrak{h} = 0$.

Proof. Write H for the analytic subgroup of G which corresponds to $\mathfrak{h}$ and let $\alpha: G \to \mathrm{Aut}\,\mathfrak{h}$ be the map which to every $g \in G$ assigns the automorphism $x \to g.x$ of $\mathfrak{h}$. Note that $\alpha(G)$ is contained in the component $(\mathrm{Aut}\,\mathfrak{h})_o$ of the identity of $\mathrm{Aut}\,\mathfrak{h}$, so by a well-known theorem $\alpha(G) = (\mathrm{Aut}\,\mathfrak{h})_o = \alpha(H)$. Assume first that the closure of $\alpha(F_R(s) \cap H)$ in $\mathrm{Aut}\,\mathfrak{h}$ is not compact. Then there is an element y in $\mathfrak{h}$ and a net $\langle g_n \mid n \in D \rangle$ in $F_R(s) \cap H$ such that $\lim \| g_n^{-1}.y \| = \infty$ or $\lim g_n^{-1}.y = 0$; we may assume that the limit $t = \lim \phi(g_n)$ exists in S. By Lemma 2 this implies that $\mathfrak{F}_R(t) \cap \mathfrak{h} \neq 0$. Let m be the minimal idempotent in the closure of $\phi(F_R(s) \cap H)$ in S. Then

$$F_R(t) \subset F_R(mt) = F_R(m) \subset F_R(s)$$

(note that $mt \in H(m)$ and that $sm = s$). Since the idempotent m is central in S, by Theorem 4, $F_R(m)$ is a closed normal subgroup of G. (If $g \in G$, $h \in F_R(m)$ then $mg^{-1}hg = g^{-1}mhg = g^{-1}mg = m$, so $g^{-1}hg \in F_R(m)$.) By our general assumption $F_R(s)$ does not contain any non-trivial normal subgroup of G, thus $F_R(m) = \{1\}$ and therefore $F_R(t) \cap H \subset F_R(m) = \{1\}$, a contradiction to $\mathfrak{F}_R(t) \cap \mathfrak{h} \neq 0$. It follows that the closure of $\alpha(F_R(s) \cap H)$ in $\mathrm{Aut}\,\mathfrak{h}$ is compact.

Now let K,A be analytic subgroups of H such that $\alpha(K)$ is a maximal compact subgroup of $(\mathrm{Aut}\,\mathfrak{h})_o$ with $\alpha(F_R(s) \cap H) \subset \alpha(K)$, A is solvable and

$$\alpha(H) = \alpha(K)\alpha(A)$$

is an Iwaswa decomposition of the semisimple linear group $\alpha(H)$ as the product of a compact group with a simply connected solvable group (cf. Helgason [7], p. 234, for example). Let K_1 be a compact subset of G with $\alpha(K_1) = K$ (such a set exists since α maps open subsets of G onto open subsets of $\alpha(H)$), let $A_1 = \alpha^{-1}\alpha(A)$. (Note that $G = K_1A_1$.) Since K_1 is compact, there is a net $<a_n | n\in D>$ in A_1 and an element $k\in K_1$ such that $<\phi(a_n)>$ converges to some element t in S and

$$s = \phi(k)t.$$

Clearly, $s\phi(G) = \phi(G)s$ if and only if $t\phi(G) = \phi(G)t$; so we may assume w.l.o.g. that $k = 1$, $s = \lim \phi(a_n) \in \overline{\phi(A_1)}$. By Theorem 4 this implies

$$s\phi(A) = \phi(A)s,$$

hence

$$aF_R(s)a^{-1} = F_R(s) \qquad \text{for all } a\in A.$$

It follows that $\alpha(F_R(s))$ is contained in the intersection

$$N = \bigcap\{\alpha(aKa^{-1}) \mid a\in A\} = \bigcap\{\alpha(hKh^{-1}) \mid h\in H\}$$

which is a compact normal subgroup of $\alpha(H)$ (and hence semisimple). Any covering group of a compact connected semisimple Lie group is again compact, so the identity component of $\alpha^{-1}(N)\cap H$ is a compact subgroup of H which is normal in G (note that $\alpha^{-1}(N)$ is normal in G). By our general assumption every compact normal subgroup of G is trivial, so $(\alpha^{-1}(N)\cap H)_o = \{1\}$ and therefore $\mathfrak{F}_R(s)\cap\mathfrak{h} = 0$. The proof is completed.

We are now ready for the proof of Theorem 5.

<u>Proof of Theorem 5.</u> By the Theorem of Levi-Malcev the Lie algebra $\mathfrak{g}$ of G is the semidirect product of its solvable radical $\mathfrak{a}$ with a semisimple subalgebra $\mathfrak{s}$. By Lemma 4 we have $\mathfrak{A}\cap\mathfrak{F}_R(s) = 0$. We first assert that this implies $[\mathfrak{A}, \mathfrak{F}_R(s)] = 0$. To show this, it is sufficient to show that $x^{-1}a^{-1}xa = 1$ for every element $x\in F_R(s)$ and every element a which lies in the analytic subgroup A corresponding to $\mathfrak{A}$. We may also assume that a lies in a neighborhood U of 1 which is symmetric and so small that $x^{-1}UxU\cap F_R(s)\cap A = \{1\}$ (note that A is generated by every neighborhood of the identity). By Lemma 2 we have $s\phi(a^{-1}xa) = \phi(g)s\phi(xa) =$

$= \phi(g)s\phi(a) = s$ for some $g \in A$. Thus $a^{-1}xa \in F_R(s)$ and therefore

$$x^{-1}a^{-1}xa \in x^{-1}UxU \cap F_R(s) \cap A = \{1\},$$

which establishes the assertion.

It follows that $\mathfrak{F}_R(s)$ is contained in the centralizer $\mathcal{C} = \{c \in \mathfrak{g} \mid [c, \mathfrak{A}] = 0\}$ of $\mathfrak{A}$. Note that $\mathfrak{A} \cap \mathcal{C}$ is an ideal of $\mathfrak{g}$ and that $\mathfrak{A} \cap \mathcal{C}$ is the solvable radical of $\mathcal{C}$, $\mathfrak{L} \cap \mathcal{C}$ a Levi complement. By the definition of $\mathcal{C}$ above, $\mathfrak{A} \cap \mathcal{C}$ is central in $\mathcal{C}$, hence $\mathcal{C}$ is the direct product

$$\mathcal{C} = (\mathcal{C} \cap \mathfrak{A}) \times (\mathcal{C} \cap \mathfrak{L}).$$

From this we conclude that $\mathcal{C} \cap \mathfrak{L}$ is a characteristic ideal of $\mathcal{C}$ and therefore an ideal of $\mathfrak{g}$.

Applying Lemma 4 we see that $\mathfrak{F}_R(s) \cap \mathcal{C} \cap \mathfrak{L} = 0$. Similar to the discussion on the preceding page we conclude that $[\mathfrak{F}_R(s), \mathfrak{L} \cap \mathcal{C}] = 0$ (to this end we only have to replace A by the analytic subgroup of G which corresponds to $\mathfrak{L}$). But this yields that $\mathfrak{F}_R(s)$ is contained in the centrum $\mathfrak{A} \cap \mathcal{C}$ of $\mathcal{C}$ and therefore $\mathfrak{F}_R(s) = 0$. The assertion follows now from Lemma 2. The proof is completed.

We finally note that Theorem 5 is not the most general result which can be achieved by the method of its proof. A still more general version will be published later.

3. Weak almost periodic compactifications which are groups with zero

The simplest type of a semitopological compactification of a non-compact locally compact topological group is the adjunction of a zero element, so that the resulting semigroup is a group with zero. It is therefore natural to ask whether there are locally compact groups whose only semitopological compactification is the one-point compactification. To put it another way - for which locally compact groups is the weak almost periodic compactification a group with zero ? (Note that this problem is equivalent to the question: for which locally compact groups is every weakly almost periodic function constant at infinity?) We shall see in a moment that this problem has a complete answer if we restrict our attention to connected groups.

Let us first note a very simple fact.

Proposition 3. Let (S,ϕ) be the weak almost periodic compactification of a locally compact topological group G, let H be a non-compact closed subgroup of G and assume that S is a group with zero. If H is either normal or open in G then $G = H$.

Proof. Suppose that H is normal in G and let e be the minimal idempotent in the closure of $\phi(H)$ in S. Then $e \neq 1$ and by Proposition 3.12 of [20] the quotient $eS/\overline{e\phi(H)}$ is isomorphic with the weak almost periodic compactification of G/H. Since S is a group with zero, we have $eS = \{e\}$, hence $G/H = \{1\}$, $G = H$.

Now suppose that H is open in G. Then by a result given in Milnes [13], the closure of $\phi(H)$ is open in S and it contains an idempotent $e \neq 1$ (since H is not compact). If there exists an element $g \in G \setminus H$ then $\overline{\phi(H)}\phi(g) \cap \overline{\phi(H)} = \emptyset$ and in particular $e\phi(g) \neq e$, a contradiction to $S = \phi(G) \cup \{e\}$. Thus $G = H$.

We now formulate the promised theorem.

Theorem 6. Let (S,ϕ) be the weak almost periodic compactification of a non-compact connected locally compact topological group G. Then the following statements are equivalent:

(i) S is a group with zero.

(ii) If N is a closed normal subgroup of G then either $G = N$ or N is compact.

(iii) The centrum C of G is compact and there is a connected simple Lie group H and a continuous homomorphism $\alpha : H \to G$ such that $\alpha(H)$ is dense in G.

Proof. The implication (i) $\Rightarrow$ (ii) is a consequence of the above proposition.

(ii)$\Rightarrow$(iii). Clearly, assertion (ii) implies that the group

C is compact; so we are left to show the existence of H and α. Every connected locally compact topological group is the projective limit of Lie groups, hence we may assume w.l.o.g. that G is a Lie group. The group G cannot be solvable; otherwise the closure N of the commutator group were compact and G/N would be a non-compact connected abelian Lie group, which certainly would contain a closed non-compact normal subgroup. It follows that the solvable radical A of G is compact, hence a central torus group. Let $\mathfrak{A}$ be the Lie algebra of A, $\mathfrak{L}$ a Levi complement to $\mathfrak{A}$ in $\mathfrak{L}$. Then $\mathfrak{G}$ is the direct product $\mathfrak{G} = \mathfrak{A} \times \mathfrak{L}$ and every ideal of $\mathfrak{L}$ is an ideal of $\mathfrak{G}$. Let $\mathfrak{H}$ be a simple ideal of $\mathfrak{G}$ and write H for the corresponding simply connected Lie group, $\alpha: H \to G$ for the induced continuous homomorphism. Since A is compact, $\mathfrak{H}$ can be choosen so that $\alpha(H)$ is non-compact (otherwise G were compact itself) and thus is dense in G, by (ii).

(iii) $\Rightarrow$ (i). By a well-known result about the closure of analytic subgroups (cf. Hochschild [10] p.210, for example) we have $G = C\,\alpha(H)$. Write $\mathfrak{L}$ for the subalgebra of $\mathfrak{G}$ which corresponds to $\alpha(H)$. The quotient group G/C is a non-compact connected simple Lie group with trivial centrum. By Theorem 6, $F_R(s)$ is a normal subgroup of G for every s in $S \setminus \phi(G)$, furthermore $\mathcal{F}_R(s) \cap \mathfrak{L} \neq 0$ by Lemma 2. (Note that the adjoint map induces a continuous homomorphism of G/C onto a closed subgroup of Aut $\mathfrak{L}$.) Thus $F_R(s) \supset \alpha(H)$ and hence $F_R(s) = \overline{\alpha(H)} = G$ for all $s \in S \setminus \phi(G)$. This means that every element in $S \setminus \phi(G)$ is a zero element and the assertion follows.

We remark that the above Theorem can be used also to provide another proof of the structure theorem for the weak almost periodic compactification of a connected simple, or, more generally, semisimple, Lie group with finite centrum, which was given first by Veech [25] (for more details, see [20]).

4. Weak almost periodic compactifications which are regular semigroups

Recall that a semigroup S is called a <u>regular semigroup</u> if we have $s \in sSs$ for every $s \in S$; it is called an <u>inverse semigroup</u> if it is the union of its maximal subgroups. An inverse semigroup is called a <u>semilattice of groups</u> if all of its idempotents are central elements. (Note that groups with zero are special cases of such semigroups.) In analogy to section 3 we pose the following problem:

<u>Problem.</u> For which locally compact topological groups is the weak almost periodic compactification [a regular semigroup] [an inverse semigroup] [a semilattice of groups] ?

We first show by a simple construction that the centrum of such a group must be compact.

<u>Construction.</u> Let G be a locally compact topological group and suppose that the centrum of G is not compact. Let $\{\alpha\}$, $\{\omega\}$ be singleton sets. On the (disjoint) union

$$S = G \cup (\{\alpha\} \times G) \cup \{\omega\}$$

we define a multiplication by the rules

$$g \circ (\alpha, g') = (\alpha, g) \circ g' = (\alpha, gg')$$

$$g \circ g' = gg'$$

$$(\alpha, g) \circ (\alpha, g') = \omega$$

for all $g, g' \in G$, and

$$\omega \circ s = s \circ \omega = \omega$$

for all $s \in S$. (Here the multiplication of G is indicated by simple iuxtaposition.) It is an easy exercise to show that this multiplication is associative; it is commutative if and only if G is abelian. The semigroup S is obviously not a regular semigroup since $(\alpha, g) \circ s \circ (\alpha, g) = \omega$ for all $s \in S$, $g \in G$. Moreover, if S is a subsemigroup of a regular semigroup S then none of the elements (α, g) can be central in S . (This follows from the fact that a central element z of a regular semigroup T is contained in the group of units of the ideal zT.)

We now provide S with a compact topology. To this end, choose a compact neighborhood V of the identity 1 of G and a countable subset $A = \{a_1, a_2, \ldots\}$ of the centrum of G such that for every $g \in G$ there are only finitely many elements a_j in A with $a_j g \in AV$. (The set A can be defined recursively: put $a_1 = 1$ and for every $n > 1$ choose a_n outside the compact set $\cup\{a_i a_j^{-1} a_k V V^{-1} \mid 1 \leq i,j,k \leq n-1\}$.) Endow the set

$$S_1 = G \cup (\{\alpha\} \times G)$$

with the topology a basis of which is formed by the open sets of G and the sets

$$(\{\alpha\} \times U) \cup \{a_n U \mid n > k\},$$

where U is open in G, $k = 1,2,\ldots$. This topology is locally compact Hausdorff and the maps

$$S_1 \times G \to S_1,\ (s,g) \to s \circ g;\ G \times S_1 \to S_1,\ (g,s) \to g \circ s,$$

are jointly continuous. Furthermore, if $s \in S_1$ and $\langle s_m \mid m \in D\rangle$ is a net in S_1 with $s \circ s_m \in S_1$ for all $m \in D$ and such that $\lim s \circ s_m$ or $\lim s_m \circ s$ exists (in S_1) then $\lim s_m$ does exist either. (The proof of these simple facts is left to the reader.) Thus, providing $S = S_1 \cup \{\omega\}$ with the unique topology making it to a one-point compactification of S_1, we get a compact semitopological semigroup. The group of units (= G) is open and dense in S, hence we have constructed a semitopological compactification of G which is not a regular semigroup.

Corollary. *Let (S,ϕ) be the weak almost periodic compactification of a locally compact topological group G. If S is a regular semigroup then the centrum of G and, more generally, the centrum of every quotient group G/N, where N is a closed normal subgroup of G, is compact.*

Similar to the situation in the preceding section we can give a satisfying solution of the above problem if G is connected.

Theorem 7. *Let* (S,ϕ) *be the weak almost periodic compactification of a connected locally compact topological group* G. *Then the following assertions are equivalent:*

(i) S *is a regular semigroup.*

(ii) S *is an inverse semigroup.*

(iii) S *is a semilattice of groups.*

(iv) S *is a finite semilattice of groups.*
(i.e. S is a semilattice of groups and has only finitely many idempotents).

(v) S *has only finitely many idempotents.*

(vi) S *has at most countably many idempotents.*

(vii) *The centrum of* G/N *is compact for every closed normal subgroup* N *of* G.

(viii) *If* G_1 *is a locally compact group and* $\alpha:G \rightarrow G_1$ *a continuous homomorphism then* $\alpha(G)$ *is closed in* G_1.

The proof of Theorem 7 is too long to be included in these short notes, so we postpone it to a forthcoming publication.

We finally give an upper bound for the number of idempotents in (v).

Theorem 8. *Let* S, ϕ, G *be as in Theorem* 7 *and suppose that one of the equivalent statements of the Theorem holds. Let* K *be a maximal compact subgroup of* G *and let* m *be the unique integer with*

$$2m \leq \dim G/K \leq 2m+1.$$

Then S *has at most* 2^m *idempotents.*

The estimate of Theorem 8 cannot be improved, as is shown by the following examples.

Example. Let G be the special linear group SL(2,$\mathbb{R}$). Then G is a simple Lie group with finite centrum, hence its weak almost periodic compactification S is equivalent to the one-point compactification of G. Obviously, $\dim G/K = 2$ and card E(S) = 2 in this case; if we replace G by its m-th power G^m then we get $\dim G^m/K^m = 2m$ and card $E(S) = 2^m$.

<u>Example.</u> Let G be the semidirect product $C \circledS T$, where T denotes the multiplicative group of all complex numbers of modulus 1; the multiplication of G is given by the formula

$$(a,b)(a',b') = (a + ba', bb').$$

Obviously, G satisfies condition (vii) of Theorem 7 and $m = 1$. The weak almost periodic compactification of G is the union $S = \phi(G) \cup M(S)$, and $M(S)$ is isomorphic with T. As in the preceding example, the m-th power G^m of G satisfies $\dim G^m/K^m = 2m$ (note that in this case $K = T$) and its weak almost periodic compactification has 2^m idempotents. We also remark that the weak almost periodic compactifications of this example are compact manifolds.

Although the decomposition $S = \phi(G) \cup M(S)$ follows immediately from Theorem 8 it is perhaps instructive to give a direct proof also. To this end, let $s \in S \setminus \phi(G)$ and choose a net $\langle g_n = (z_n, t_n) | n \in D \rangle$ in G with $\lim \phi(g_n) = s$. We may assume w.l.o.g. that $t = \lim t_n$ exists. Since $s \notin \phi(G)$ we have $\lim \| z_n \| = \infty$. Choose a net $\langle h_n | n \in D \rangle$ in T such that

$$\lim \| (h_n - 1) z_n \| = 1;$$

clearly this implies $\lim h_n = 1$ and we may also assume that the limit

$$z = \lim (h_n - 1) z_n$$

exists. Now

$$\lim (z_n, t_n)^{-1} (0, h_n)(z_n, t_n) = \lim (t_n^{-1}(h_n - 1) z_n, h_n) = (\bar{t}^{1} z, 1)$$

and therefore, since the multiplication of S is jointly continuous at $(s,1)$,

$$s = \lim \phi(0, h_n) \phi(z_n, z_n) =$$

$$= \lim \phi(z_n, t_n) \phi((z_n, t_n)^{-1} (0, h_n)(z_n, t_n)) = s\phi((\bar{t}^{1} z, 1)).$$

It follows that $F_R(s)$ is not trivial and, since $N = \{(z,1) | z \in C\}$ is the minimal non-trivial normal subgroup of G, $s = se$, where e is the minimal idempotent in $\overline{\phi(N)}$. Also, $F_R(e) \supset N$, hence, with $H = \{(0,t) | t \in T\}$,

$$eS = e\overline{\phi(G)} = e\overline{\phi(H)} = e\phi(H) = H(e).$$

We conclude $eS = M(S)$ and the assertion follows.

R e f e r e n c e s

1. Berglund, J.F. and K.H. Hofmann: Compact semitopological semigroups and weakly almost periodic functions. Lecture Notes in Mathematics No.42. Springer-Verlag, Berlin-Heidelberg-New York 1967.

2. Bourbaki, N.: General topology I, II. Herrmann; Paris 1966.

3. Brown, G. and W. Moran: Idempotents of compact monothetic semigroups. Proc.London Math. Soc.(3), 22 (1971) 203-216.

4. Chou, C.: On the size of the set of left invariant means on a semigroup.Proc.AMS 23 (1969) 199-205.

5. Clifford, A.H., and G.B. Preston: The algebraic theory of semigroups I, II. Math. Surveys 7. Amer.Math. Soc., Providence, I., 1961.

6. Ellis, R.: Locally compact transformation groups. Duke Math.J. 24 (1957) 119-126.

7. Helgason, S.: Differential geometry and symmetric spaces. Academic Press, New York 1962.

8. Helmer, D.: Joint continuity of affine semigroup actions. Semigroup Forum 21(1980)153 - 165.

9. ———— : Continuity of semigroup actions. Semigroup Forum 23 (1981) 153-188.

10. Hochschild, G.: La structure des groupes de Lie. Dunod, Paris 1968.

11. Lawson, J.D.: Joint continuity in semitopological semigroups. Illinois J.Math. 18(1974) 275-285.

12. ———— : Additional notes on continuity in semitopological semigroups. Semigroup Forum 12 (1976) 265-280.

13. Milnes, P.: On the extension of continuous and almost periodic functions. Pacific J.Math. 56(1975) 187-192.

14. Montgomery, D. and L. Zippin: Topological transformation groups, 2nd ed. Interscience Tracts in Pure and Applied Mathematics 1. Krieger, New York 1974.

15. Namioka, I.: Separate continuity and joint continuity Pac.J.Math. 51 (1974), 515-531.

16. Ruppert, W.: Rechtstopologische Halbgruppen. J.reine angew.Math. 261 (1973), 123-133.

17. ———— : Notes on compact semigroups with identity. Semigroup Forum 14(1977) 199-234, 16(1978) 199-200.

18. ———— : The structure of semitopological monoids on compact connected manifolds. Math. Z.170 (1980) 15-42.

19. ———— : On semitopological compactifications of non-abelian groups. Semigroup Forum 21(1980) 383-389.

20. ———— : On semitopological compactifications of non-abelian groups, to appear in Illinois. J.Math. 26(1982).

21. Ryll-Nardzewsky, R.: On fixed points of semigroups of endomorphisms of linear spaces. Roceedings of the Fifth Berkely Symposium on Mathematical Statistics and Probability (1965) II, Part I. Theory of Probability. Univ. of Calif.Press. Berkely, California.

22. Troallic, J.-P.: Fonctions à valeurs dans des espaces fonctionels généraux: théorèmes de R.Ellis et de I. Namioka. C.R. Acad. Sc. Paris A 287 (1978) 63-66.

23. ———— : Une approche nouvelle de la presque-periodicité faible. Semigroup Forum 22(1981) 247-255.

24. Veech. W.A.: A fixed point theorem - free approach to weak almost periodicity. Trans.Amer.Math. Soc. 177(1973) 353-362.

25. ———— : Weakly almost periodic functions on semi-simple Lie groups. Monatsh. Math. 88(1979) 55-68.

Institut für Mathematik
u. angew. Statistik der
Universität für Boden-
kultur

A-1180 Wien, Gregor Mendelstr.33

SEMIGROUPES SEMITOPOLOGIQUES ET PRESQUE-PERIODICITE

Jean-Pierre Troallic

1. INTRODUCTION

Dans le présent travail, nous établissons - en nous appuyant sur un théorème de I.Namioka [13] - de nouvelles propriétés des semigroupes semitopologiques. Ces propriétés nous permettent de retrouver simplement divers résultats obtenus antérieurement par W.F.Eberlein [6], K.Deleeuw et I.Glicksberg [5], K.Jacobs [9], J.F.Berglund et K.H.Hofmann [2], C.Ryll-Nardzewski [14].

On remarquera que le théorème de point fixe de C.Ryll-Nardzewski - indispensable dans l'approche originale de ces problèmes - n'est pas utilisé ici.

Ce travail prolonge celui présenté dans [16].

2. PRELIMINAIRES

NOTATIONS 2.1. Soient Y un espace topologique et $\mathcal{C}(Y, \mathbb{R})$ l'ensemble des applications continues et bornées de Y dans $\mathbb{R}$. On note $\mathcal{C}_u(Y, \mathbb{R})$ l'espace métrique obtenu en munissant $\mathcal{C}(Y, \mathbb{R})$ de la métrique de la convergence uniforme. On note $\mathcal{C}_s(Y, \mathbb{R})$ l'espace topologique obtenu en munissant $\mathcal{C}(Y, \mathbb{R})$ de la topologie de la convergence simple.

Le résultat suivant de Namioka [13] [15] est l'outil de base de ce travail.

THEOREME 2.2. *Soient X et Y des espaces compacts et $\varphi : X \to \mathcal{C}_s(Y, \mathbb{R})$ une application continue. Alors l'ensemble des points de continuité de l'application $\varphi : X \to \mathcal{C}_u(Y, \mathbb{R})$ est dense dans X.*

REMARQUE 2.3. Le théorème 2.2 peut - de façon équivalente - s'énoncer ainsi : Soient X et Y des espaces compacts et $f : X \times Y \to \mathbb{R}$ une application séparément continue (c'est à dire une application continue en chacune de ses variables). Alors il existe une partie A de X dense dans X telle que f soit continue en tout point de $A \times Y$. En effet, on a la propriété élémentaire suivante :

LEMME 2.4. (cf par exemple [3]). Soient X un espace topologique, Y un espace compact et $\varphi : X \to \mathcal{C}(Y, \mathbb{R})$ une application. Soit $f : X \times Y \to \mathbb{R}$ l'application définie par $f(x,y) = \varphi(x)(y)$. Soit $x \in X$. Alors l'application $\varphi : X \to \mathcal{C}_u(Y, \mathbb{R})$ est continue en x si et seulement si f est continue en tout point de $\{x\} \times Y$.

3. ACTIONS DE SEMIGROUPES SEMITOPOLOGIQUES

Les théorèmes 3.2 et 3.6 de cette section sont essentiels dans ce travail. L'un et l'autre assurent l'existence de points de continuité jointe pour des actions séparément continues de semigroupes. Le théorème 3.2 est un résultat connu de J.D.Lawson [10] [11] ; nous en donnons une démonstration qui simplifie de manière substantielle celle donnée dans [11] .

DEFINITIONS 3.1. a) Un semigroupe est un ensemble non vide S muni d'une loi interne associative. S'il existe dans S un élément neutre, on dit que S est un monoïde.

Un semigroupe semitopologique (resp. topologique) est un semigroupe muni d'une topologie rendant la multiplication séparément continue (resp. continue).

b) Soient S un semigroupe et Y un ensemble. On dit que S agit dans Y si l'on s'est donné une application de $S \times Y$ dans Y, qui à (s,y) associe sy, telle que (st)y = s(ty) pour tous s et t dans S et tout y dans Y.

Supposons que S et Y soient munis d'une topologie. On dit que S agit séparément continûment (resp. continûment) dans Y si l'application de $S \times Y$ dans Y qui à (s,y) associe sy est séparément continue (resp. continue).

THEOREME 3.2. *Soit S un monoïde muni d'une topologie compacte rendant les translations à gauche continues. Soit $\pi : S \times Y \to Y$ une action séparément continue de S dans un espace compact Y. Alors, pour tout élément inversible s de S et tout y de Y, l'application π est continue en (s,y).*

Démonstration. Soient s un élément inversible de S et $y \in Y$. Montrons que Π est continue en (s,y). On peut supposer que pour tout z de Y on a sz = z (la vérification est aisée). Pour tout $f \in \mathcal{C}(Y, \mathbb{R})$, soit X_f l'espace compact obtenu en munissant $\{L_t(f) \mid t \in S\}$ de la topologie de la convergence simple dans Y ($L_t(f)$ étant l'élément de $\mathcal{C}(Y, \mathbb{R})$ défini par $L_t(f)(z) = f(tz)$). Soit A_f l'ensemble des points de continuité de l'application d'inclusion de X_f dans $\mathcal{C}_u(Y, \mathbb{R})$. Il résulte de 2.2 que f est adhérent à A_f dans X_f. Soit A la réunion des A_f. A est dense dans $\mathcal{C}_s(Y, \mathbb{R})$ et pour tout $g \in A$, l'application de $S \times Y$ dans $\mathbb{R}$ qui à (t,z) associe g(tz) est continue en (s,y)(cf 2.4). On en déduit que Π est continue en (s,y).

<u>COROLLAIRE 3.3.</u> <u>Soit</u> S <u>un monoïde semitopologique compact. Alors, pour tout élément inversible</u> s <u>de</u> S, <u>la multiplication est continue en tout point de</u> $(\{s\} \times S) \cup (S \times \{s\})$.

Démonstration. Soit s un élément inversible de S. Le semigroupe S agit séparément continûment dans lui-même par translations à gauche donc, d'après 3.2, la multiplication est continue en tout point de $\{s\} \times S$.

Par symétrie, la multiplication est également continue en tout point de $S \times \{s\}$.

<u>RAPPELS 3.4 (cf par exemple</u> [5]). Soit S un semigroupe semitopologique compact. Alors : a) Il existe dans S au moins un idempotent, c'est à dire un élément a tel que $a^2 = a$.

b) Une partie non vide A de S est un idéal à gauche (resp. à droite) de S si $SA \subset A$ (resp. $AS \subset A$). Un idéal à gauche (resp. à droite) minimal de S est un élément minimal dans l'ensemble ordonné par inclusion des idéaux à gauche (resp. à droite) de S.

Tout idéal à gauche (resp. à droite) de S contient un idéal à gauche (resp. à droite) minimal de S et tout idéal à gauche (resp. à droite) minimal de S est un fermé de S.

c) Tout idéal à gauche (resp. à droite) minimal I de S contient au moins un idempotent a et a est élément neutre à droite (resp. à gauche) de I.

<u>DEFINITION 3.5.</u> Soit S un semigroupe semitopologique. On dit que S est topologiquement simple à gauche (resp. à droite) si pour tout s de S, Ss (resp. sS) est dense dans S.

<u>THEOREME 3.6.</u> *Soient S un semigroupe semitopologique compact et I un idéal à gauche minimal de S. S'il existe dans S un semigroupe topologiquement simple à gauche dense, l'application de S $\times$ I dans I qui à (s,x) associe sx est continue. (L'énoncé "à droite" correspondant en découle par symétrie).*

Démonstration. Soient $f \in \mathcal{C}(I, \mathbb{R})$ et $\varphi : I \longrightarrow \mathcal{C}_s(S, \mathbb{R})$ l'application continue définie par $\varphi(x)(s) = f(sx)$. Montrons que l'application $\varphi : I \longrightarrow \mathcal{C}_u(S, \mathbb{R})$ est continue (cf 2.4.). Soit $\varepsilon > 0$ et soit A_ε l'ensemble des x de I tels que $\{u \mid u \in I, \sup_{s \in S} |\varphi(x)(s) - \varphi(u)(s)| < \varepsilon\}$ soit un voisinage de x dans I. Nous devons démontrer que $A_\varepsilon = I$.

Remarquons tout d'abord que Int A_ε (intérieur de A_ε dans I) est non vide. En effet, $A_{\frac{\varepsilon}{2}}$ est inclus dans Int A_ε (immédiat) et, compte-tenu du théorème 2.2, $A_{\frac{\varepsilon}{2}}$ est non vide.

Soit $x \in I$; montrons que $x \in A_\varepsilon$. Désignons par T un semigroupe topologiquement simple à gauche dense dans S. T est dense dans S et Sx = I, par conséquent, Tx est dense dans I. Soit $t \in T$ tel que tx appartienne à A_ε et soit ω l'ensemble des u de I tels que $\sup_{s \in S} |\varphi(tx)(s) - \varphi(tu)(s)| < \varepsilon$. ω est un voisinage de x dans I et, compte-tenu de ce que S t = S, ω est inclus dans l'ensemble des u de I tels que $\sup_{s \in S} |\varphi(x)(s) - \varphi(u)(s)| < \varepsilon$. Par conséquent, $x \in A_\varepsilon$.

<u>COROLLAIRE 3.7.</u> *Soit S un semigroupe semitopologique compact. Tout idéal à gauche (resp. à droite) minimal I de S est un semigroupe topologique compact.*

Démonstration. En effet, I est un semigroupe semitopologique compact (cf 3.4) et pour tout $s \in I$, on a Is = I (resp. sI = I).

<u>REMARQUES 3.8.</u> a) Il découle de 3.7 (ou de 3.3) que tout groupe semitopologique compact est un groupe topologique. Plus généralement, tout groupe semitopologique localement compact est un groupe topologique (théorème de R.Ellis [7]). Ce théorème s'obtient aisément à partir de 3.3 (cf [10]).

b) La propriété 3.7 est obtenue dans [2] par une autre voie.

4. STRUCTURE DU NOYAU DE SEMIGROUPES SEMITOPOLOGIQUES.

Dans cette section, nous démontrons - en nous appuyant sur le théorème 3.6 - que le noyau d'un semigroupe semitopologique compact S est un groupe topologique compact dès lors que S contient un monoïde topologiquement simple à gauche dense. Cette propriété est établie dans [2] en recourant au théorème de point fixe de C.Ryll-Nardzewski ; au contraire, notre démonstration n'a recours à aucun théorème de point fixe répondant ainsi à une question soulevée par J.F.Berglund et K.H. Hofmann [1] [2] .

Démontrons tout d'abord - en nous appuyant sur le théorème de J.D.Lawson - que la dissymétrie de l'énoncé n'est qu'apparente :

THEOREME 4.1. *Soit S un semigroupe semitopologique compact. Les conditions suivantes sont équivalentes :*

a) *Il existe dans S un monoïde topologiquement simple à gauche dense.*

b) *Il existe dans S un monoïde topologiquement simple à droite dense.*

Démonstration. Démontrons que a) implique b).

Soit T un monoïde topologiquement simple à gauche dense dans S. Par continuité, l'élément neutre e de T est élément neutre de S. Soit $U = \{ s \mid s \in S,\ sS = S \}$. Montrons que le sous-monoïde U de S est dense dans S et topologiquement simple à droite.

Montrons que U est dense dans S. Il suffit de montrer que $T \subset \overline{U}$. Soit $t \in T$. Tt est dense dans T donc dans S. Soit (s_α) une suite généralisée de points de T telle que $e = \lim s_\alpha t$. Pour tout α, on a $Ss_\alpha = S$. Soit $r_\alpha \in S$ tel que $r_\alpha s_\alpha = e$. S étant compact, on peut supposer que la suite généralisée (r_α) converge vers un point r de S. Compte-tenu du théorème 3.3 de J.D.Lawson, on a

$$\lim (r_\alpha s_\alpha t) = t = \lim r_\alpha \lim s_\alpha t = r.$$

Le point t est limite de la suite généralisée (r_α) de points de U ; par conséquent il est dans $\overline{U}$.

Montrons que U est topologiquement simple à droite. Soit $s \in U$. Le sous-monoïde U de S est dense dans S et sS = S, par conséquent on a $\overline{sU} = S$. Il en résulte que sU est dense dans U.

Nous avons démontré que a) implique b). Par symétrie, b) implique a).

DEFINITION 4.2. Soit S un semigroupe. Une partie non vide A de S est un idéal bilatère de S si $SA \subset A$ et $AS \subset A$. L'intersection des idéaux bilatères de S est appelée le noyau de S.

PROPOSITION 4.3. (*cf par exemple* [5]). *Soit S un semigroupe semitopologique compact. S'il existe dans S un seul idéal à gauche minimal I et un seul idéal à droite minimal J, le noyau K de S est un groupe et l'on a I = J = K.*

THEOREME 4.4. *Soit S un semigroupe semitopologique compact. S'il existe dans S un monoïde topologiquement simple à gauche ou à droite dense, le noyau de S est un groupe topologique compact.*

Démonstration. Désignons par T un monoïde topologiquement simple à gauche dense dans S (cf 4.1). Soit e l'élément neutre de T. Par continuité, e est élément neutre de S.

Soit I un idéal à gauche minimal de S et soient u et v des idempotents de I (cf 3.4). Montrons que u = v. Il en résultera que S admet un seul idéal à droite minimal. Soit (t_α) une suite généralisée de points de T telle que $\lim t_\alpha = u$. Pour tout α, on a $St_\alpha = S$. Soit $s_\alpha \in S$ tel que $s_\alpha t_\alpha = e$. S étant compact, on peut supposer que la suite généralisée (s_α) converge dans S. Compte-tenu du théorème 3.6 et de ce que uu = uv (cf 3.4), on a

$$u = (\lim s_\alpha)\ (\lim t_\alpha u) = (\lim s_\alpha)\ (\lim t_\alpha v) = v.$$

Nous avons démontré que S admet un seul idéal à droite minimal. Par symétrie (cf 4.1), S admet également un seul idéal à gauche minimal. Le noyau K de S est donc un groupe (cf 4.3). Munissons K de la topologie induite par S ; K est un groupe topologique compact (cf 3.4 et 3.8).

COROLLAIRE 4.5. *Soit S un semigroupe semitopologique compact. S'il existe dans S un groupe dense, le noyau de S est un groupe topologique compact.*

5. TOPOLOGIES FAIBLE ET FORTE SUR DES SEMIGROUPES D'OPERATEURS.

Soit S un monoide faiblement presque-périodique d'opérateurs sur un espace de Banach E. Le résultat principal de cette section est le suivant : Tout élément de S inversible dans S est un point denté de S dans $\mathcal{L}(E)$ pour la topologie forte des opérateurs. En particulier, compacité faible et compacité forte sont équivalentes pour les groupes d'opérateurs sur E (résultat bien connu de K.Deleeuw et I.Glicksberg [5]).

NOTATIONS ET RAPPELS 5.1. Soit E un espace de Banach et soit $\mathcal{L}(E)$

l'ensemble des opérateurs (linéaires continus) sur E. On note $\mathcal{L}_w(E)$ (resp. $\mathcal{L}_s(E)$) l'espace topologique obtenu en munissant $\mathcal{L}(E)$ de la topologie faible (resp. forte) des opérateurs. Pour les opérations usuelles, $\mathcal{L}_w(E)$ et $\mathcal{L}_s(E)$ sont des espaces localement convexes séparés. L'enveloppe convexe fermée d'une partie de $\mathcal{L}(E)$ est la même dans $\mathcal{L}_w(E)$ et dans $\mathcal{L}_s(E)$. L'enveloppe convexe fermée d'une partie relativement compacte de $\mathcal{L}_w(E)$ (resp. $\mathcal{L}_s(E)$) est une partie compacte de $\mathcal{L}_w(E)$ (resp. $\mathcal{L}_s(E)$).

Pour la composition des opérateurs, $\mathcal{L}_w(E)$ et $\mathcal{L}_s(E)$ sont des semigroupes semitopologiques. Soit S un semigroupe d'opérateurs sur E. Un point x de E est faiblement presque-périodique (resp. presque-périodique) par rapport à S si son orbite $S(x) = \{s(x) \mid s \in S\}$ est faiblement relativement compacte (resp. relativement compacte) dans E. Si tout point de E est faiblement presque-périodique (resp. presque-périodique), on dit que S est faiblement presque-périodique (resp. presque-périodique). Si S est équicontinu, l'ensemble des éléments faiblement presque-périodiques (resp. presque-périodiques) de E est un sous-espace de Banach de E stable par rapport à S. Rappelons enfin que S est faiblement presque-périodique (resp. presque-périodique) si et seulement si son adhérence dans $\mathcal{L}_w(E)$ (resp. $\mathcal{L}_s(E)$) est un semigroupe semitopologique (resp. topologique) compact pour la topologie faible (resp. forte) des opérateurs (cf par exemple [5]).

<u>THEOREME 5.2.</u> <u>Soit</u> S <u>un monoïde faiblement presque-périodique d'opérateurs sur un espace de Banach</u> E <u>et soit</u> s <u>un élément de</u> S <u>inversible dans</u> S. <u>Alors les voisinages faibles et forts de</u> s <u>dans</u> S <u>coïncident</u>.

Démonstration. Tout voisinage faible de s dans S est bien sûr un voisinage fort de s dans S. Soit (s_α) une suite généralisée de points de S convergeant vers s pour la topologie faible des opérateurs. Montrons qu'elle converge vers s pour la topologie forte. Il en résultera que tout voisinage fort de s dans S est un voisinage faible de s dans S. Le monoïde S est équicontinu (d'après le théorème de Banach-Steinhaus), par conséquent, $A = \{\gamma \circ t \mid \gamma \in E',\ \|\gamma\| \leq 1,\ t \in S\}$ est une partie équicontinue du dual topologique E' de E. Soit $\overline{A}$ l'adhérence de A dans E' pour la topologie faible $\sigma(E',E)$. Munissons $\overline{A}$ de la topologie faible ; $\overline{A}$ est un espace compact. Soit $\overline{S}$ l'adhérence de S dans $\mathcal{L}_w(E)$. Munissons $\overline{S}$ de la topologie faible des opérateurs et de la loi interne notée . définie par $s.t = t \circ s$; $\overline{S}$ est un monoïde semitopologique compact. A tout couple (t,θ) de $\overline{S}\times\overline{A}$ associons l'élément $\theta\circ t$ de $\overline{A}$; on définit ainsi une action séparément continue de $\overline{S}$ dans $\overline{A}$. Soit $x\in E$. On a, compte-tenu de 3.2 et 2.4 :

$$\lim_{\alpha}\ \sup_{\theta\in\overline{A}} |(\theta\circ s_\alpha)(x) - (\theta\circ s)(x)| = 0.$$ On en déduit que

$$\lim_{\alpha}\ \sup_{\gamma\in E',\ \|\gamma\|\leq 1} |\gamma(s_\alpha(x) - s(x))| = 0,$$ autrement dit que

$\lim_{\alpha} \| s_\alpha(x) - s(x) \| = 0$ (théorème de Hahn-Banach). La suite généralisée (s_α) converge donc vers s pour la topologie forte des opérateurs.

<u>COROLLAIRE 5.3.</u> <u>Soit</u> G <u>un groupe faiblement presque-périodique d'opérateurs sur un espace de Banach</u>. <u>Alors sur</u> G, <u>les topologies faible et forte des opérateurs coïncident</u>.

<u>COROLLAIRE 5.4.</u> <u>Soit</u> G <u>un groupe d'opérateurs sur un espace de Banach</u>. <u>Si</u> G <u>est compact pour la topologie faible des opérateurs</u>, G <u>est compact pour la topologie forte des opérateurs</u> (<u>et réciproquement</u>).

<u>NOTATIONS ET DEFINITION 5.5.</u> Soit E un espace localement convexe séparé. Pour toute partie A de E, on note co(A) (resp. $\overline{co}$(A)) l'enveloppe convexe (resp. l'enveloppe convexe fermée) de A dans E.

Soit A une partie non vide de E. Un point a de A est un point denté de A dans E si pour tout voisinage V de 0 dans E, $a\notin \overline{co}(A-(a+V))$.

<u>THEOREME 5.6.</u> <u>Soit</u> S <u>un monoïde faiblement presque-périodique d'opérateurs sur un espace de Banach</u> E. <u>Alors tout élément de</u> S <u>inversible dans</u> S <u>est un point denté de</u> S <u>dans</u> $\mathcal{L}_s(E)$.

Démonstration. Soit s un élément de S inversible dans S. $\overline{co}$(S) est un semigroupe semitopologique compact pour la topologie faible des opérateurs (cf 5.1). Il est immédiat que l'élément neutre e de S est élément neutre de $\overline{co}$(S) et que s est un élément de $\overline{co}$(S) inversible dans $\overline{co}$(S).

Vérifions que s est un point extrémal de $\overline{co}$(S) (cette propriété est connue [2]). Soient $s_1, s_2 \in \overline{co}(S)$ et $\alpha\in\,]0,1[$ tels que $s = \alpha s_1 + (1-\alpha)s_2$. Soit t un point de $\overline{co}$(S). On a $s\circ t$ qui est égal à $\alpha(s_1\circ t) + (1-\alpha)(s_2\circ t)$. Si t est un point extrémal de $\overline{co}$(S), $s\circ t$ est un point extrémal de $\overline{co}$(S) et par conséquent, $s_1\circ t = s_2\circ t$. Il en résulte, compte-tenu du théorème de Krein-Milman, que $s_1\circ t$ est égal à $s_2\circ t$ pour tout t de $\overline{co}$(S). En particulier, on a $s_1\circ e = s_2\circ e$, donc on a $s_1 = s_2$.

Montrons que s est un point denté de S dans $\mathcal{L}_s(E)$. Soit V un voisinage de 0 dans $\mathcal{L}_s(E)$. D'après le théorème 5.2, $(s+V) \cap \overline{co}(S)$ est un voisinage de s dans $\overline{co}(S)$ pour la topologie faible des opérateurs, autrement dit, s n'appartient pas à l'adhérence de $\overline{co}(S)-(s+V)$ dans $\mathcal{L}_w(E)$. Il en résulte que $s \notin \overline{co}(\overline{co}(S)-(s+V))$. En effet dans le cas contraire, s serait un point extrémal de $\overline{co}(\overline{co}(S)-(s+V))$ donc un point adhérent à $\overline{co}(S)-(s+V)$ dans $\mathcal{L}_w(E)$ (cf [4] page 108). Le point s de S est par conséquent un point denté de $\overline{co}(S)$ dans $\mathcal{L}_s(E)$ donc un point denté de S dans $\mathcal{L}_s(E)$.

6. PROPRIETES DE DECOMPOSITION

Dans cette section, nous nous appuyons sur les théorèmes 4.4 et 5.4 pour obtenir des propriétés de décomposition du type de celles de W.F.Eberlein [6] , K.Jacobs [9] , K.Deleeuw et I.glicksberg [5] , C.Ryll-Nardzewski [14] . L'approche originale nécessitait le théorème de point fixe de C.Ryll-Nardzewski ; au contraire, notre approche ne fait appel à aucun théorème de point fixe (cf également [17]).

Dans le théorème qui suit, l'élément neutre du monoïde d'opérateurs considéré est supposé être l'application identique.

<u>THEOREME 6.1.</u> <u>Soit</u> S <u>un monoïde faiblement presque-périodique d'opérateurs sur un espace de Banach</u> E <u>et soit</u> A <u>le sous-espace de Banach de</u> E <u>constitué par les éléments de</u> E <u>presque-périodiques par rapport à</u> S. <u>Soit</u> N <u>l'ensemble des</u> x <u>de</u> E <u>tels que</u> 0 <u>soit faiblement adhérent à l'orbite</u> $S(x) = \{ s(x) \mid s \in S \}$. <u>Alors, si</u> S <u>est topologiquement simple à gauche ou à droite pour la topologie faible des opérateurs,</u> N <u>est un sous-espace de Banach de</u> E <u>stable par rapport à</u> S <u>et</u> E <u>est somme directe de</u> A <u>et</u> N.

Démonstration. Soit $\overline{S}$ le semigroupe semitopologique compact obtenu en munissant l'adhérence de S dans $\mathcal{L}_w(E)$ de la topologie faible des opérateurs. Il résulte des théorèmes 4.4 et 5.4 que le noyau K de $\overline{S}$ est un groupe topologique compact pour la topologie forte des opérateurs. Soit e_K le projecteur continu élément neutre de K. Les sous-espaces de Banach Ker e_K et Im e_K de E sont supplémentaires ; par conséquent, pour établir 6.1, il suffit de montrer que N = Ker e_K, que Ker e_K est stable par rapport à S et que A = Im e_K.

a) Démontrons que $N = \text{Ker } e_K$ et que $\text{Ker } e_K$ est stable par rapport à S.

Soient $x \in E$ et $\overline{S(x)}$ l'adhérence faible dans E de $S(x)$. Il est immédiat que $\overline{S(x)} = \overline{S}(x)$. Par conséquent, si $e_K(x) = 0$, $0 \in \overline{S(x)}$. Inversement, supposons que $0 \in \overline{S(x)}$. Soit $s \in \overline{S}$ tel que $s(x) = 0$ et soit k l'inverse dans K de $e_K \circ s$.

On a $e_K(x) = (k \circ (e_K \circ s))(x) = ((k \circ e_K) \circ s)(x) = 0$. Par conséquent, on a $N = \text{Ker } e_K$. Le fait que $\text{Ker } e_K$ soit stable par rapport à S découle de ce que pour tout $s \in S$, on a $s \circ e_K = e_K \circ s$ ($= e_K \circ s \circ e_K$).

b) Démontrons que $A = \text{Im } e_K$.

Soit $x \in E$. e_K étant un projecteur, $x \in \text{Im } e_K$ si et seulement si $e_K(x) = x$.

Supposons que $e_K(x) = x$. On a alors $S(x) \subset \overline{S}(x) = K(x)$. K étant compact pour la topologie forte des opérateurs, $K(x)$ est un compact de E. Par conséquent, $x \in A$. Inversement, supposons que x soit presque-périodique par rapport à S. Le monoïde T d'opérateurs sur A induit par S est presque-périodique. Son adhérence $\overline{T}$ dans $\mathcal{L}_s(A)$ est par conséquent un monoïde topologique compact pour la topologie forte des opérateurs. Soit T_1 (resp. T_2) l'ensemble des éléments de $\overline{T}$ inversibles à gauche (resp. à droite) dans $\overline{T}$. La multiplication dans $\overline{T}$ étant continue, T_1 et T_2 sont des fermés de $\overline{T}$. Compte-tenu de 4.1, T_1 et T_2 sont denses dans $\overline{T}$. Par conséquent, on a $T_1 = T_2 = \overline{T}$, autrement dit, $\overline{T}$ est un groupe. L'application identique de A dans A étant le seul idempotent de $\overline{T}$, on a $e_K(x) = x$.

<u>NOTATIONS ET RAPPELS 6.2.</u> Soient S un monoïde semitopologique et C(S) l'espace de Banach, pour la norme uniforme, des applications continues et bornées de S dans $\mathbb{R}$ (ou $\mathbb{C}$). Pour tout $s \in S$, on note L_s (resp. R_s) l'opérateur de translation à gauche (resp. à droite) sur C(S) défini par $L_s(f)(t) = f(s\,t)$ (resp; $R_s(f)(t) = f(t\,s)$). Une fonction f de C(S) est dite faiblement presque-périodique (resp. presque-périodique) si elle est faiblement presque-périodique (resp. presque-périodique) par rapport au monoïde $\{L_s \mid s \in S\}$ ou - ce qui est équivalent [8] (resp. [12]) - faiblement presque-périodique (resp. presque-périodique) par rapport au monoïde $\{R_s \mid s \in S\}$. On note W(S) (resp. A(S)) le sous-espace de Banach de C(S) constitué par les fonctions faiblement presque-périodiques (resp. presque-périodiques). W(S) et

A(S) sont stables par rapport aux translations L_s et R_s. On a $A(S) \subset W(S)$ On montre également que W(S) et A(S) sont stables pour le produit de fonctions et que $|f| \in W(S)$ (resp. A(S)) dès que $f \in W(S)$ (resp. A(S)).

LEMME 6.3. *Soit S un monoïde semitopologique et soit $f \in W(S)$. Alors, dans l'espace de Banach W(S), 0 est faiblement adhérent à $\{L_s(f) \mid s \in S\}$ (resp. $\{R_s(f) \mid s \in S\}$) si et seulement si $0 \in \overline{co}\{L_s(|f|) \mid s \in S\}$ (resp. $0 \in \overline{co}\{R_s(|f|) \mid s \in S\}$).*

Démonstration. Il suffit bien sûr d'établir la propriété pour les translations à gauche. Remarquons tout d'abord que 0 est faiblement adhérent à $\{L_s(f) \mid s \in S\}$ si et seulement si 0 est faiblement adhérent à $\{L_s(|f|) \mid s \in S\}$.
Soit $K_f = \overline{co}\{L_s(|f|) \mid s \in S\}$. Si 0 est faiblement adhérent à $\{L_s(|f|) \mid s \in S\}$, 0 appartient à K_f d'après le théorème de Mazur. Inversement, supposons que 0 appartienne à K_f. K_f est un convexe faiblement compact de W(S) et 0 est un point extrémal de ce convexe ; par conséquent, 0 est faiblement adhérent à $\{L_s(|f|) \mid s \in S\}$ (cf [4] page 108).

THEOREME 6.4. *Soit S un monoïde semitopologique topologiquement simple à gauche ou à droite. Alors :*

a) Pour tout f de W(S), 0 appartient à $\overline{co}\{L_s(|f|) \mid s \in S\}$ si et seulement si 0 appartient à $\overline{co}\{R_s(|f|) \mid s \in S\}$.

b) Soit N(S) l'ensemble des f de W(S) tels que 0 appartienne à $\overline{co}\{L_s(|f|) \mid s \in S\}$ (ou à $\overline{co}\{R_s(|f|) \mid s \in S\}$). Alors N(S) est un sous-espace de Banach de W(S) stable par rapport aux translations L_s et R_s et W(S) est somme directe de A(S) et N(S).

Démonstration. Soit S_1 (resp. S_2) le monoïde faiblement presque-périodique des translations à gauche (resp. à droite) sur l'espace de Banach W(S). Le monoïde S_i est topologiquement simple à gauche ou à droite pour la topologie faible des opérateurs (i = 1,2). Soit $N_1(S)$ (resp. $N_2(S)$) l'ensemble des f de W(S) tels que 0 appartienne à $\overline{co}\{L_s(|f|) \mid s \in S\}$ (resp. 0 appartienne à $\overline{co}\{R_s(|f|) \mid s \in S\}$). Compte-tenu de 6.1 et 6.3, $N_i(S)$ est un sous-espace de Banach de W(S) stable par rapport à S_i et W(S) est somme directe de A(S) et $N_i(S)$ (i = 1,2).

Pour obtenir 6.4, il reste à démontrer que $N_1(S) = N_2(S)$. Soit $\overline{S_i}$ le semigroupe semitopologique compact obtenu en munissant l'adhérence de S_i dans $\mathcal{L}_w(W(S))$ de la topologie faible des opérateurs. Soit K_i

le noyau de $\overline{S_i}$; K_i est un groupe. Soit e_{K_i} l'opérateur idempotent sur W(S) élément neutre de K_i ; on a

$A(S) = \mathrm{Im}\, e_{K_i} = \{f \mid f \in W(S),\ e_{K_i}(f) = f\}$ et $N_i(S) = \mathrm{Ker}\, e_{K_i}$ (cf 6.1 et 6.3). Montrons que $e_{K_1} = e_{K_2}$. (Il en résultera que $N_1(S) = N_2(S)$).

Tout élément de S_1 commute avec tout élément de S_2 et $\mathcal{L}_w(W(S))$ est un semigroupe semitopologique ; par conséquent e_{K_1} et e_{K_2} commutent.

Si $f \in W(S)$, on a

$$e_{K_1}(f) = (e_{K_2} \circ e_{K_1})(f) = (e_{K_1} \circ e_{K_2})(f) = e_{K_2}(f)$$

En particulier, on retrouve le théorème bien connu suivant [6] [5] [14] :

COROLLAIRE 6.5. *Soit G un groupe semitopologique. Alors :*

a) *Pour tout f de W(G), 0 appartient à* $\overline{co}\{L_g(|f|) \mid g \in G\}$ *si et seulement si 0 appartient à* $\overline{co}\{R_g(|f|) \mid g \in G\}$.

b) *Soit N(G) l'ensemble des f de W(G) tels que 0 appartienne à* $\overline{co}\{L_g(|f|) \mid g \in G\}$ (*ou à* $\overline{co}\{R_g(|f|) \mid g \in G\}$). *Alors N(G) est un sous-espace de Banach de W(G) stable par rapport aux translations* L_g *et* R_g *et W(G) est somme directe de A(G) et N(G).*

REFERENCES

[1] Berglund, J.F., *Problems about semitopological semigroups*, Semigroup Forum 19 (1980), 373-383.

[2] Berglund, J.F., and K.H.Hofmann, *Compact Semitopological semigroups and weakly almost periodic functions*, Lecture Notes in Mathematics 42, Berlin-Heidelberg-New York 1967.

[3] Bourbaki, N., *Topologie Générale*, Chap. 10. Paris, Hermann, 1967 (Act. scient. et ind., 1084 ; Bourbaki, 10).

[4] Choquet, G., *Lectures on analysis*, vol. 2, Benjamin, New York and Amsterdam, 1969.

[5] Deleeuw, K., and I.Glicksberg, *Applications of almost periodic compactifications*, Acta Math. 105 (1961), 63-97.

[6] Eberlein, W.F., *Abstract ergodic theorems and weakly almost periodic functions*, Trans. A.M.S. 67 (1949), 217-240.

[7] Ellis, R., *Locally compact transformation groups*, Duke Math. J.

24 (1957), 119-125.

[8] Grothendieck, A., Critères de compacité dans les espaces fonctionnels généraux, Amer. J. Math. 74 (1952), 168-186.

[9] Jacobs, K., Ergodentheorie und fastperiodische Funktionen auf Halbgruppen, Math. Z. 64 (1956), 298-338.

[10] Lawson, J.D., Joint continuity in semitopological semigroups, Illinois J. Math. 18 (1974), 275-285.

[11] Lawson, J.D., Additional notes on continuity in semitopological semigroups, Semigroup Forum 12 (1976), 265-280.

[12] Loomis, L., An introduction to abstract harmonic analysis, Van Nostrand, New York, 1953.

[13] Namioka, I., Separate continuity and joint continuity, Pacific J. of Math. 51 (1974), 515-531.

[14] Ryll-Nardzewski, C., On fixed points of semigroups of endomorphisms of linear spaces, Proc. Fifth Berkeley Sympos. Math. Statis. and Probability (Berkeley, Calif., 1965-66), Vol. 2, Part 1, 55-61.

[15] Troallic, J.P., Espaces fonctionnels et théorème de I.Namioka, Bull. Soc. Math. France 107 (1979), 127-137.

[16] Troallic, J.P., Une approche nouvelle de la presque-périodicité faible, Semigroup Forum 22 (1981), 247-255.

[17] Veech, W.A., A fixed point theorem-free approach to weak almost periodicity, Trans. A.M.S. 177 (1973), 353-362.

Département de Mathématiques
Faculté des Sciences et des Techniques
Université de Rouen
76130 Mont Saint Aignan
France

Preface to the papers on the algebraic and combinatorial theories of semigroups, delivered at the Conference on Semigroups in Oberwolfach 1981, May 24 through May 30.

This part of the proceedings comprises those papers with an emphasis on the algebraic or combinatorial theories of semigroups. It contains 12 contributions; there were further 28 talks at the conference whose subjects belong into this area; most of the latter have appeared elsewhere or will do so, soon.

The variety of currently existing research directions in the area of "non-topological" semigroups is reflected by the wide range of subjects treated in this part of the conference and the proceedings.

Combining these papers into well-defined groups would be rather artifical; nevertheless, there are, of course, quite a few connections, which we shall try to exhibit in the sequel.
The following remarks should be considered a brief survey of the papers included rather than an evaluation of the current state of the respective areas.

The paper of F. Migliorini and J. Szép presents a new approach to a natural decomposition theory of semigroups, of finite ones, in particular, which is based on earlier work by Szép. Radical theory of monoids is represented here in a paper by L.C.A. van Leeuwen. T. Imaoka's paper deals with fundamental regular ∗-semigroups; the author determines the maximal regular ∗-subsemigroups of a ∗-semigroup and the set of projections of a fundamental regular ∗-semigroup, and he characterizes the minimum right precongruence on a regular ∗-semigroup. J.D.Ph. Meldrum studies regular semigroups of endomorphisms of groups; surprisingly, semigroups of endomorphisms of groups, a most natural subject to consider, have not been treated widely in the literature yet.

Semigroups acting on sets - known as S-acts, S-systems, S- sets, etc. - have always been an important tool in semigroup theory. U. Knauer's paper which extends results due to Skornjakov, Dorofeeva, Kilp, Fountain, Knauer and Petrich is devoted to homological characterizations of monoids by properties of finitely generated S-acts and their right ideals. It is exhibited in the paper by W. Lex that a generalization of S-acts with groupoids instead semigroups leads to results which specialize to semigroups and groups in an interesting manner.
The last two papers as well as the following ones are related to the theory of semigroups of right, say, quotients starting with classical ones up to those consisting of classes of partial S-mappings with respect to right quotient filters, a concept due to C.V.Hinkle (cf. the Survey by H.J. Weinert, Semigroup Forum 19 (1980), 1-78). Thus J.K. Luedeman's paper develops a torsion theory for S - systems of monoids with zero, which exhibits a new construction of the semigroups of quotients, however, restricted to so-called special right quotient filters.
A general concept of the latter ones, comparing different supplementary conditions on right quotient filters, introduced in order to obtain minimal axiom systems as well as counter-examples in this context, is the scope of the paper by U. Hebisch.

In H.J. Weinert's paper the hierarchy of semigroups of right quotients (Q_r-semigroups) of an arbitrary semigroup (S,.) according to his survey article mentioned above is applied to algebras (S,+,.), subject to the assumptions that (S,.) is a semigroup and that the right distributive law (a+b) c = ab+ac holds, only. In fact, the more general subject is an S-groupoid M_S, i.e., a groupoid (M,+) such that the semigroup (S,.) acts on M correspondingly. Then in most cases, each Q_r-semigroup (T,.) of (S,.) has exactly one addition such that (T,+,.) is again an algebra as (S,+,.). This theory developed in different steps includes the well-known ring theoretical constructions of R.E. Johnson and Y. Utumi as special cases for instance, which have been the starting point of all extensions by right quotients beyond the classical ones.

H.J. Bandelt considers free objects in the variety generated by rings and distributive lattices and exhibits a connection between regular semigroups, considerations as given in the papers by Knauer and Weinert.

H. Jürgensen studies a generalization of Bruck-Reilly- extensions of semigroups and characterizes the situation when such a semigroup is totally disjunctive; the results have an interpretation in terms of formal languages. Free monoids and languages of primitive words associated with certain sets of natural numbers are the subject of the paper by A.D. Paradis and G. Thierrin; they characterize the situation when these sets are disjunctive and exhibit properties of the syntactic monoids of these languages; these results allow for a new equivalent formulation of the twin primes conjecture.

Helmut Jürgensen und Hanns Joachim Weinert

FREE OBJECTS IN THE VARIETY GENERATED BY RINGS AND DISTRIBUTIVE LATTICES

Hans-J. Bandelt *

Rings and distributive lattices are semirings with regular and commutative addition, whence they can be regarded as semirings equipped with the unary operation of additive inversion. We shall show that within this framework the least variety containing all rings and distributive lattices is the variety of "semirings which are lattices of rings" in the sense of Bandelt and Petrich [1]. The proof of this is based on a description of the free objects in the latter variety.

A semiring $(S,+,\cdot)$ is a system with associative addition and multiplication connected by distributivity of multiplication over addition. Let $\mathcal{S}$ be the variety of all semirings $(S,+,\cdot,-)$ with an additional unary operation $-$ such that the following axioms are satisfied:

$$a + b = b + a \quad ,$$

$$-a + 2a = a \quad ,$$

$$-2a + a = -a \quad ;$$

here we use the notation na as a shorthand for $a + a + \dots + a$ (n times) and write $-na$ for $n(-a)$. In other words, $(S,+,\cdot,-)$ is a member of $\mathcal{S}$ if and only if $(S,+,\cdot)$ is a semiring and $(S,+,-)$ is a commutative regular semigroup. Any such semigroup is a semilattice of abelian groups, where $-a$ is the additive inverse of $a \in S$ in the additive group it belongs to. The zero of the latter group is the idempotent

$$\bar{a} = -a + a \quad .$$

For every member of $\mathcal{S}$ the following equations hold:

$$\overline{a+b} = \bar{a} + \bar{b} \quad ,$$

$$\overline{ab} = \bar{a}\bar{b} = \bar{a}b = a\bar{b} \quad .$$

* Research partially supported by NSERC Grant A 2985

Certainly, the variety $\mathcal{R}$ of all rings and the variety $\mathcal{D}$ of all distributive lattices are subvarieties of $\mathcal{S}$. As a member of $\mathcal{S}$ every distributive lattice is endowed with the dummy operation $-$, viz. the identity mapping. Henceforth we denote the objects in $\mathcal{S}$ simply by S, R, D, etc. , and do not mention the operations $+, \cdot, -$ explicitly. For the varieties $\mathcal{R}$ and $\mathcal{D}$ their Mal'cev product $\mathcal{R} \circ \mathcal{D}$ (with respect to $\mathcal{S}$) is defined as follows (cf. p.339 of [2]). A semiring $S \in \mathcal{S}$ belongs to $\mathcal{R} \circ \mathcal{D}$ if and only if S admits a congruence ρ such that each ρ-class is a subring and S/ρ is a distributive lattice. The following fact was established in [1]: $\mathcal{R} \circ \mathcal{D}$ is a subvariety of $\mathcal{S}$, given by the three axioms

(A1) $\qquad a\bar{a} = \bar{a}$,

(A2) $\qquad \bar{a}b = b\bar{a}$,

(A3) $\qquad a + \bar{a}b = a$.

The variety $\mathcal{R} \circ \mathcal{D}$ contains all subdirect products of rings and distributive lattices. Examples show that there are subdirectly irreducible members of $\mathcal{R} \circ \mathcal{D}$ which are neither rings nor lattices (see [1]). From the Theorem below one gets that every free object in $\mathcal{R} \circ \mathcal{D}$ is a subdirect product of a (free) ring and a (free) distributive lattice. Consequently, every member of $\mathcal{R} \circ \mathcal{D}$ is a homomorphic image of a subdirect product of a ring and a distributive lattice. Hence, as one would expect, $\mathcal{R} \circ \mathcal{D}$ is the join of $\mathcal{R}$ and $\mathcal{D}$, that is, $\mathcal{R} \circ \mathcal{D}$ is the least variety containing $\mathcal{R}$ and $\mathcal{D}$.

The construction of the free objects can be described as follows. A ring R and a distributive sublattice Δ of its ideal lattice $\mathcal{I}(R)$ give rise to a subobject

$$F = \{(a,\alpha) \mid a \in \alpha \in \Delta\}$$

of the direct product $R \times \Delta$ in $\mathcal{S}$ if one defines operations $+, \cdot, -$ componentwise, i.e.

$$(a,\alpha) + (b,\beta) = (a + b, \alpha + \beta) ,$$

$$(a,\alpha) \cdot (b,\beta) = (ab, \alpha \cap \beta) ,$$

$$-(a,\alpha) = (-a,\alpha) .$$

Now, if R is a free ring and if Δ is the least sublattice of $\mathcal{I}(R)$ containing all principal ideals generated by some free generator of R, then F is indeed a free object in $\mathcal{R} \circ \mathcal{D}$:

<u>THEOREM</u>. <u>Let</u> R <u>be the free ring on the set</u> X <u>and let</u> Δ <u>be the sublattice of the ideal lattice</u> $\mathcal{I}(R)$ <u>generated by all principal ideals</u> (x) <u>of</u> R <u>where</u> $x \in X$. <u>Then</u> $F = \{(w,\alpha) \mid w \in \alpha \in \Delta\}$ <u>is the free object on</u> X <u>in the variety</u> $\mathcal{R} \circ \mathcal{D}$.

<u>COROLLARY</u>. $\mathcal{R} \circ \mathcal{D}$ <u>is the join of</u> $\mathcal{R}$ <u>and</u> $\mathcal{D}$.

We now proceed to the proof of the Theorem. First some notation. The <u>monomials</u> y on X are the elements of the free semigroup X^+ on X and are expressed as products

$$y = \prod_j x_j = x_1 x_2 \dots x_n$$

of <u>indeterminates</u> $x_j \in X$. The free ring R on X is the set of all linear combinations

$$w = \sum_i k_i y_i = \sum_i k_i \cdot \prod_j x_{ij}$$

of monomials $y_i \in X^+$ with non - zero integer coefficients k_i, including the empty sum to represent 0. To each element $w \neq 0$ of R one associates an ideal $\delta w \in \Delta$ as follows. For an indeterminate x let δx be the principal ideal (x), and for a monomial $y = x_1 x_2 \dots x_n$ let δy be the intersection of all δx_j $(j = 1,\dots,n)$. Finally, if $w = \sum k_i y_i$ is any element of $R - \{0\}$, then let δw be the join $\sum \delta y_i$ of the ideals δy_i in the lattice $\mathcal{I}(R)$. It is easily seen that $w \in \delta w$, and moreover, that an element $v \in R$ belongs to the ideal δw if and only if the set of indeterminates of each monomial y of v contains the set of indeterminates of some monomial y_i of w. Therefore the image of $R - \{0\}$ under δ is a sublattice of $\mathcal{I}(R)$, whence $\delta(R - \{0\}) = \Delta$. Since $\delta v \subseteq \delta w$ for $v,w \neq 0$ in R if and only if the set of indeterminates of each monomial occurring in v contains the set of indeterminates of some monomial occurring in w, we see that Δ is nothing else but the free distributive lattice on the set $\{(x) \mid x \in X\}$, the latter being a copy of X.

Since for $w \neq 0$ in R the ideal δw is the least member of Δ containing w, we infer that

$$F = \{(w,\alpha) \in R \times \Delta \mid \text{either } w = 0 \text{ or } \delta w \subseteq \alpha\} .$$

Hence every element (w,α) of F with $w \neq 0$ can be written as

$$(w,\alpha) = (w,\delta w) + (0,\alpha) .$$

The elements $(0,\alpha)$ form an isomorphic copy of Δ, which is (freely) generated by $\{(0,\delta x) \mid x \in X\}$. For each $x \in X$ we have

$$(0,\delta x) = (x,\delta x) + (-x,\delta x) = \overline{(x,\delta x)} .$$

If $w = \sum_i k_i \prod_j x_{ij}$, then by the definition of δ we get

$$(w,\delta w) = \sum_i k_i \prod_j (x_{ij}, \delta x_{ij}) .$$

Therefore F is generated by the set $\{(x,\delta x) \mid x \in X\}$ (with respect to the fundamental operations $+, \cdot, -$).

Now let f be any mapping of $\{(x,\delta x) \mid x \in X\}$ into a member S of $\mathcal{R} \circ \mathcal{D}$. We wish to show that f extends to a homomorphism of F into S. For $w = \sum_i k_i \prod_j x_{ij} \neq 0$ in R we define

$$f(w,\delta w) = \sum_i k_i \prod_j f(x_{ij}, \delta x_{ij}) ,$$

as necessary. Further, for each $x \in X$ we put

$$f(0,\delta x) = \overline{f(x,\delta x)} .$$

Thus f maps the set of free generators of $\{0\} \times \Delta$ into the distributive lattice $\overline{S} = \{\overline{a} \mid a \in S\}$. Since Δ is free, this mapping can be extended to a homomorphism $\alpha \mapsto f(0,\alpha)$ of Δ into $\overline{S}$. Finally, define $f(w,\alpha)$ for $0 \neq w \in \alpha$ by

$$f(w,\alpha) = f(w,\delta w) + f(0,\alpha) .$$

First we have to check that for $\alpha = \delta w$ this does not conflict with the definition of $f(w,\delta w)$ as given above. Since $a \mapsto \overline{a}$ is a homomorphism of S onto the lattice $\overline{S}$, we see that

$$\overline{f(w,\delta w)} = \sum_i \prod_j \overline{f(x_{ij}, \delta x_{ij})} = \sum_i \prod_j f(0,\delta x_{ij})$$

$$= f(0, \sum_i \prod_j \delta x_{ij}) = f(0,\delta w) .$$

Then

$$f(w,\delta w) = f(w,\delta w) + \overline{f(w,\delta w)} = f(w,\delta w) + f(0,\delta w) ,$$

as required. It remains to prove that $f : F \to S$ is a homomorphism. Recall that its restriction to $\{0\} \times \Delta$ is indeed a homomorphism. Now let $v \in \alpha$ and $w \in \beta$ be non-zero elements of R. Denote by u the sum of all monomials ky with integer coefficient k such that y appears in v and w as ky and (-k)y, respectively. Then v and w can be written as $v = v' + u$ and $w = w' - u$, so that $\delta(v + w) = \delta v' + \delta w'$. By the choice of u, all the monomials of v' and w' appear in v + w with some non-zero coefficients. Now an easy computation shows that $f(v+w, \delta(v+w))$ is equal to the sum $f(v',\delta v') + f(w',\delta w')$. Then

$$\begin{aligned} f(v,\alpha) + f(w,\beta) &= f(v,\delta v) + f(w,\delta w) + f(0,\alpha+\beta) \\ &= f(v',\delta v') + f(w',\delta w') + f(u,\delta u) + f(-u,\delta u) + f(0,\alpha+\beta) \\ &= f(v+w, \delta(v+w)) + f(0,\alpha+\beta) \\ &= f(v+w, \alpha+\beta) \quad . \end{aligned}$$

Since additive inverses in S are unique, f is also a homomorphism with respect to the operation - . To see that f preserves multiplication, first observe that

$$f[(v,\delta v)(w,\delta w)] = f(vw,\delta(vw)) = f(v,\delta v) \cdot f(w,\delta w)$$

for monomials v and w. Using distributivity of multiplication over addition, this equality carries over to the case of arbitrary $v,w \neq 0$ in R. From this and the formula $\overline{a}b = \overline{a}\overline{b}$ in S we get

$$\begin{aligned} f(v,\alpha) \cdot f(w,\beta) &= [f(v,\delta v) + f(0,\alpha)] \cdot [f(w,\delta w) + f(0,\beta)] \\ &= f(v,\delta v)f(w,\delta w) + f(0,\alpha)\overline{f(w,\delta w)} + \overline{f(v,\delta v)}f(0,\beta) + f(0,\alpha\beta) \\ &= f(vw,\delta(vw)) + f(0,\alpha\beta) \\ &= f(vw,\alpha\beta) \quad . \end{aligned}$$

This completes the proof that f is a homomorphism. We conclude that F is freely generated by $\{(x,\delta x) \mid x \in X\}$. Since the latter is a copy of X, we have shown that F is the free object on X.

One final remark. The Theorem and the Corollary remain true if, for any integer $n \geqq 2$, we substitute $\mathcal{R}$ by the variety $\mathcal{R}_n$ of all rings whose characteristics divide n. Indeed, the above proof carries over to this case provided that all integer coefficients are read modulo n.

REFERENCES

1. H.-J. Bandelt and M. Petrich, Subdirect products of rings and distributive lattices, Proc. Edinburgh Math. Soc. 25 (1982), 155-171.

2. G. Grätzer, Universal algebra, 2nd ed., Springer - Verlag, 1979.

Fachbereich Mathematik/Informatik
Universität Oldenburg
D-2900 Oldenburg

ON RIGHT QUOTIENT FILTERS OF SEMIGROUPS

U. Hebisch

§ 1 Introduction

In order to generalize considerations about semigroups of right quotients due to McMorris [7], the concept of a right quotient filter $\mathcal{F}$ (briefly Q_r-filter) of a semigroup $S = S^0$ was introduced by Hinkle [4]. According to Botero de Meza [2], a Q_r-filter $\mathcal{F}$ can be defined in the same way for any semigroup S (cf. Def. 2.1), and for each such $\mathcal{F}$ there exists the right quotient semigroup $Q_r(S,\mathcal{F})$ of S with respect to $\mathcal{F}$. For this concept, corresponding statements and generalizations see [2], [4], [8] and [9]. The case of McMorris, called McMorris Q_r-semigroup in [9], is that with the Q_r-filter $\mathcal{F} = S^\Delta$ consisting of all dense right ideals of S.

Sometimes, in particular to transfer results on $Q_r(S,S^\Delta)$ to other $Q_r(S,\mathcal{F})$ (e.g. [4], [5], [6], [9]), one needs the concept of special Q_r-filters (cf. Def. 2.2). In this context it is useful to deal also with weak Q_r-filters (cf. Def. 2.4) and with bases of Q_r-filters (cf. Def. 2.7). In § 2 various characterizations of these concepts are considered and proved to be equivalent. Moreover, necessary and sufficient conditions are given such that a subset $\mathcal{B}$ of the set $\mathcal{S}$ of all right ideals of a semigroup S generates a (weak) Q_r-filter $\mathcal{F} = \langle\langle \mathcal{B} \rangle\rangle$ of S, or even a special one, as a base of $\mathcal{F}$.

According to different definitions of a special Q_r-filter $\mathcal{F}$, certain subsets $\mathcal{F}_a, \mathcal{F}_b, \mathcal{F}_c$ and $\mathcal{F}^{sp}$ of $\mathcal{S}$ are defined for each weak Q_r-filter $\mathcal{F}$ of S and investigated in § 3. In particular, one obtains $\mathcal{F} \subseteq \mathcal{F}_a = \mathcal{F}_c = \langle\langle \mathcal{F}_b \rangle\rangle \subseteq \mathcal{F}^{sp}$, where $\mathcal{F}_a = \mathcal{F}_c$ is again a weak Q_r-filter of S and $\mathcal{F}^{sp}$ the smallest special one containing $\mathcal{F}$. This implies again characterizations for special Q_r-filters. In this context, Thm. 3.2 and Thm. 3.5

contain corresponding results which already have been announced without proofs in Thm. 11.4 of [9].

§ 2 Concepts and equivalent conditions

For an arbitrary semigroup S denote by $\mathcal{S}$ the set of all right ideals of S and for any $A \in \mathcal{S}$ by $\Lambda_S(A,S) = \Lambda(A,S) = \{f\colon A \to S \mid f(as) = f(a)s \text{ for all } a \in A,\ s \in S\}$ the set of all S-mappings on A. Further write $f|B$ for the restriction of any $f \in \Lambda(A,S)$ to a nonempty subset $B \subseteq A$ and $1_A \in \Lambda(A,S)$ for the identical mapping on A. Finally let $\Lambda_i(S,S) = \{\lambda_x \mid x \in S\} \subseteq \Lambda(S,S)$ be the semigroup of all inner left-translations on S, i.e. $\lambda_x(s) = xs$ for all $x,s \in S$. Following Weinert [9] one may use

DEFINITION 2.1 A (nonempty) subset $\mathcal{F} \subseteq \mathcal{S}$ is called right quotient filter (Q_r-filter) of S iff it satisfies

(1) $A \in \mathcal{F},\ A \subseteq B \in \mathcal{S} \Rightarrow B \in \mathcal{F}$,

(2) $A,B \in \mathcal{F} \Rightarrow A \cap B \in \mathcal{F}$ and

(3) $A \in \mathcal{F},\ f \in \Lambda(A,S) \Rightarrow f^{-1}(A) \in \mathcal{F}$.

Since $(f|A\cap B)^{-1}(A\cap B) = (A\cap B) \cap f^{-1}(A\cap B) \subseteq f^{-1}(A\cap B) \subseteq f^{-1}(B)$ and $A \cap B = (1_A)^{-1}(B)$ hold for all $A,B \in \mathcal{S}$, $f \in \Lambda(A,S)$, an equivalent definition for a Q_r-filter $\mathcal{F}$ of S (corresponding to that of Hinkle [3]) is given by (1) and

(4) $A,B \in \mathcal{F},\ f \in \Lambda(A,S) \Rightarrow f^{-1}(B) \in \mathcal{F}$.

The most important Q_r-filters of a semigroup S, according to the notion of [2] or [9], consist of all dense, all $\cap$-large or all weakly $\cap$-large ideals of S defined respectively by

$$S^{\Delta} = \{A \in \mathcal{S} \mid \text{for all } x_1 \neq x_2,\ x \text{ of } S \text{ there is some } s \in S \text{ such that } x_1 s \neq x_2 s \text{ and } xs \in A\} \ (\text{if } S^{\Delta} \neq \emptyset),$$

$$S^{\lambda} = \{A \in \mathcal{S} \mid \text{for all } B \in \mathcal{S} \text{ with } |B| \geq 2 \text{ also } |B \cap A| \geq 2\}$$

and

$$\overline{S} = \{A \in \mathcal{S} \mid xS \cap A \neq \emptyset \text{ for all } x \in S\}.$$

Also the following concept of a special Q_r-filter (sometimes, e.g. in [5], called idempotent) originates from Hinkle [4]:

DEFINITION 2.2 A Q_r-filter $\mathcal{F}$ of S is called special iff it satisfies

(a) $\forall A \in \mathcal{S}$: if there is a $C \in \mathcal{F}$ such that $\lambda_c^{-1}(A) \in \mathcal{F}$ for all $c \in C$ then $A \in \mathcal{F}$

or equivalently (cf. Lemma 2.3)

(b) $\forall C \in \mathcal{F}$: if $A_c \in \mathcal{F}$ for all $c \in C$ then also $\tilde{C} = \cup\{cA_c \mid c \in C\} \in \mathcal{F}$.

In [6], Luedeman established the axiom

(c) $\forall A \in \mathcal{S}$: if there is $C \in \mathcal{F}$ such that $A_c \in \mathcal{F}$ and $cA_c \subseteq A$ for all $c \in C$ then $A \in \mathcal{F}$

to define a special Q_r-filter $\mathcal{F}$ of S. All these definitions are equivalent due to

LEMMA 2.3 *For any nonempty subset $\mathcal{F} \subseteq \mathcal{S}$ the following statements are equivalent:*

i) (1) $\wedge$ (a), ii) (1) $\wedge$ (b), iii) (c).

Proof i) $\Rightarrow$ ii): If $C \in \mathcal{F}$ and $A_c \in \mathcal{F}$ for all $c \in C$, then $\tilde{C} = \cup\{cA_c \mid c \in C\} \in \mathcal{S}$ and $\lambda_c^{-1}(\tilde{C}) \supseteq A_c \in \mathcal{F}$ for all $c \in C$. Therefore (1) gives $\lambda_c^{-1}(\tilde{C}) \in \mathcal{F}$ for all $c \in C$ and (a) implies $\tilde{C} \in \mathcal{F}$ which shows (b).
ii) $\rightarrow$ iii): If $A \in \mathcal{S}$, $C \in \mathcal{F}$ and $A_c \in \mathcal{F}$ such that $cA_c \subseteq A$ for all $c \in C$, then $\tilde{C} = \cup\{cA_c \mid c \in C\} \subseteq A$ and $\tilde{C} \in \mathcal{F}$ because of (b). Since (1) gives $A \in \mathcal{F}$ one gets (c).
iii) $\Rightarrow$ i): Obviously (c) implies $S \in \mathcal{F}$, hence (1) follows from $aS \subseteq A \subseteq B$ for all $a \in A \in \mathcal{F}$, again by (c). Now assume $A \in \mathcal{S}$ and $C \in \mathcal{F}$ such that $\lambda_c^{-1}(A) \in \mathcal{F}$ for all $c \in C$. Then $A_c = \lambda_c^{-1}(A)$ satisfies $cA_c \subseteq A$ for all $c \in C$, hence (c) implies $A \in \mathcal{F}$. This proves (a).

Note that for any semigroup S the Q_r-filters S^Δ (if $S^\Delta \neq \emptyset$) and $\overline{S}$ are always special, whereas S^λ may be special or not; for necessary and sufficient conditions cf. Thm. 11.3 in [9].

Corresponding to the restriction of $\Lambda(A,S)$ to $\Lambda_i(A,S) = \{\lambda_x|A \mid x \in S\} \subseteq \Lambda(A,S)$ Weinert in [10] gave the

DEFINITION 2.4 A (nonempty) subset $\mathcal{F} \subseteq \mathcal{S}$ is called weak Q_r-filter $\mathcal{F}$ of S, iff it satisfies (1), (2) and

(3') $A \in \mathcal{F},\ x \in S \Rightarrow \lambda_x^{-1}(A) = \{s \in S \mid xs \in A\} \in \mathcal{F}$.

(In this context cf. [8], Chap. VI. § 4 and § 5, for similar concepts on rings). Since $(\lambda_x|A)^{-1}(A) = \lambda_x^{-1}(A) \cap A$ holds for all $A \in \mathcal{S}$, $x \in S$, a weak Q_r-filter $\mathcal{F}$ can also be defined by (1), (2) and

(3") $A \in \mathcal{F},\ x \in S \Rightarrow (\lambda_x|A)^{-1}(A) \in \mathcal{F}$.

Clearly, each Q_r-filter $\mathcal{F}$ of S is a weak Q_r-filter of S, but not conversely (cf. Expl. 3.3). Nevertheless, the concepts of a special Q_r-filter and a special weak Q_r-filter, i.e. a weak Q_r-filter satisfying (a), (b) or (c), coincide according to (cf. [1], Prop. 1, in the case $S = S^1$)

PROPOSITION 2.5 Let S be a semigroup. For any (nonempty) subset $\mathcal{F}$ of $\mathcal{S}$, the axioms (3') and (a) together imply all the other axioms (1), (2), (3), (3"), (4), (b), and (c). Moreover, one has

$(3')\wedge(a) \Leftrightarrow (4)\wedge(c) \Leftrightarrow (3)\wedge(c) \Leftrightarrow (3")\wedge(c) \Leftrightarrow (3')\wedge(c)$

and the corresponding equivalences resulting from

$(c) \Leftrightarrow (1)\wedge(a) \Leftrightarrow (1)\wedge(b)$ (cf. Lemma 2.3).

On the other hand, all the axioms (a), (b), (2), (3), (3"), and (4) together do neither imply (1) nor (3').

Proof If $A \in \mathcal{F}$ then (3') shows $\lambda_a^{-1}(A) = S \in \mathcal{F}$ for any $a \in A$. Hence for $A \subseteq B \in \mathcal{S}$ one has $\lambda_a^{-1}(B) = \lambda_a^{-1}(A) = S \in \mathcal{F}$ and so (a) implies (1). Lemma 2.3 gives (b) and (c). To show (4) let $A, B \in \mathcal{F}$ and $f \in \Lambda(A,S)$. From (3') one obtains

$$\lambda_a^{-1}(f^{-1}(B)) = \{s \in S \mid as \in f^{-1}(B)\}$$
$$= \{s \in S \mid f(as) = f(a)s \in B\} = \lambda_{f(a)}^{-1}(B) \in \mathcal{F}$$

for all $a \in A \in \mathcal{F}$. This yields $f^{-1}(B) \in \mathcal{F}$ by (a). From (4), the remaining axioms follow trivially by $(4) \Rightarrow (3) \Rightarrow (3")$.

In particular, one has $(3')\wedge(a) \Rightarrow (4)\wedge(c) \Rightarrow (3)\wedge(c) \Rightarrow (3")\wedge(c)$. Since obviously $(3")\wedge(1) \Rightarrow (3')$, Lemma 2.3 allows to continue those implications by $(3")\wedge(c) \Rightarrow (3')\wedge(c) \Rightarrow (3')\wedge(a)$, which shows the second part of the proposition. The last statement is proved by

Example 2.6 Let $(S,\cdot)$ be the semigroup defined by the left hand multiplication table

$\cdot$	0	a	b	1
0	0	0	0	0
a	0	a	a	a
b	0	a	1	b
1	0	a	b	1

	λ_0	λ_a	λ_b	λ_1
0	0	0	0	0
a	0	a	a	a
b	0	a	1	b
1	0	a	b	1

Clearly, $\mathcal{S} = \{\{0\}, A = \{0,a\}, S\}$ and $\Lambda(S,S) = \{\lambda_0, \lambda_a, \lambda_b, \lambda_1\}$ given by the right hand table above. If one takes $\mathcal{F} = \{A\}$, (1) and (3') fail to be true since $\lambda_0^{-1}(A) = \lambda_a^{-1}(A) = S \notin \mathcal{F}$. Clearly (2) and (3") hold. Since $f(0) = f(00) = f(0)0 = 0$ and $f(a) = f(aa) = f(a)a \in A$ for each $f \in \Lambda(A,S)$, one gets $\Lambda(A,S) = \Lambda_i(A,S)$ and this implies (3) and (4). Finally $\mathcal{F}$ satisfies (a), since $\lambda_0^{-1}(\{0\}) = \lambda_0^{-1}(S) = S \notin \mathcal{F}$, and (b), because $\tilde{C} = 0A \cup aA = A \in \mathcal{F}$ for the only possibility to choose $C = A \in \mathcal{F}$.

Sometimes it is useful to deal with "bases" of (weak) Q_r-filters (cf. [4], [9], [10]).

DEFINITION 2.7 A (nonempty) subset $\mathcal{B} \subseteq \mathcal{S}$ is called a base of a (weak) Q_r-filter $\mathcal{F}$ of S, iff $\mathcal{F} = \langle\langle \mathcal{B} \rangle\rangle$ holds for the set $\langle\langle \mathcal{B} \rangle\rangle = \{A \in \mathcal{S} \mid B \subseteq A \text{ for some } B \in \mathcal{B}\}$.

Now the following statements are straightforward to prove.

PROPOSITION 2.8 *Let S be any semigroup. If $\mathcal{B} \neq \emptyset$ is a subset of $\mathcal{S}$, then*

i) *$\mathcal{B}$ is a base of a Q_r-filter of S iff*

(B2) $A,B \in \mathcal{B} \Rightarrow C \subseteq A \cap B$ *for some* $C \in \mathcal{B}$ *and*

(B3) $A \in \mathcal{B},\ f \in \Lambda(A,S) \Rightarrow C \subseteq f^{-1}(A)$ *for some* $C \in \mathcal{B}$ *hold,*

or equivalently iff

(B4) $A,B \in \mathcal{B},\ f \in \Lambda(A,S) \Rightarrow C \subseteq f^{-1}(B)$ *for some* $C \in \mathcal{B}$ *holds.*

ii) *$\mathcal{B}$ is a base of a weak Q_r-filter of S iff* (B2) *and*

(B3') $A \in \mathcal{B},\ x \in S \Rightarrow C \subseteq \lambda_x^{-1}(A)$ *for some* $C \in \mathcal{B}$ *hold.*

iii) *$\mathcal{B}$ is a base of a special Q_r-filter of S iff* (B3') *and*

(Ba) $A \in \mathfrak{J}$, $C \in \mathcal{B}$ *and* $\lambda_c^{-1}(A) \in \mathcal{B}$ *for all* $c \in C \Rightarrow B \subseteq A$ *for some* $B \in \mathcal{B}$.

Here one may replace (Ba) *by axioms* (Bb) *or* (Bc) *corresponding to* (b) *and* (c) (cf. [10]).

§ 3 Results concerning special Q_r-filters

To consider interrelations between special Q_r-filters of a semigroup S and arbitrary (weak) Q_r-filters of S, firstly we note

LEMMA 3.1 *Each weak* Q_r*-filter* $\mathcal{F}$ *of* S *is contained in a unique smallest special* Q_r*-filter* $\mathcal{F}^{sp}$ *of* S, *given by*

$$\mathcal{F}^{sp} = \cap\{\mathcal{G} \mid \mathcal{G} \text{ is a special } Q_r\text{-filter of } S \text{ and } \mathcal{F} \subseteq \mathcal{G}\}.$$

Clearly, $\mathcal{F}$ *is a special* Q_r*-filter of* S *iff* $\mathcal{F} = \mathcal{F}^{sp}$.

Proof From the proof of Theorem 5.9 in [9] one gets $\mathcal{F} \subseteq \overline{S}$ for the special Q_r-filter $\overline{S}$ of all weakly $\cap$-large right ideals of S. Since the intersection of a family of special Q_r-filters is again a special Q_r-filter (cf. [10], Lemma 3), $\mathcal{F}^{sp}$ as defined above is the unique smallest special Q_r-filter of S satisfying $\mathcal{F} \subseteq \mathcal{F}^{sp}$.

To give more statements in this context, for each weak Q_r-filter $\mathcal{F}$ of a semigroup S we define the following subsets of $\mathfrak{J}$.

$\mathcal{F}_a = \{A \in \mathfrak{J} \mid$ there is $C \in \mathcal{F}$ such that $\lambda_c^{-1}(A) \in \mathcal{F}$ for all $c \in C\}$,

$\mathcal{F}_b = \{\tilde{C} \in \mathfrak{J} \mid$ there are $C \in \mathcal{F}$ and $A_c \in \mathcal{F}$ for all $c \in C$ such that $\tilde{C} = \cup\{cA_c \mid c \in C\}\}$ and

$\mathcal{F}_c = \{A \in \mathfrak{J} \mid$ there are $C \in \mathcal{F}$ and $A_c \in \mathcal{F}$ for all $c \in C$ such that $cA_c \subseteq A\}$.

They clearly correspond to the above characterizations (a), (b) and (c) of a special Q_r-filter such that $\mathcal{F}$ satisfies these axioms iff $\mathcal{F}_a \subseteq \mathcal{F}$ or $\mathcal{F}_b \subseteq \mathcal{F}$ or $\mathcal{F}_c \subseteq \mathcal{F}$ hold, respectively. (In [9], § 11, $\mathcal{F}_b$ was denoted by $\tilde{\mathcal{F}}$.) More statements on these subsets are given in

<u>THEOREM 3.2</u> <u>Let $\mathcal{F}$ be a weak Q_r-filter of a semigroup</u> S. <u>Then $\mathcal{F}_a$ and $\mathcal{F}_c$ coincide and $\mathcal{F}_a$ is again a weak Q_r-filter of</u> S <u>satisfying</u> $\mathcal{F} \subseteq \mathcal{F}_a = \mathcal{F}_c \subseteq \mathcal{F}^{sp}$. <u>But $\mathcal{F}_a = \mathcal{F}_c$ need not be a Q_r-filter of</u> S, <u>even not if $\mathcal{F}$ is one.</u>

<u>In general, $\mathcal{F}_b$ is neither a weak Q_r-filter of</u> S <u>nor $\mathcal{F}_b$ contains $\mathcal{F}$, but $\mathcal{F}_b$ is a base of the weak Q_r-filter</u> $\mathcal{F}_a = \mathcal{F}_c$.

<u>In particular, $\mathcal{F}$ is a special Q_r-filter of</u> S <u>iff</u> $\mathcal{F} = \mathcal{F}_a$ <u>or</u> $\mathcal{F} = \mathcal{F}_c$ <u>or</u> $\mathcal{F} = \langle\langle \mathcal{F}_b \rangle\rangle$.

<u>Proof</u> If $A \in \mathcal{F}_a$, there is $C \in \mathcal{F}$ such that $\lambda_c^{-1}(A) \in \mathcal{F}$ for all $c \in C$. Then $A_c = \lambda_c^{-1}(A)$ is in $\mathcal{F}$ and therefore $cA_c \subseteq A$ for all $c \in C$, hence $A \in \mathcal{F}_c$. Conversely, let be $C \in \mathcal{F}$ and $A_c \in \mathcal{F}$ for all $c \in C$ such that $cA_c \subseteq A$ for some $A \in \mathcal{S}$. Now $\lambda_c^{-1}(A) \supseteq A_c \in \mathcal{F}$ implies $\lambda_c^{-1}(A) \in \mathcal{F}$ for all $c \in C$. Hence $\mathcal{F}_a = \mathcal{F}_c$ is proved.

Clearly, $\mathcal{F}_a$ satisfies (1). To prove (2) for $\mathcal{F}_a$ let be $A_1, A_2 \in \mathcal{F}_a$ and $C_1, C_2 \in \mathcal{F}$ such that $\lambda_c^{-1}(A_i) \in \mathcal{F}$ for all $c \in C_i$. Then one has $C = C_1 \cap C_2 \in \mathcal{F}$ and $\lambda_c^{-1}(A_1 \cap A_2) = \lambda_c^{-1}(A_1) \cap \lambda_c^{-1}(A_2) \in \mathcal{F}$ for all $c \in C$. Finally, if $A \in \mathcal{F}_a$ and $C \in \mathcal{F}$ such that $A_c \in \mathcal{F}$ and $cA_c \subseteq A$ for all $c \in C$, define $D = \lambda_x^{-1}(C) \in \mathcal{F}$ for any $x \in S$ and $A_d = A_c$ if $c = xd \in C$ for $d \in D$. Then one gets $A_d \in \mathcal{F}$ and $xdA_d = cA_c \subseteq A$ which shows $dA_d \subseteq \lambda_x^{-1}(A)$ for all $d \in D$. Hence $\lambda_x^{-1}(A) \in \mathcal{F}_a$ and $\mathcal{F}_a$ satisfies (3'), i.e. $\mathcal{F}_a$ is a weak Q_r-filter of S.

$\mathcal{F} \subseteq \mathcal{F}_a$ comes from $\lambda_a^{-1}(A) = S \in \mathcal{F}$ for all $a \in A \in \mathcal{F}$. Obviously $\mathcal{F} \subseteq \mathcal{F}_a \subseteq \mathcal{G}$ holds for any special Q_r-filter $\mathcal{G}$ of $\mathcal{S}$ satisfying $\mathcal{F} \subseteq \mathcal{G}$. This implies $\mathcal{F}_a \subseteq \mathcal{F}^{sp}$.
For each $\tilde{C} = \cup\{cA_c \mid c \in C\} \in \mathcal{F}_b$ such that $C \in \mathcal{F}$, $A_c \in \mathcal{F}$ one has $cA_c \subseteq \tilde{C}$ for all $c \in C$. Therefore $\lambda_c^{-1}(\tilde{C}) \supseteq \lambda_c^{-1}(cA_c) \supseteq A_c \in \mathcal{F}$ for all $c \in C \in \mathcal{F}$ shows $\tilde{C} \in \mathcal{F}_a$ and so $\langle\langle \mathcal{F}_b \rangle\rangle \subseteq \langle\langle \mathcal{F}_a \rangle\rangle = \mathcal{F}_a$ holds.
Conversely, for any $A \in \mathcal{F}_a$ there are $C \in \mathcal{F}$ and $A_c \in \mathcal{F}$ for all $c \in C$ such that $cA_c \subseteq A$. Now $\tilde{C} = \cup\{cA_c \mid c \in C\} \subseteq A$ and $\tilde{C} \in \mathcal{F}_b$ imply $A \in \langle\langle \mathcal{F}_b \rangle\rangle$, thus $\langle\langle \mathcal{F}_b \rangle\rangle = \mathcal{F}_a$.

The last statement of Theorem 3.2 is obvious and the remaining negative statements are proved by the following

Example 3.3 Consider the subsemigroup $S = \{2^n \mid n \in \mathbb{N}\}$ of $(\mathbb{N},\cdot)$. As for each semigroup, $\mathcal{F} = \{S\}$ is a Q_r-filter of S. One easily checks $\mathcal{F}_a = \mathcal{F}_c = \{2S, S\}$ and $\mathcal{F}_b = \{2S\}$, and that, according to the general statement above, $\mathcal{F}_a = \mathcal{F}_c = \langle\langle \mathcal{F}_b \rangle\rangle$ is a weak Q_r-filter of S. Thus $\mathcal{F}_b$ is not even a weak Q_r-filter of S, since $\mathcal{F}_b \subset \mathcal{F}_a$, and does not contain $\mathcal{F}$. Finally, since $f\colon 2S \to S$ defined by $f(x) = \frac{x}{2} \in S$ belongs to $\Lambda(2S,S)$ and satisfies $f^{-1}(2S) = 4S \notin \mathcal{F}_a = \mathcal{F}_c$, the latter violates (3); hence $\mathcal{F}_a = \mathcal{F}_c$ is not a Q_r-filter of S.

In this context it should be noted, that the mapping $\mathcal{F} \to \mathcal{F}_a$ may yield an infinite chain of weak Q_r-filters of a semigroup S,

$$\mathcal{F} \subset \mathcal{F}_a \subset (\mathcal{F}_a)_a \subset \dots ,$$

even if the weak Q_r-filter $\mathcal{F}$ contains a special Q_r-filter $\mathcal{K}$ of S. This is shown by

Example 3.4 Consider $S = S^1 = \{2^n \mid n \in \mathbb{N}_o\}$ as a subsemigroup of $(\mathbb{N},\cdot)$. Then $\mathcal{K} = \{S\}$ is a Q_r-filter of S and a special one by $\mathcal{K}_b = \{\cup\{sS \mid s \in S\}\} = \{S\} = \mathcal{K}$. Straightforward computations show that $\mathcal{F} = \{2S, S\}$ is a weak Q_r-filter of S, yielding the weak Q_r-filters

$$\mathcal{F}^{(m)} = (\mathcal{F}^{(m-1)})_a = \{2^{2^{m+1}-1}S, \dots, S\},$$

hence the ascending chain $\mathcal{K} \subset \mathcal{F} \subset \mathcal{F}_a \subset \dots \subset \mathcal{F}^{sp} = \bigcup_{m \in \mathbb{N}} \mathcal{F}^{(m)}$.

Coming back to the so called standard Q_r-filters S^Δ (if $S^\Delta \neq \emptyset$), S^λ and $\overline{S}$ of a semigroup S, it was already mentioned that only S^λ need not be special. But, contrasting the general situation described in Theorem 3.2, one has

THEOREM 3.5 Let $\mathcal{F} = S^\lambda$ be the Q_r-filter consisting of all $\cap$-large right ideals of a semigroup S. Then $\mathcal{F}_a = \mathcal{F}_c$ is a Q_r-filter of S.

Proof Since $\mathcal{F}_a = \langle\langle \mathcal{F}_b \rangle\rangle$, it remains to proove (B3) of Proposition 2.8 for $\mathcal{B} = \mathcal{F}_b$. So take $A = \{cA_c \mid c \in C\} \in \mathcal{F}_b$ for any $C \in \mathcal{F}$, $A_c \in \mathcal{F}$ and $f \in \Lambda(A,S)$. There exists (cf. [9], Thm. 5.14) a maximal extension $\hat{f} \in \Lambda(\hat{A},S)$ of f such that $A \subseteq \hat{A} \in S^\lambda = \mathcal{F}$ and $\hat{f}|A = f$. Now $C \cap \hat{A} \in \mathcal{F}$ and $B = \hat{f}^{-1}(C \cap \hat{A}) \cap C \in \mathcal{F}$.

Therefore $B \subseteq \hat{A}$ and $\hat{f}(b) = c \in C$ for all $b \in B$. Using this, one can define $A_b = A_{\hat{f}(b)} = A_c \in \mathcal{F}$ for each $b \in B$. Since $B \subseteq C$, it follows $\tilde{B} = \cup \{bA_b \mid b \in B\} \in \mathcal{F}_b$ and $\tilde{B} \subseteq A$. Thus

$$f(\tilde{B}) = \hat{f}(\tilde{B}) = \cup \{\hat{f}(b)A_b \mid b \in B\}$$
$$= \cup \{cA_c \mid c = \hat{f}(b),\ b \in B\} \subseteq \cup \{cA_c \mid c \in C\} = A$$

implies $f^{-1}(A) \supseteq \tilde{B} \in \mathcal{F}_b$ and (B3) is proved.

REFERENCES

[1] Allouch, D., Filtre sur un monoide fini, Semigroup Forum 18 (1979), 27-32.

[2] Botero de Meza, M.M., Zur Theorie allgemeiner Rechtsquotientenhalbgruppen und Rechtsquotientenhalbringe, Dissertation, Technische Universität Clausthal 1975.

[3] Hinkle jr., C.V., Semigroups of right quotients of a semigroup which is a semilattice of groups, Semigroup Forum 5 (1972), 167-173.

[4] Hinkle jr., C.V., Generalized semigroups of quotients, Trans. Amer. Math. Soc. 183 (1973), 87-117.

[5] Luedeman, J., The generalized translational hull of a semigroup, Semigroup Forum 9 (1974), 76-83.

[6] Luedeman, J., Torsion theories and semigroups of quotients, Dept. Math. Sciences, Clemson Univ. 1979.

[7] McMorris, F.R., The maximal quotient semigroup, Semigroup Forum 4 (1972), 360-364.

[8] Stenström, B., Rings of Quotients, Springer, New York 1975.

[9] Weinert, H.J., S-sets and semigroups of quotients, Semigroup Forum 19 (1980), 1-78.

[10] Weinert, H.J., On special right quotient filters of semigroups, Lecture Notes in Mathematics 855, Semigroups, 211-221, Springer, New York 1981.

Institut für Mathematik
Technische Universität Clausthal
D-3392 Clausthal-Zellerfeld, Germany

SOME REMARKS ON FUNDAMENTAL REGULAR *-SEMIGROUPS

Teruo Imaoka

In this paper, firstly we shall determine a maximal regular *-subsemigroup of a given *-semigroup. Secondly, we shall characterize the set of projections of a fundamental regular *-semigroup by introducing a P-groupoid. Finally, we shall study a precongruence on a regular *-semigroup S which is introduced by McAlister [4]. We shall show that the minimum right precongruence on S is the *-equivalence generated by $\{(a,b) \in S\times S: V(a) \cap V(b) \neq \square\}$.

1. Introduction

A semigroup S with a unary operation $*: S \to S$ is called a *-*semigroup* if it satisfies

(i) $$(x^*)^* = x,$$

(ii) $$(xy)^* = y^*x^*.$$

If a *-semigroup S satisfies

(iii) $$xx^*x = x,$$

it is called a *regular* *-*semigroup*. Let S be a *-semigroup. An idempotent e in S is called a *projection* if $e^* = e$. By P(S) we denote the set of all projections of S. Let θ be a relation on S. If $(a,b) \in \theta$ implies $(a^*,b^*) \in \theta$, we call it a *-*relation* on S. A (semigroup) homomorphism $\Phi: S \to T$ is called a *-*homomorphism* if $x^*\Phi = (x\Phi)^*$ for all $x \in S$. The following result is basic and we use it frequently throughout this paper.

RESULT 1.1. ((i) due to [5] and the others due to [3]). *Let S be a regular *-semigroup, and let P and E be the sets of projections and idempotents, respectively, of S. Then we* have the following conclusions:

(i) each L-class and each R-class contain one and only one projection,

(ii) $E = P^2$,

(iii) a^*ea is a projection for any $a \in S$ and $e \in P$,

(iv) the maximum idempotent-separating congruence μ on S is also a *-congruence, and μ is given by

$$\mu = \{(a,b) \in S\times S: a^*ea = b^*eb \text{ and } aea^* = beb^* \text{ for all } e \in P\}.$$

The notation and terminology are those of [2] and [3], unless otherwise stated.

2. Maximal regular *-subsemigroups

The following theorem is the analogue of Theorem 14 [1].

THEOREM 2.1. Let S be a *-semigroup with the set of projections P, and let Q be a subset of P. Then we have the following conclusions:

(i) S has a regular *-subsemigroup with Q as its set of projections if and only if $efe \in Q$ and ef is an idempotent for all $e,f \in Q$.

(ii) In that case, there exists the largest such regular *-subsemigroup, namely, $Q^c = \{a \in S: aa^*a = a$ and $aa^*, a^*a, aea^*, a^*ea \in Q$ for all $e \in Q\}$.

Proof. To see (i), it is sufficient to show that $\langle Q \rangle$, the *-subsemigroup generated by Q, is a regular *-subsemigroup if $efe \in Q$ and ef is an idempotent for all $e,f \in Q$. Let $x = e_1e_2 \ldots e_n$, $e_i \in Q$, be any element of $\langle Q \rangle$. We shall show that $xx^*x = x$ and $xx^* \in Q$ by using induction on n. It is obvious that the statement is true for $n = 1$. Let $n > 1$ and assume that the statement is true for $n-1$. Then

$$\begin{aligned} x &= e_1e_2\ldots e_n \\ &= e_1(e_2\ldots e_n)(e_n\ldots e_2)(e_2\ldots e_n) \\ &= e_1(e_2\ldots e_n)(e_n\ldots e_2)e_1(e_2\ldots e_n)(e_n\ldots e_2)(e_2\ldots e_n), \end{aligned}$$

since $(e_2 \dots e_n)(e_n \dots e_2) \in Q$, and hence $e_1(e_2 \dots e_n)(e_n \dots e_2)$ is an idempotent,

$$= e_1(e_2 \dots e_n)(e_n \dots e_2)e_1 e_1(e_2 \dots e_n)$$
$$= xx^*x,$$

and

$$xx^* = e_1(e_2 \dots e_n)(e_n \dots e_2)e_1 \in Q.$$

Thus, $\langle Q \rangle$ is a regular *-subsemigroup of S.

To see (ii), it is sufficient to show that Q^C is closed under the multiplication of S. Let a and b be any elements of Q^C. Since a^*abb^* is an idempotent, we have

$$ab(ab)^*ab = abb^*a^*ab = a(a^*abb^*)(a^*abb^*)b$$
$$= a(a^*abb^*)b = ab.$$

By the definition of Q^C, it is clear that $ab(ab)^*$, $(ab)^*ab$, $abe(ab)^*$ and $(ab)^*eab$ are all contained in Q for all $e \in Q$, and hence Q^C is closed. Thus we have the theorem.

3. Fundamental regular *-semigroups

The author has given a construction of a fundamental regular *-semigroup in [3]. In his paper [6], Yamada has given another construction of a fundamental regular *-semigroup and characterized its set of idempotents and projections by introducing the concepts of "a fundamental regular warp" and "a P-set". In this section, we shall give another characterization of its set of projections.

Let P be a set. Let θ be a mapping of P into $\mathcal{T}_P$, the full transformation semigroup on P, such that θ satisfies the following axioms:

(P1) $e\theta_e = e$,

(P2) $\theta_e\theta_e = \theta_e$,

(P3) $e\theta_f\theta_e = f\theta_e$,

(P4) $\theta_e\theta_f\theta_e = \theta_{f\theta_e}$,

(P5) $\theta_e\theta_f\theta_e\theta_f = \theta_e\theta_f$,

where θ_e is the image of $e(\in P)$ by θ. Define a partial product on P by

$$ef = \begin{cases} e\theta_f & \text{if } e\theta_f = f\theta_e, \\ \text{undefined} & \text{otherwise.} \end{cases}$$

We call such a groupoid P a P-groupoid with respect to θ, or simply say that $(P;\theta)$ is a P-groupoid. It immediately follows from (P1) that ee is always defined in P and $e^2 = e$. From (P2), (P4) and (P5) one obtains: if ef is defined in P then fe is also defined, $ef = fe$ and $\theta_e\theta_f = \theta_f\theta_e = \theta_{ef}$.

REMARK. Let θ satisfy (P1), (P4) and (P5). Then θ is injective if and only if θ satisfies (P3).

LEMMA 3.1. *Let e, f and g be elements of a P-groupoid $(P;\theta)$ such that ef and fg are defined in P. If either (ef)g or e(fg) is defined in P, so is the other and $(ef)g = e(fg)$. In that case, we simply denote (ef)g by efg.*

Proof. Suppose that ef, fg and (ef)g are defined in P. Then $ef = e\theta_f$, $fg = g\theta_f$ and $ef\theta_g = g\theta_{ef}$. Now,

$$e\theta_{fg} = e\theta_f\theta_g = (ef)\theta_g = g\theta_{ef} = g\theta_f\theta_e = fg\theta_e.$$

Thus e(fg) is also defined in P, and $e(fg) = e\theta_{fg} = e\theta_f\theta_g = (ef)g$.

Next, we define a relation $\leq$ on P by

$$e \leq f \iff ef \text{ is defined in P and } ef = fe = e,$$
$$\text{that is, } e\theta_f = f\theta_e = e.$$

Then $\leq$ is a partial order on P. For, it is obvious that $\leq$ is reflexive and anti-symmetric. Suppose that $e \leq f$ and $f \leq g$. Then ef, fg and e(fg) are defined in P, and so is (ef)g. Thus, $eg = (ef)g = e(fg) = ef = e$. Similarly, we have $ge = e$. Then $e \leq g$, and hence $\leq$ is transitive. For each $e \in P$, let $\langle e\rangle = \{f \in P : f \leq e\}$. Then we can easily see that $\langle e\rangle = P\theta_e$.

Let A and B be subsets of P-groupoids $(P;\theta)$ and $(Q;\eta)$, respectively. For $x, y \in A$, "$xy \in A$" means "xy is defined in P and xy is contained in A". A mapping $\alpha: A \to B$ is called a P-homomorphism if for $x,y \in A$, (i) $xy \in A$ implies $(x\alpha)(y\alpha) \in B$ and $(xy)\alpha = (x\alpha)(y\alpha)$, and (ii) $(x\theta_y)\alpha = (x\alpha)\eta_{y\alpha}$. If a P-homomorphism $\alpha: A \to B$ is bijective and $(x\alpha)(y\alpha) \in B$ implies $xy \in A$, we call α a P-isomorphism. In that case, we say that A is P-isomorphic to B, and denote it by $A \simeq B$.

For a P-groupoid $(P;\theta)$, let $\mathcal{U} = \{(e,f) \in P\times P: \langle e\rangle \simeq \langle f\rangle\}$. For each $(e,f) \in \mathcal{U}$, let $T_{e,f}$ be the set of all P-isomorphisms of $\langle e\rangle$ onto $\langle f\rangle$, and let $T_{(P;\theta)} = \{(\theta_e\alpha, \theta_f\alpha^{-1}): \alpha \in T_{e,f}, (e,f) \in \mathcal{U}\}$. Let us define a multiplication and a unary operation on $T_{(P;\theta)}$ by

$$(\theta_e\alpha, \theta_f\alpha^{-1})(\theta_g\beta, \theta_h\beta^{-1}) = (\theta_e\alpha\theta_g\beta, \theta_h\beta^{-1}\theta_f\alpha^{-1}),$$

$$(\theta_e\alpha, \theta_f\alpha^{-1})^* = (\theta_f\alpha^{-1}, \theta_e\alpha).$$

Then it is clear that $T_{(P;\theta)} \subset \mathcal{T}_P \times \mathcal{T}_P^{op}$, where $\mathcal{T}_P^{op}$ is the dual semigroup of $\mathcal{T}_P$. We shall show that $T_{(P;\theta)}$ is a fundamental regular *-semigroup. Let $(\theta_e\alpha, \theta_f\alpha^{-1})$ and $(\theta_g\beta, \theta_h\beta^{-1})$ be any elements of $T_{(P;\theta)}$. Then $\mathrm{ran}(\theta_e\alpha) = P\theta_f$, and $\mathrm{ran}(\theta_e\alpha\theta_g\beta) = P\theta_f\theta_g\beta$. Now,

$$P\theta_f\theta_g = P\theta_f\theta_g\theta_f\theta_g \subset P\theta_g\theta_f\theta_g \subset P\theta_f\theta_g.$$

Then $\mathrm{ran}(\theta_e\alpha\theta_g\beta) = P\theta_{f\theta_g}\beta = P\theta_{f\theta_g\beta}$. Let $i = g\theta_f\alpha^{-1}$, $j = f\theta_g\beta$ and $\gamma = (\alpha|\langle i\rangle)(\theta_g|\langle g\theta_f\rangle)(\beta|\langle f\theta_g\rangle)$. Then it is clear that γ is a P-isomorphism of $\langle i\rangle$ onto $\langle j\rangle$. We shall show that $\theta_e\alpha\theta_g\beta = \theta_i\gamma$. For any $x \in P$,

$$\begin{aligned}
x\theta_i\gamma &= x\theta_{ei}\gamma && \text{since } i \leq e,\\
&= (x\theta_e)\theta_i(\alpha|\langle i\rangle)(\theta_g|\langle g\theta_f\rangle)(\beta|\langle f\theta_g\rangle)\\
&= (x\theta_e\alpha)\theta_{i\alpha}(\theta_g|\langle g\theta_f\rangle)(\beta|\langle f\theta_g\rangle)\\
&= (x\theta_e\alpha)\theta_{g\theta_f}\theta_g(\beta|\langle f\theta_g\rangle)\\
&= (x\theta_e\alpha)\theta_f\theta_g\theta_f\theta_g\beta\\
&= (x\theta_e\alpha)\theta_f\theta_g\beta\\
&= x\theta_e\alpha\theta_g\beta, && \text{since } x\theta_e\alpha \leq f.
\end{aligned}$$

Thus we have $\theta_e\alpha\theta_g\beta = \theta_i\gamma$. Similarly, $\theta_h\beta^{-1}\theta_f\alpha^{-1} = \theta_j\gamma^{-1}$, and hence $(\theta_e\alpha, \theta_f\alpha^{-1})(\theta_g\beta, \theta_h\beta^{-1}) = (\theta_i\gamma, \theta_j\gamma^{-1}) \in T_{(P;\theta)}$.

Since $T_{(P;\theta)} \subset T_P \times T_P^{op}$, $T_{(P;\theta)}$ is a semigroup. It is clear that $T_{(P;\theta)}$ is a regular *-semigroup with $\{(\theta_e, \theta_e): e \in P\}$ as its set of projections. By Result 1.1 (iv) and the fact that θ is injective, we can easily see that $T_{(P;\theta)}$ is fundamental. Thus we have the first part of the following theorem.

THEOREM 3.2. *Let* $(P;\theta)$ *be a P-groupoid. Then* $T_{(P;\theta)}$, *constructed above, is a fundamental regular *-semigroup with* $\{(\theta_e, \theta_e): e \in P\} = Q$, *say, as its set of projections. If we define a mapping* $\eta: Q \to T_Q$ *by*

$$(\theta_e, \theta_e)\eta_{(\theta_f, \theta_f)} = (\theta_{e\theta_f}, \theta_{e\theta_f}),$$

then $(Q;\eta)$ *is a P-groupoid and it is P-isomorphic to* $(P;\theta)$.

*Conversely, any fundamental regular *-semigroup is *-isomorphic to a full regular *-subsemigroup of* $T_{(P;\theta)}$ *for some P-groupoid* $(P;\theta)$.

Proof. It is obvious that $(Q;\eta)$ is P-isomorphic to $(P;\theta)$. We shall prove the second half of the theorem. Let S be a regular *-semigroup with the set of projections P. For each $e \in P$, define a mapping $\theta_e: P \to P$ by

$$f\theta_e = efe \quad \text{for any } f \in P.$$

It is clear that $(P;\theta)$ becomes a P-groupoid. For any $a \in S$, let ϱ_a and λ_a be mappings of P into P defined by

$$e\varrho_a = a^*ea,$$
$$e\lambda_a = aea^*.$$

It is clear that $\varrho_{ab} = \varrho_a\varrho_b$, $\lambda_{ab} = \lambda_b\lambda_a$ and that $\varrho_e = \lambda_e = \theta_e$ for all $e \in P$. For $a \in S$, we denote aa^* and a^*a by e and f, respectively. Then it is clear that mappings $\alpha: x \mapsto a^*xa$ and $\alpha': y \mapsto aya^*$ are mutually inverse P-isomorphisms of $\langle e\rangle$ onto $\langle f\rangle$ and $\langle f\rangle$ onto $\langle e\rangle$, respectively. Hence $(\varrho_a, \lambda_a) = (\theta_e\alpha, \theta_f\alpha^{-1}) \in T_{(P;\theta)}$. Since $\varrho_{ab} = \varrho_a\varrho_b$ and $\lambda_{ab} = \lambda_b\lambda_a$, a mapping $\xi: a \mapsto (\varrho_a, \lambda_a)$ is a *-homomorphism of S into $T_{(P;\theta)}$. Assume that $a\xi = b\xi$. Then $a^*ea = b^*eb$ and $aea^* = beb^*$ for all $e \in P$. By the Result 1.1 (iv), we have that $a \mu b$. Since S is fundamental, we have $a = b$. It is obvious that $S\xi$ is a full *-subsemigroup of $T_{(P;\theta)}$.

4. Precongruences

Let S be a regular *-semigroup. Let $\sigma = \{(a,b) \in S\times S: ab^*a = a$ and $ba^*b = b\}$. It is clear that σ is a reflexive and symmetric *-relation on S. By π_0 we denote the equivalence on S generated by $\{(a,b) \in S\times S: V(a) \cap V(b) \neq \Box\}$, where V(a) is the set of all inverses of a.

LEMMA 4.1. (i) π_0 *is the transitive closure of* σ.
(ii) *For any* $a \in S$ *and* $a' \in V(a)$, *we have that* $aa^* = aa'(a')^*a$.
(iii) *For* $a,x,y \in S$ *and* $a' \in V(a)$, $(xaa', yaa') \in \sigma$ *implies* $(xa,ya) \in \sigma$.
(iv) For $a,x,y \in S$ *and* $a' \in V(a)$, $(xaa',y) \in \sigma$ *implies* $(xaa', y(a')^*a') \in \sigma$.

Proof. (i) It is obvious that $\sigma^t \subset \pi_0$. Let a and b be elements of S such that $V(a) \cap V(b) \neq \Box$. Then there exists c in $V(a) \cap V(b)$. Since $aca = a$ and $cac = c$, we have $a(c^*)^*a = a$ and $c^*a^*c^* = c^*$. Thus $(a,c^*) \in \sigma$. Similarly, we have $(b,c^*) \in \sigma$, and so $(a,b) \in \sigma^t$. Then we have $\pi_0 \subset \sigma^t$.
(ii) It is clear that $aa' \mathcal{R}\, a \,\mathcal{R}\, aa^*$. Since each $\mathcal{R}$-class contains one and only one projection, we have
$aa^* = aa'(aa')^* = aa'(a')^*a^*$.
(iii) Let $(xaa', yaa') \in \sigma$. Then $xaa'(yaa')^*xaa' = xaa'$ and $yaa'(xaa')^*yaa' = yaa'$. Now,

$$\begin{aligned} xa(ya)^*xa &= xaa^*y^*xa \\ &= xaa'(a')^*a^*y^*xaa'a \qquad \text{by (ii)}, \\ &= xaa'a = xa. \end{aligned}$$

Similarly, we have $ya(xa)^*ya = ya$, and hence $(xa,ya) \in \sigma$.
(iv) Suppose that $(xaa',y) \in \sigma$. Then $xaa'y^*xaa' = xaa'$ and $y(xaa')^*y = y$. Now,

$$\begin{aligned} xaa'(y(a')^*a')^*xaa' &= xaa'(a')^*a'y^*xaa' \\ &= xaa'y^*xaa' = xaa'. \end{aligned}$$

Similarly, we have $y(a')^*a'(xaa')^*y(a')^*a' = y(a')^*a'$, and so $(xaa',y(a')^*a') \in \sigma$.

An equivalence relation ϱ on a regular *-semigroup S is called a <u>right precongruence</u> if it satisfies the following conditions:

(i) $V(a) \cap V(b) \neq \square$ implies $(a,b) \in \varrho$,

(ii) $(xaa',yaa') \in \varrho$ if and only if $(xa,ya) \in \varrho$,

for any $a,b,x,y \in S$ and $a' \in V(a)$. We denote the minimum right precongruence on S by π.

<u>LEMMA 4.2.</u> $\pi = \pi_0$.

<u>Proof.</u> Since π_0 is the minimum equivalence on S which satisfies the condition (i) above, we have $\pi_0 \subset \pi$. Let $(xaa',yaa') \in \pi_0$. It follows from Lemma 4.1 (i) that there exist $z_0, z_1, \ldots, z_n$ in S such that $xaa' = z_0 \,\sigma\, z_1 \,\sigma \ldots \sigma\, z_n = yaa'$. By Lemma 4.1 (iv), we have $xaa' \,\sigma\, z_1(a')^*a' \,\sigma \ldots \sigma\, yaa'(a')^*a' = yaa'$. By Lemma 4.1 (iii), we have $xa \,\sigma\, z_1(a')^* \,\sigma \ldots \sigma\, ya$, and hence $(xa,ya) \in \pi_0$. By the similar argument, we have that $(xaa'a,yaa'a) = (xa,ya) \in \pi_0$ implies $(xaa',yaa') \in \pi_0$. Thus π_0 is a right precongruence on S, and hence $\pi = \pi_0$.

By using Theorem 2.2 and Proposition 3.8 [4], we have the following corollary.

<u>COROLLARY 4.3.</u> *Let S be a regular *-semigroup with the set of projections P. Then we have*

(i) *S strongly divides the direct product of a fundamental regular *-semigroup and a group if and only if* $\mu \cap \pi_0 = \iota$, *where* ι *is the identity relation on S,*

(ii) *the *-subsemigroup generated by P is self-conjugate, that is,* $a'\langle P\rangle a \subset \langle P\rangle$ *for any* $a \in S$ *and* $a' \in V(a)$.

<u>LEMMA 4.4.</u> *Let* Φ *be a *-homomorphism of a regular *-semigroup S to a *-semigroup T. Then* $\mathrm{Im}\Phi$ *is a regular *-subsemigroup of T. Moreover, for any projection e in* $\mathrm{Im}\Phi$, $e\Phi^{-1}$ *is a regular *-subsemigroup of S.*

Proof. The first part is clear. Let e be any projection in ImΦ. Let a be any element of $e\Phi^{-1}$. Since Φ is a *-homomorphism, $a^*\Phi = (a\Phi)^* = e^* = e$. Then $a^* \in e\Phi^{-1}$, and hence $e\Phi^{-1}$ is a regular *-subsemigroup of S.

THEOREM 4.5. *Let S be a regular *-semigroup with the set of projections P. Let τ_0 be a relation on S defined by*

$$a\,\tau_0\,b \Leftrightarrow a^*\langle P\rangle b \subset \langle P\rangle,$$

*and let $\tau = \tau_0^t$. Then τ is the minimum group *-congruence on S.*

Proof. Is it clear that τ_0 is a symmetric *-relation on S. By Corollary 4.3 (ii), τ_0 is a reflexive congruence, and hence τ is a *-congruence. It is obvious that $(e,f) \in \tau$ for all $e,f \in P$. By the lemma above, S/τ is a regular *-semigroup with only one projection. It follows from Result 1.1 (ii) that S/τ has only one idempotent, and hence it is a group.

Let ϱ be any group [*-]congruence on S. Let $(a,b) \in \tau_0$. Then $a^*\langle P\rangle b \subset \langle P\rangle$. Let $1_{S/\varrho}$ be the identity of S/ϱ. Since $\langle P\rangle \subset 1_{S/\varrho}$, we have $(a\varrho)^{-1}b\varrho = (a\varrho)^{-1}1_{S/\varrho}(b\varrho) = 1_{S/\varrho}$. Then $a\varrho = b\varrho$, and hence $\tau \subset \varrho$.

5. Structure theorem

Let S be a regular *-semigroup and π_0 the *-equivalence on S defined in § 4. For any a in S, we denote the π_0-class containing a by [a]. For each $a \in S$, we define a mapping Φ_a by

$$\mathrm{dom}(\Phi_a) = \{[x]: x \in Saa^*\},$$

and

$$[x]\Phi_a = [xa] \quad \text{for all } [x] \in \mathrm{dom}(\Phi_a).$$

Since π_0 is a right precongruence, Φ_a is a one-to-one mapping of $\{[x]: x \in Saa^*\}$ onto $\{[x]: x \in Sa^*a\}$, with inverse Φ_{a^*}.

THEOREM 5.1. Let $\Phi: S \to I_{S/\pi_0}$ be a mapping defined by $a\Phi = \Phi_a$ for all $a \in S$. Then Φ is a [*-]homomorphism of S into I_{S/π_0} such that $\Phi \circ \Phi^{-1}$ is the minimal inverse semigroup [*-]congruence on S.

Proof. Let a and b be any elements of S. It follows from Construction 1.13 [4] that $\mathrm{dom}(\Phi_a\Phi_b) \supset \mathrm{dom}(\Phi_{ab})$ and that $[x]\Phi_a\Phi_b = [x]\Phi_{ab}$ for any $[x] \in \mathrm{dom}(\Phi_{ab})$. We remark that $\mathrm{dom}(\Phi_a\Phi_b) = (\mathrm{ran}(\Phi_a) \cap \mathrm{dom}(\Phi_b))\Phi_a^{-1}$. Let [z] be any element of $\mathrm{ran}(\Phi_a) \cap \mathrm{dom}(\Phi_b)$. Then there exist $x \in Sa^*a$ and $y \in Sbb^*$ such that $[x] = [y] = [z]$. By Lemma 4.1 (i), there exist $t_0, t_1, \ldots, t_n$ in S such that $x = t_0 \,\sigma\, t_1 \,\sigma \ldots \sigma\, t_n = y$. Since $ybb^* = y$, it follows from Lemma 4.1 (iv) that $xbb^* \,\sigma\, t_1bb^* \,\sigma \ldots \sigma\, y$. Then

$$[z]\Phi_a^{-1} = [xbb^*]\Phi_{a^*} = [xa^*abb^*a^*] \in \mathrm{dom}(\Phi_{ab}),$$

since $xa^*a = x$. Thus, $\mathrm{dom}(\Phi_a\Phi_b) \subset \mathrm{dom}(\Phi_{ab})$, and hence we have $\Phi_a\Phi_b = \Phi_{ab}$. It is clear that $S\Phi$ is an inverse subsemigroup of I_{S/π_0}.

Let Π be the [*-]congruence on S generated by $\{(a,b) \in S\times S: V(a) \cap V(b) \neq \square\}$. It is clear that Π is the minimal inverse semigroup [*-]congruence on S. For each $a \in S$, let ψ_a be a mapping defined by

$$\mathrm{dom}(\psi_a) = \{x\Pi : x \in Saa^*\},$$

and

$$(x\Pi)\psi_a = (xa)\Pi \quad \text{for all } x \in Saa^*.$$

It is clear that ψ_a is an element of $I_{S/\Pi}$. Let $\psi: S \to I_{S/\Pi}$ be a mapping defined by $a\psi = \psi_a$ for any $a \in S$. Then ψ is a *-homomorphism. Since $\pi_0 \supset \Pi$, there exists a *-homomorphism $\nu: S\Phi \to S\psi$ such that $\Phi\nu = \psi$. On the other hand, S/Π is isomorphic to $S\psi$. Then $\Phi \circ \Phi^{-1}$ is the minimal inverse semigroup [*-]congruence on S.

COROLLARY 5.2. Let S be a regular *-semigroup with the set of projections P. Then S is embeddable into the direct product of the maximal inverse semigroup homomorphic image and $T_{(P;\theta)}$ if and only if $\mu \cap \pi_0 = \iota$.

References

[1] T.E. Hall, On regular semigroups, J. Algebra 24 (1973), 1-24.

[2] J.M. Howie, An introduction to semigroup theory, Academic Press, London, 1976.

[3] T. Imaoka, On fundamental regular * semigroups, Mem. Fac. Sci., Shimane Univ. 14 (1980), 19-23.

[4] D.B. McAlister, Regular semigroups, fundamental semigroups and groups, J. Austral. Math. Soc. (Series A) 29 (1980), 475-503.

[5] T.E. Nordahl and H.E. Scheiblich, Regular * semigroups, Semigroup Forum 16 (1978), 369-377.

[6] M. Yamada, On regular *-semigroups, II. Fundamental regular *-semigroups, to appear.

Department of Mathematics
Shimane University
Matsue, Japan

TOTAL DISJUNKTIVE VERALLGEMEINERTE BRUCK-REILLY-ERWEITERUNGEN VON HALBGRUPPEN UND FORMALE SPRACHEN

H. Jürgensen

1. Vorbemerkungen

Diese Arbeit befaßt sich mit der Beschreibung von Halbgruppen, in denen jedes Element disjunktiv ist. Offenbar ist jede derartige Halbgruppe einfach oder O-einfach. Für Halbgruppen mit O gibt ein Satz von Schein [7] eine allgemeine Charakterisierung an: Es gilt nämlich, daß genau dann jedes Element einer Halbgruppe mit O disjunktiv oder S-disjunktiv ist, wenn die Halbgruppe h-einfach ist, d.h., keine nichttrivialen Kongruenzen besitzt (vgl. [3]). Jedoch auch in diesem Falle sind für die allgemeine Situation darüber hinaus nur wenig Details bekannt. In [3] leiteten wir einige Einzelheiten für den Fall von Rees-Matrix-Halbgruppen mit O über kürzbaren einfachen Halbgruppen her.

Der Fall der Halbgruppen ohne O ist noch weiter von einer Lösung entfernt. Bisher konnte die Situation nur für Rees-Matrix-Halbgruppen über kürzbaren einfachen Halbgruppen einigermaßen zufriedenstellend geklärt werden; die erhaltene Charakterisierung steht in gewissem Zusammenhang mit Schützenberger-Darstellungen und Fragen der Erweiterungstheorie [3]. Jedoch schon in [3] wiesen wir darauf hin, daß es Halbgruppen gibt, in denen jedes Element disjunktiv ist, die aber weder Rees-Matrix-Halbgruppen über kürzbaren einfachen Halbgruppen noch h-einfach sind: Die bizyklische Halbgruppe ist ein derartiges Beispiel ([8] zeigt, daß sogar jede bi-einfache inverse Halbgruppe total disjunktiv ist). Diese Bemerkung bildet den Ausgangspunkt der vorliegenden Arbeit.

Bekanntlich ist die bizyklische Halbgruppe der Grundbaustein einer ganzen Klasse von Halbgruppen, die man aus ihr mit Hilfe einer bestimmten Erweiterungsoperation erzeugen kann, welche im wesentlichen auf Bruck und Reilly zurückgeht (vgl. [2]). Wir werden eine leicht verallgemeinerte Version dieser Erweiterungsoperation einführen (diese Verallgemeinerung findet man, wie wir inzwischen feststellten, schon bei Justin [5] angegeben) und dann zunächst Charakterisierungen der Greenschen Relationen herleiten. Mit ihrer Hilfe werden wir dann diejenigen Bruck-Reilly-Erweiterungen beschreiben, in denen jedes Element disjunktiv ist. Die Arbeit schließt mit einigen Hinweisen zur Interpretation dieser Ergebnisse im Rahmen der Theorie der formalen Sprachen.

2. Definitionen, Notation

S sei eine Halbgruppe; S^1 sei gleich S, wenn S ein Monoid ist, und gleich $S \cup \{1\}$ sonst, wobei $1 \notin S$ ist und 1 als Einselement auf S^1 operiert. Mit E(S) sei die Menge der idempotenten Elemente von S bezeichnet. S_1, S_2 seien Teilmengen von S^1, und $L \subseteq S_1 \cap S_2$. S_1, S_2, L definieren die Äquivalenz $P_L^{(S_1,S_2)}$ auf S_1 folgendermaßen:

$$\forall y,z \in S_1 : \left[y \equiv z \quad (P_L^{(S_1,S_2)}) \Longleftrightarrow \left[\forall u,v \in S_2 : uyv \in L \Leftrightarrow uzv \in L \right]\right].$$

Wie üblich, lassen wir bei einelementigen Mengen häufig die Mengenklammern weg, schreiben also z.B. $P_x^{(S_1,S_2)}$ statt $P_{\{x\}}^{(S_1,S_2)}$.

<u>2.1. Definition.</u> L heißt (S_1,S_2)-*disjunktiv*, wenn $P_L^{(S_1,S_2)}$ die Diagonale auf S_1 ist.

Wie in [3] interessieren wir uns in dieser Arbeit für die folgenden Spezialfälle:

(1) <u>$S_1 = S$, $S_2 = S^1$:</u> Wir nennen L *disjunktiv*, wenn L (S,S^1)-disjunktiv ist. Die Äquivalenzrelation $P_L^{(S,S^1)}$ ist eine Kongruenzrelation auf S - in der Sprachentheorie die syntaktische Kongruenz von L - , die als P_L^1 geschrieben werden soll. P_L^1 ist die gröbste Kongruenz auf S, die L saturiert.

(2) <u>$S_1 = S_2 = S$:</u> Wir nennen L S-*disjunktiv*, wenn L (S,S)-disjunktiv ist. Auch die Äquivalenzrelation $P_L^{(S,S)}$ ist eine Kongruenzrelation auf S - die Hauptkongruenz zu L im

Sinne von Dubreil u.a. (vgl. [1], § 10) -, die wir mit P_L bezeichnen wollen. Es gilt $P_L^1 \subseteq P_L$, d.h., aus S-Disjunktivität folgt Disjunktivität.

Eine Halbgruppe S wird als *total* [S-]*disjunktiv* bezeichnet, wenn jedes ihrer Elemente [S-]disjunktiv ist. Aus [3] übernehmen wir die folgende Aussage:

2.2. Lemma. *S sei eine total disjunktive oder total S-disjunktive Halbgruppe. Dann ist S einfach oder O-einfach.*

Für Halbgruppen mit O gilt:

2.3. Satz [6,3]. *S sei eine Halbgruppe mit O.*
Die folgenden Aussagen sind äquivalent:

(1) *S ist total disjunktiv.*

(2) *S ist O-einfach und O ist disjunktiv.*

(3) *S ist h-einfach.*

(4) *S ist total S-disjunktiv.*

3. Elementare Eigenschaften von Bruck-Reilly-Erweiterungen

In diesem Abschnitt wird zunächst eine natürliche Verallgemeinerung der Bruck-Reilly-Erweiterungskonstruktion definiert. Danach charakterisieren wir die Greenschen Relationen $\mathcal{R}$, $\mathcal{L}$, $\mathcal{J}$, $\mathcal{H}$ und leiten notwendige und hinreichende Bedingungen dafür her, daß eine Bruck-Reilly-Erweiterung einfach oder bieinfach ist.

B sei die bizyklische Halbgruppe, d.h., die Halbgruppe $\mathbb{N}_0 \times \mathbb{N}_0$, wobei $\mathbb{N}_0 = \mathbb{N} \cup \{0\}$ ist, mit der folgendermaßen definierten Multiplikation:

$$(m_1,n_1) \quad (m_2,n_2) = (m_1-n_1+\max(n_1,m_2), n_2-m_2+\max(n_1,m_2)).$$

Zur Vereinfachung der Notation schreiben wir, wie üblich,

$$z \dot{-} y = \begin{cases} 0 & \text{für } y > z, \\ z - y & \text{für } y \leq z, \end{cases}$$

und es ist

$$(m_1,n_1)(m_2,n_2) = (m_1 + (m_2 \dot{-} n_1),\ n_2 + (n_1 \dot{-} m_2)).$$

H sei eine beliebige Halbgruppe, Endo(H) sei die Endomorphismenhalbgruppe von H, und es sei $\vartheta \in$ Endo(H). Für $h \in H$ und $i \in \mathbb{N}_0$ vereinbaren wir

$$h\vartheta^i = \begin{cases} h & , \quad \text{falls } i = 0, \\ h\vartheta & , \quad \text{falls } i = 1, \\ (h\vartheta)\vartheta^{i-1}, & \quad \text{falls } i > 1. \end{cases}$$

Mit BR(H, ϑ) bezeichnen wir das Gruppoid mit der Trägermenge $\mathbb{N}_0 \times H \times \mathbb{N}_0$ und der durch

$$(m_1,h_1,n_1)\ (m_2,h_2,n_2)$$
$$= (m_1 + (m_2 \dot{-} n_1),\ h_1\vartheta^{m_2 \dot{-} n_1}\ h_2\vartheta^{n_1 \dot{-} m_2},\ n_2 + (n_1 \dot{-} m_2))$$

definierten Multiplikation.

Man rechnet nach, daß diese Multiplikation assoziativ ist. Damit ist BR(H, ϑ) eine Halbgruppe.

3.1. Definition. BR(H, ϑ) ist die *Bruck-Reilly-Erweiterung* von H bezüglich ϑ.

Bruck-Reilly-Erweiterungen im üblichen Sinne bilden einen Spezialfall dieser Definition: Bei jenen ist H ein Monoid und ϑ ein Homomorphismus von H in seine Einheitengruppe [2].

Zur Vereinfachung der Notation vereinbaren wir für $q \in$ BR(H, ϑ) die Schreibweise

$q = (m_q, h_1, n_q)$ mit $m_q \in \mathbb{N}_o$, $h_q \in H$, $n_q \in \mathbb{N}_o$. Seien nun $u,v,y \in BR(H,\vartheta)$. Man weist die folgenden Beziehungen nach:

$$m_{uyv} = m_u - n_u + m_y - n_y + \tau(u,y,v),$$

$$n_{uyv} = n_v - m_v + \tau(u,y,v),$$

$$h_{uyv} = h_u \vartheta^{\phi(u,y,v)} h_y \vartheta^{\psi(u,y,v)} h_v \vartheta^{\chi(u,y,v)}$$

mit

$$\tau(u,y,v) = \max(n_y - m_y + \max(n_u, m_y), m_v)$$
$$= \max(n_y + (n_u \dot{-} m_y), m_v)$$

$$\phi(u,y,v) = \tau(u,y,v) - n_u + m_y - n_y$$

$$\psi(u,y,v) = \tau(u,y,v) - n_y,$$

$$\chi(u,y,v) = \tau(u,y,v) - m_v.$$

Wir werden im weiteren immer wieder die folgenden Fälle zu unterscheiden haben:

	Fall (a) $n_u \leq m_y$ $m_v \leq n_y$	Fall (b) $n_u \leq m_y$ $m_v > n_y$	Fall (c) $n_u > m_y$ $m_v \leq n_y - m_y + n_u$	Fall (d) $n_u > m_y$ $m_v > n_y - m_y + n_u$
$\tau =$	n_y	m_v	$n_y - m_y + n_u$	m_v
$\phi =$	$m_y - n_u$	$m_v - n_u + m_y - n_y$	0	$m_v - n_u + m_y - n_y$
$\psi =$	0	$m_v - n_y$	$n_u - m_y$	$m_v - n_y$
$\chi =$	$n_y - m_v$	0	$n_y - m_y + n_u - m_v$	0
$m_{uyv} =$	$m_u - n_u + m_y$	$m_u - n_u + m_y - n_y + m_v$	m_u	$m_u - n_u + m_y - n_y + m_v$
$n_{uyv} =$	$n_v - m_v + n_y$	n_v	$n_v - m_v + n_y - m_y + n_u$	n_v

Man beachte, daß die zu den Fällen b und d gehörigen Spalten übereinstimmen.

3.2. Lemma. Sei $S = BR(H, \vartheta)$ und $x, y \in S$.

Es gilt $x \in SyS$ genau dann, wenn

$$h_x \in \bigcup_{c \in \mathbb{N}_0} H\, h_y\, \vartheta^c\, H$$

ist.

Beweis:

Falls $x = uyv$ für $u, v \in S$, dann gilt trivialerweise

$$h_x = h_{uyv} = h_u \vartheta^{\phi(u,y,v)}\; h_y\; \vartheta^{\psi(u,y,v)}\; h_v \vartheta^{\chi(u,y,v)}$$

$$\in \bigcup_{c \in \mathbb{N}_0} H\, h_y\, \vartheta^c\, H.$$

Zum Beweis der Umkehrung sei nun $c \in \mathbb{N}_0$ und

$$h_x \in H\, h_y\, \vartheta^c H$$

gegeben.

Für c = 0 betrachten wir den Fall a. Man kann dann etwa $m_u = m_x$, $n_u = m_y$, $m_v = n_y$, $n_v = n_x$ wählen, was für $\phi(u,y,v)$ und $\chi(u,y,v)$ die kleinstmöglichen Werte, nämlich 0, ergibt. Man erhält $h_{uyv} = h_u h_y h_v$. Nach Voraussetzung können h_u und h_v so gewählt werden, daß $h_{uyv} = h_x$ ist.

Man beachte, daß unter der Voraussetzung $c = \psi(u,y,v)$ nur der Fall a für c = 0 anwendbar ist; daß die Fälle b und c ausgeschlossen sind, ist nach Definition klar; für den Fall d folgert man aus $m_v > n_y - m_y + n_u$ und $m_v - n_y = c = 0$, daß $m_y > n_u$ gelten müßte.

Sei jetzt $c \geq 1$. Falls wir $\psi(u,y,v) = c$ verlangen, ist der Fall a generell ausgeschlossen; c = 1 schließt den Fall d und $m_x < c$ den Fall b aus.

Betrachten wir den Fall c: Sei $m_u = m_x$, $n_u = m_y + c$, $m_v = n_y + c$, $n_v = n_x$. Man erhält $m_{uyv} = m_x$, $n_{uyv} = n_x$, $\phi(u,y,v) = 0$, $\psi(u,y,v) = c$, $\chi(u,y,v) = 0$. Nach Voraussetzung können h_u, $h_v \in H$ geeignet gewählt werden, so daß

$$h_x = h_{uyv} = h_u h_y \vartheta^c h_v \text{ gilt.} \qquad \square$$

Wir betrachten die im Beweis des Lemma nicht untersuchten, weil zum Beweis überflüssigen, Fälle:

Im Falle d mit $c > 1$ erhält man eine Lösung mit der Parameterwahl $m_u = m_x - 1$, $n_u = m_y + c - 1$, $m_v = n_y + c$, $n_v=n_x$. Dies setzt natürlich $m_x > 0$ voraus. Damit berechnet man $m_{uyv} = m_x$, $n_{uyv} = n_x$, $\phi(u,y,v) = 1$, $\psi(u,y,v) = c$, $\chi(u,y,v) = 0$. Ist nun $h_x \in H \quad h_y \vartheta^c H$, so kann man $h_u \in H$ und $h_v \in H$ geeignet wählen, so daß x = uyv gilt.

Wir stellen die Ergebnisse zusammen

3.3. Lemma. Es sei $x,y \in BR(H,\vartheta)$, $h_x \in \bigcup_{c \in \mathbb{N}_0} Hh_y\vartheta^c H$.

Damit für $u,v \in BR(H,\vartheta)$ *gilt* $uyv=x$, *ist notwendig und hinreichend, daß*

$$m_u = m_x - \phi,\ n_u = m_y - \phi + \psi$$

$$m_v = n_y - \chi + \psi,\ n_v = n_x - \chi$$

$$h_x \in H\vartheta^{\phi} h_y \vartheta^{\psi} H \vartheta^{\chi}$$

ist, wobei ϕ,ψ,χ *entsprechend der folgenden Tabelle zu wählen sind:*

Fall a	Fall b	Fall c	Fall d
$\phi,\chi \in \mathbb{N}_0$	$\phi,\psi \in \mathbb{N}$,	$\phi = 0, \psi \in \mathbb{N}$,	$\psi \in N, \psi \in \mathbb{N}$,
$\psi = 0$	$\chi = 0$	$\chi \in \mathbb{N}_0$,	$\chi = 0$,
$\phi \leq m_x, m_y$	$\phi \leq m_x$,	$\psi \leq n_x$,	$\psi \leq m_x$
$\chi \leq n_x, n_y$	$0 \leq \psi-\phi \leq m_y$	$\chi - \psi \leq n_y$	$\phi < \psi$.

Aus 3.3 berechnet man ferner die einseitigen Analoga von 3.2 und 3.3 , z.B.:

3.4. Lemma. Es seien $x,y \in BR(H,\vartheta) = S$. *Es gilt* $x \in yS$ *genau dann, wenn* $m_x \geq m_y$ *und* $h_x \in h_y \vartheta^{m_x - m_y} H$ *ist. Dafür, daß* $x = yv$ *mit* $v \in S$ *gilt, ist notwendig und hinreichend, daß* v *gemäß der folgenden Tabelle gewählt wird:*

	(Fall a) $m_v \leq n_y$	(Fall b) $m_v > n_y$
m_v	$n_y - \beta$	$m_x - m_y + n_y$
n_v	$n_x - \beta$	n_x
h_{yv}	$h_y \vartheta^0 h_v \vartheta^\beta$	$h_y \vartheta^{m_x - m_y} h_v \vartheta^0$
Bedin-gungen	$\beta \in \mathbb{N}_0, \beta \leq n_y, n_x,$ $m_x = m_y$ $h_x \in h_y H \vartheta^\beta$	$m_x > m_y$ $h_x \in h_y \vartheta^{m_x - m_y} H$

Die duale Version von 3.4 ist analog.

Durch Zusammenfassen der Aussagen erhält man die folgende Charakterisierung der Greenschen Relationen auf BR(H, ϑ).

3.6. Satz. Sei S = BR(H, ϑ) und $x, y \in S$. Mit $\mathcal{R}_H, \mathcal{L}_H, \ldots$ seien die Greenschen Relationen auf H bezeichnet. Es gilt:

(1a) $x \leq_{\mathcal{R}} y \Longleftrightarrow \left[x = y \vee (m_x \geq m_y \wedge h_x \in h_y \vartheta^{m_x - m_y} H) \right]$.

(1b) $x \mathcal{R} y \Longleftrightarrow \left((m_x = m_y \wedge h_x \mathcal{R}_H h_y) \right)$.

(2a) $x \leq_{\mathcal{L}} y \Longleftrightarrow \left(x = y \vee (n_x \geq n_y \wedge h_x \in H h_y \vartheta^{n_x - n_y}) \right)$.

(2b) $x \mathcal{L} y \Longleftrightarrow \left((n_x = n_y \wedge h_x \mathcal{L}_H h_y) \right)$.

(3a) $x \leq_{\mathcal{J}} y \Longleftrightarrow (x \leq_{\mathcal{R}} y \vee x \leq_{\mathcal{L}} y \vee h_x \in \bigcup_{c \in \mathbb{N}_0} H h_y \vartheta^c H)$.

(4) $x \mathcal{H} y \Longleftrightarrow (m_x = m_y \wedge n_x = n_y \wedge h_x \mathcal{H}_H h_y)$.

(5) $x \mathcal{D} y \Longleftrightarrow h_x \mathcal{D}_H h_y$.

3.7. Korollar. Sei S = BR(H, ϑ). S ist genau dann bi-einfach, wenn H bi-einfach ist. S ist genau dann einfach, wenn

$$H = \bigcup_{c \in \mathbb{N}_0} H h \vartheta^c H$$

für alle $h \in H$ gilt.

Die für Bruck-Reilly-Erweiterungen im üblichen Sinne bekannten Aussagen erhält man als Spezialfälle.

Falls H Monoid und ϑ Homomorphismus von H in die Einheitengruppe von H ist, so ist $H = H h\vartheta H$ für alle $h \in H$, und BR(H, ϑ) - im üblichen Sinne genommen - ist einfach (vgl. [2]). Falls H einfache Halbgruppe ist, so ist BR(H, ϑ) für beliebige ϑ einfach.

3.8. Satz. Sei S = BR(H, ϑ).

(1) S ist genau dann ein Monoid, wenn H ein Monoid ist. In diesem Falle ist $(0,1_H,0)$ das Einselement von S, wobei 1_H das Einselement von H ist.

(2) $E(S) = \{(n,e,n) \mid n \in \mathbb{N}_0, e \in E(H)\}$.

(3) S ist genau dann regulär, wenn H regulär ist.

(4) S ist genau dann invers, wenn H invers ist.

Beweis: Der Beweis der Aussagen (2)-(4) ist identisch mit dem der Aussagen c,d von Proposition V.6.10 in [2]. Wir beweisen nur (1): Falls H Monoid ist, so ist $(0,1_H,0)$ offensichtlich Einselement von S und S ist Monoid.

Sei nun umgekehrt S Monoid mit Einselement (m_1,h_1,n_1).

Für alle $x \in S$ gilt also

$$x = (m_1, h_1, n_1)\,(m_x, h_x, n_x)$$
$$= (m_1 + (m_x \dot{-} n_1),\ h_1\,\vartheta^{m_x \dot{-} n_1}\ h_x\,\vartheta^{n_1 \dot{-} m_x},\ n_x + (n_1 \dot{-} m_x))$$
$$= (m_x, h_x, n_x).$$

Folglich muß $(n_1 \dot{-} m_x) = 0$, also $(m_x \dot{-} n_1) = m_x - n_1$ gelten.

Daraus ergibt sich $m_1 = n_1 = 0$ und

$$h_x = h_1\,\vartheta^{m_x}\,h_x.$$

Dual erhält man

$$h_x = h_x\,h_1\,\vartheta^{n_x}$$

für alle m_x, n_x, h_x. Es folgt, daß h_1 Einselement von H sein muß. □

4. Total disjunktive Bruck-Reilly-Erweiterungen von Monoiden

Im folgenden sei $S = BR(H, \vartheta)$. Wir wollen in diesem Abschnitt unter der Voraussetzung, daß H ein Monoid ist, diejenigen S charakterisieren, die total disjunktiv sind. Zur Vorbereitung dieses Ergebnisses und derjenigen des nächsten Abschnitts führen wir noch etwas Notation ein und beweisen einige Lemmata, in denen wir allerdings nicht voraussetzen, daß H Monoid ist.

Es sei

$$-: BR(H, \vartheta) \to B: x \mapsto \bar{x} = (m_x, n_x)$$

und

$$\Delta x = \Delta \bar{x} = m_x - n_x.$$

Wir sagen, x sei *disjunktiv* (oder S-*diskunktiv*) für y,z, wenn $y \nparallel z\ (P_x^1)$ (bzw. $y \nparallel z\ (P_x)$) gilt.

4.1. Satz. H sei ein Monoid, $S = BR(H, \vartheta)$ für $\vartheta \in$ Endo H. Genau dann ist S total disjunktiv, wenn H einfach und total disjunktiv ist.

Diesen Satz beweisen wir mit Hilfe mehrerer Hilfssätze, von denen einige später zu seiner Verallgemeinerung weiter benötigt werden.

4.2. Lemma. Für $\bar{x}$ B, y,z, ε BR(H, ϑ), $y \neq z$, $\bar{y} = \bar{z}$ existieren u,v BR(H, ϑ) mit $uyv \neq uzv$. $\overline{uyv} = \overline{uzv} = \bar{x}$ genau dann, wenn es

$$(\alpha,\beta,\gamma) \in \mathrm{Ind}_1 \ (\bar{x},\bar{y})$$

mit $h_u\vartheta^{\alpha}h_y\vartheta^{\beta}h_v\vartheta^{\gamma} \neq h_u\vartheta^{\alpha}h_z\vartheta^{\beta}h_v\vartheta^{\gamma}$

für geeignete h_u, $h_v \in H$ gibt. Dabei ist

$$\mathrm{Ind}_1(\bar{x},\bar{y}) = \left\{ (\alpha,\beta,\gamma) \;\middle|\; \begin{array}{l} 0 \leq \alpha \leq m_x, m_y,\ \beta = 0, 0 \leq \gamma \leq n_x, n_y \\ \text{oder} \\ 1 \leq \alpha \leq m_x,\quad \alpha \leq \beta \leq m_y + \alpha, \gamma = 0 \\ \text{oder} \\ \alpha = 0,\ 1 \leq \beta \leq n_x,\ 0 \leq \gamma \leq n_y + \beta \\ \text{oder} \\ 1 \leq \alpha \leq m_x,\ \alpha < \beta,\ \gamma = 0. \end{array} \right\}$$

Beweis: Lemma 3.3. □

Falls H ein Monoid ist, wähle man $h_u = h_v = 1_H$, $\alpha=\beta=\gamma=0$. Damit haben wir gezeigt:

4.3. Lemma. H sei ein Monoid $x \in BR(H, \vartheta)$ ist genau dann für alle $y,z \in BR(H, \vartheta)$ mit $y=z$, $\bar{y}=\bar{z}$ disjunktiv, wenn h_x in H disjunktiv ist.

Für Monoide H ergibt Lemma 4.3 sofort, daß BR(H, ϑ) nur dann total disjunktiv ist, wenn H total disjunktiv und damit einfach oder O-einfach ist.

<u>4.4. Lemma. Zu</u> $\bar{y}$, $\bar{z} \in B$ <u>gibt es</u> $\bar{u}, \bar{v} \in B$ <u>mit</u> $\overline{uyv} = \overline{uzv}$ <u>genau dann, wenn</u> $\Delta\bar{y} = \Delta\bar{z}$ <u>ist.</u>

<u>Beweis:</u> Nachrechnen mit Lemma 3.3. □

Für $x,y,z \in BR(H,\vartheta)$ mit $\Delta\bar{y} \neq \Delta\bar{z}$ folgt daher aus $uyv = x$ immer $uzv \neq x$. Damit gilt:

<u>4.5. Lemma. Damit jedes</u> $x \in BR(H,\vartheta)$ <u>für alle</u> $y,z \in BR(H,\vartheta)$ <u>mit</u> $\Delta\bar{y} \neq \Delta\bar{z}$ <u>S-disjunktiv ist, ist notwendig und hinreichend, daß</u> H <u>einfach ist.</u>

<u>Beweis:</u> Die Behauptung muß insbesondere für $\bar{x} = (0,0)$ erfüllt sein. Zur Bestimmung von $u,v \in BR(H,\vartheta)$ mit $uyv = x$ (oder $uzv = x$) kommt daher gemäß Lemma 4.2 nur das Tripel $(\alpha,\beta,\gamma) = (0,0,0)$ in Frage. Für jedes $h_x, h_y, h_z \in H$ muß also $h_x \in Hh_yH$ oder $h_x \in Hh_zH$ sein. H ist also einfach oder O-einfach; weil die Aussage aber insbesondere für $h_y = h_z = 0$ gelten müßte, folgt, daß H kein O-Element besitzen kann. Ist nun umgekehrt H einfach, so gilt a fortiori

$$H = \bigcup_{\beta \in \mathbb{N}_0} H h_y \vartheta^\beta H$$

für alle $h_y \in H$, und man findet mit Lemma 4.2 aus $(\alpha,\beta,\gamma) = (0,\beta,0)$ Elemente $u,v \in BR(H,\vartheta)$ mit $uyv=x$. Wegen $\Delta\bar{y} = \Delta\bar{z}$ ist $uzv \neq x$. □

Es bleibt der Fall $\bar{y} \neq \bar{z}$, $\Delta\bar{y} = \Delta\bar{z}$ zu betrachten. Wegen des Lemmas 4.5 dürfen wir generell voraussetzen, daß H einfach ist. Eine genaue Analyse der Möglichkeiten gemäß Lemma 3.3 zeigt:

4.6. Lemma. H sei einfach. Zu jedem $x,y,z \in BR(H,\vartheta)$ mit $\bar{y} \neq \bar{z}$, $\Delta\bar{y} = \Delta\bar{z}$, $m_y < m_z$ existieren $u,v \in BR(H,\vartheta)$ mit $uyv = x$, $\overline{uzv} \neq \bar{x}$.

Beweis: Die u,v gemäß Behauptung werden bestimmt durch die Wahl von $h_u, h_v \in H$ und $(\alpha,\beta,\gamma) \in Ind_1(\bar{x},\bar{y})$, $(\alpha,\beta',\gamma) \notin Ind_1(\bar{x},\bar{z})$ mit $\beta' = m_y - m_z + \beta = n_y - n_z + \beta$, wobei

$$h_x = h_u\, \vartheta^\alpha\, h_y\, \vartheta^\beta\, h_v\, \vartheta^\gamma$$

ist. Sei

$$Ind_2(x,y,z) = \left\{ (\alpha,\beta,\gamma) \;\middle|\; \begin{array}{l} (\alpha,\beta,\gamma) \in Ind_1(\bar{x},\bar{y}), \\ (\alpha,m_y-m_z+\beta,\gamma) \notin Ind_1(\bar{x},\bar{z}) \end{array} \right\}.$$

Das Tripel (0,0,0) liegt in $Ind_2(\bar{x},\bar{y},\bar{z})$. Es gilt nämlich einerseits bei dieser Wahl von (α,β,γ)

$$(m_x,m_y)\ (m_y,n_y)\ (n_y,n_x) = (m_x,n_x) = \bar{x}$$

und andererseits wegen $m_y < m_z$ und $\Delta\bar{y} = \Delta\bar{z}$

$$(m_x,m_y)(m_z,n_z)(n_y,n_x) = (m_x+m_z-m_y, n_x+n_z-n_y) \neq \bar{x}$$

Geeignete Wahl von $h_u, h_v \in H$ ergibt daher mit

$m_u = m_x$, $n_u = m_y$, $m_v = n_y$, $n_v = n_x$:

$uyv = x$ und $\overline{uzv} \neq \bar{x}$. □

Aus der Kombination von 4.3, 4.5 und 4.6 folgt die Behauptung von Satz 4.1. □

5. Total S-disjunktive Bruck-Reilliy-Erweiterungen von Halbgruppen

In diesem Abschnitt können wir ohne Beschränkung der Allgemeinheit voraussetzen, daß H kein Monoid ist (andernfalls liegt der Fall des Kapitels 4 vor). Lemma 4.5 bleibt auch in der vorliegenden Situation gültig, so daß H notwendigerweise einfach ist, wenn BR(H, ϑ) total S-disjunktiv ist. Damit ist auch Lemma 4.6 gültig, und wir können uns darauf beschränken, die Situation von Lemma 4.2 für den Fall, daß H kein Monoid ist, weiterzuverfolgen.

Man beachte, daß die Aussage von 4.2 insbesondere auch für $\bar{x} = (0,0)$ gelten muß. Damit ergibt sich

$$\mathrm{Ind}_1((0,0),\bar{y}) = \{(0,0,0)\},$$

und es ist ferner

$$\mathrm{Ind}_1(\bar{x},\bar{y}) \supseteq \mathrm{Ind}_1((0,0),\bar{y})$$

für alle $\bar{y}$. Es folgt, daß die Wahl von h_u, $h_v \in H$ so möglich sein muß, daß

$$h_{uyv} = h_u h_y h_v = h_x \longleftrightarrow h_{uzy} = h_u h_z h_v \neq h_x$$

gilt. Damit ist gezeigt:

<u>5.1. Satz.</u> H <u>sei eine Halbgruppe</u>. S =BR(H, ϑ) <u>ist genau dann total</u> S-<u>disjunktiv</u>, <u>wenn</u> H <u>einfach und total</u> H-<u>disjunktiv ist</u>.

6. Total disjunktive Bruck-Reilly-Erweiterungen von Halbgruppen

Wir wollen voraussetzen, daß H kein Monoid ist, da sonst $BR(H, \vartheta)^1 = BR(H, \vartheta)$ wäre und somit S-Disjunktivität und Disjunktivität äquivalent wären. Wir werden im Prinzip der Fallunterscheidung aus Abschnitt 4 folgen und die nötigen Modifikationen andeuten.

6.1. Lemma. Seien $y, z \in BR(H, \vartheta)$, $y \neq z$, $\bar{y} = \bar{z}$ *und* $\bar{x} \in B$. *Es existieren* $u, v \in BR(H, \vartheta)^1$ *mit* $\overline{uyv} = \overline{uzv} = \bar{x}$ *und* $uyv = x \leftrightarrow uzv \neq x$ *genau dann, wenn eine der folgenden Bedingungen erfüllt ist:*

(a) *Es gibt* $(\alpha, \beta, \gamma) \in \mathrm{Ind}_1(\bar{x}, \bar{y})$ *und*
$h_u, h_v \in H$ *mit* $h_u \vartheta^\alpha h_y \vartheta^\beta h_v \vartheta^\gamma = h_x \leftrightarrow h_u \vartheta^\alpha h_z \vartheta^\beta h_v \vartheta^\gamma \neq h_x$.

(b) *Es gibt* $(\alpha, \beta) \in \mathrm{Ind}_1^L(\bar{x}, \bar{y})$ *und* $h_u \in H$ *mit*
$h_u \vartheta^\alpha h_y \vartheta^\beta = h_x \leftrightarrow h_u \vartheta^\alpha h_z \vartheta^\beta \neq h_x$.

(c) *Es gibt* $(\beta, \gamma) \in \mathrm{Ind}_1^R(\bar{x}, \bar{y})$ *und* $h_v \in H$ *mit*
$h_y \vartheta^\beta h_v \vartheta^\gamma = h_x \leftrightarrow h_z \vartheta^\beta h_v \vartheta^\gamma \neq h_x$.

(D) $\bar{y} = \bar{z} = \bar{x}$ *und* $x = y \leftrightarrow x \neq z$.

Dabei sind

$$\mathrm{Ind}_1^L(\bar{x}, \bar{y}) = \left\{ (\alpha, \beta) \;\middle|\; \begin{array}{l} 0 \leq \alpha \leq m_x, m_y, \quad \beta = 0 \\ \text{oder} \\ \alpha = 0, \; \beta = n_x - n_y \end{array} \right\}$$

und

$$\mathrm{Ind}_1^R(\bar{x}, \bar{y}) = \left\{ (\beta, \gamma) \;\middle|\; \begin{array}{l} 0 \leq \gamma \leq n_y, n_x, \quad \beta = 0 \\ \text{oder} \\ \gamma = 0, \; \beta = m_x - m_y \end{array} \right\}$$

Der Fall (b) ist nur für $n_x \geq n_y$ *und (c) nur für* $m_x \geq m_y$ *möglich.*

Beweis: Umformulierung von Lemmata 3.4, 3.5 und dem zu 3.5 Dualen. □

Die Mengen $\mathrm{Ind}_1(\bar{x},\bar{y})$, $\mathrm{Ind}_1^L(\bar{x},\bar{y})$, $\mathrm{Ind}_2^R(\bar{x},\bar{y})$ wachsen offenbar mit "wachsendem" $\bar{x}$ ($\bar{x}$ kleiner als $\bar{x}'$, falls $m_x < m_{x'}$, $n_x \leq n_{x'}$, oder $m_x \leq m_{x'}$, $n_x < n_{x'}$). Folglich genügt es in der Situation von 6.1, den Fall $\bar{x} = (0,0)$ zu behandeln. Es ist

$$\mathrm{Ind}_1((0,0),\bar{y}) = \{(0,0,0)\},$$
$$\mathrm{Ind}_1^L((0,0),\bar{y}) = \{(0,0)\},\ n_y = 0,$$
$$\mathrm{Ind}_1^R((0,0),\bar{y}) = \{(0,0)\},\ m_y = 0.$$

Aus den Bedingungen (a)-(c) von 6.1 wird unter diesen Voraussetzungen:

(a') $\exists\, h_u, h_v \in H : h_u h_y h_v = h_x \leftrightarrow h_u h_z h_v \neq h_x$,

(b') $\exists\, h_u \in H : h_u h_y = h_x \leftrightarrow h_u h_z \neq h_x$,

(c') $\exists\, h_v \in H : h_y h_v = h_x \leftrightarrow h_z h_v \neq h_x$,

(d') $h_y = h_x \leftrightarrow h_z \neq h_x$.

Insbesondere kann $n_y \neq 0$, $m_y \neq 0$ sein. Dann sind die Fälle (b)-(d) nicht möglich, und es folgt:

6.2. Lemma. Damit jedes $x \in BR(H,\vartheta)$ für alle $y,z \in BR(H,\vartheta)$ mit $y \neq z$, $\bar{y} = \bar{z}$ disjunktiv ist, ist notwendig und hinreichend, daß H total H-disjunktiv ist.

Aus 6.2 folgt, falls $S = BR(H,\vartheta)$ total disjunktiv ist, daß H einfach oder 0-einfach ist. Die zweite Möglichkeit wird durch folgendes Analogon zu Lemma 4.5 ausgeschlossen:

6.3. Lemma. Damit jedes $x \in BR(H,\vartheta)$ für alle $y,z \in BR(H,\vartheta)$ mit $\Delta\bar{y} \neq \Delta\bar{z}$ disjunktiv ist, ist notwendig und hinreichend, daß H einfach ist.

Beweis: Analog Lemma 4.5, wobei wie im Beweis zu 6.2 $m_y > 0$, $n_y > 0$, $m_z > 0$, $n_z > 0$ zu wählen sind. □

Wegen Lemma 6.3 ist auch Lemma 4.6 anwendbar. Es folgt:

6.4. Satz. H sei eine Halbgruppe. $S = BR(H, \vartheta)$ ist genau dann total disjunktiv, wenn H einfach und total H-disjunktiv ist.

Die Sätze 4.1, 5.1 und 6.4 zusammenfassend, erhält man:

6.5. Korollar. Die folgenden Aussagen für $S = BR(H, \vartheta)$ sind äquivalent:

(1) S ist total disjunktiv,

(2) S ist total S-disjunktiv,

(3) H ist einfach und total H-disjunktiv.

7. Spezialfälle von Bruck-Reilly-Erweiterungen

In diesem Abschnitt diskutieren wir kurz einige spezielle Situationen, die sich aus der besonderen Wahl von H oder ϑ ergeben können.

Falls H Halbgruppe mit O ist, so ist $S = BR(H, \vartheta)$ genau dann total [S-]disjunktiv, wenn H=O ist.

Falls H endlich und $S = BR(H, \vartheta)$ total [S-]disjunktiv ist, muß H vollständig einfach und total H-disjunktiv sein. Nach [3] ist H eine Rees-Matrix-Halbgruppe $\mathfrak{M}[G;I,\Lambda;P]$, deren beide Schützenberger-Darstellungen treu sind.

In [3] wurde mit den Rees-Matrix-Halbgruppen über kürzbaren einfachen Halbgruppen zwar eine Konstruktion für eine unendliche Klasse von total [S-]disjunktiven Halbgruppen angegeben. Diese Konstruktion ist jedoch nicht rekursiv fortsetzbar. Im Gegensatz dazu erlauben die Bruck-Reilly-Erweiterungen eine rekursive Anwendung der Konstruktion: Man beginne mit einer beliebigen total H_o-disjunktiven einfachen Halbgruppe H_o und $\vartheta_o \in$ Endo H_o. $BR(H_o, \vartheta_o) = H_1$ ist dann total H_1-disjunktiv und einfach. Allgemein sei dann $H_{i+1} = BR(H_i, \vartheta_i)$ für einen beliebig gewählten Endomorphismus $\vartheta_i \in$ Endo H_i.

Die Struktur von B läßt allerdings erwarten, daß die Endomorphismen von Bruck-Reilly-Erweiterungen spezielle Eigenschaften haben. Es gilt

7.1. Satz. Für $\vartheta_1 \in$ Endo $BR(H, \vartheta_o)$ gilt:

(a) $|BR(H, \vartheta_o)\vartheta_1| = 1$

oder

(b) $BR(H, \vartheta_o)\vartheta_1$ ist einer Unterhalbgruppe von H isomorph,

oder

(c) $\overline{x\vartheta_1} = \bar{x}$ und $\exists\, \hat{\vartheta} \in$ Endo H mit $h_x\vartheta_1 = h_x\hat{\vartheta}$.

Beweis: Da jedes nicht-triviale homomorphe Bild von B zyklische Gruppe ist, B aber $\mathcal{H}$-trivial ist, folgt, daß entweder $BR(H, \vartheta_o)\vartheta_1$ einer Unterhalbgruppe von H isomorph ist oder daß $\overline{x\vartheta_1} = \bar{x}$ für alle $x \in BR(H, \vartheta_o)$ gelten muß. Im zweiten Falle wirkt ϑ_1 wie ein Endomorphismus von H auf die mittlere Komponente. □

In der Tat sind alle Situationen aus 7.1 möglich.
Sei etwa

$$H = \mathcal{Z}_2 \times \mathcal{Z}_3 = \{0,1\} \times \{0,1,2\}$$

das direkte Produkt der zyklischen Gruppen der Ordnungen 2

und 3, ϑ_o die identische Abbildung,

$$\hat{\vartheta} : H \to H : (a,b) \mapsto (0,b).$$

Es ist $\hat{\vartheta} \in$ Endo H und $H\hat{\vartheta} \simeq \mathfrak{Z}_3 = \{0,1,2\}$.

Die Abbildung

$$\vartheta_1 : BR(H,\vartheta_o) \to BR(H,\vartheta_o) : (m,h,v) \mapsto (0,h\hat{\vartheta},0)$$

ist ein Endomorphismus der Art (b). Andererseits ist

$$\vartheta_1' : BR(H,\vartheta_o) \to BR(H,\vartheta_o) : (m,h,n) \mapsto (m,h\hat{\vartheta},n)$$

von der Art (c).

8. Interpretation für die Theorie der formalen Sprachen

Im Hinblick auf die Frage, wie die bisherigen Aussagen für Bruck-Reilly-Erweiterungen im Sinne der Theorie der formalen Sprachen zu interpretieren sind, ist es zunächst sinnvoll, für Bruck-Reilly-Erweiterungen eine abstrakte Präsentation anzugeben.

8.1. Satz. H sei durch die Präsentation $\langle G,R\rangle$ mit Erzeugendenmenge G und Relationenmenge R definiert. Sei $\vartheta \in$ Endo H, und für jedes $g \in G$ sei eine Darstellung von $g\vartheta$ als Wort über G gegeben. a,b,e seien neue Symbole. Dann wird $S = BR(H^1, \vartheta)$ durch die folgende Präsentation $\langle G',R'\rangle$ definiert:

$$G' = G \cup \{a,b,e\},$$

$$R' = R \cup \bar{R},$$

wobei $\bar{R}$ aus den folgenden Relationen besteht ($h \in G$):

$$he = eh = h$$
$$ab = e$$
$$hb = bh\vartheta$$
$$ah = h\vartheta a$$
$$ae = ea = a$$
$$be = eb = b$$
$$ee = e$$

Ferner $e = 1_H$, *falls* H *ein Monoid ist*.

Beweis: G'^+ sei die Menge aller Wörter über G'. Wir betrachten die Abbildung

$$\phi : G' \to \mathbb{N}_0 \times G \times \mathbb{N}_0,$$

die durch

$$h \mapsto (0,h,0),\ h \in G,$$

$$e \mapsto \begin{cases} (0,e,0), & \text{falls } H \neq H^1, \\ (0,1_H,0) & \text{sonst,} \end{cases}$$

$$a \mapsto \begin{cases} (0,e,1) & \text{falls } H \neq H^1 \\ (0,1_H,1) & \text{sonst} \end{cases}$$

$$b \mapsto \begin{cases} (1,e,0) & \text{falls } H \neq H^1 \\ (1,1_H,0) & \text{sonst} \end{cases}$$

gegeben wird. Offensichtlich induziert ϕ einen Homomorphismus der von <G',R'> definierten Halbgruppe T auf $BR(H^1, \vartheta)$. Andererseits sieht man leicht, daß jedes Element von T genau eine der folgenden "Normalformen" besitzt: $b^m h a^n$ und eventuell $b^m e a^n$ mit $h \in H$, $n, m \in \mathbb{N}_0$. Damit folgt die Injektivität der Abbildung ϕ. □

Das Relationensystem vereinfach sich natürlich, wenn H ein Monoid ist. In diesem Falle wählt man

$$G' = G \cup \{a,b\}$$

$$R' = R \cup \bar{R}$$

mit $\bar{R}$ bestehend aus:

$$\left.\begin{array}{l} a\,b = 1_H \\ a\,h = h\vartheta\,a \\ h\,b = b\,h\vartheta \end{array}\right\} \quad h \in G.$$

Diesen einfacheren Fall wollen wir im Hinblick auf formale Sprachen weiter untersuchen. Der von $x \in BR(H, \vartheta)$ erzeugte Rechtsoperand ist in Abbildung 1 dargestellt.

Dabei sind die Rechtecke als Darstellung von Übergangsgraphen für Operationen aus H zu interpretieren, spezifiziert durch die mit $H\vartheta^i$ beschrifteten Schleifen.

Interpretiert man nun diesen Operanden als Akzeptor $\mathfrak{A}$ mit Zustandsmenge xS, Startzustand x, Stoppzustand x, Eingabealphabet G' und durch die Rechtstranslationen definierter Übergangsfunktion, so hat man global die Struktur einer Zählermaschine zum Akzeptieren von $\bar{x} \in B$, jedoch an jeder Stelle versehen mit einer aus der Halbgruppe H abzuleitenden speziellen Übergangsstruktur. Dabei bewirkt die Eingabe a ein Weiterschalten des Zählers. Ist der Zähler bei n, so entspricht die Übergangsstruktur für Eingabesymbole aus G der regulären Darstellung von $H\vartheta^n$; mit wachsendem n kann der Akzeptor die Elemente von G* also "immer schlechter unterscheiden". b bewirkt ein Zurücksetzen des Zählers. Damit folgt:

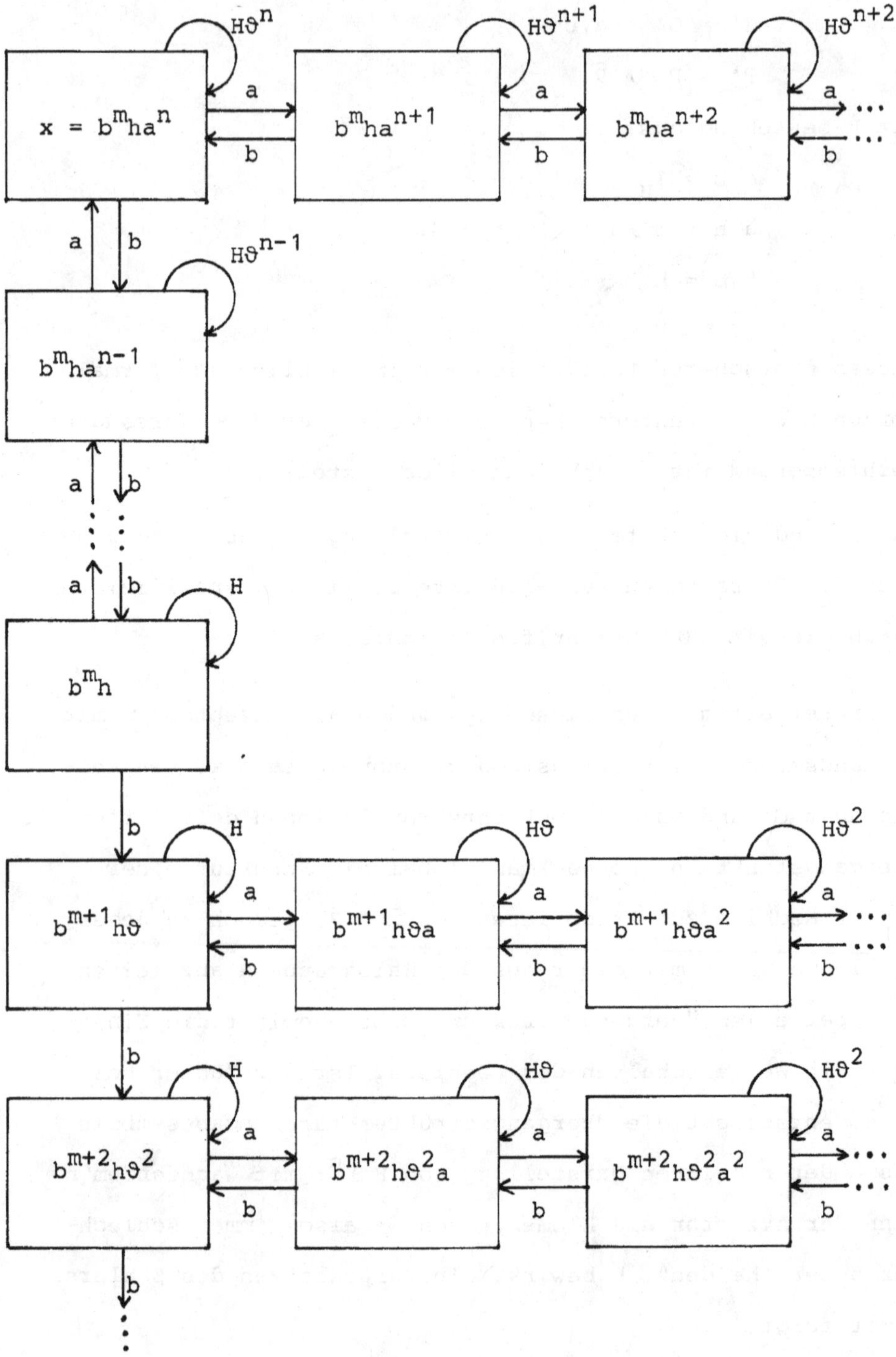

x = b^m h a^n
b^m h a^{n+1}
b^m h a^{n+2}
b^m h a^{n-1}
b^m h
b^{m+1} hϑ
b^{m+1} hϑa
b^{m+1} hϑa^2
b^{m+2} hϑ^2
b^{m+2} hϑ^2 a
b^{m+2} hϑ^2 a^2
Hϑ^n
Hϑ^{n+1}
Hϑ^{n+2}
Hϑ^{n-1}
H
Hϑ
Hϑ^2
a
b

ABBILDUNG 1

8.2. Lemma. H sei eine endliche einfache total H-disjunktive Halbgruppe, S = BR(H, ϑ) und x ε S. Jede Sprache L, deren syntaktische Halbgruppe S und deren Bild in S das Element x ist, ist eine deterministische kontextfreie Sprache.

Beweis: Es sei <G',R'> die Präsentation von S gemäß Satz 8.1, <G,R> die Präsentation von H.

Es ist ein deterministischer Kellerautomat $\mathfrak{A}$ zu definieren. Ohne wesentliche Einschränkung der Allgemeinheit wollen wir x = h ε H voraussetzen. Andere x wird man durch geeignetes Vorbesetzen des Kellers behandeln.

Weil H endlich ist, ist die Folge

$$\vartheta^0, \vartheta^1, \vartheta^2, \ldots$$

schließlich periodisch, etwa

$$\vartheta^0 \neq \vartheta^1 \neq \vartheta^2 \neq \ldots \neq \vartheta^r \neq \vartheta^{r+1} \neq \ldots \neq \vartheta^{r+p} = \vartheta^r.$$

Als Kelleralphabet wählt man die Menge

$$\{\#, 0, 1, 2, \ldots, r+p-1\}$$

Eas Eingabealphabet von $\mathfrak{A}$ sei G', die Zustandsmenge sei H $\dot{\cup}$ 1 $\dot{\cup}$ O. Startzustand sei 1, Stoppzustand sei x, Anfangsinschrift auf dem Keller sei O. Es muß jetzt die Wirkung der Eingabe beschrieben werden.

Eingabe	oberstes Kellersymbol	Wirkung auf Keller	Zustand h'
h	#	keine	h'h
	i	keine	$h'h\,\vartheta^i$
a	#	push:0, falls $r + p - 1 = 0$, sonst 1.	h'
	i	push: i+1, falls $i < r+p-1$, sonst r	h'
b	#	keine	0
	i	pop;	h'
e	#	keine	h'
	i	keine	h'

$\mathbb{A}$ ist offenbar ein deterministischer Kellerakzeptor für eine der im Lemma genannten Sprachen. Jede andere derartige Sprache (für festes S und x) geht aus dieser durch inverse Homomorphismen hervor [6] und ist daher ebenfalls deterministisch-kontext-frei. □

Wir wollen jetzt wie in Abschnitt 7 eine Folge $H = (H_o, H_1, H_2, \ldots)$ von Halbgruppen und $\Theta = (\vartheta_o, \vartheta_1, \vartheta_2 \ldots)$ von Endomorphismen betrachten, so daß gilt: H_o ist endliche, einfache, total H_o-disjunktive Halbgruppe, $\vartheta_i \in \text{Endo}\, H_i$, $H_{i+1} = BR(H_i, \vartheta_i)$. Ferner sei für jedes i ein $x_i \in H_i$ fest gewählt.

<u>8.3. Satz.</u> $k \in \mathbb{N}_o \cup \{\infty\}$ <u>sei minimal mit der Eigenschaft, daß</u> $H_k \vartheta_k \nsubseteq H_o$. <u>Dann gilt für jede Sprache</u> L_i <u>mit</u> H_i <u>als syntaktischer Halbgruppe und</u> x_i <u>als Bild in</u> H_i:

(1) <u>Für</u> i = 0 <u>ist</u> L_i <u>rational</u>.

(2) <u>Für</u> i = 1,2,...,k-1 <u>ist</u> L_i <u>deterministisch kontext-frei</u>.

(3) <u>Für</u> i = k, k+1,... <u>ist</u> L_i <u>nicht kontext-frei, aber kontext-abhängig</u>.

<u>Beweis:</u> Die Aussage (1) ist bekannt. Zum Beweis der Aussage (2) hat man das Lemma 8.2 zu verallgemeinern; dazu beachtet man, daß, sobald ein Symbol a_j aus H_j mit $j \leq i$ auf dem Keller steht, alle Eingaben aus H_l mit $l < j$ nur auf die Zustandsmenge wirken. Im obersten Kellersymbol muß nur die Kellersituation für alle $l \leq j$ registriert sein. Diese offensichtliche Modifikation sei dem Leser überlassen. Es ist also nur die Aussage (3) noch zu beweisen. Ohne wesentliche Beschränkung der Allgemeinheit dürfen wir k = 1 voraussetzen. Ferner gilt die Aussage a fortiori, wenn sie schon für $|H_o| = 1$ gilt. Dies sei also ebenfalls vorausgesetzt. Schließlich können wir dann auch o.B.d.A.

$$\mathbf{x} = (0,0,1,0,0)$$

wählen. Nach Voraussetzung ist $H_1 \vartheta_1 \nsubseteq H_o = \{1\}$, also $\vartheta_1 = id_{H_1}$ mit Satz 7.1.

Als Präsentation für H_1 erhält man

$$G_1 = \{a_1,b_1\},\ R_1 = \{a_1b_1 = 1\}$$

und für H_2

$$G_2 = \{a_2,b_2,a_1,b_1\}$$

$$R_2 = \{a_1b_1 = 1,\ a_2b_2 = 1,\ a_1a_2 = a_2a_1,\ a_1b_2 = b_2a_1,\ b_1a_2 = a_2b_1,\ b_1b_2 = b_2b_1\}.$$

Das Urbild L von 1 über dem Alphabet G_2 ist also das Mischprodukt[1] $D = D(a_1,b_1) ш D(a_2,b_2)$ der Dycksprachen $D(a_1,b_1)$, $D(a_2,b_2)$. Es ist zu zeigen, daß D nicht kontextfrei, aber kontextabhängig ist.

Daß D kontext-abhängig ist, sieht man leicht: Man konstruiert in naheliegender Weise einen linear beschränkten (sogar deterministischen) Akzeptor für D[2].

Zum Beweis, daß D nicht kontext-frei ist, verwenden wir das "pumping lemma": Für jede kontextfreie Sprache L existiert eine Konstante q, so daß für jedes Wort $r \varepsilon L$ mit $|r| \geqslant q$ eine nicht-triviale Aufteilung $uvwxy = r$ mit $uv^nwx^ny \varepsilon L$ für alle $n \varepsilon \mathbb{N} \cup \{0\}$ existiert, wobei aber $|vx| \geqslant 1$, $|vwx| \leqslant q$ gilt. Man nehme an, daß D kontextfrei ist, und q sei die Konstante aus dem "pumping lemma". Als $r \varepsilon D$ betrachte man

$$a_1^q a_2^q b_1^q b_2^q \ .$$

Sei schließlich $r = uvwxy$ eine Aufteilung, für die $|vx| \geqslant 1$ und $uv^nwx^ny \varepsilon D$ für $n = 0,1,2,...$ gilt. Enthält v ein a_i, so muß vwx auch ein b_i enthalten, und $|vwx| > q$. Enthält aber v kein a_i, aber ein b_i, so ist $vwx \varepsilon b_1^* b_2^*$ und folglich $uv^nwx^ny \notin D$ für $n > 1$. Schließlich ist noch möglich, daß v leer ist. Dann ist aber x nicht leer; enthält es ein b_i, so auch ein a_i, und $|vwx| > q$; enthält es kein b_i, aber

---

1) Für Sprachen L_1, L_2 ist

$$L_1 ш L_2 = \left\{ w \;\middle|\; \begin{array}{l} \exists n \varepsilon \mathbb{N},\ \exists u_1,...,u_n,\ \exists v_1,...,v_n: \\ w = u_1v_1u_2v_2...u_nv_n,\ u_1u_2...u_n \varepsilon L_1, v_1v_2..v_n \varepsilon L_2 \end{array} \right\}$$

2) Dies gilt natürlich auch ohne die speziellen Annahmen.

ein a_i, so ist $uv^nwx^ny \notin D$ für $n > 1$. Damit ist gezeigt, daß D nicht kontextfrei ist. □

Literatur

[1] A.H. Clifford, G.B. Preston: The algebraic theory of semigroups. Amer. Math. Soc., Providence, Rh.I.,1964, Vol. II, 1971.

[2] J.M. Howie: An introduction to semigroup theory, Academic Press, London, 1976.

[3] H. Jürgensen, G. Thierrin: Semigroups with each element disjunctive. Semigroup Forum 21 (1980/81),127-141.

[4] H. Jürgensen: Disjunktive Teilmengen inverser Halbgruppen. Arch. Math. (Brno) XV (1979), 205-208.

[5] J. Justin: Sur une construction de Bruck et Reilly. Semigroup Forum 3 (1971/72), 148-155.

[6] J. Sakarovitch: Monoides syntactiques et langages algébriques. Thèse 3ème cycle, Paris 1976.

[7] B.M. Schein: Homomorphisms and subdirect decompositions of semigroups. Pacific J. Math. 17 (1966), 529-547.

[8] B.M. Schein: A remark concerning congruences on [O-]bisimple inverse semigroups. Semigroup Forum 3 (1971/72), 80-93.

H. Jürgensen
Institut für theoretische Informatik
Technische Hochschule Darmstadt
Alexanderstr. 24
D - 6100 Darmstadt

CHARACTERIZATION OF MONOIDS BY PROPERTIES OF FINITELY GENERATED RIGHT ACTS AND THEIR RIGHT IDEALS

Ulrich Knauer

There exist quite a few papers describing monoids by different properties of their categories of right (left) acts or subclasses there of (often called homological classifications of monoids). Especially the properties of being torsion free, flat, projective, and free have been used for homological classifications in different combinations by Skornjakov [10], Dorefeeva [2], Kilp [6,7], Fountain [3], and in [8]. The purpose of this paper is to extend the results of the cited literature to the remaining cases. So all together there are given necessary and sufficient conditions on a monoid S in order that, for example, all flat right S-acts are free. These conditions are given for all meaningful variants of these concepts also considering the combinations where "all" is replaced by "all finitely generated" or "all cyclic" and "S-acts" is replaced by "S-ideals". On the way some restricted concept of perfectness of monoids is discussed (in Section 2).

In the following S will always stand for a monoid. A right S-act is a set A on which S acts unitarily from the right in the usual way, that is to say

$a(rs) = (ar)s$, $a1 = a$ ($a \in A, r, s \in S$, 1 the identity of S).

1. BASIC DEFINITIONS AND RESULTS

We recall the definitions of torsion free, flat, projective and free right S-acts. They can be found in [8], for example, and are given in their elementary non-homological versions. They originate in the theory of rings and modules.

1.1 DEFINITIONS. A right S-act is <u>torsion free</u> if as = bs, with $a,b \in A$ and s a right cancellable element of S, implies a = b.
A right S-act A is <u>flat</u> if as = bt, with $a,b \in A$ and $s,t \in S$, implies the existence of elements $c \in A$ and $s',t' \in S$ such that cs' = a, ct' = b and s's = t't. Morever, if a = b there exists $s' \in S$ such that cs' = a and s's = s't.
A right S-act A is <u>projective</u> if $A \cong \coprod e_i S$ for $e_i^2 = e_i \in S$, where the coproduct is the disjoint union.
A right S-act is <u>free</u> if $A \cong \coprod S$, S being considered as a right act.

For the definitions of the Green's relations $\mathcal{D}, \mathcal{R}, \mathcal{L}$ see for example [1].

1.2 PROPOSITION ([7]). *For any right S-act, we have the following implications:*

free → projective → flat → torsion free.

Next we give some elementary properties of flat and projective cyclic right S-acts.

1.3 LEMMA. *A cyclic right S-act xS is flat if and only if for xs = xt, $s,t \in S$ there exists $r \in S$ with xr = x and rs = rt.*
PROOF. Necessity . Assume that xs = xt for $s,t \in S$. As xS is flat there exists $z \in xS$, i.e. z = xv, $v \in S$ and there exists $s' \in S$ such that zs' = xvs' = x and s's = s't. Then vs's = vs't and with r = vs' the condition is fulfilled. Sufficiency is trivial.

1.4 LEMMA. *If the cyclic right S-act xS is flat then for any $s \neq t \in S$ with xs = xt there exists a chain $q_0, q_1, \ldots$ of not necessarily distinct elements $q_i \in S$, $i \in \mathbb{N}$ such that $xq_i = x$, $q_i q_{i-1} = q_i$, $i = 1,2\ldots$, and $q_0 s = q_0 t$.*
PROOF. The statement follows with 1.3 when applying the definition of flatness subsequently to xs = xt which im-

plies $q_0 s = q_0 t$ and $xq_0 = x$, then to $xq_0 = x$ and so on.

1.5 LEMMA. *If there exists a chain* $q_0, q_1, \dots$ *of elements* $q_i \in S$, $i \in \mathbb{N}$, *with* $q_i q_{i-1} = q_i$ *for all* $i = 1,2,\dots$, *consider the submonoid* $Q \subset S$ *generated by the set* $\{1, q_0, q_1, \dots\}$. *Define a relation* ρ *on* S *by* $s \rho t$, $s,t \in S$, *if there exist* $p,q \in Q$ *with* $s = pr$, $t = qr$, $r \in S$. *Denote by* $\sim$ *the transitive closure of* ρ. *Then* $xS = S/\sim$ *is a flat cyclic right* S-*act*.
PROOF. By construction $\sim$ is a right congruence on S, so (by 11.6 of [1]) xS is a cyclic right S-act. Assume that $xu = xv$, that is $u \sim v$, $u, v \in S$. Then we have for $i_j, j_j, k_j, l_j, m_j \dots \in \mathbb{N}$, $j = 1,2,\dots$,

$$u = q_{i_1} \dots q_{i_1} u_1 \rho\, q_{j_1} \dots q_{j_k} u_1 = q_{k_1} \dots q_{k_h} u_2 \;\rho\; q_{l_1} \dots q_{l_k} u_2$$

$$= \rho \dots \rho\; q_{m_1} \dots q_{m_i} u_m = v$$

where the subscripts of the elements q are (not necessarily strictly) growing. Assume that q_k is the element with the highest subscript in this line. Remark that if $c \rho d$ then $qc \rho qd$ and $c \rho qd$ for $q \in Q$, $c,d \in S$ since Q is a monoid. Now left multiplication with q_{k+1} gives

$$q_{k+1} u = q_{k+1} u_1 \;\rho\; q_{k+1} u_1 = q_{k+1} u_2 \;\rho\; q_{k+1} u_2 =$$

$$= \rho \dots \rho\; q_{k+1} u_m = q_{k+1} v.$$

So we have shown that $q_{k+1} u = q_{k+1} v$ where $xq_{k+1} = x$. That is, xS is a flat cyclic right S-act.

1.6 LEMMA. *A cyclic right* S-*act* xS *is projective if and only if there exists* $f^2 = f \in S$ *such that by* $\psi(x) = f$ *there is defined an isomorphism of right acts* $\psi: xS \to fS$. *Then* $xf = x$.
PROOF. Necessity. Let xS be projective. Then there exists $e^2 = e \in S$ such that $eS \cong xS$ with $\varphi: eS \to xS$ an isomorphism of acts. Let $\varphi(e) = xr$, $\varphi^{-1}(x) = er'$, $r, r' \in S$. Then $f = rer'$ is idempotent as $er'r = \varphi^{-1}(x)r = \varphi^{-1}(xr) = \varphi^{-1}\varphi(e) = e$ and thus $rer'rer = rer'$. Now define $\psi: xS \to fS$ by $\psi(x) = f$. Then ψ is well defined as $xu = xv$, $u,v \in S$ implies

$rer'u = r\varphi^{-1}(x)u = r\varphi^{-1}(xu) = r\varphi^{-1}(xv) = r\varphi^{-1}(x)v =$

$= rer'v$.

Moreover, ψ is injective as $rer'u = rer'v$, $u,v \in S$ implies $xu = xrer'u = xrer'v = xv$, since $xr = \varphi(e) = \varphi(e)e = xre$ and $xrr' = \varphi(e)r' = \varphi\varphi^{-1}(x) = x$. By definition ψ is surjedtive and a homomorphism.

Finally $xf = \psi^{-1}(f)f = \psi^{-1}(f) = x$.

Sufficiency is obvious.

2. SOME MODIFICATIONS OF PERFECT MONOIDS

We know that a monoid S is (right) perfect if and only if all flat right S-acts are projective [3]. The analogue of the original definition of perfectness coming from ring theory says that a monoid S is right perfect if every right S-act has a projective cover [5]. It turns out that in restricting the existence of a projective cover to cyclic right S-acts, monoids with this property form a wider class then those over which all flat cyclic right S-acts are projective.

First we restate 1.5 of [5] in the terminology of [3].

2.1 PROPOSITION. *Every cyclic right S-act has a projective cover if and only if*

(D) *every right unitary subsemigroup of S has minimal right ideal generated by an idempotent.*

Then Lemma 5 of [3] reads as follows.

2.2 LEMMA. *If every cyclic right S-act has a projective cover then every flat cyclic right S-act is projective.*

For the sake of completeness we also repeat Lemma 4 of [3].

2.3 LEMMA. *If S satisfies the minimum condition for principal left ideals, then every flat cyclic S-act is projec-*

tive.

The next example shows that the converses are not true in general.

2.4 EXAMPLE. Consider the monoid $(N,\cdot)$. As it is left cancellative it follows from the definition of flatness that all flat cyclic right S-acts are free (cf. 1.3) and thus projective. But (D) is not fulfilled as $\mathbb{N}$ itself does not have a minimal right ideal. So by 2.1 not every cyclic right $(\mathbb{N},\cdot)$-act has a projective cover. And of course the minimum condition for principal left ideals is not fulfilled either as the chain $\mathbb{N} \supset 2\mathbb{N} \supset 4\mathbb{N} \supset \ldots$ shows.

Analogous considerations can be made in restricting perfectness to the ideals of a monoid. Call a monoid S (right) <u>ideal perfect</u> if every right ideal of S has a projective cover. It turns out that ideal perfect monoids are not characterized by the condition that flat right ideals are projective. For principal ideals the situation is similar to cyclic acts. A condition similar to condition (D) in [3] (cf. 2.1) can be found and the following result corresponds to the result for cyclic acts (cf. 2.1).

2.5 PROPOSITION. <u>Every principal right ideal of S has a projective cover if and only if</u>
(D') <u>for</u> $s \in S$ <u>the monoid of right identities of</u> s <u>contains a minimal right ideal generated by an idempotent</u>.
PROOF. Necessity. For $s \in S$ consider $B = \{t \in S \mid st = s\}$. Using the proof of 1.5 in [5], we get (D').
Sufficiency. Let sS be a right ideal, $s \in S$ and $\varphi : S \to sS$ the canonic epimorphism. Then $B = \varphi^{-1}(S)$ contains a minimal right ideal rB generated by an idempotent r. Then as in the proof of 1.5 in [5] rS is a projective cover of sS.

Before we turn to ideal perfect monoids we need an ideal version of 1.2 in [5].

2.6 LEMMA. *Let S be a monoid. Equivalent are*

(i) *Every locally cyclic right ideal of S is cyclic.*

(ii) *S has no ascending chain of principal right ideals.*

(iii) *Flat right ideals of S are coproducts of principal flat right ideals.*

PROOF. (i) → (iii) is trivial since flat right ideals are locally cyclic.

(iii) → (ii): Consider a chain $s_1S \subset s_2S \subset \ldots$, $s_i \in S$, of right ideals in S. Then $\bigcup_{i\in I} s_iS$ is indecomposable. By [5] $\bigcup_{i\in I} s_iS$ is a direct limit of free acts which by [10] is flat. By construction $\bigcup_{i\in I} s_iS$ is an ideal, so it is cyclic. Thus (ii) is fulfilled.

(ii) → (i): Let $A \subset S$ be a locally cyclic right ideal of S which is not cyclic. For $a_1, a_2 \in A$ there exists $s_1 \in A$ such that $a_1, a_2 \in s_1S$. Now take $a_3 \in A$, $a_3 \notin s_1S$, then there exists $s_2 \in A$ such that $a_3, s_1 \in s_2S$.

Continuation of this procedure gives an infinite ascending chain of right ideals in S. This is a contradiction to (ii). Consequently, A must be cyclic.

Now we can characterize ideal perfect monoids.

2.7 THEOREM. *Let S be a monoid. Equivalent are*

(i) *Every locally cyclic right ideal of S is cyclic and every right principal ideal of S has a projective cover.*

(ii) *The ascending chain condition* (ACC) *for right principal ideals of S and* (D').

(iii) *S is ideal perfect*, *i.e. every right ideal of S has a projective cover.*

PROOF. The equivalence of (i) and (ii) is shown by 2.6 and 2.5. The equivalence of (i) and (iii) is proved as in 1.1 of [5].

2.8 COROLLARY. *If all principal right ideals of S have a projective cover then all flat principal right ideals of S are projective.*

Different from the situation for acts (cf. [3]) there exist monoids such that (D') is not fulfilled even if the flat principal ideals are projective and the ACC for principal right ideals is fulfilled. This is shown by the next example.

2.9 EXAMPLE. Consider the monoid Z generated by the following mappings

$P = \{p_i : \mathbb{Z} \to \mathbb{Z} \mid i \in \mathbb{N}\}$,

$N = \{n_i : \mathbb{Z} \to \mathbb{Z} \mid i \in \mathbb{N}\} \cup \{n_1'\}$, and $1_{\mathbb{Z}}$.

We define these mappings as follows ($k \in \mathbb{Z}$, arguments written on the left):

$$kp_i = \begin{cases} k+1 & \text{if } k \geq 0 \\ k & \text{if } k < 0 \end{cases},$$

$$kn_i = \begin{cases} 0 & \text{if } k > 0 \\ k-1 & \text{if } k \leq 0 \end{cases},$$

and $kn_1' = \begin{cases} -2 & \text{if } k = 0 \\ kn_1 & \text{otherwise} \end{cases}$.

Then one gets the following equations:

$p_i p_j = p_{i+j}$,

$n_i n_j = n_{i+j}$,

$n_i p_j = p_i$,

$n_1' p_i = n_1'$,

$$kp_i n_j = \begin{cases} 0 & \text{if } k = 0 \\ kn_j & \text{otherwise,} \end{cases}$$

$p_i n_1' = p_i n_1$ (since $0p_i n_1 = in_1 = 0 = in_1' = 0p_i n_1'$),

$n_i n_1' = n_i n_1$,

$$kn_1'n_i = \begin{cases} (-1)n_1n_i & \text{if } k = 0 \\ kn_1n_i & \text{otherwise} \end{cases}, \text{ for all } i,j \in \mathbb{N},$$

and $n_1'n_1' = n_1'n_1$.

For $x \in S$ assume that xZ is flat. If $xs = st$ then there exist $r,q \in Z$ such that $rs = rt$, $xr = r, qr = q$. Now if $s \neq t \in P$ or $s \in P, t = 1$ it follows that $r \in N$, but then q does not exist. If $s \in P$ and $t \in N$ or $s \neq t \in N \setminus \{n_1'\}$ then r does not exist. But if $s = n_1$, $t = n_1'$ then $n_i n_1 = n_i n_1'$ but q does not exist. Finally if $s = n_1$, $t = n_1'$ then $n_i n_1 = n_i n_1'$ but q does not exist. Thus $s = t$ consequently over Z all flat cyclic right ideals are free. Now $B = P \cup \{1_{\mathbb{Z}}\}$ is the monoid of right identities of every $n \in N$. The existence of the infinite chain $p_1B \supset p_2B \supset \dots$ of right ideals in B shows that B has no minimal right ideal, that is, (D') is not fulfilled. Obviously Z fulfills the ACC for principal right ideals.

3. CHARACTERIZATION OF MONOIDS BY IDEALS

3.1 REMARK. All right ideals of a monoid S are torsion free. And consequently all (all finitely generated, all principal) torsion free right ideals of S are free (projective, flat) if and only if all (all finitely generated, all principal) right ideals of S are free. By the structure of projective right acts it becomes clear that a projective right ideal is free if and only if it is free as right act. Consequently all projective right ideals are free if and only if $e \mathcal{D} 1$ for all $e^2 = e \in S$.

Before we recall some results of [6,7] and [2], we define some sorts of weakened cancellability of elements in monoids.

3.2 DEFINITION. An element $u \in S$ is called <u>left semi cancellable</u> if for $us = ut$, $s,t \in S$ there exists $r \in S$ such that ur = u and rs = rt. An element $u \in S$ is called <u>left</u> e-<u>cancellable</u> if there exists $e \in S$ such that us = ut, s, $t \in S$, implies ue = u and es = et.

So obviously every left e-cancellable element is left semi cancellable and any idempotent e is left e-cancellable.

Now we can state 1.3 for ideals als follows.

3.3 COROLLARY. *A principal right ideal aS, $a \in S$, is flat if and only if aS = uS, $u \in S$ where u is left semi cancellable.*

3.4 LEMMA. *A right ideal of S is generated by a left cancellable element if and only if it is generated by a left e-cancellable element and $e \mathcal{D} 1$, $e^2 = e \in S$.*

PROOF. Let aS be such that a is left cancellable. Then a S = 1S, that is a is left 1-cancellable. Now consider uS where u is left e-cancellable, $e^2 = e \in S$. By Lemma 1 of [2] we get that uS = eS. As $e \mathcal{D} 1$ we have $e \mathcal{R} a$, $a \mathcal{L} 1$, consequently eS = aS and a is left cancellable.

3.5 THEOREM.

(i) *All principal right ideals of S are (projective) free if and only if every principal right ideal of S is generated by a left (e-) cancellable element (for $e^2 = e \in S$).*

(ii) *All principal right ideals of S are flat if and only if every principal right ideal of S is generated by a left semi cancellable element.*

(iii) *All finitely generated right ideals of S are free (projective, flat) if and only if all principal right ideals of S are free (projective, flat) and*

  (c) *all finitely generated right ideals of S are coproducts of principal right ideals (or equivalently, all non-comparable principal right ideals are disjoint).*

(iv) *All right ideals of S are free (projective) if and*

<u>only if all finitely generated right ideals of</u> S <u>are free</u> (<u>projective</u>) <u>and the</u> ACC <u>for principal right ideals of</u> S <u>is fulfilled</u>.

(v) <u>All right ideals of</u> S <u>are flat if and only if all finitely generated right ideals of</u> S <u>are flat</u>.

PROOF. The proofs are in [2], [7], or [6], (ii) is a reformulation of Prop. 2 in [7] using 3.2. For (i) see also 3.4.

We recall ([2]) that a monoid, all (finitely generated) right ideals of which are projective is called (<u>semi-</u>) <u>hereditary</u>. A monoid all principal right ideals of which are projective is called <u>right</u> PP <u>monoids</u> (cf. [4]).

3.6 EXAMPLES.

1. In the monoid $(\mathbb{N},.)$ all principal right ideals are generated by left cancellable elements but (c) is not fulfilled, take for example the right ideal $2\mathbb{N}\cup 3\mathbb{N}$ Thus in $(\mathbb{N},.)$ all principal right ideals are free but not all finitely generated right ideals are free.

2. Let $B = \{(m,n) \mid n, m \in \mathbb{N}_0\}$ be the bicyclic monoid with the multiplication $(m,n)(k,l) = (m+k-\min(n,k), n+l-\min(n,k))$, $k, l, m, n \in \mathbb{N}_0$, cf [1]. Then B fulfills (c) and the ACC for principal right ideals. Moreover $(m,n)B$ is generated by (m,o) which is left cancellable. Consequently all right ideals of B are free.

3. Consider $R = G \cup (\mathbb{N},.)$ where G is a group and $g_1g_2 \in G$, $n_1n_2 \in \mathbb{N}$, $gn = ng = n$, for all $g_1, g_2, g \in G$, $n_1, n_2, n \in \mathbb{N}$. If $us = ut$ in R then either $s = t$ or $u \in \mathbb{N}$, $s, t \in G$. But then $u1_{\mathbb{N}} = u$ and $1_{\mathbb{N}}s = 1_{\mathbb{N}}t$. So in the first case u is left 1_R-cancellable and in the second case u is left $1_{\mathbb{N}}$-cancellable. Thus all elements of R are left e-cancellable and consequently all principal right ideals of R are projective, but not all of them are flat as no $n \in \mathbb{N}$ is left cancellable, and any right ideal gen-

erated by $n \in \mathbb{N}$, $n \neq 1_{\mathbb{N}}$, is generated only by n.

4. In [7] there is given an example of a monoid such that all (principal) ideals are flat but not projective.

3.7 THEOREM.

(i) All flat principal right ideals of S are (projective) free if and only if every principal right ideal generated by a left semi cancellable element of S is generated by a left (e-) cancellable element (for some $e^2 = e \in S$).

(ii) All flat finitely generated right ideals of S are free (projective) if and only if all flat principal right ideals of S are free (projective).

(iii) All flat right ideals of S are free (projective) if and only if all flat principal right ideals of S are free (projective) and the ACC for principal right ideals of S is fulfilled.

PROOF. Using Lemma 1 of [2] and 3.4, we can prove (i) immediately. For (ii) recall that by [11] every finitely generated right S-act is a coproduct of flat cyclic right S-acts. For the proof of (iii) use the equivalence of (ii) and (iii) in 2.6.

3.8 EXAMPLES.

1. Let G be a group and consider $S = G \cup (\mathbb{N} \setminus \{1\}$, with multiplication $g_1 g_2 \in G$, $n_1 n_2 \in \mathbb{N} \setminus \{1\}$, and $ng = gn = n$ for all $g_1, g_2, g \in G$, $n_1, n_2, n \in \mathbb{N} \setminus \{1\}$. If $us = ut$ then either $s = t$ or $u \in \mathbb{N} \setminus \{1\}$. In the latter case u is not left semi cancellable. So all left semi cancellable elements of S are left cancellable. Consequently all finitely generated flat right ideals are free.

2. Let $S = \{0, 1, 2, 3\}$ with multiplication table

	0	1	2	3
0	0	0	0	0
1	0	1	2	3
2	0	2	0	2
3	0	3	2	1

.

Then 0 is left 0-cancellable, 1 and 3 are left cancellable, and 2 is not left semi cancellable. Consequently all flat right ideals are projective, not all of them are free, as 0 is not left cancellable. Moreover not all right ideals are flat, as 2S = {0,2} is not generated by a left semi cancellable element.

4. CHARACTERIZATION OF MONOIDS BY FINITELY GENERATED AND CYCLIC ACTS

This section characterizes monoids over which finitely generated or cyclic acts fulfill the meaningful implications between the concepts torsion free, flat, projective and free. Firstly we recall the results which imply that the regarded monoid is trivial.

4.1 THEOREM. *The following conditions on a monoid* S *are equivalent*

(i) *All cyclic right* S-*acts are free*.
(ii) *All cyclic right* S-*acts are projective*.
(iii) *All cyclic right* S-*acts are flat*.
(iv) *All torsion free cyclic right* S-*acts are free*.
(v) *All torsion free cyclic right* S-*acts are projective*.
(vi) *All torsion free cyclic right* S-*acts are flat*.
(vii) S = {1} *or* S = {0,1}.

PROOF. (cf. [5], which contains the portions without torsion free). Obviously (vii) implies anything else. Conversely it is clear that (i) → (ii) → (iii), (iv) → (v) → (vi), (i) → (iv), (ii) → (v), and (iii) → (vi). In Theorem 4 of [5] the implication (iii) → (vii) is proved. To prove the implication (vi) → (vii), assume that there exists $a \in S$, $a \neq 1,0$. Consider the submonoid of S generated by {1,a}. Define a relation ρ by $s\rho t$ if $s = a^n r$, $t = a^m r$, $r \in S$ and let $\sim$ be the transitive closure of ρ. Then $S/\sim$ = xS is a right S-act generated by the class of 1. We show that xS is torsion free. Assume $ut \sim vt$ where $t \in S$ is right cancellable. Then this equivalence is given as

$$ut = a^{n_1}u_1 \rho a^{n_2}u_1 = a^{n_3}u_2 \rho a^{n_4}u_2 = \rho \cdots \rho a^{n_i}u_i = vt$$

If $n_2 > n_1$ then

$$a^{n_2-n_1}ut = a^{n_2}u_1 \rho a^{n_2}u_1 = \cdots,$$

that is we get

$$a^{n_2-n_1}ut = a^{n_3}u_2 \rho \cdots.$$

If $n_2 > n_1$ then $ut = a^{n_1}u_1 \rho a^{n_1}u_1$, so we get

$$ut = a^{n_3}u_2 \rho \cdots.$$

Proceeding this way one finally gets $a^p ut = a^q vt$ for suitable p and q. By hypothesis t is right cancellable and then $a^p u = a^q v$. By definition of $\sim$ this implies $u \sim v$ which shows that xS is torsion free. By (vi) xS then is flat. So $xa = x$ implies the existence of $r \in S$ such that $xr = x$ and $ra = r$. By the procedure applied to the situation $ut \sim vt$ we get from $r \sim 1$ that $r = a^q$ and thus $a^{q+1} = a^q$. But then no $a \neq 1$ is right cancellable. Consequently every cyclic right S-act is torsion free. So using the contraposition of (vii) we have shown the implication (vi) $\to$ (iii). But as (iii) $\to$ (vii) we have a contradition which completes the proof.

When considering finitely generated right S-acts one gets the analogue of 4.1 as in [5], cf. [8].

4.2 COROLLARY. _The following conditions on a monoid_ S _are equivalent_

(i) _All finitely generated right_ S-_acts are free_.
(ii) _All finitely generated right_ S-_acts are projective_.
(iii) _All finitely generated right_ S-_acts are flat_.
(iv) _All torsion free finitely generated right_ S-_acts are free_.

(v) _All torsion free finitely generated right S-acts are projective._

(vi) _All torsion free finitely generated right S-acts are flat._

(vii) S = {1}.

PROOF. Obviously (vii) implies anything else, and anything else implies S = {1} or S = {0,1} according to 4.1. If S = {0,1} then as in [6] the three element non-cyclic right S-act {0,a,b} with {0,a} ≅ {0,b} ≅ S as right S-acts, is torsion free but not flat.

Corollary 4.2 also answers the question left open in [8]:

4.3 COROLLARY. _All torsion free right S-acts are flat if and only if_ S = |1|.

4.4 REMARK. The difference between 4.1 and 4.2 of course disappears if monoids and acts with zero are considered in the corresponding category.

The following result extends 3.2 of [8] to cyclic acts using a different proof.

4.5 THEOREM. _All cyclic right S-acts are torsionfree if and only if all right cancellable elements of_ S _are right invertible._

PROOF. Sufficiency follows from 2.2 of [8].

Necessity. Let $s \in S$ be right cancellable but not right invertible. Then s is not left invertible since otherwise $s's = 1$ implies $ss's = s$ and by right cancellation $ss'=1$. Then the ideal SsS in particular gives a right congruence $\sim$ on S and $A := S/\sim$ is a cyclic right S-act generated by the class of the identity of S. As $1 \notin SsS$ we have in A that $a \not\sim 1s = s$, but $1s \sim 1s^2$ as $s, s^2 \in SsS$. That is A is not torsion free.

From this follows

4.6 COROLLARY. _All finitely generated right S-acts are torsionfree if and only if all right cancellable elements_ *of S are right invertible.*

4.7 COROLLARY. *The following conditions on S are equivalent*

(i) *All projective right S-acts are free.*

(ii) *All projective finitely generated right S-acts are free.*

(iii) *All cyclic projective right S-acts are free.*

(iv) *All projective right ideals of S are free.*

(v) *All projective finitely generated right ideals of S are free.*

(vi) *All projective principal right ideals of S are free.*

(vii) $e \mathcal{D} 1$ *for all idempotents* e *of* S.

PROOF. (i) $\to$ (ii) $\to$ (iii) $\to$ (iv) $\to$ (v) $\to$ (vi) is trivial.
(vii) $\leftrightarrow$ (i) is proved in [8].
(vi) $\to$ (i) since a projective right S-act is isomorphic to the disjoint union of projective principal right ideals of S.

Now we turn to the remaining cases connecting flat, projective, and free acts.

4.8 LEMMA. *If all flat cyclic right S-acts are projective, then*

(FP_1) *For all infinite chains* $q_0, q_1, \ldots \in S$ *with* $q_i q_{i-1} = q_i$, $i = 1,2,\ldots$, *there exists* $m \in \mathbb{N}$ *such that* $q_m q_i = q_m$ *for all* $i = 0,1,\ldots$.

PROOF. Let $q_0, q_1, \ldots \in S$ be a chain with $q_i q_{i-1} = q_i$ for $i = 1,2,\ldots$. Define $S/\sim\; = xS$ as in 1.5. Then xS is a flat cyclic right S-act by 1.5. By hypothesis xS is projective, so by 1.6 there exists an isomorphism of right S-acts $\psi : xS \to fS$, with $\psi(x) = f$, $f^2 = f \in S$. From $xq_i = x$ it follows that $fq_i = f$ for all $i = 0,1,\ldots$, and we have $xf = x$. This implies $f = q_{i_1} \ldots q_{i_n} u_1 \,\rho\, q_{j_1} \ldots q_{j_m} u_1 =$
$= q_{k_1} \ldots q_{k_l} u_2 \,\rho = \ldots \rho\, q_{l_1} \ldots q_{l_i} u_l = 1$, where $u_j \in S$.
Let m be greater than all subscripts of all these elements q. Then left multiplication by q_m gives $q_m f = q_m$.

Now these two properties of f imply that
$q_m q_i = q_m f q_i = q_m f = q_m$ for all $i \in \mathbb{N}$.

4.9 LEMMA. *Let* $M \subset S$ *be a set of elements with the property:*
For $e_1, \cdots e_n, f_1, \cdots f_m \in M$ *there exists* $e \in M$ *such that*

$$ee_1 \cdots e_n = ef_1 \cdots f_m.$$

Consider the submonoid $Q \subset S$ *generated by* $M \cup \{1\}$. *Define a relation* ρ *on* S *by* $s \rho t$ *if* $s = pr$, $t = qr$ *for* $p, q \in Q$, $r \in S$.
Let $\sim$ *be the transitive closure of* ρ *on* S.
Then $S/\sim \; = xS$ *is a flat cyclic right* S-*act*.

PROOF. As $\sim$ by definition is a right congruence, xS is a cyclic right S-act.
Assume that $xu = xv$, that is $u \sim v$, $u, v \in S$.
Then for $e_{i_j}, e_{j_j}, e_{k_j}, e_{l_j}, \cdots \in M$, $j = 1, \cdots$; and $u_i \in S$ we have

$$u = e_{i_1} \cdots e_{i_n} u_1 \,\rho\, e_{j_1} \cdots e_{j_m} u_1 = e_{k_1} \cdots e_{k_l} u_2 \,\rho\, e_{l_1} \cdots e_{l_i} u_2 =$$
$$= \rho \cdots \rho\, e_{m_1} \cdots e_{m_j} = v.$$

By hypothesis there exists e_i with $e_i e_{i_1} \cdots e_{i_n} = e_i e_{j_1} \cdots e_{j_m}$. Then

$$e_i u = e_i e_{i_1} \cdots e_{i_n} u_1 \,\rho\, e_i e_{j_1} \cdots e_{j_m} u_1 = e_i e_{k_1} \cdots {}_{k_l} u_1 \,\rho = $$
$$= \cdots e_i v.$$

By hypothesis there exists e_j with $e_j e_i e_{k_1} \cdots e_{k_l} = e_j e_i e_{l_1} \cdots e_{l_i}$. Then

$$e_j e_i u = \cdots = e_j e_i e_{l_1} \cdots e_{l_i} = \rho \cdots = e_j e_i v$$

Proceeding in this manner we finally get $e_m \cdots e_j e_i u = e_m \cdots e_j e_i v$. That is $ru = rv$ for $r = e_m \cdots e_j e_i$ and thus $xr = x$.
Consequently xS is flat.

Now we can formulate the desired characterization of monoids over which all flat cyclic right S-acts are projective. The conditions are not very elegant but they reflect conditions on S-acts into S.

4.10 THEOREM. <u>All flat cyclic right</u> S-<u>acts are projective if and only if</u>

(FP_1) <u>For all (infinite) chains</u> $q_0, q_1, \cdots$ <u>with</u> $q_i\, q_{i-1} = q_i$, $i = 1,2,\cdots, q_i \in S$ <u>there exists</u> $m \in \mathbb{N}$ <u>such that</u> $q_m q_i = q_m$ <u>for all</u> $i = 0,1,\cdots$.

<u>and</u>

(FP_2) <u>For any set</u> M <u>of idempotents of</u> S <u>with the property "for</u> $e_1,\cdots,e_n,f_1,\cdots,f_m \in M$ <u>there exists</u> $f \in M$ <u>such that</u> $fe_1\cdots e_n = ff_1\cdots f_m$"

<u>the subsemigroup of</u> S <u>generated by</u> M <u>contains a left zero</u>

PROOF. Sufficiency. Let xS be a flat cyclic right S-act. Then for all $q_0 \in S$ with $xq_0 = x$ consider chains $q_0, q_1, \cdots$ with $q_i q_{i-1} = q_i$, $i = 1,2,\cdots$ of elements $q_i \in S$ and $xq_i = x$. Let M be the set of idempotents received from these chains $q_0, q_1, \cdots$ by (FP_1). Then for any $e \in M$ one has $xe = x$ by(FP_1). M has the property required for (FP_2). For take $e_1,\cdots,e_n,f_1,\cdots f_m \in M$ then $xe_1\cdots e_n = xf_1\cdots f_m = x$. As xS is flat, there exists a chain $q_0, q_1, \cdots$ as above such that $q_0 e_1\cdots e_n = q_0 f_1 \cdots f_m$ and by definition of M and with (FP_1) the set M contains q_m such that $q_m q_i = q_m$ for all $i = 0,1,\cdots$, and $q_m^2 = q_m$ for some m. Then $q_m e_1 \cdots e_n = q_m f_1 \cdots f_m$. Using ($FP_2$) the subsemigroup of S generated by M contains a left zero element e for which consequently $xe = x$. Show that $xS \cong eS$. By defining $\varphi(x) = e$ one gets a mapping, as $xs = xt$ implies $es = et$ by definition of M and e. This mapping is injective as $es = et$ implies $xs = xes = xet = xt$. By definition φ is surjective and a homomorphism of the right S-act xS onto the right S-act eS. So xS is projective.

Necessity. By 4.8 the hypothesis implies (FP_1). Let M be a set of idempotents of S which fulfills the condition required for (FP_2). Then by 4.9 we define xS which is flat. Then, by hypothesis xS is projective. By 1.6 there exists $e^2 = e \in S$ such that $\psi(x) = e$ defines an isomorphism of right S-acts, $\psi : xS \to eS$. Now $xe = x$, thus for any $f \in M$ we get $e = \psi(x) = \psi(xf) = \psi(x)f = ef$. To get the left zero element for M which belongs to the subsemigroup generated by M proceed as follows.

Since $xe = x$, that is $e \sim 1$ one has the existence of elements $e_{i_j} \in M$, $u_i \in S$, such that

$$e = e_{i_1} \cdots e_{i_n} u_1 \rho e_{j_1} \cdots e_{j_m} u_1 = e_{k_1} \cdots e_{k_l} u_2 \rho e_{l_1} \cdots e_{l_i} u_2 =$$
$$= \rho \cdots \rho e_{m_1} \cdots e_{m_i} u_m = 1.$$

As in the proof of 4,9 this gives $e_m \cdots e_j e_i e = e_m \cdots e_j e_i$. Set $e_m \cdots e_j e_i = f$, this element belongs to the semigroup generated by M. Then f is a left zero element for M, as $f = fe$ and thus $ff' = fef' = fe = f$ from the above for any f' M. So f is the left zero wanted.

4.11 COROLLARY. If all flat cyclic right S-acts are projective then S does not have an infinite descending chain of idempotents with respect to their natural order.

4.12 COROLLARY. All flat finitely generated right S-acts are projective if and only if all flat cyclic right S-acts are projective.

PROOF. By [10] flat finitely generated right S-acts are coproducts of cyclic S-acts.

4.13 REMARK. (FP_1) in particular is fulfilled in S if we have the minimal condition for cyclic left ideals (M_L) in S, consider $Sq_0\ Sq_1 \cdots$. (FP_2) in particular is fulfilled in S if we know that every right unitary subsemigroup of S contains a minimal right ideal generated by an idempotent (D).

Any of these two conditions (M_L) or (D) together with the ACC for cyclic subacts (A) is equivalent to saying that all flat right S-acts are projective (cf [3]).

Next we give examples of monoids which fulfill none, either one, or both of the conditions (FP_1), (FP_2).

4.14 EXAMPLES.

1. Consider the bycyclic monoid $R = \{(m,n) \mid m,n \in \mathbb{N}\}$ with multiplication
$(m,n)(k,l) = (m + k - \min(n,k), n+l - \min(n,k))$, $k,l,m,n \in \mathbb{N}_0$
R fulfills neither (FP_1) nor (FP_2).
The element (0,0) is the identity and idemptents are of the form (n,n), $n \in \mathbb{N}_0$. To get a right congruence $\sim$ on R define the class of the identity as $[(0,0)]_\sim = \{(i,i) \mid i \in \mathbb{N}_0\}$. Then $S/\sim = xS$ can be represented by
$xS = \{x, x(0,1), x(0,2), \dots\}$
where $x(n,m) = x(n+i, m+i)$ for all $n,m,e \in \mathbb{N}_0$ and these are the only elements $s,t \in R$ with $sx = xt$. Now xS is flat as $x(n,m) = x(n+i, m+i)$ implies $(n+i,n+i)(n,m) = (n+i, n+i)(n+i, m+i)$, $x(n+i, n+i) = x$ (cf. 1.3). But xR is not projective as for any idempotent $(j,j) \in R$ the right R-act $(j,j)R$ is not isomorphic to xR when defining $\psi : xR \to (j,j)R$ by $\psi(x) = (j,j)$ (cf. 1.6). Indeed $(j,j)(j+1, j+1) = (j+1, j+1) \neq (j,j)$ but $x(j+1,j+1) = x$, so ψ is not well defined. In this case the set of idempotents $\{(0,0),(1,1),(2,2),\dots\}$ does not have a left zero element, that is (FP_2) is not fulfilled. At the same time this chain of idempotents has the property required for (FP_1). So (FP_1) is not fulfilled either.

2. The following monoid fulfills (FP_2) but not (FP_1).
Consider mappings $q_i : [\, i,\infty[\to [i,\infty[\subset \mathbb{R}$

such that $q_i(r) = \begin{cases} r & \text{if } r \geq i+1 \\ i + \frac{r-i}{2} & \text{if } i \leq r < i+1 \end{cases}$ $i = 0,1,2,\dots$.

Then q_i is not idempotent but $q_i q_{i-1} = q_{i-1} q_i = q_i$ using the usual composition of partial mappings.
Consider the monoid $S = \{q_1^n \mid i=0,1,\dots,n=1,2,\dots\} \cup \{1_{\mathbb{R}}\}$.

Then in S condition (FP_2) is trivially fulfilled but (FP_1) is not true by definition of S. So by the equivalence class $[1]_\sim$ of the identity, generated by $\{1, q_1, q_2, \ldots\}$, we get a flat cyclic S-act $S/\sim\ = xS = \{x, xq_0, xq_0^2, \ldots\}$ which is not projective.

3. Any right zero semigroup with externally adjoint identity trivially fulfills (FP_1), but obviously not (FP_2).

4. Any left zero semigroup with externally adjoint identity obviously fulfills (FP_1) and (FP_2).

4.15 COROLLARY. *All flat cyclic right S-acts are free if and only if* (FP_1), (FP_2) *and* $e \mathcal{D} 1$ *for all idempotents* $e \in S$.

4.16 COROLLARY. *All flat finitely generated right S-acts are free if and only if all flat cyclic right S-acts are free.*

4.17 COROLLARY. *All flat finitely generated right S-acts are free if S is left cancellative and in particular if S is a group.*

Next we show that left cancellativety is not a necessary condition on S such that all flat cyclic right S-acts are free.

4.18 EXAMPLE. Take Z of 2.9. Then Z fulfills the conditions of 4.15, as (FP_2) and $e \mathcal{D} 1$ for all idempotents are vacuously fulfilled, and so is (FP_1). Consequently all flat cyclic right Z-acts are free. In fact, this has already been showed directly in 2.9 if taking x to be a generator of a cyclic right Z-act. But as $n_i n_1 = n_i n_1'$ the monoid Z is not left cancellative.

5. OVERVIEW

We tabulate the results of sections 3 and 4 and for the sake of completeness include the table of [8].

5.1 Right ideals

	free	projective	flat
principal fin. gen. projective all	$e\mathfrak{D}_1$		
principal fin. gen. flat all	pri generated by left semi canc. are gen. by left canc. & ACC for pri	pri generated by left semi canc. are gen. by left e-canc. & ACC for pri	
principal fin. gen. all	all pri gen. by left canc. el. & (c) & ACC for pri	all pri gen. by left e-canc. el. & (c) & ACC for pri	all pri gen. by left semi canc. & (c)

5.2 Right S-acts

		free	projective	flat	torsion free
cyclic fin. gen. all	projective	5.1			
cyclic fin. gen.	flat	(FP_1) (FP_2), $e \mathcal{D} 1$	(FP_1) (FP_2)		
all		group	(A), (D)		
cyclic fin. gen. all	torsion free	$\vee$ $S = \{0,1\}$ $S = \{1\}$	$\vee$ $S = \{0,1\}$ $S = \{1\}$	$\vee$ $S = \{0,1\}$ $S = \{1\}$	
cyclic fin. gen. all		$\vee$ $S = \{0,1\}$ $S = \{1\}$	$\vee$ $S = \{0,1\}$ $S = \{1\}$	$\vee$ $S = \{0,1\}$ $S = \{1\}$	right canc. el. are right inv.

REFERENCES

1. Clifford, A.H., G.B. Preston, The algebraic theory of semigroups, I, II, Amer. Math. Soc., Providence (1961), (1967).

2. Dorofeeva, M., Hereditary and semihereditary monoids, Semigroup Forum 4 (1972) 301-311.

3. Fountain, J., Perfect semigroups, Proc. Edinburg Math. Soc., 20 (1976) 87-93.

4. Fountain, J., Right PP Monoids with central idempotents, Semigroup Forum 13 (1977) 229-337.

5. Isbell, J., Perfect monoids, Semigroup Forum 2 (1971) 95-118.

6. Kilp, M., On the homological classification of monoids, Sibir. Matem. Ž. 13(1972)3, 578-586 (in Russian).

7. Kilp, M., On the homological classification of monoids by properties of their left ideals, Acta et commentationes Universitatis Tartuensis 336(1974) 178-188 (in Russian).

8. Knauer, U., M. Petrich, Characterization of monoids by torsion free, flat, projective, and free acts, Arch. Math., 36(1981)4, 289-294.

9. Petrich, M., Structure of regular semigroups, Cahiers Mathematiques 11 Montpellier (1977).

10. Skornjakov, L.A., On homological classification of monoids, Sibir. Matem. Ž. 10(1969) 1139-1143 (in Russian).

11. Stenström, B., Flatness and localization over monoids, Math. Nachr. 48(1971) 315-334.

FB 6 Mathematik
Universität
D 2900 Oldenburg

REMARKS ON A MONOID RADICAL THEORY

L.C.A. van Leeuwen

Let V be a fixed variety of monoids. A radical ϱ in V is a mapping which assigns to each $A \in V$ a congruence ϱA on A and satisfies the following two conditions for all $A \in V$:

(I) $\varphi(\varrho A) \subseteq \varrho(\varphi A)$ for every homomorphism φ from A,

(II) $\varrho(A/\varrho A) = \omega$ (cf. H.-J. Hoehnke [1]).

DEFINITION. A radical ϱ in V is a strict radical if it satisfies

$$\varrho A = \omega \text{ iff } (\forall B \leq A)\ \varrho B \neq \iota$$

(cf. L. Márki, R. Mlitz & R. Strecker [3]).

A class of monoids is called the radical class or the semisimple class with respect to a radical ϱ if it consists of those monoids A for which $\varrho A = \iota$ or $\varrho A = \omega$, respectively.

The above notation $(\forall B \leq A)$ means: for all non-trivial submonoids of A. In the same way $(\forall \varphi A)$ means: for all non-trivial homomorphic images of A.

Suppose that a variety V of monoids and a subclass $\mathbb{C}$ of V are given. Then we define the classes $R\mathbb{C}$ and $S\mathbb{C}$ by

$$R\mathbb{C} = \{A \in V \mid (\forall \varphi A)\ \varphi A \notin \mathbb{C}\},$$
$$S\mathbb{C} = \{A \in V \mid (\forall B \leq A)\ B \notin \mathbb{C}\}.$$

For any $\mathbb{C} \in V$ we have that $R\mathbb{C}$ is closed under homomorphic images and $S\mathbb{C}$ is closed under taking submonoids.

The monoid E as submonoid or homomorphic image is called trivial. We consider E as contained in all radical and all semisimple classes. Further, supposing that $E \in \mathbb{C}$, for any $A \in V$ we define

$\mathbb{C}A = \vee\{B \leq A \mid B \in \mathbb{C}\} \leq A,$

$A\mathbb{C} = \cap\{\kappa \mid \kappa \in \mathrm{Con}\, A,\ A/\kappa \in \mathbb{C}\} \in \mathrm{Con}\, A.$

Given an arbitrary family $(B_i \mid i \in I)$ of submonoids of A, $\bigvee_{i \in I} B_i$ denotes the submonoid of A generated by the subset $\bigcup_{i \in I} B_i$. For any monoid A, Con A stands for its congruence lattice.

We call a radical class or a semisimple class a <u>strict radical class</u> and a <u>strict semisimple class</u> respectively if the class in question belongs to a strict radical, according to [3].
The monoid radical theory, as presented in [3], is quite analogous to the corresponding theory for Ω-groups (cf. [2] or [4]).

<u>THEOREM 1. Let</u> $\mathbb{R}$ <u>be a strict radical class and let</u> $\mathbb{S}$ <u>be a strict semisimple class. Then</u> $\mathbb{R} = RS\mathbb{R}$ <u>and</u> $\mathbb{S} = SR\mathbb{S}$.

<u>Proof.</u> First we show that $\mathbb{R} = RS\mathbb{R}$. Let $A \in \mathbb{R}$ and let φA be a non-trivial homomorphic image of A. Suppose $\varphi A \in S\mathbb{R}$, then $(\forall B \leq \varphi A)\ B \notin \mathbb{R}$. But $A \in \mathbb{R}$ implies $(\forall \varphi A)\ \exists B \leq \varphi A$, $B \in \mathbb{R}\setminus\{E\}$ by Theorem 3.1.b [3]. Hence $\varphi A \notin S\mathbb{R}$. Then $A \in RS\mathbb{R}$ by definition.
Conversely, let $A \in RS\mathbb{R}$, then A has no non-trivial homomorphic image in $S\mathbb{R}$. Let φA be an arbitrary non-trivial homomorphic image of A, then $\varphi A \notin S\mathbb{R}$ implies $\exists B \leq \varphi A$, $B \in \mathbb{R}$, where B is a non-trivial submonoid of A. Now Theorem 3.1.b [3] implies $A \in \mathbb{R}$.
Secondly we prove that $\mathbb{S} = SR\mathbb{S}$. Let $A \in \mathbb{S}$ and let B be a non-trivial submonoid of A. By Theorem 3.1.a [3] we have that $B \notin R\mathbb{S}$. Then $A \in SR\mathbb{S}$ by definition.
Conversely, let $A \in SR\mathbb{S}$. This means that A has no non-trivial submonoids in $R\mathbb{S}$, i.e. $(\forall B \leq A)\ B \notin R\mathbb{S}$. Hence $A \in \mathbb{S}$ by Theorem 3.1.a [3].

In Theorem 3.6 Márki, Mlitz and Strecker describe how a strict radical ϱ can be determined from its radical class or semisimple class. We show

<u>THEOREM 2. Let ϱ be an arbitrary strict radical in V, $\mathbb{R}_\varrho$ and $\$_\varrho$ be the corresponding radical class and semisimple class. Then $\mathbb{R}_\varrho = R\$_\varrho$ and $\$_\varrho = S\mathbb{R}_\varrho$.</u>

<u>Proof.</u> By Theorem 3.1 [3], $A/\kappa \in \$_\varrho$ holds if and only if $(\forall C \leq A/\kappa)\ C \notin \mathbb{R}_\varrho$ and on the other hand $A/\kappa \in S\mathbb{R}_\varrho$ means that $(\forall C \leq A/\kappa)\ C \notin \mathbb{R}_\varrho$ by the definition of $S\mathbb{R}_\varrho$. Then $\$_\varrho = S\,\mathbb{R}_\varrho$. This implies that $R\$_\varrho = RS\mathbb{R}_\varrho$ or $R\$_\varrho = \mathbb{R}_\varrho$ (Theorem 1), which completes the proof.

For any $A \in V$ we have

$$A\$_\varrho = \cap\{\kappa \mid \kappa \in \mathrm{Con}\,A,\ A/\kappa \in \$_\varrho\}.$$

Since $\$_\varrho$ ist hereditary (Corollary 3.2 [3]), we have that

$A/\kappa \in \$_\varrho$ if and only if $(\forall B \leq A)\ B/\kappa \in \$_\varrho$, hence

$A\$_\varrho = \cap\{\kappa \mid \kappa \in \mathrm{Con}\,A,\ (\forall B \leq A)\ B/\kappa \in \$_\varrho\}$. Also

$\varrho A = \cap\{\kappa \mid \kappa \in \mathrm{Con}\,A,\ (\forall B \leq A)\ B/\kappa \in \$_\varrho\}$ (Theorem 3.6 [3]).

Hence, for any $A \in V$, we get $\varrho A = A\$_\varrho$ for a strict radical ϱ and its corresponding semisimple class $\$_\varrho$. We cannot expect a similar nice property with respect to ϱA and $\mathbb{R}_\varrho A$. First of all, ϱA is a congruence on A and $\mathbb{R}_\varrho A$ is a submonoid of A. Now we introduce the following notation: for any submonoid $B \leq A$, $\langle B\rangle_A$ stands for the congruence generated by B in A, i.e. for the smallest congruence which contains B in one class (in the class of 1); for any congruence ϱ in a monoid, $[1]_\varrho$ denotes the class of 1 under ϱ.
Then one might ask to characterize the strict radicals ϱ in V having the property: $\varrho A = \langle\mathbb{R}_\varrho A\rangle_A$ for all $A \in V$. These are the so-called <u>tight</u> radicals ([3], Definition 3). A characterization has been given in [3], section 4. Or one can impose the condition: $[1]_{\varrho A} = \mathbb{R}_\varrho A$ for a strict radical ϱ and all $A \in V$. This leads to a torsion theory and a strict radical satisfying this last condition is also considered in [3], section 5. The connection between these conditions is given by the property: $\varrho A = \langle[1]_{\varrho A}\rangle_A$ for a strict radical ϱ and every $A \in V$ [cf. [3], Theorem 3.12]. Hence a torsion theory implies that the strict radical is

tight and Corollary 5.3 in [3] implies that a torsion theory is a stronger condition than that the strict radical be tight.

<u>THEOREM 3.</u> <u>Let</u> $\mathbb{S} \subseteq V$ <u>be a semisimple class of a strict radical in</u> V. <u>Then the following are equivalent:</u>

1) $\mathbb{S}$ <u>is a torsionfree class in</u> V.

2) γ') <u>for all</u> $A \in V$, $[1]_{A\mathbb{S}} \in R\mathbb{S}$.

<u>Proof.</u> 1) ⇒ 2) (see Theorem 5.8 in [3]).
2) ⇒ 1) The condition β): for all $A \in \mathbb{S}$, $(\forall B \leq A)\ B \notin R\mathbb{S}$ is valid for all strict semisimple classes and $\mathbb{S}$ is closed under subdirect products, since any semisimple class has this property. By β): $A/A\mathbb{S}$ has no non-trivial submonoids in $R\mathbb{S}$, moreover $R\mathbb{S}$ is homomorphically closed, so $(R\mathbb{S})A \subseteq [1]_{A\mathbb{S}}$. By γ') we get $[1]_{A\mathbb{S}} \in R\mathbb{S}$ implies $[1]_{A\mathbb{S}} \subseteq (R\mathbb{S})A$. So $(R\mathbb{S})A = [1]_{A\mathbb{S}}$ for all $A \in V$. Suppose now that $\mathbb{S}$ is the semisimple class of a strict radical ϱ, i.e. $\mathbb{S} = \mathbb{S}_\varrho$. Then $R\mathbb{S} = R\mathbb{S}_\varrho = \mathbb{R}_\varrho$, where $\mathbb{R}_\varrho$ is the radical class of ϱ (Theorem 2). Hence $(R\mathbb{S})A = [1]_{A\mathbb{S}}$ implies $\mathbb{R}_\varrho A = [1]_{\varrho A}$, since $A\mathbb{S} = A\mathbb{S}_\varrho = \varrho A$ for a strict radical ϱ. So we get a strict radical ϱ satisfying (*), which is a torsion theory (see [3], Theorem 5.1). This completes the proof.

From the proof of Theorem 3 we can see that the condition: $(R\mathbb{S})A = [1]_{A\mathbb{S}}$ is essential in proving that the semisimple class $\mathbb{S}$ is a torsionfree class.

<u>THEOREM 4.</u> <u>Let</u> $\mathbb{S} \subseteq V$ <u>be a semisimple class of a strict radical in</u> V. <u>Then the following are equivalent:</u>

1) $\mathbb{S}$ <u>is a torsionfree class in</u> V.

2) $(R\mathbb{S})A = [1]_{A\mathbb{S}}$ <u>for all</u> $A \in V$.

<u>Proof.</u> 1) ⇒ 2) according to the proof of Theorem 3.
In order to prove 2) ⇒ 1) we have to show that γ') holds by Theorem 3. Now $(R\mathbb{S})A \in R\mathbb{S}$ by Corollary 3.9 [3], hence $[1]_{A\mathbb{S}} \in R\mathbb{S}$.
In [3], Theorem 5.8 the authors prove that a subclass

$\mathcal{S} \subseteq V$ is a torsionfree class if and only if $\mathcal{S}$ satisfies $\alpha)$, $\beta)$, $\delta)$, $\varepsilon)$, $\zeta)$, $\eta)$, where $\delta)$ means: for all $A \in V$, $A\mathcal{S} = <[1]_{A\mathcal{S}}>_A$, $\varepsilon)$ is the closure under extensions and $\zeta)$ means: for all $A \in V$, $[1]_{([1]_{A\mathcal{S}})\mathcal{S}} \triangleleft_N A$ (normal submonoid) whereas $\eta)$ reads: for all $A \in V$, $[1]_{A\mathcal{S}}/\sigma \in \mathcal{S}$ where $\sigma = <[1]_{([1]_{A\mathcal{S}})\mathcal{S}}>_A$. It is also remarked that if the variety V satisfies some weak congruence extension property for all $A \in V$ and all $B \triangleleft_N A$, then $\eta)$ can be omitted. The condition $\delta)$ describes how the maximal radical submonoid determines the radical congruence, which is trivial for Ω-groups. The remaining conditions $\alpha)$, $\beta)$, $\varepsilon)$, $\zeta)$ are just the analogues of those which characterize semisimple classes of Ω-groups (see [2], Theorem 10).

Now we want to add a further condition

$\vartheta)$ $[1]_{([1]_{A\mathcal{S}})\mathcal{S}} = [1]_{A\mathcal{S}}$ for all $A \in V$.

The corresponding analogue in [2] is condition (3), and $\zeta)$ corresponds to (4). In addition $\alpha)$ is the closure under subdirect products, which corresponds to (b*) in [2]; $\beta)$ corresponds to the regularity (condition (A*) in [2]). The closure under extensions is condition (c) = (c*) in [2]. In [2] it is shown that any class $\mathcal{X}$ of Ω-groups satisfying (A*), (b*) and (c) has the property, that (2*) $(U\mathcal{X}(A) = (A)\mathcal{X})$, (3) and (4) are equivalent. The above characterization of semisimple classes of Ω-groups is via (A*), (b*), (c) and (4). So another characterization is via (A*), (b*), (c) and (3). This suggests that there is a similar characterization of a torsionfree class in V by $\alpha)$, $\beta)$, $\varepsilon)$ and $\vartheta)$.

THEOREM 5. *For a subclass $\mathcal{S} \subseteq V$ the following conditions are equivalent:*

1) *$\mathcal{S}$ is a torsionfree class in* V.

4) $\alpha)$ *$\mathcal{S}$ is closed under subdirect products,*

$\beta)$ *for all* $A \in \mathcal{S}$, $(\forall B \leq A)\ B \notin R\mathcal{S}$,

$\varepsilon)$ *$\mathcal{S}$ is closed under extensions,*

$\vartheta)$ *for all* $A \in V$, $[1]_{([1]_{A\mathcal{S}})\mathcal{S}} = [1]_{A\mathcal{S}}$.

Proof. 1) ⇒ 4): Since 1) ⇒ α), β), ε), δ), γ') ⇒ ϑ) (see the proof of Theorem 5.8 [3], part (2) ⇒ (3)).
4) ⇒ 1): By ϑ) we get $[1]_{([1]_{A\$})\$} = [1]_{A\$} \triangleleft_N A$, so this gives ζ). Also η) is trivial.

References

[1] H.-J. Hoehnke, Radikale in allgemeinen Algebren, Math. Nachr. 32 (1966), 347-383.

[2] L.C.A. van Leeuwen and R. Wiegandt, Radicals, semisimple classes and torsion theories, Acta Math. Acad. Sci. Math. Hung. 36 (1980), 37-47.

[3] L. Márki, R. Mlitz & R. Strecker, Strict radicals of monoids, Semigroup Forum 21 (1980), 27-66.

[4] R. Mlitz, Radicals and semisimple classes of Ω-groups, Proc. Edinburgh Math.Soc., 23 (1980), 37-42.

Mathematisch Instituut
Rijksuniversiteit te Groningen
Postbus 800
NL - 9700 Groningen

SIMPLE ACTS [1)]

W. Lex

An act, as we understand it, is nothing but a triplet (G,M,f) consisting of a non-empty set G, any set M and an arbitrary mapping f from G×M into M. Of course f alone already determines an act but this notation seems to be more suggestive. So an act is essentially a unary algebra or what we call in German an "äußere Verknüpfung".

Examples of acts in our sense abound: various sets of mappings of a set into itself (especially permutation groups), semi-automata or the act (G,G,·) for any groupoid (G,·).

My aim here is not to develop a theory of simple acts but to show by some examples that this fairly general concept of an act is still strong enough to derive relatively concrete results: via the description of the group of automorphisms of a simple act (theorem 3) we shall determine the monoid of endomorphisms of a simple act (thm.5) and thus gain some information about the right nucleus of any groupoid without non-trivial left congruences (proposition 7), which will be specialized to semigroups and abelian quasigroups (corollary 8). Further we will obtain a characterization of faithful groupoid acts with a commutative groupoid (thm.9). - For more information about acts, especially with an operating semigroup, cf. e.g. [1], [2], [6], [9], [10] and [5].

Let me begin by introducing some notation and concepts. - We usually write $_GM$ or even only M for (G,M,f), especially $_GG$ for the above mentioned act (G,G,·) belonging to a groupoid (G,·), and just αx for $f(\alpha,x)$.

1) The results of this note are part of the author's "Habilitationsschrift" [3]. - I would like to express my gratitude here to Prof. Dr. Weinert for his encouragement and continual interest.

For the following let $M = {}_G M = (G,M,f)$ be any act. - Very often the operating set G carries an additional structure, e.g. $G = (G,\cdot)$ is a groupoid. If we then postulate that the operation is connected with the groupoid multiplication in the usual way, i.e.

$$\alpha(\beta x) = (\alpha\beta)x \qquad (\alpha,\beta \in G;\ x \in M),$$

then we call M a groupoid act and write $M = {}_G M$.

If $U \subseteq M$ and $U = (G,U,f|U\times U)$ is an act then U is said to be a subact of M, in short $U \leq M$; let

$$U(M) \leftrightharpoons \{U \mid U \leq M\}.$$

There are always trivial subacts: the empty act ${}_G\emptyset$ and M itself; let

$$U'(M) \leftrightharpoons U(M)\setminus\{{}_G\emptyset, M\}.$$

M is called irreducible if $U'(M) = \emptyset$ and reducible otherwise.- As usual for $a \in M$ let $Ga = \{\alpha a \mid \alpha \in G\}$, further

$$G^{n+1}a \leftrightharpoons G(G^n a) \quad (n \in \mathbb{N})^{2)} \quad \text{with } G^1 a = Ga$$

and

$$\langle a\rangle \leftrightharpoons \bigcup_{n\in\mathbb{N}} G^n a .$$

Thus ${}_G\langle a\rangle$ is the smallest subact of M "generated" by a, but $\langle a\rangle$ need not contain a. Let us say that M is strictly cyclic with respect to a if $a \in M$ and $\langle a\rangle = M$. - An element t of M is called a trap of M if $Gt = \{t\}$.

Let $1_A = \{(a,a) \mid a \in A\}$ for any set A. - An equivalence relation $\equiv$ on M is said to be a congruence of M if it is compatible with the operation of M, i.e. if

$$x \equiv y \ \Rightarrow\ \forall\alpha \in G:\ \alpha x \equiv \alpha y$$

holds for all $x,y \in M$. Let us denote by $C(M)$ the set of all congruences of M and by $C'(M)$ the set of all non-trivial congruences of M, i.e.

$$C'(M) \leftrightharpoons C(M)\setminus\{1_M, M\times M\}.$$

M is called simple if $C'(M) = \emptyset$.

2) $\mathbb{N}$ stands for the set of positive integers!

For later application we note

PROPOSITION 1: Let $M = {}_G M$ be a simple act and $|M| > 2$. Then M is either irreducible or M has exactly one trap and is strictly cyclic with respect to every other element.

Proof: Let T denote the set of all traps of M.

Since $U'(M) = \emptyset$ and $T \neq \emptyset$ exclude each other it suffices to prove the assertion with "or" instead of "either-or", which will be done as contrapositive. Thus let M be reducible and

(1) $|T| \neq 1 \quad \vee \quad \exists a \in M \setminus T: \langle a \rangle \neq M.$

First let $T = \emptyset$. Because of the reducibility of M there is a ${}_G U \in U'(M)$. Should $|U| = 1$ then $GU = U$ in contradiction to $T = \emptyset$. Therefore we have $|U| > 1$ and

$$(U \times U) \cup 1_{M \setminus U} \in C'(M),$$

hence the non-simplicity of M.

If $|T| > 1$ then for every $U \subseteq T$ with $|U| = 2$ one immediately obtains ${}_G U \in U'(M)$ and thus as before M is not simple.

For $|T| = 1$ because of (1) there is an $a \in M \setminus T$ with ${}_G\langle a \rangle \in U'(M)$ and $|\langle a \rangle| > 1$ and the non-simplicity of M follows as before.

Let $\mathbb{P}$ denote the set of all primes and let $\mathbb{P}_1 = \mathbb{P} \cup \{1\}$.- For any set A we write T(A) for the set of all transformations of A, i.e. mappings from A into A, and P(A) for the set of all permutations on A, i.e. bijective transformations of A; further let $T(A) \doteqdot (T(A), \circ)$ stand for the full transformation monoid of A and $P(A) \doteqdot (P(A), \circ)$ for the full permutation group on A.

To describe the group of automorphisms of a simple act we shall use

LEMMA 2: A group G of permutations has a non-trivial intransitive subgroup iff G possesses a non-trivial subgroup.[3)]

---

3) This lemma does not seem to be explicitly formulated elsewhere - I am greatly indebted to Prof. Dr. H. Wielandt for some remarks on this subject.

Proof: Let $G = (G,\circ)$ be a permutation group on M, further $g \coloneqq |G|$ and $m \coloneqq |M|$. As one direction is trivial let $g \notin \mathbb{P}_1$. If G operates intransitively on M then naturally every subgroup of G does also; therefore G is assumed to be transitive. - For every transitive group $(T,\circ) \leq P(M)$ one has

(2) $|T| \geq m$.

If $m \in \mathbb{N}$ then $G \leq P(M)$ thus implies $g|m!$ and $m \leq g$. As $g = qrh$ with $q,r \in \mathbb{P}$ and $h \in \mathbb{N}$ it follows that there is a $p \in \mathbb{P}$ with $p|g$ and $p < m$. Hence Cauchy's theorem ensures the existence of a non-trivial subgroup of G acting intransitively on M because of (2).

Let M be infinite, $\gamma \in G\setminus\{1_M\}$ and $\langle\gamma\rangle$ the subgroup generated by γ. If $|\langle\gamma\rangle| \in \mathbb{N}$ then $\langle\gamma\rangle$ is a non-trivial intransitive subgroup of G.

Thus let $|\langle\gamma\rangle| \notin \mathbb{N}$ and $a \in M$, further $a_\nu \coloneqq \gamma^\nu(a)$ for $\nu \in \mathbb{Z}$. If there are integers k,ℓ with $a_k = a_\ell$ and $k > \ell$ then $\langle\gamma^n\rangle$ with $n = k-\ell$ is an infinite cyclic group which fixes a because $\gamma^k(a) = \gamma^\ell(a)$ implies $\gamma^n(a) = a$; therefore $\langle\gamma^n\rangle$ is a non-trivial intransitive subgroup of G. - In case such a pair (k,ℓ) does not exist one has a non-trivial subgroup in $\langle\gamma^2\rangle$ whose intransitivity follows from $\varphi(a) \neq a_1$ for all $\varphi \in \langle\gamma^2\rangle$.

An endomorphism of an act $M = {}_GM$ is a transformation h of M with $h(\alpha x) = \alpha h(x)$ for all $\alpha \in G$ and all $x \in M$; let End M denote the set of endomorphisms of M. An element of

$$\operatorname{Aut} M \coloneqq (\operatorname{End} M) \cap P(M)$$

is called an automorphism of M. - Of course the endomorphisms of M form a monoid with respect to concatenation $\circ$, and the automorphisms a group.

I am greatly indebted to Prof. Dr. G. Thierrin who kindly pointed out to me that the following theorem for finite unitary monoid acts had already appeared as prop.9, p.348, in his work [9] - which also contains some interesting characterizations of simple automata.

THEOREM 3: The group of automorphisms of a simple act is simple and abelian and all simple abelian groups can be obtained in this way.

Proof: Let $M = {}_GM$ be any act and $A \leftrightharpoons (A,\circ)$ with $A = \mathrm{Aut}\, M$. We contrapose one half of the assertion.

Thus let $|A| \notin \mathbb{P}_1$. Then A has a non-trivial subgroup and, as a permutation group on M by lemma 2, also a non-trivial intransitive subgroup $(U,\circ)$. Hence the equivalence relation

$$\{(x,y) \mid \exists \varphi \in U: \varphi(x) = y\}$$

lies in $C'(M)$ and M is not simple.

Since for a simple abelian group $G = (G,\cdot)$ the act ${}_GG$ is simple and G isomorphic to $(\mathrm{Aut}\,{}_GG,\circ)$ - cf. [4], Korollar 1, p.100 - one gets every simple abelian group as the automorphism group of an appropriate act.

As an easy consequence we have

COROLLARY 4: The centre of a primitive permutation group is either trivial or cyclic of prime order.

Proof: Let $G = (G,\circ)$ be a permutation group primitively operating on M, thus the act $M = {}_GM$ is simple, and let C be the centre of G. For $\varphi \in C$ and $\alpha \in G$ one has

$$\varphi(\alpha x) = (\varphi\alpha)x = (\alpha\varphi)x = \alpha(\varphi x) \quad (x \in M),$$

hence $\varphi \in \mathrm{Aut}\, M$ and so $C \subseteq \mathrm{Aut}\, M$, which proves the assertion by theorem 3.

Again using theorem 3 we can now determine the endomorphism monoid of a simple act:

THEOREM 5: The monoid of endomorphisms of a simple act $M = {}_GM$ is for an irreducible M a group and for a reducible M either a group with zero - if $|M| > 2$ or if $M = \{a,b\}$ with $a \neq b$ and $Ga = \{a\}$, $Gb \neq \{b\}$ - or the full transformation monoid of a set with 2 elements - if $|M| = 2$ and $Gx = \{x\}$ for $x \in M$.

Proof: Let $M = {}_GM$ be a simple act and T the set of traps of M, further $E = (E,\circ)$ with $E = \mathrm{End}\, M$ and $A = (A,\circ)$ with

$A = \mathrm{Aut}\, M$. - If $|M| \in \{0,1\}$ then $|E| = 1$, hence $E = A$ is a trivial group; thus for the following let $|M| > 1$.

Let M be irreducible. Then the assumption of the existence of an $h \in E$ with $hM = \{a\}$ gives

$$Ga = Gh(a) = hGa = \{a\},$$

hence ${}_G\{a\} \in U'(M)$ which contradicts the irreducibility of M. Therefore if $|M| = 2$ we have $E \subseteq A$ and if $|M| > 2$ for a non-injective $g \in E$ in

$$\{(x,y) \mid g(x) = g(y)\} \in C'(M),$$

a contradiction to the simplicity of M.

Since for every $f \in E$ the image fM forms a subact ${}_G fM$ and as M is irreducible it follows that every $f \in E$ is surjective and thus bijective, which proves that alway $E \subseteq A$ and hence that $E = A$ is a group.

Let M be reducible. With $|M| > 2$ proposition 1 yields $T = \{b\}$; with $|M| = 2$ we assume also $T = \{b\}$. Then for $h = (x \mapsto b)$ we have

$$h(\alpha x) = b = \alpha b = \alpha h(x) \qquad (\alpha \in G,\ x \in M),$$

hence $h \in E$. Because of $T = \{b\}$ there is no $c \in M \setminus T$ with $(x \mapsto c) \in E$ and for $f \in E \setminus \{h\}$ one thus gets $|fM| > 1$. It follows that $fM = M$, with $|M| = 2$ trivially and with $|M| > 2$ because otherwise

$$(fM \times fM) \cup 1_{M \setminus fM} \in C'(M)$$

which contradicts the simplicity of M. The injectivity of f results as in the irreducible case and thus $f \in A$ for every $f \in E \setminus \{h\}$, which implies $E = A \cup \{h\}$. - Because $T = \{a\}$ one has $f(a) = a$ for $f \in E$ and for $x \in M$ hence $(f \circ h)(x) = a$ thus $f \circ h = h$, on the other hand as $(h \circ f)(x) = a$ also $h \circ f = h$ holds; therefore E is a group with zero: A^0. - Thus the case of a reducible M with $|M| = 2 = |T|$ remains since $T = \emptyset$ with $|M| = 2$ means just the irreducibility of M, hence $Ga = \{a\}$ and $Gb = \{b\}$ with $\{a,b\} = M$. Then however each mapping from M into M is an endomorphism of M and thus $E = T(M)$.

Before applying this result to the act ${}_G G$ of an arbitrary roupoid $(G,\cdot)$ in proposition 7 we have to introduce further oncepts and to state some elementary facts. - Let $\mathcal{G} = (G,\cdot)$ e any groupoid and

$$N_r(\mathcal{G}) = \{a \in G \mid \forall x,y \in G: (xy)a = x(ya)\}$$

he right nucleus of $\mathcal{G}$; further analogously $N_\ell(\mathcal{G})$ and $N_m(\mathcal{G})$ he left and the middle nucleus of $\mathcal{G}$, respectively. The ucleus of $\mathcal{G}$,

$$N(\mathcal{G}) = N_\ell(\mathcal{G}) \cap N_m(\mathcal{G}) \cap N_r(\mathcal{G}),$$

nd the just defined seminuclei of $\mathcal{G}$ are subsemigroups of $\mathcal{G}$ nless they are empty.

EMMA 6: There are 10 groupoids of order 2 up to isomorphism: semigroups and the following (which are described by their ayleytables without the marginal rows, i.e. $\frac{1}{2}$ and 1 2):

$\mathcal{G}_1$	$\mathcal{G}_2$	$\mathcal{G}_3$	$\mathcal{G}_4$	$\mathcal{G}_5$
1 1	1 2	2 1	2 2	2 2
2 1	1 1	2 1	1 1	2 1

f R_ν denotes the right nucleus of $\mathcal{G}_\nu$ with $\nu = 1,\dots,5$ then $R_1 = \{1\} = R_2$ and $R_\nu = \emptyset$ for $\nu = 3,4,5$.

The lemma can be proved by the description of all possible ases.

In analogy to a concept introduced in [7], p.145, for semigroups we define somewhat more generally: a groupoid $(G,\cdot)$ is called left reductive with respect to U if $U \subseteq G$ and

$$\forall a,b \in U \ (\varrho_a = \varrho_b \Rightarrow a = b)$$

where ϱ_c denotes the right (inner) translation by c on G, i.e. $x \mapsto xc$, and just left reductive in the case of left reductivity with respect to G. - A groupoid $\mathcal{G}$ is said to be left congruence free, lcf for short, if $\mathcal{G}$ possesses no nontrivial left congruences.

PROPOSITION 7: Let $\mathcal{G} = (G,\cdot)$ be a lcf groupoid and $R = N_r(\mathcal{G})$. Then R is empty or $(R,\cdot)$ is a group with zero $\mathcal{H}^0 = (H^0,\cdot)$ with $|H| \in \mathbb{P}_1$ or a left or right zero semigroup or zero semigroup of order 2, further the case of the one- or two-sided zero semigroup only occurs if $|G| = 2$.

<u>Except</u> <u>for</u> <u>the</u> <u>left</u> <u>zero</u> <u>or</u> <u>zero</u> <u>semigroup</u> <u>with</u> 2 <u>elements</u> G <u>is</u> <u>left</u> <u>reductive</u> <u>with</u> <u>respect</u> <u>to</u> R.

<u>Proof:</u> $M = {}_G G$ is a simple act. Let $E = (\mathrm{End}\,M, \circ)$ and $\mathcal{D}$ be the semigroup, with respect to concatenation, of right multiplications by elements of R; further let $R = (R, \cdot)$ and $R \neq \emptyset$.

First we consider the case $|G| > 2$. If G is not left reductive with respect to R, i.e. if there are $a,b \in R$ with $a \neq b$ and $xa = xb$ for all $x \in G$, then

$$(\{a,b\} \times \{a,b\}) \cup 1_{G \setminus \{a,b\}} \in C'(M)$$

since $|G| > 2$, which contradicts the assumption. Therefore R is antiisomorphic to $\mathcal{D}$ (cf. [4], Satz, a), p. 95). - Because of theorem 5 and $|G| > 2$ we have E as a group with zero, H^0, or as a group H with $|H| \in \mathbb{P}_1$ and $\mathcal{D}$ as a subsemigroup of E (cf. Satz l.c.) thus of the same structure. This also holds for R as R and $\mathcal{D}$ are antiisomorphic.

For $|G| = 2$ we shall use lemma 6: let M be

1. irreducible. Then G is either a left zero semigroup or a group or isomorphic to G_1, G_3, G_4 or G_5 and thus R is a left zero semigroup or a group of order 2 or 1 or R is empty. If G is not a left zero semigroup, the left reductivity of G with respect to R results from the fact that then G always has a "left unit with respect to R", i.e. an $e \in G$ with $ey = y$ for all $y \in R$.

2. reducible. Then G is either a right zero semigroup or a zero semigroup or a group with zero or isomorphic to G_2 and therefore R is one of the mentioned semigroups or a trivial group. It follows, as in the irreducible case, that G is left reductive with respect to R unless G is a zero semigroup.

In the case $|G| = 1$, finally, R is a trivial group. - Since every G turns out to be per definitionem left reductive with respect to $\emptyset$ all is proved.

Before specializing proposition 7 in corollary 8 let me give the following definitions: a groupoid $G = (G, \cdot)$ is called <u>left</u> <u>quasigroup</u> <u>with</u> <u>respect</u> <u>to</u> U, if $\emptyset \neq U \subseteq G$ and

$_a \in P(G)$ for all $a \in U$ where $\varrho_a = x \mapsto xa$ as before, and
eft quasigroup if G is a left quasigroup with respect to G.-
et the corresponding "right concepts" be completely ana-
ogously defined. - A left and right quasigroup is called
uasigroup.

From proposition 7 we get by specialization

OROLLARY 8: a) A semigroup is lcf - and thus also without
on-trivial right congruences and v.v. - iff it is of order 2
cf. lemma 6) at the most or a cyclic group of prime order.[4]

) A monoid is lcf iff it is either a group of prime order
r a trivial group.

) A band is lcf iff it is either a trivial group or a tri-
ial group with zero or a left or right zero semigroup of
rder 2.

) A semilattice is without non-trivial congruences iff it
s a trivial group or a trivial group with zero.

) A left quasigroup with respect to its non-trivial right
ucleus is lcf iff it is a group of prime order or a left
ero semigroup of order 2.

) A commutative quasigroup with a non-trivial nucleus is
ithout non-trivial congruences iff it is a group of prime
rder.

Proof: By proposition 7 a lcf semigroup S with more than 2
elements is a group with zero or a simple abelian group.
But should S be a group with zero, say $(G^0,\cdot)$, then the act
$_{G^0}G^0$ would not be simple - in contradiction to the left
congruence freedom of S - since

$$(\{0\}\times\{0\}) \cup (G\times G) \in C'(S).$$

Thus S is a group of prime order.

If $S = (S,\cdot)$ on the other hand is a group with $|S| \in \mathbb{P}$
or a semigroup with $|S| \leq 2$, then S is without non-trivial
congruences and, because of the commutativity in case $|S| > 2$,
even lcf, which proves a).

4) This result was essentially already published in [8].

The statements b) to d) by lemma 6 follow immediately from what we have just shown.

Proposition 7 and lemma 6 imply e) analogously to a) and one gets f) at once as a special case of e).

Before we apply theorem 5 to groupoid acts in theorem 9 we have to introduce more generally for acts a notion which is well known in module theory: an act $M = {}_G M$ is called faithful if

$$\forall \alpha,\beta \in G \quad (\forall x \in M\colon \alpha x = \beta x \Rightarrow \alpha = \beta)\,.$$

THEOREM 9: Let $G = (G,\cdot)$ be an abelian groupoid and $M = {}_G M$ a faithful act with $|M| > 2$. Then M is simple iff G is a group with $|G| \in \mathbb{P}\setminus\{2\}$ and M strictly cyclic.

Proof: Let $E = (E,\circ)$ and $E = \operatorname{End} M$, further λ_α the operation on M induced by $\alpha(\in G)$, i.e. $\lambda_\alpha \mathrel{\hat=} (x \mapsto \alpha x)$, and $L \mathrel{\hat=} (\{\lambda_\alpha \mid \alpha \in G\},\circ)$. With $\alpha,\gamma \in G$ and $x \in M$ one gets

$$\lambda_\alpha(\gamma x) = (\alpha\gamma)x = \gamma\lambda_\alpha(x)$$

and thus $\lambda_\alpha \in E$, hence $L \leq E$. As M is faithful we have $G \simeq L$.

First let M be simple: by proposition 1 M is strictly cyclic and by assumption a groupoid act and thus $|G| \geq |M| > 2$. The simplicity of M implies by theorem 5 and $|M| > 2$ that E, and thus also G, is a simple abelian group or a group with zero with respect to such a group. Thus let $G = H^0$ with a cyclic group $H = (H,\cdot)$ of prime order. With $\langle a\rangle = M$ one has $M = Ha \cup \{0a\}$. Should $0a \in Ha$ there would be an $\alpha \in H$ with $\alpha a = 0a$, thus $Ga = \{0a\}$, which contradicts $|M| > 2$. Hence

$$(\{0a\}\times\{0a\}) \cup (Ha \times Ha) \in C'(M)\,.$$

in contradiction to the simplicity of M. Therefore G is a group with $|G| \in \mathbb{P}\setminus\{2\}$.

On the other hand let G be such a group, M strictly cyclic with respect to, say, a and $\equiv\, \in C(M)$. This induces according to

$$\alpha \sim \beta \;\Leftrightarrow\; \alpha a \equiv \beta a$$

a left congruence, and, as G is abelian, even a congruence, n G. Since G is simple $\sim$ is trivial, thus $\sim = G\times G$ or $\sim = 1_G$. n the first case, because M is strictly cyclic, it follows hat $\equiv = M\times M$ and in the second that $\equiv = 1_M$, which proves M o be simple.

With a trivial additional consideration for the case $|M| \leq 2$ theorem 9 immediately yields the well known

OROLLARY 10: An abelian permutation group is primitive iff t is trivial or of prime order, and transitive.

REFERENCES

1] Deussen, P., Halbgruppen und Automaten, Berlin, Heidelberg, New York 1971.

2] Eilenberg, S., Automata, Languages, and Machines, B. New York, London 1976.

[3] Lex, W., Akte, Habilitationsschrift, Clausthal-Zellerfeld 1980.

[4] Lex, W., Beziehungen zwischen Rechtstranslationen eines Gruppoids (G,·) und den Endomorphismen des Aktes $_G G$, Mitt. math. Sem. Gießen 149 (1981), 91-102.

[5] Lex, W. and R. Wiegandt, Torsion Theory for Acts, Studia Sci. Math. Hungar., to appear.

[6] Skornjakov, L.A., Axiomatizability of a class of injective polygons, (Russian), Trudy Sem. Petrovsk. 4 (1978), 233-239.

[7] Thierrin, G., Sur les automorphismes intérieurs d'un demi-groupe réductif, Commentarii math. Helvet. 31 (1956), 145-151.

[8] Thierrin, G., Une caractérisation des groupes d'ordre premier, International Congress of Mathematicians, Stockholm 1962, Abstracts of Short Communications, Section 2, 55.

[9] Thierrin, G., Simple Automata, Kybernetika (Prague) 5 (1970), 343-350.

[10] Weinert, H.J., S-sets and semigroups of quotients, Semigroup Forum 19 (1980), 1-78.

Institut für Mathematik
Technische Universität Clausthal
D-3392 Clausthal-Zellerfeld, Germany

TORSION THEORIES AND SEMIGROUPS OF QUOTIENTS

John K. Luedeman

In this paper we present the definitions and basic properties of the formation of an S-system of quotients for unitary centered S-systems with zero over a monoid with zero. This leads to a new construction of the semigroup of quotients of a monoid S with 0 and a determination of some of its basic properties. We also develop a special torsion theory for S-systems which demonstrates the importance of the torsion congruence of Hinkle [3] and stresses the difference between the semigroup of quotients of S and the ring of quotients of a ring R. These results are used to develop a primary decomposition theory for semigroups.

1. RIGHT QUOTIENT FILTERS

In this paper, S will always be a semigroup with identity 1 and zero 0. Each right S-system $M_S=M$ is assumed to be unitary (i.e., M1=M) and centered (i.e., m0=0s=0 for the zero 0 of M.) By Hom(M,N) we denote the set of all S-homomorphisms $f:M_S \to N_S$ (i.e., all mappings $f:M_S \to N_S$ satisfying $f(ms) = f(m)s$). In particular, $f \in \mathrm{Hom}(M,N)$ is called "0-restricted" if $f^{-1}(0) = \{0\}$.

Each S-congruence κ on a right S-system $N_S=N$ defines a right S-system $N/\kappa=(N/\kappa)_S$ in an obvious way. Let, in particular, M be an S-subsystem of N, denoted by $M \subseteq N$. Then an S-congruence κ on N is given by $x\kappa y$ iff $x, y \in M$ or $x=y$, and we shall write N/M in this case rather than N/κ. As a special case we have S/A for each right ideal A of S, considering both as right S-systems. Finally w denotes the universal congruence and id denotes the identity for each S-system.

<u>DEFINITION</u>: A nonempty collection Σ of right ideals of S is a <u>right quotient filter</u> (rqf) if Σ satisfies

Q1) if $A,B \in \Sigma$ and $f \in \text{Hom}(A,S)$, then $f^{-1}(B) \in \Sigma$.

Q2) if $I \subseteq S$ and $J \in \Sigma$, and for each $a \in J$ there is $T_a \in \Sigma$ with $a\,T_a \subseteq I$, then $I \in \Sigma$.

We remark that Hinkle [5] calls such filters "special" right quotient filters, and has noted that in the presence of condition Q3

Q3) if $A \in \Sigma$ and $A \subseteq B \subseteq S$, then $B \in \Sigma$,

{Q1,Q2} is equivalent to {Q1,Q2',Q3} where condition Q2' is as follows:

Q2') if $J \in \Sigma$ and to each $a \in J$ we associate $T_a \in \Sigma$, then $\bigcup_{a \in J} a\,T_a \in \Sigma$.

We remark that (1) Q2 $\Rightarrow$ Q2' but (2) Q1 and Q2' do not imply Q2 as shown in the following example do to Weinert [11].

Example: Let S be the commutative monoid whose Cayley table is:

•	o	a	b	c
o	o	o	o	o
a	o	a	b	c
b	o	b	a	c
c	o	c	c	c

where $J = \{o,c\}$ and $\Sigma = \{J\}$. Then Hom $(J,S) = \{f_1,f_2\}$ where f_1 maps J to {o} and f_2 is the insertion map. Condition Q1 is satisfied since A=B=J must hold. Likewise Q2' is satisfied since $Jo \cup Jc = J \in \Sigma$. However since Q2 implies that $S \in \Sigma$, Q2 does not hold.

<u>PROPOSITION 1</u>: Let Σ be a rqf on S. Then the following conditions are satisfied by Σ.

(1) if $A \in \Sigma$, $A \subseteq B \subseteq S$, then $B \in \Sigma$ (Condition Q3),

(2) if $J \in \Sigma$ and $s \in S$, then $s^{-1}J=\{t \in S \mid st \in J\} \in \Sigma$,

(3) if $A, B \in \Sigma$, then $A \cap B \in \Sigma$.

Proof: (1) By Q2, for all $a \in A$, $aA \subset A \subset B$ and $A \in \Sigma$ so $B \in \Sigma$.

(2) Let $f:S \to S$ be defined by $f(t) = st$, then for

$J \in \Sigma$, $f^{-1}(J) = s^{-1}J \in \Sigma$ by Q1.

(3) Let $f: A \to S$ be the insertion map, then $f^{-1}(B) = A \cap B \in \Sigma$ by Q1.

For any right S-system M and rqf Σ, define a relation τ_M on M by $m\tau_M m'$ there is $A \in \Sigma$ with $ma = m'a$ for all $a \in A$. It is easily seen τ_M is an S-congruence on M and is called the <u>special torsion congruence given by</u> Σ.

Now let $\mathcal{M}$ be the category of right S-systems and $(\mathcal{M},\tau)$ be the collection of all pairs (M,τ_M) for $M \in \mathcal{M}$.

<u>PROPOSITION 2</u>: If $f \in \mathrm{Hom}(M,N)$, then $(f\times f)\tau_M \subseteq \tau_N$.

Proof: Let $m\tau_M m'$, then if $A \in \Sigma$ with $ma = m'a$, then $f(ma) = f(m'a)$ or $f(m)a = f(m')a$ for all $a \in A$. Thus $f(m)\tau_N f(m')$.

Thus $(\mathcal{M},\tau)$ becomes a category in the obvious way.

<u>THEOREM 3</u>: (1) Let $M \subseteq N$, then $\tau_M = \tau_N \cap (M\times M)$.

(2) If $f \in \mathrm{Hom}(M,N)$, $f^{-1}(0) = \{0\}$, $\tau_N = w$, then $\tau_M = w$.

(3) $\tau_{(M/\tau_M)} = \mathrm{id}$.

Proof: (1) follows from the definition of τ_M, and (3) is implied by Q2'. Finally, to verify (2), let $x \in M$, then $(fx, f0) \in \tau_N$ so there is $A \in \Sigma$ with $f(x)a = f(xa) = f(0a) = f(0)a = 0$ for all $a \in A$. Now $f^{-1}(0) = \{0\}$ so $xa = 0$ for all $a \in A$ and $x\tau_M 0$. Since $x \in M$ was arbitrary, $x\tau_M 0$ for all $x \in M$ so $\tau_M = w$.

<u>DEFINITION</u>: For $M_S \subseteq N_S$, let $M \subseteq' N$ denote that for all $n \in N$, $n^{-1}M = \{s\in S \mid ns \in M\} \in \Sigma$.

<u>PROPOSITION 4</u>: The following conditions are equivalent:

(1) $A \in \Sigma$

(2) $\tau_{S/A} = w$

(3) $A \subseteq' S$

Proof: (1) $\Rightarrow$ (2): Let $s \in S$, then $s^{-1}A \in \Sigma$ and $s(s^{-1}A) \subseteq A$ so $\bar{s}(\bar{s}^{-1}A) = \bar{0}$ in S/A. Thus $\tau_{S/A} = w$.

(2) $\Rightarrow$ (3): Let $s \in S$, if $s \in A$ then $sS \subseteq A$ and $s^{-1}A \in \Sigma$. If $s \notin A$, then $\bar{s} \in S/A$, $\bar{s} \neq \bar{0}$ so since $\tau_{S/A} = w$, there is $B \in \Sigma$ with $\bar{s}B = \bar{0}$ or $sB \subseteq A$. Thus $A \subseteq' S$.

(3) $\Rightarrow$ (1): If for $s \in S$ there is $B_s \in \Sigma$ with $sB_s \subseteq A$, then since $S \in \Sigma$, $A \in \Sigma$ by Q2.

<u>COROLLARY 5</u>: $M \subseteq' N$ if and only if $\tau_{N/M} = w$.

<u>PROPOSITION 6</u>: Let $f \in Hom(M,N)$, then the following conditions are equivalent:

(1) $B \subseteq' N \Rightarrow f^{-1}(B) \subseteq' M$

(2) $\tau_N = w,\ f^{-1}(0) = \{0\} \Rightarrow \tau_M = w$

Proof: (2) $\Rightarrow$ (1): If $B \subseteq' N$, then $\tau_{N/B} = w$. Now f yields $\hat{f} \in Hom(M/f^{-1}(B), N/B)$ which satisfies $f^{-1}(0) = \{0\}$. Thus $\tau_{M/f^{-1}(B)} = w$ or $f^{-1}(B) \subseteq' M$.

(1) $\Rightarrow$ (2): Let $\tau_N = w$, then $(0_N) \subseteq' N$. Now if f satisfies $f^{-1}(0) = \{0\}$, then $(0_M) \subseteq' M$ and $\tau_M = w$.

<u>PROPOSITION 7</u>: Q2' is equivalent to the condition:

$$M \subseteq' N \text{ and } N \subseteq' K \Rightarrow M \subseteq' K.$$

Proof: $\Rightarrow$: Let $k \in K$, then $k^{-1}N \in \Sigma$ and if $ka \in N$, there is $B_a \in \Sigma$ with $(ka)B_a \subseteq M$. By Q2', $\bigcup_{a \in k^{-1}N} aB_a \in \Sigma$ so $M \subseteq' K$.

$\Leftarrow$: Let $J \in \Sigma$ and for $a \in J$ let $B_a \in \Sigma$. Then $\bigcup aB_a \subseteq' J$ and $J \subseteq' S$ so $\bigcup aB_a \subseteq' S$ and $\bigcup aB_a \in \Sigma$. Thus Q2' holds.

<u>DEFINITION</u>: Let (M,τ) denote a category whose objects are pairs (M,τ_M) where M is a right S-system and τ_M is any S-congruence on M. The morphisms of (M,τ) are the S-homomorphisms $f: M \to N$ which must satisfy $(f \times f)(\tau_M) \subseteq \tau_N$. Moreover, we require that (M,τ) satisfy the following conditions:

(M1) $\tau_{N/M} = w$ and $\tau_{K/N} = w \Rightarrow \tau_{K/M} = w$.

(M2) $\tau_{M/\tau_M} = id$.

In this situation τ is called a <u>torsion congruence</u> on M. If (M,τ) also satisfies

(M3) Let $f: (M,\tau_M) \to (N,\tau_N)$ be 0-restricted, then $\tau_N = w \Rightarrow \tau_M = w$,

then τ is called a <u>special torsion congruence</u>.

We have previously shown that given a right quotient filter Σ, we can obtain a category (M,τ) by $\tau_M = \{(x,y) \mid xa = ya$ for $a \in A \in \Sigma\}$.

Now suppose a special torsion congruence (M,τ) is given. Let

$$\Sigma' = \{A \subseteq S \mid \tau_{S/A} = w\}.$$

THEOREM 8: Σ' is a right quotient filter for S.

Proof: (Q1): Let $f \in \mathrm{Hom}(A,S)$ with $A \in \Sigma'$ and let $B \in \Sigma'$, then $\tau_{S/B} = w$. Since $\hat{f} \in \mathrm{Hom}(A/f^{-1}(B), S/B)$ is given by $\hat{f}(\bar{a}) = \overline{f(a)}$, then $\hat{f}$ is 0-restricted so $\tau_{A/f^{-1}(B)} = w$. By (M3), since $\tau_{S/A} = w$, $\tau_{S/f^{-1}(B)} = w$ or $f^{-1}(B) \in \Sigma'$.

(Q2'): Let $J \in \Sigma'$ so that $\tau_{S/J} = w$. Moreover, suppose that for all $a \in J$, there is $T_a \in \Sigma'$, and let $B = \cup aT_a$. For each $\bar{s} \in J/B$, let $f: S/T_s \to J/B$ be given by left multiplication by s. Then since $\tau_{S/T_s} = w$, $\bar{1}\tau_{S/T_s}\bar{0}$ so $\bar{s}\tau_{J/B}\bar{0}$. Thus $\tau_{J/B} = w$ and so $\tau_{S/B} = w$ by (M3). Hence $B \in \Sigma$ and Q2' is satisfied.

(Q3): Let $A \in \Sigma$ and $A \subseteq B$. Then the canonical epimorphism $f: S/A \to S/B$ makes $\tau_{S/B} = (f\times f)(\tau_{S/A}) = w$.

Let Σ be given, form (M,τ) from Σ and then form Σ' from (M,τ).

THEOREM 9: $\Sigma = \Sigma'$

Proof: If $A \in \Sigma$, then $A \subseteq' S$ so $\tau_{S/A} = w$ or $A \in \Sigma'$. Conversely, if $A \in \Sigma'$ then $\tau_{S/A} = w$ so $A \subseteq' S$ or $A \in \Sigma$ by Proposition 4.

NOTATION: In the rest of this paper, the rqf determining the special torsion congruence τ given by Σ, will be denoted by Σ_τ.

2. TORSION THEORIES

A torsion theory for S is a pair $(\underline{\underline{T}},\underline{\underline{F}})$ of classes of S-systems such that

(1) $\mathrm{Hom}(T,F) = 0$ for all $T \in \underline{\underline{T}}$ and $F \in \underline{\underline{F}}$; and

(2) $\underline{\underline{T}}$ and $\underline{\underline{F}}$ are maximal classes having property (1).

The S-systems in $\underline{\underline{T}}$ are called torsion systems and those in $\underline{\underline{F}}$ are called torsion-free.

Now any class $\underline{\underline{C}}$ generates a torsion theory by

$$\underline{\underline{F}} = \{F \mid \mathrm{Hom}(C,F) = 0 \text{ for all } C \in \underline{\underline{C}}\} \text{ and}$$
$$\underline{\underline{T}} = \{T \mid \mathrm{Hom}(T,F) = 0 \text{ for all } F \in \underline{\underline{F}}\}.$$

We characterize those classes of S-systems which may appear as torsion classes for some torsion theory; such a class we call a torsion class. (We assume that classes are closed under isomorphism.)

<u>PROPOSITION 1</u>: A class of S-systems is a torsion class if and only if it is closed under quotients, disjoint unions, and extensions.

Proof: A class $\underline{\underline{C}}$ is closed under extensions if whenever $N \subset M$ and $N, M/N \in \underline{\underline{C}}$, then $M \in \underline{\underline{C}}$.

Now suppose $(\underline{\underline{T}},\underline{\underline{F}})$ is a torsion theory, then $\underline{\underline{T}}$ is clearly closed under quotients and since $\mathrm{Hom}(\dot{\cup} M_i, F) \cong \Pi\mathrm{Hom}(M_i,F)$, $\dot{\cup} M_i \in \underline{\underline{T}}$ if each $M_i \in \underline{\underline{T}}$. Finally let N and M/N be torsion, and $\alpha: M \to F$ where $F \in \underline{\underline{F}}$. Then since $N \in \underline{\underline{T}}$, $\alpha|_N = 0$ so $\bar{\alpha}: M/N \to F$ defined by $\bar{\alpha}(\bar{m}) = \alpha(m)$ for $m \notin N$ is an S-homomorphism. But $\bar{\alpha} = 0$ so $\alpha = 0$ and $M \in \underline{\underline{T}}$.

Conversely, assume $\underline{\underline{C}}$ to be closed under quotients, disjoint unions and extensions. Let $(\underline{\underline{T}},\underline{\underline{F}})$ be the torsion theory generated by $\underline{\underline{C}}$, then $\underline{\underline{C}} \subseteq \underline{\underline{T}}$. Now let $T \in \underline{\underline{T}}$, then $\mathrm{Hom}(T,F) = 0$ for all $F \in \underline{\underline{F}}$. Let $C = \cup\{M \subseteq T | M \in \underline{\underline{C}}$. Then $\dot{\cup} M \in \underline{\underline{C}}$ and $C = \dot{\cup} M/\sim \; \in \underline{\underline{C}}$ where $\sim$ is the kernel of the canonical epimorphism $\dot{\cup} M \to C$. Now to show that $C = T$ it suffices to show that the canonical epimorphism $\alpha: T \to T/C$ is zero or equivalently, that $T/C \in \underline{\underline{F}}$. Now suppose $\gamma: C' \to T/C$ where $C' \in \underline{\underline{C}}$, then $\mathrm{im}\,\gamma$ is a subsystem of T/C and if $\gamma \neq 0$, we obtain a subsystem B of T with $C \subsetneq B$ and $B/C \cong C'/\ker\gamma \in \underline{\underline{C}}$ so $B \in \underline{\underline{C}}$ which contradicts the maximality of C. Thus $\gamma = 0$ so $T/C \in \underline{\underline{F}}$, and $T = C$. Hence each $M \in \mathcal{M}$ has a maximal torsion subsystem $C \in T$. Denote C by $t(M)$.

Now let Σ be a right quotient filter and form $(\mathcal{M},\tau)$. Let

$$\underline{\underline{T}} = \{M | \tau_M = w\}.$$

Then $\underline{\underline{T}}$ is a torsion class by the preceding theorem. Moreover,

$$\underline{\underline{F}} = \{M | \tau_M{}^{\#}(0) = 0\},$$

where $\tau_M{}^{\#}(0)$ is the 0-class of τ_M. We call $\underline{\underline{T}}$ the torsion class generated by Σ.

<u>THEOREM 2</u>: Let $\underline{\underline{T}}$ be the torsion class under $(\mathcal{M},\tau)$. Then $\underline{\underline{T}}$ satisfies

(*) Let $f \in \mathrm{Hom}(M,N)$ be 0-restricted, then $N \in \underline{\underline{T}}$ implies $M \in \underline{\underline{T}}$.

Proof: Let $m \in M$, then there is $A \in \Sigma$ with $f(ma) = 0$ for all $a \in A$. Since $f(0) = 0$ and f is 0-restricted, then $mA=0$ so $m\tau_M 0$ or $\tau_M = w$ so $M \in \underline{\underline{T}}$.

<u>DEFINITION</u>: A torsion theory $(\underline{\underline{T}},\underline{\underline{F}})$ is <u>hereditary</u> if whenever $N \subseteq M$ and $M \in \underline{\underline{T}}$, then $N \in \underline{\underline{T}}$.

Notice that the torsion class generated by Σ is hereditary.

<u>THEOREM 3</u>: $(\underline{\underline{T}},\underline{\underline{F}})$ is a hereditary torsion theory if and only if $\underline{\underline{F}}$ is closed under injective hulls.

Proof: $\Rightarrow$: Let $T \in \underline{\underline{T}}$, $F \in \underline{\underline{F}}$ and $E(F)$ be the injective hull of F. Then if α: $T \to E(F)$ is not the zero map, then $\alpha(T) \neq 0$. But if $N = \{t \in T | \alpha(t) \in F\}$, then $N \in \underline{\underline{T}}$ since $(\underline{\underline{T}},\underline{\underline{F}})$ is hereditary, thus $\alpha|_N$: $N \to F$ is not the zero map since $\alpha(T) \cap F \neq 0$, and so we have a contradiction.

$\Leftarrow$: Let $N \subseteq M$ and $M \in \underline{\underline{T}}$. Then consider the commutative diagram

$$\begin{array}{ccc} N & \hookrightarrow & M \\ \downarrow \alpha & & \downarrow \beta \\ N/t(N) & \hookrightarrow & E(N/t(N)) \end{array}$$

where $N/t(N) \in \underline{\underline{F}}$ and so $E(N/t(N)) \in \underline{\underline{F}}$ and β exists since $E(N/t(N))$ is injective. Thus $\beta = 0$ so $\alpha = 0$ and $N = t(N)$ so $N \in \underline{\underline{T}}$.

<u>DEFINITION</u>: A torsion theory $(\underline{\underline{T}},\underline{\underline{F}})$ is <u>special</u> if $f \in \mathrm{Hom}(M,N)$ is 0-restricted and $N \in \underline{\underline{T}}$, imply $M \in \underline{\underline{T}}$.

Note that a special torsion theory is hereditary since the insertion mapping is 0-restricted.

We have seen that if we start with a right quotient filter Σ, we obtain a special torsion theory $(\underline{\underline{T}},\underline{\underline{F}})$ where

$$\underline{\underline{T}} = \{M | \tau_M = w\},$$
$$t(M) = \tau_M{}^{\#}(0),$$
$$\text{and } \underline{\underline{F}} = \{M | t(M) = 0\}.$$

Next, let

$$\Sigma' = \{A | S/A \in \underline{\underline{T}}\}.$$

<u>THEOREM 4</u>: Σ' is a right quotient filter for S.

Proof: (Q1): Let f: $A \to S$ and $B \in \Sigma'$, then $\bar{f}$: $A/f^{-1}(B) \to S/B$ is 0-restricted and $S/B \in \underline{\underline{T}}$, thus $A/f^{-1}(B) \in \underline{\underline{T}}$. Now $S/A \in \underline{\underline{T}}$ and since $\underline{\underline{T}}$ is closed under extensions, $S/f^{-1}(B) \in \underline{\underline{T}}$ so $f^{-1}(B) \in \Sigma'$.

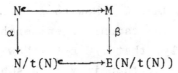

(Q2): Let $J \in \Sigma$, $I \subseteq S$ and $a^{-1}I \in \Sigma$ for $a \in J$. Let T_a denote $a^{-1}I$, then $S/T_a \in \underline{\underline{T}}$ for all $a \in J$ so $\cup S/T_a \in \underline{\underline{T}}$. Now let $\lambda: \cup S/T_a \to J/\cup aT_a$ by $\lambda|_{S/T_a} = \lambda_a$, then λ maps onto $J/\cup aT_a$ so $J/\cup aT_a \in \underline{\underline{T}}$. Since $S/J \in \underline{\underline{T}}$, then $S/\cup aT_a \in \underline{\underline{T}}$. But $\cup aT_a \subsetneq I$ so $S/\cup aT_a$ maps onto S/I so $S/I \in \underline{\underline{T}}$. Thus Q2 is satisfied.

Recall that $T \in \underline{\underline{T}}$ iff for $t \in T$ there is $A \in \Sigma$ with $tA = 0$. Now begin with Σ, generate $\underline{\underline{T}}$, and in turn obtain Σ'.

<u>THEOREM</u> <u>5</u>: $\Sigma = \Sigma'$.

Proof: Let $A \in \Sigma$, then for all $s \in S$, $s^{-1}A \in \Sigma$ and $s(s^{-1}A) \subseteq A$ so $S/A \in \underline{\underline{T}}$ and $A \in \Sigma'$.

Conversely, if $A \in \Sigma'$, then $S/A \in \underline{\underline{T}}$ so for all $s \in S$, there is $T_s \in \Sigma$ with $sT_s \subseteq A$, thus $A \in \Sigma$ and $\Sigma = \Sigma'$.

Next begin with a special torsion class $\underline{\underline{T}}$, obtain Σ, and then generate $\underline{\underline{T}}'$.

<u>THEOREM</u> <u>6</u>: $\underline{\underline{T}} = \underline{\underline{T}}'$.

Proof: Let $T \in \underline{\underline{T}}$, $t \in T$ and let $A = \{s \mid ts = 0\}$. Then $\lambda_t: S/A \to T$ which is 0-restricted and since $T \in \underline{\underline{T}}$, $S/A \in \underline{\underline{T}}$ so $A \in \Sigma$ and $T \in \underline{\underline{T}}'$.

Conversely, let $T \in \underline{\underline{T}}'$ and $F \in \underline{\underline{F}}$ with $\alpha: T \to F$. If $t \in T$ and $\alpha(t) \neq 0$, then $tS \subset T$ so $tS \in \underline{\underline{T}}'$ since $\underline{\underline{T}}'$ is hereditary. Thus there is some $A \in \Sigma$ with $tA = 0$. Thus $S/A \in \underline{\underline{T}}$ and if $\alpha|_{tS}: tS \to F$ and $\lambda_t: S/A \to tS$ then $\alpha \circ \lambda_t = 0$ but λ_t is onto so $\alpha = 0$.

3. RADICALS

In section 2, we noted that M_S was torsion free if $\tau_M^{\#}(0) = 0$. We also noted that each M_S had a maximal torsion subsystem

$$t(M) = \{x \in M \mid xA = 0 \text{ for some } A \in \Sigma\}.$$

Thus (1) $t(M) \subseteq M$, and

(2) if $f \in \mathrm{Hom}(M,N)$, then $f(t(M)) \subseteq t(N)$.

<u>THEOREM</u> <u>1</u>: t also satisfies the following properties:

(1) $t(t(M)) = t(M)$,

(2) $t(M/t(M)) = 0$,

and (3) if $f \in \mathrm{Hom}(M,N)$ is 0-restricted and $t(N) = N$, then $t(M) = M$.

Proof: (1) Clearly $t(t(M)) \subseteq t(M)$. But since $t(M)$ is torsion, $t(t(M)) = t(M)$.

(2) If $x \in t(M/t(M))$, then there is $A \in \Sigma$ with $xA \subseteq t(M)$. Thus for each $a \in A$, there is $B_a \in \Sigma$ with $xaB_a = 0$. Thus $\bigcup aB_a \in \Sigma$ and $x(\bigcup aB_a) = 0$ so $x \in t(M)$ and $t(M/t(M)) = 0$.

(3) Let $x \in M$, then there is $A \in \Sigma$ with $f(x)A = f(xA) = 0$. Since f is 0-restricted, $xA = 0$ and $x \in t(M)$. Thus $t(M) = M$.

<u>DEFINITION</u>: A mapping t: ${}_SM \to {}_SM$ satisfying

(1) $t(M) \subseteq M$

(2) if $f \in Hom(M,N)$ then $f(t(M)) \subseteq t(N)$,

(3) $t(t(M)) = t(M)$

(4) $t(M/t(M)) = 0$, and

(5) if $f \in Hom(M,N)$ is 0-restricted and $t(N) = N$, then $t(M) = M$

is called a <u>special</u> <u>idempotent</u> <u>radical</u> for ${}_SM$.

<u>THEOREM 2</u>: A special idempotent radical t for ${}_SM$ satisfies

$$(*) \quad \text{if } M \subseteq N \text{ then } t(M) = t(N) \cap M.$$

Proof: If $i: M \to N$ is the insertion map, then $t(M) \subseteq t(N) \cap M$. Conversely, $i: t(N) \cap M \to t(N)$ is 0-restricted and $t(t(N)) = t(N)$ so $t(t(N) \cap M) = t(N) \cap M$. But $t(N) \cap M \subset M$ so $t(t(N) \cap M) = t(N) \cap M \subset t(M)$ so $t(M) = t(N) \cap M$.

<u>PROPOSITION 3</u>: $\underline{\underline{T}} = \{M | t(M) = M\}$ is a torsion class.

Proof: Let $\underline{\underline{T}}'$ be the torsion class generated by $\underline{\underline{T}}$, then $\underline{\underline{T}} \subseteq \underline{\underline{T}}'$. Now let $M \in \underline{\underline{T}}'$, then $M/t(M) \in \underline{\underline{F}}'$. Thus if $\alpha: M \to M/t(M)$ is the canonical homomorphism, $\alpha = 0$ so $M/t(M) = 0$ or $M = t(M)$. Hence $M \in \underline{\underline{T}}$ and $\underline{\underline{T}} = \underline{\underline{T}}'$.

Given a <u>special</u> <u>idempotent</u> <u>radical</u> t for ${}_SM$, let

$$\Sigma = \{A \subseteq S | t(S/A) = S/A\}.$$

<u>THEOREM 4</u>: Σ is a right quotient filter for S.

Proof: (Q1): Let $f: A \to S$ and $B \in \Sigma$, then $t(S/B) = S/B$ and $f: A/f^{-1}(B) \to S/B$ is 0-restricted so $t(A/f^{-1}(B)) = A/f^{-1}(B)$. However $t(S/A) = S/A$ so $t(S/f^{-1}(B)) = S/f^{-1}(B)$ and $f^{-1}(B) \in \Sigma$ since $\underline{\underline{T}} = \{M|t(M) = M\}$ is a torsion class.

(Q2): Again, since $\underline{\underline{T}} = \{M|t(M) = M\}$ is a torsion class, Σ satisfies Q2 as in the proof of Theorem 2.4.

Since $\underline{\underline{T}}$ is a special torsion class, Σ determines $\underline{\underline{T}}$ and so determines t.

Let τ be a special torsion congruence, $\Sigma = \{A \subseteq S | S/A$ is torsion$\}$ and let Σ give the torsion congruence σ. Then for any right S-system M, define $t(M) = \{x \varepsilon M | x \tau_M 0\}$ and $s(M) = \{x \varepsilon M | x \sigma_M 0\}$. Then if $x \varepsilon t(M)$, let $A = \{s \varepsilon S | xs = 0\}$, then $S/A \mapsto xS$ given by $1 \to x$ is 0-restricted and since xS is τ-torsion, $A \varepsilon \Sigma$ so $x \varepsilon s(M)$.

Conversely, if $m \varepsilon s(M)$ and $A \varepsilon \Sigma$ with $mA = 0$, then $\{s \varepsilon S | ms = 0\} = B \varepsilon \Sigma$ and $f: S/B \to mS$ is 0-restricted. Since mS is σ-torsion, S/B is σ-torsion and so τ-torsion so mS is τ-torsion and $m \varepsilon t(M)$. Thus, $s(M) = t(M)$.

Thus we have proved

<u>THEOREM 5</u>: Σ uniquely determines the special idempotent radical t and conversely.

4. THE MODULE OF QUOTIENTS

Let Σ be a right quotient filter on S and τ be its associated special torsion congruence. In the preceding sections, we have seen that a right quotient filter determines a special torsion theory $(\underline{\underline{T}},\underline{\underline{F}})$, a special torsion congruence τ_M, and a special idempotent radical t. In this section, we shall see that the torsion free class is not what we really need. Recall that M is torsion free if $\tau_M^{\#}(0) = t(M) = 0$.

<u>DEFINITION</u>: M_S is <u>strongly torsion free</u> if $\tau_M = id$.

E_S is <u>Σ-injective</u> if $N/M \varepsilon \underline{\underline{T}}$ and $f: M \to E$, then f has an extension $\hat{f}: N \to E$.

E_S is <u>faithfully Σ-injective</u> iff $\hat{f}$ is unique.

E_S is <u>weakly Σ-injective</u> if each $f \varepsilon Hom(A,E)$ where $A \varepsilon \Sigma$ has an extension $\hat{f} \varepsilon Hom(S,E)$.

<u>THEOREM 1</u>: E is faithfully <u>Σ-injective</u> if and only if E is Σ-injective and $\tau_E = id$.

Proof: $\Rightarrow$: If $x \tau_E y$, then for some $A \varepsilon \Sigma$, $xa = ya$ for all $a \varepsilon A$. Thus if $f: A \to E$ is given by $f(a) = xa = ya$, then f has both λ_x and λ_y as extensions. Hence $x = x1 = y1 = y$ and $\tau_E = id$.

$\Leftarrow$: Let $N/M \varepsilon \underline{\underline{T}}$ and $g, h \varepsilon Hom(N,E)$ satisfy $g|_M = h|_M$. Then if $gn \neq hn$, there is $A \varepsilon \Sigma$ with $g(na)=h(na)$

for all $a \in A$ and $nA, mA \subseteq M$ since $N/M \in \underline{\underline{T}}$. Thus $gn \tau_E hn$ so $gn = hn$ and $g = h$. Hence E is faithfully Σ-injective.

<u>THEOREM 2</u>: Let $F \subseteq E$, $L = E/F$, $t(L) = 0$ and E be Σ-injective, then F is Σ injective.

Proof: Let $N/M \in \underline{\underline{T}}$ and consider the commutative diagram:

$$\begin{array}{ccc} M & \longrightarrow & N \\ \alpha \downarrow & & \downarrow \hat{\alpha} \\ F & \longrightarrow & E \end{array}$$

Now if $n \in N$, there is $A \in \Sigma$ with $\hat{\alpha}(nA) = \hat{\alpha}(n)A \subseteq F$. Since $N/M \in \underline{\underline{T}}$, if $\hat{\alpha}(n) \notin F$, $\hat{\alpha}(n) \in t(L) = 0$ so $\hat{\alpha}(n) = 0$ or $\hat{\alpha}(n) \in F$. Thus $\hat{\alpha}(N) \subseteq F$, and F is Σ-injective.

Recall that each S-system M has a unique (up to isomorphism over M) injective hull E(M) [2]. M is <u>essential</u> in E(M) in the sense that any S-congruence τ on E(M) whose restriction to M is the identity must be the identity on E(M). We next show that each M_S has a unique faithfully Σ-injective extension. To this end let M_S be strongly torsion free, and let $E = E(M)$, then $\tau_E = id$ since $\tau_M = id$. Now let $M \subseteq N \subseteq E$ with $N/M = t(E/M)$. Then $t(E/N) = t((E/M)/t(E/M)) = 0$ so E/N is torsion free. By Theorem 2, since E is Σ-injective and E/N is torsion free, N is Σ-injective but since $\tau_N \subseteq \tau_E$, $\tau_N = id$ so N is faithfully Σ-injective.

Thus we have proven

<u>THEOREM 3</u>: Let Σ be a right quotient filter. If M_S is strongly torsion free, then there is a faithfully Σ-injective S-system N containing M such that N/M is torsion.

The S-system N just constructed is an $\cap$-large extension of M because of the way it was constructed. Since we shall need to use this later we isolate this fact as

<u>LEMMA 3</u>: Let $t(X) = 0$, $Y \subseteq X$ and $t(X/Y) = X/Y$, then Y is $\cap$-large in X.

Proof: Let $0 \neq x \in X$, then there is $A \in \Sigma$ with $xA \subseteq Y$. If $xA = 0$, then $t(X) \neq 0$. Thus $xA \neq 0$ and Y is $\cap$-large in X.

Let $E_\Sigma(M)$ denote the S-system N of Theorem 2.

<u>PROPOSITION 4</u>: $E_\Sigma(M)$ is the unique up to isomorphism over M faithfully Σ-injective extension of M with $E_\Sigma(M)/M$ torsion.

Proof: Let E'_Σ be another such extension. Then we have the commutative diagram

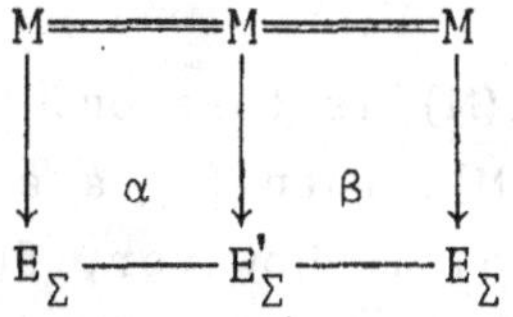

and by the uniqueness of the mappings, $\beta\alpha = 1_{E_\Sigma}$ and $\alpha\beta = 1_{E'_\Sigma}$.

We now give a construction for $E_\Sigma(M)$ and in so doing, construct an S-system of quotients for M. Again let M be strongly torsion free and let $\Omega = \{(A,f): f \in \text{Hom}(A,M), A \in \Sigma\}$. Define $(A,f)\ \theta\ (B,g)$ if there is some $C \in \Sigma$, $C \subseteq A \cap B$ and f and g agree on C. Set $Q_\Sigma(M)=\Omega/\theta$. Clearly θ is an equivalence relation on Ω. Denote the θ class of (A,f) by $[A,f]$.

For $s \in S$, $[A,f] \in Q_\Sigma(M)$, let $B = s^{-1}A \in \Sigma$ and define $g \in \text{Hom}(B,M)$ by $g(b) = f(sb)$. Then $[B,g] \in Q_\Sigma(M)$ and $[B,g]=[A,f]s$. This gives a binary operation $Q_\Sigma(M)xS \to Q_\Sigma(M)$. With this operation, $Q_\Sigma(M)$ becomes a unitary S-system with 0 since for $s, t \in S$,

$[A,f]st = [B,g]$ where $B = s^{-1}(t^{-1}A) \in \Sigma$

$[A,f]t = [C,h]$ where $C = t^{-1}A \in \Sigma$, and

$[C,h]s = [D,k]$ where $D = s^{-1}C \in \Sigma$,

thus $D = B$ and for $d \in D$,

$g(d) = f((ts)d)$ and

$k(d) = h(sd) = f(t(sd)) = g(d)$.

Thus $(D,k)\ \theta\ (B,g)$ or $[A,f](ts) = ([A,f]t)s$.

For $m \in M$, define $j(m): S \to M$ by $j(m)(s) = ms$. Then $[S,j(m)] \in Q_\Sigma(M)$. Let $i: M \to Q_\Sigma(M)$ be defined by $i(m) = [S,j(m)]$.

PROPOSITION 5: i is an S-homomorphism.

Proof: For $s \in S$, $m \in M$, $i(ms) = [S,j(ms)]$ where for $t \in S$, $j(ms)(t) = (ms)t$. But $i(m)s = [S,j(m)]s = [S,k]$ where $k(t) = j(m)(st) = (m)(st)$. Thus for $t \in S$, $k(t) = j(ms)t$ so $i(m)s = i(ms)$.

PROPOSITION 6: ker i = id.

Proof: For $n,m \in M$, $i(m) = i(n)$ implies $[S,j(m)] = [S,j(n)]$ so for all $s \in S$, $ms = j(m)(s) = j(n)(s) = ns$ so $m = n$ since $\tau_M = id$.

<u>PROPOSITION 7</u>: $Q_\Sigma(M)/i(M)$ is torsion.

Proof: Let $[A,f] \in Q_\Sigma(M)$, then for $a \in A$, $[A,f]a = [S,g]$ where $g(s) = f(as) = f(a)s$. However, $f(a) \in M$ and $i(f(a)) = [S,j(f(a))]$ with $i(f(a)) \in i(M)$. Thus $[A,f]a = i(f(a)) \in i(M)$.

Note that if $[A,f]A = 0$, then $f(a) = 0$ for all $a \in A$ so $[A,f] = 0$. Thus for $[A,f] \neq 0$, $0 \neq [A,f]A \subseteq i(M)$ so <u>$i(M)$ is $\cap$-large in $Q_\Sigma(M)$</u>.

We remark that since $\tau_M = id$, M is isomorphic to i(M) so $Q_\Sigma(M)$ is $\cap$-large over M.

<u>PROPOSITION 8</u>: $Q_\Sigma(M)$ is strongly torsion free.

Proof: Suppose $A \in \Sigma$ and $[B,f]a = [C,g]a$ for all $a \in A$. Let $D_a = a^{-1}B \cap a^{-1}C \in \Sigma$, then $[B,f]a = [C,g]a$ and $f(ad) = g(ad)$ for all $d \in D_a$. But $T = \bigcup aD_a \in \Sigma$ and $f(t) = g(t)$ for all $t \in T$ so $[B,f] = [C,g]$.

<u>THEOREM 9</u>: Let $\tau_E = id$, then if E is weakly Σ-injective, E is Σ-injective.

Proof: Let N/M be torsion and $f \in Hom(M,E)$. We assume that M is maximal with respect to extension of f. Let $n \in N/M$ and $A = n^{-1}M \in \Sigma$, then define $\alpha: A \to E$ by $\alpha(a) = f(na)$. Since E is weakly Σ-injective, α has an extension $\hat{\alpha}: S \to E$. Define $g: M \cup nS \to E$ by $g(m) = f(m)$ and $g(ns) = \hat{\alpha}(s)$. Then if $ns \in M$, $g(ns) = f(ns) = \alpha(s) = \hat{\alpha}(s)$. Moreover, g is well defined for if $ns = nt$, then $g(ns) = \hat{\alpha}(s)$, $g(nt) = \hat{\alpha}(t)$ but for some $B \in \Sigma$, $sB, tB \subseteq A$ since $A \in \Sigma$. Thus for $b \in B$, $g(nsb) = f(nsb) = f(ntb) = g(ntb)$ so $(g(ns), g(nt)) \in \tau_E = id$. Hence $g(ns) = g(nt)$ and g is well defined. This contradicts the maximality of M so E is Σ-injective.

<u>REMARK</u>: $Q_\Sigma(M)$ is weakly Σ-injective by construction and is strongly torsion free so $Q_\Sigma(M)$ is Σ-injective and so faithfully Σ-injective. Thus $Q_\Sigma(M) = E_\Sigma(M)$.

Now let M be an arbitrary S-system, then M/τ_M is strongly torsion free. Define the module of quotients of M with respect to Σ to be $Q_\Sigma(M) = Q_\Sigma(M/\tau_M)$. The insertion map i: $M \to Q_\Sigma(M)$ is $M \to M/\tau_M \to Q_\Sigma(M/\tau_M)$. Moreover, i is monic if and only if $\tau_M = \text{id}$.

Let M and M' be S-systems with $M \xrightarrow{f} M'$, then we have a natural mapping $M/\tau_M \xrightarrow{f'} M'/\tau_{M'}$. By the construction of Q_Σ, there is a unique homomorphism f_Σ: $Q_\Sigma(M) \to Q_\Sigma(M')$ such that the diagram

$$\begin{array}{ccc} M & \xrightarrow{\quad f \quad} & M' \\ {\scriptstyle i}\downarrow & & \downarrow{\scriptstyle i} \\ Q_\Sigma(M) & \xrightarrow{\quad f_\Sigma \quad} & Q_\Sigma(M') \end{array}$$

commutes. Thus $M \to Q_\Sigma(M)$ and $f \to f_\Sigma$ forms a covariant functor.

5. THE SEMIGROUP OF QUOTIENTS

Let Σ be a right quotient filter. Since for any S-system M, τ_M is described by $x\tau_M y$ if and only if there is $A \in \Sigma$ with $xa = ya$ for all $a \in A$, then if $m \in M$ and $a\tau_S b$, then $ma\tau_M bm$. Thus, if M is strongly torsion free, then M is an S/τ_S-system.

Now form $Q_\Sigma(S)$ which contains S/τ_S and which is an S/τ_S system.

<u>THEOREM 1</u>: $Q_\Sigma(S)$ is a semigroup in such a way that S/τ_S is a subsemigroup and that the structure of $Q_\Sigma(S)$ as a right S-system is induced by the semigroup structure. Moreover, this semigroup structure is unique in the sense that if T is a semigroup containing S and $\tau_{T/S} = w$ then T/τ_T is isomorphic over S/τ_S to a subsemigroup of $Q_\Sigma(S)$.

Proof: Let $S_o = S/\tau_S$. Then if $[A,f] \in Q_\Sigma(S)$, $s \mapsto [A,f]s$ is an S-mapping from S_o to $Q_\Sigma(S)$. First view $Q_\Sigma(S)$ as an extension of S_o with $Q_\Sigma(S)/S_o$ as a torsion system, and secondly as a faithfully Σ-injective system. Thus the mapping $s \mapsto [A,f]s$ extends uniquely to a map γ: $Q_\Sigma(S) \to Q_\Sigma(S)$. If $[B,g] \in Q_\Sigma(S)$, then $[A,f]\,[B,g]$ is defined to be $\gamma[B,g]$. It is easy (but tedious) to verify that this is the required semigroup structure on $Q_\Sigma(S)$ and shows why the semigroup structure is unique.

If E is a faithfully Σ-injective system, then $\tau_E = \text{id}$ so that E is an S/τ_S-system. By exactly the same argument as above, we have

COROLLARY 2: If E is a faithfully Σ-injective S-system, then the S-system structure extends uniquely to give E the structure of a right $Q_\Sigma(S)$-system.

Of course, $Q_\Sigma(S)$ is called the semigroup of quotients of S with respect to Σ. In the following three theorems we apply our characterization of $Q_\Sigma(S)$ to determine some categorical properties of $Q_\Sigma(S)$.

THEOREM 3: Let Σ be a right quotient filter on S. The following assertions are equivalent:

(1) Every $Q_\Sigma(S)$-system is strongly torsion free,

(2) $i(A)Q_\Sigma(S) = Q_\Sigma(S)$ for every $A \in \Sigma$,

(3) every $Q_\Sigma(S)$-system is faithfully Σ-injective,

(4) $M_S \otimes Q_\Sigma(S)$ is isomorphic to $Q_\Sigma(M)$.

Proof: (1) $\Rightarrow$ (2): We have $i(A) \subseteq i(S) \subseteq Q_\Sigma(S)$ with $i(S)/i(A)$ and $Q_\Sigma(S)/i(S)$ torsion. Thus $Q_\Sigma(S)/i(A)$ is torsion. Since $i(A) \subseteq i(A)Q_\Sigma(S)$, $Q_\Sigma(S)/i(A)Q_\Sigma(S)$ is torsion. However, this is also a $Q_\Sigma(S)$-system and so is torsion free. Thus $Q_\Sigma(S)/i(A)Q_\Sigma(S) = 0$ or $i(A)Q_\Sigma(S) = Q_\Sigma(S)$.

(2) $\Rightarrow$ (3): Let E be a $Q_\Sigma(S)$-system. Thus if $a\tau_S b$, then $xa = xb$ for all $x \in E$ so $(\lambda_x \times \lambda_x)\tau_S \subseteq \mathrm{id}$. Hence if $a \in A$ and $A \in \Sigma$, $xa = xi(a)$. Thus if $(x,y) \in \tau_E$ and $xa=xa$ for all $a \in A \in \Sigma$, then $xi(a)\eta = yi(a)\eta$ for all $i(a)\eta \in i(A)Q_\Sigma(S)$. Since $Q_\Sigma(S) = i(A)Q_\Sigma(S)$, $x\xi = y\xi$ for all $\xi \in Q_\Sigma(S)$. Since $1 \in Q_\Sigma(S)$ and E is unitary, $x=y$ and $\tau_E=\mathrm{id}$. (Thus (2) $\Rightarrow$ (1).)

Now this shows that E is strongly torsion free so we need only show that E is weakly Σ-injective. To this end let $A \in \Sigma$ and $f \in \mathrm{Hom}(A,E)$. Then $\ker f \supseteq \tau_A$ so we have $\bar{f} \in \mathrm{Hom}(i(A),E)$. Since $Q_\Sigma(S) = i(A)Q_\Sigma(S)$, $1 = i(a)\xi$ for some $\xi \in Q_\Sigma(S)$. Now there is some $B \in \Sigma$ with $\xi B \subseteq i(S)$ so let $C = A \cap B \in \Sigma$ and $e = f(a)\xi = \bar{f}(i(a))\xi \in E$. Then if $c \in C$,

$$ec = f(a)(\xi c) = \bar{f}(i(a))(\xi c) = f(a\xi c)$$

since $\xi c \in i(S)$. From $1 = i(a)\xi$ we obtain $i(c) = i(a)\xi c$ so

$$\begin{aligned} f(c) = \bar{f}(i(c)) &= \bar{f}[i(a)\xi c] \\ &= \bar{f}(i(a))\xi c \\ &= f(a)\,\xi\, c \\ &= ec \end{aligned}$$

so $f|_C$ has an extension to $\hat{f}: S \to E$.

Now let $a \in A$. Then there is $T \in \Sigma$ with $aT \subseteq C$ so for $t \in T$, $\hat{f}(a)t = \hat{f}(at) = f|_C(at) = f(at) = f(a)t$ so $(\hat{f}(a), f(a)) \in \tau_E = id$. Thus $\hat{f}$ is an extension of f and E is weakly Σ-injective.

(3) ⇒ (4): Let $M_S \otimes Q_\Sigma(S)$ be the usual tensor product which is a $Q_\Sigma(S)$-system by $(m \otimes \lambda)\xi = m \otimes (\lambda\xi)$. By (3), $M_S \otimes Q_\Sigma(S)$ is a faithfully Σ-injective S-system. Let $j: M \to M_S \otimes Q_\Sigma(S)$ be defined by $j(m) = m \otimes 1$. Since $M_S \otimes Q_\Sigma(S)$ is strongly torsion free, $\ker j \supseteq \tau_M$. But $Q_\Sigma(M)$ is Σ-injective so there is $k: M_S \otimes Q_\Sigma(S) \to Q_\Sigma(M)$ with $jk = i: M \to Q_\Sigma(M)$. Thus $\ker j = \tau_M$, and the induced mapping $j: M/\tau_M \to M_S \otimes Q_\Sigma(S)$ is monic.

Now $M_S \otimes Q_\Sigma(S)/j(M)$ is torsion for if $m \otimes \xi \in M_S \otimes Q_\Sigma(S)$ there is some $A \in \Sigma$ with $\xi A \subseteq i(S)$, and for $a \in A$, $(m \otimes \xi)a = m(\xi a) \otimes 1$. Thus $M_S \otimes Q_\Sigma(S)$ is the Σ-injective hull of $j(M)$ and so k is an isomorphism.

(4) ⇒ (1): Let M be a $Q_\Sigma(S)$-system. Let $f: M_S \otimes Q_\Sigma(S) \to M$ be given by $f(m \otimes \xi) = m\xi$. Then f is clearly an $Q_\Sigma(S)$-epimorphism. Moreover, if $m\alpha = m\beta$, let $A \in \Sigma$ with $\alpha A, \beta A \in i(S)$. Then for $a \in A$, $m\alpha a = m\beta a$ and

$$\begin{aligned}(m \otimes \alpha)a &= m \otimes (\alpha a) = m\alpha a \otimes 1 \\ &= m\beta a \otimes 1 = m \otimes \beta a \\ &= (m \otimes \beta)a\end{aligned}$$

Then since $M \otimes Q_\Sigma(S) \cong Q_\Sigma(M)$ is strongly torsion free, $m \otimes \alpha = m \otimes \beta$ so f is one-to-one. Thus $M \cong M \otimes Q_\Sigma(S) \cong Q_\Sigma(M)$ is strongly torsion free.

DEFINITION: A right quotient filter Σ which has any of the properties listed in Theorem 3 is said to have property T (for tensor).

Define Σ to be noetherian if Σ has the following property: if $A_1 \subset A_2 \subset \ldots$ is an ascending chain of right ideals of S with $\cup A_i \in \Sigma$, then $A_n \in \Sigma$ for some n.

THEOREM 4: Let Σ be a right quotient filter for S. The following conditions are equivalent:

(1) The direct sum of faithfully Σ-injective S-systems is faithfully Σ-injective,

(2) Q_Σ commutes with direct sums,

(3) Σ is noetherian.

Proof: Let $\{M_\alpha : \alpha \in M\}$ be a collection of right S-systems. By their direct sum $\amalg M_\alpha$ we mean the subset of ΠM_α consisting of all $(m_\alpha) \in \Pi M_\alpha$ for which $\{\alpha | m_\alpha \neq 0\}$ is finite. Then $\amalg M_\alpha$ is a right S-system under component wise multiplication.

(1) $\Rightarrow$ (2): Let $\{M_\alpha : \alpha \in M\}$ be a family of strongly torsion free S-systems, then by (1), $\amalg Q_\Sigma(M_\alpha)$ is faithfully Σ-injective and there is an obvious embedding of $\amalg M_\alpha$ into $\amalg Q_\Sigma(M_\alpha)$. Now for all $\alpha \in M$, $Q_\Sigma(M_\alpha)/M_\alpha$ is torsion, and $\amalg Q_\Sigma(M_\alpha)/\amalg M_\alpha \cong \amalg Q_\Sigma(M_\alpha)/M_\alpha$ so $\amalg Q_\Sigma(M_\alpha)/\amalg M_\alpha$ is torsion. Since $\amalg Q_\Sigma(M_\alpha)$ is faithfully Σ-injective and $\amalg Q_\Sigma(M_\alpha)/\amalg M_\alpha$ is torsion, $\amalg Q_\Sigma(M_\alpha) \cong Q_\Sigma(\amalg M_\alpha)$.

(2) $\Rightarrow$ (3): Let $A_1 \subset A_2 \subset \ldots$ be an ascending chain of right ideals with $D = \cup A_i \in \Sigma$. Consider the obvious mapping $f\colon D \to \amalg(S/A_i) \subset \amalg Q_\Sigma(S/A_i)$. Since $D \in \Sigma$, f has an extension $f'\colon S \to \amalg Q_\Sigma(S/A_i)$, because $\amalg Q_\Sigma(S/A_i) \cong Q_\Sigma(\amalg S/A_i)$ is Σ-injective. Let $\xi = f'(1) \in Q_\Sigma(\amalg S/A_i)$, then there is an h with $\xi_n = 0$ for $n \geq h$. Thus since $f(d) = \xi d$, $(\xi d)_h = \xi_h d = 0$ and so $D \subset A_h$ and $A_h \in \Sigma$.

(3) $\Rightarrow$ (1): Let $\{E_\alpha | \alpha \in M\}$ be a family of faithfully Σ-injective S-systems. Then $\amalg E_\alpha$ is strongly torsion free since each E_α is torsion free. To show that E_α is Σ-injective we need only show that $\amalg E_\alpha$ is weakly Σ-injective. Let $D \in \Sigma$ and consider

$$\begin{array}{ccc} D & \xrightarrow{f} & \amalg E_\alpha . \\ & & \downarrow \pi_\beta \\ & & E_\beta \end{array}$$

Then for each $\beta \in M$ there is some $\xi_\beta \in E$ with $(f(d))_\beta = \xi_\beta d$. Let $\xi = (\xi_\alpha) \in \Pi E_\alpha$ and $f(d) = \xi d$. We will show that $\xi_\alpha = 0$ for all $\alpha \geq h$ for some h, thus $\xi \in \amalg E_\alpha$. Let $X = \{\alpha | \xi_\alpha \neq 0\}$. Let $C = \{\alpha_1, \alpha_2, \ldots\}$ be any countable subset of X. Let $A_n = \{d \in D\colon \xi_{\alpha_i} d = 0 \text{ for } i \geq n\}$. Since $f(d) \in \amalg E_\alpha$, $[f(d)]_\alpha = \xi_\alpha d = 0$ for all but a finite number of α. Thus $A_1 \subset A_2 \subset A_3 \subset \ldots$ is an ascending chain of right ideals and $D = \cup A_i \in \Sigma$. Thus $A_h \in \Sigma$ for some h, so

$\xi_{\alpha_n} A_h = 0$ for $n \geq h$. But E_{α_n} is torsion free so $\xi_{\alpha_n} = 0$ for $n \geq h$. Thus C is finite and as C was arbitrary, X is finite and $\xi S \subset \Pi E_\alpha$.

REMARK: In the proof of (3) $\Rightarrow$ (1) we need only show that $t(E_\alpha) = 0$ for all α, not that $\tau_{E_\alpha} = id$ for all α.

Finally, we close by discussing the right ideal structure of $Q_\Sigma(S)$.

THEOREM 5: Let Σ be a right quotient filter with property T. Then every right ideal of $Q_\Sigma(S)$ is generated by a right ideal of S/τ_S. In particular, if S is right noetherian, $Q_\Sigma(S)$ is also.

Proof: Let I be a right ideal of $Q_\Sigma(S)$, and let $A = I \cap i(S) \subset i(S)$. Then I/A is torsion so $I/AQ_\Sigma(S)$ is also torsion. By property T, every $Q_\Sigma(S)$-system is torsion free so $I/AQ_\Sigma(S)$ is both torsion and torsion free and so is 0. Thus $I = AQ_\Sigma(S)$.

6. THE TORSION CONGRUENCE GENERATED BY AN S-SYSTEM

Let M_S be an S-system. There is at least one torsion congruence ρ with $\rho_M = id$, namely $\rho = id$. In general, there are other such ρ and we will consider the class of all such ρ. For an essential extension E of M, $\rho_E = id$ if and only if $\rho_M = id$. Let $V(M) = \{\rho | \rho_M = id\}$ then $V(M) = V(E)$. Since V(0) consists of all torsion congruences on M_S, we will consider only nonzero S-systems.

THEOREM 1: Let E be a nonzero injective S-system. For any M_S, define τ_M to be the intersection of the kernels of all S-homomorphisms for M to E. Then

(1) τ is a torsion congruence on M_S

(2) $\tau \in V(E)$ and for any torsion congruence ρ, $\rho \in V(E)$ if and only if $\rho < \tau$.

Proof: (1) Clearly τ_M is an S-congruence on M and if $f \in Hom_S(M',M)$, let $(x,y) \in \tau_{M'}$. We must show that $(fx,fy) \in \tau_M$. Let $g: M \to E$ be arbitrary, then $gf: M' \to E$ so $gfx = gfy$ or $(fx,fy) \in \ker g$ so $(fx,fy) \in \tau_M$.

Let $f: K/M \to F$, $f(\bar{x}) \neq 0$. Then $x \notin N$ since Hom $(N/M,F) = 0$. Thus since $f: K/M \to F$ is defined and $f|_{N/M} = 0$, $\hat{f}: K/N \to F \neq 0$. (Thus (M1) is satisfied.)

Let K be a right S-system and consider $\alpha: K/\tau_K \to E$. Then we have, for each such α, the commutative diagram

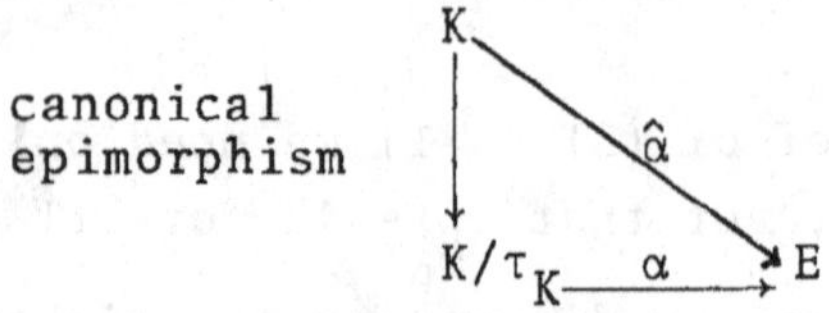

Let $[k_1]\tau_{(K/\tau_K)}[k_2]$, then $[k_1]\alpha = [k_2]\alpha$ so $\hat{\alpha}(k_1) = \hat{\alpha}(k_2)$. Since each $\beta: K \to E$ factors through K/τ_K, $[k_1]\tau_{(K/\tau_K)}[k_2]$ if and only if $k_1\tau_K k_2$ so $\tau_{(K/\tau_K)} = \mathrm{id}$, and (M2) is satisfied.

(2) Since $1: E \to E$ is an S-homomorphism, $\tau_E = \mathrm{id}$. If $\rho < \tau$ then $\rho_E < \tau_E = \mathrm{id}$ so $\rho_E = \mathrm{id}$. Let $M \in M_S$, then for $g: M \to E$, $(g\times g)(\rho_m) \subseteq \rho_E = \mathrm{id}$ so $\rho_M \subset \ker g$ or $\rho_M \subseteq \tau_M$ and so $\rho < \tau$.

For any $0 \neq M \in M_S$, the above theorem shows that V(M) has a largest element τ obtained by using the injective hull E of M. This torsion congruence will be denoted by τ^M and is called the <u>torsion congruence generated</u> by M. We set $\tau^0 = w$.

Note that the torsion congruence τ^M given by M may not necessarily be a special torsion congruence nor given by a rqf Σ. However, we do have the following results.

<u>THEOREM 2</u>: Let μ be a special torsion congruence. Then $\mu = \cap\{\tau^M \mid \mu_M = \mathrm{id}\}$.

Proof: Since $\mu_M = \mathrm{id}$, $\mu \leq \tau^M$ for all such M so $\mu \leq \cap\{\tau^M \mid \mu_M = \mathrm{id}\}$.

Let $T \in M_S$ and $(t_1,t_2) \in \cap\{\tau^M \mid \mu_M = \mathrm{id}\}$. Since T/μ_T is strongly torsion free, if ε is the canonical epimorphism $\varepsilon: T \to T/\mu_T$, then $\varepsilon(t_1) = \varepsilon(t_2)$ or $t_1\mu_T t_2$ so $\cap\{\tau^M \mid \mu_M = \mathrm{id}\} = \mu$.

<u>THEOREM 3</u>: Let σ be a special torsion congruence given by Σ_σ. Then if ρ is any other special torsion congruence given by some Σ_ρ and $\sigma \leq \rho \leq \tau^{(S/\sigma_S)}$, then $\rho = \sigma$.

Proof: First since $\sigma_{(S/\sigma_S)} = \mathrm{id}$, $\sigma \leq \tau^{(S/\sigma_S)}$.

Let $\Sigma_1 = \{A \subseteq S: \mathrm{Hom}(S/A, S/\sigma_S) = 0\}$, then $\Sigma_1 = \Sigma_\sigma$ for if $A \notin \Sigma_1$, then there is some $f: S/A \to S/\sigma_S$, $f \neq 0$. Let

$t = f(1) \neq 0$, then $tA = 0$, and $(t,0) \notin \sigma_S$ so $A \notin \Sigma_\sigma$. Conversely, if $A \notin \Sigma_\sigma$, define $f\colon S/A \to S/\sigma_S$ by $f(1) = [1]_{\sigma_S}$. Since $A \notin \Sigma_\sigma$, $f \neq 0$ so $A \notin \Sigma_1$.

Next let ρ be given by Σ_ρ and $\sigma \leq \rho \leq \tau^{(S/\sigma_S)}$. Then $\Sigma_\rho = \{A \subseteq S \mid \rho_{S/A} = w\}$ but $A \in \Sigma_\rho \Rightarrow \tau_{S/A}^{(S/\sigma_S)} = w \Rightarrow \mathrm{Hom}(S/A, S/\sigma_S) = 0 \Rightarrow A \in \Sigma_\sigma$. Conversely, $A \in \Sigma_\sigma \Rightarrow \sigma_{S/A} = w \Rightarrow A \in \Sigma_\rho$. Since $\Sigma_\sigma = \Sigma_\rho$, $\sigma = \rho$.

7. Prime Congruences

Let S be a commutative monoid with zero.

DEFINITION: A congruence ρ on S is prime if every non-zero element of S/ρ is cancellative.

An ideal B of S is completely prime if $ab \in B$ implies $a \in B$ or $b \in B$.

Let B be a completely prime ideal of S. Define a relation μ_B on arbitrary M by $x\mu_B y$ if and only if there is some element $d \in S \setminus B$ with $xd = yd$. It is easily seen that μ_B is an S-congruence and is a special torsion congruence with associated left quotient filter $\Sigma = \{A \subseteq S \mid A \cap (S/B) \neq 0\}$. In this connection see also [1] and [10]. μ_B is not necessarily a prime congruence on S. However, we will show that μ_B is related to a minimal prime congruence on S.

LEMMA 1: Let $\{\rho_i\} = C$ be a chain of prime congruences. Then $\rho = \cap \rho_i$ is a prime congruence.

Proof: Let $t \in S$ satisfy $t\not\rho 0$, thus $t\not\rho_i 0$ for some i. Moreover let $xt\rho yt$, thus $xt\rho_j yt$ for all j. However, if $x\not\rho y$, there is some j with $x\not\rho_j y$. If $\rho_i \subset \rho_j$, then $xt\rho_i yt$ implies that $x\rho_i y$ so $x\rho_j y$, a contradiction. On the other hand, if $\rho_j \subset \rho_i$ then $t\not\rho_i 0$ so $xt\rho_j yt$ implies $x\rho_j y$, another contradiction. Thus $x\rho y$.

THEOREM 2: Let α be a congruence on S, then α is contained in a minimal prime congruence μ.

Proof: By Lemma 1 and Zorn's Lemma, the theorem is valid.

When γ is a prime congruence on S, the zero class of γ is a completely prime ideal of S. Call this ideal B. Form a new congruence ρ_B on S by $\rho_B = B \times B \cup \{(x,y) \mid$ for some $t \notin B$, $xt = yt$; $x,y \notin B\}$. Note that if $\alpha\colon S \to S/B$

is the canonical epimorphism, then $\alpha^{-1}(\mu_B) = \rho_B$ where μ_B is the special torsion congruence applied to S/B. Clearly, ρ_B is a prime congruence on S since $S/\rho_B \cong \frac{S/B}{\mu_B}$.

THEOREM 3: Let B be a completely prime ideal of S. If γ is a prime congruence on S with zero class B, then $\rho_B \subset \gamma$.

Proof: First if $x\rho_B 0$ then $x \in B$ so $x\gamma 0$.

Next, if $x\rho_B y$, then there is $t \notin B$ with $xt = yt$. Thus $xt\gamma yt$ but $t \notin B$ so $x\gamma y$.

COROLLARY 4: Given a completely prime ideal B, ρ_B is the smallest prime congruence on S with zero class B.

DEFINITION: A congruence β on S is a primary congruence if the elements of S/β are either nilpotent or cancellative.

An ideal Q of S is primary if whenever $ab \in Q$ then $a \notin Q$ implies $b^n \in Q$ for some n.

Given a primary ideal Q of S, let $\sqrt{Q} = \{x \in S | x^n \in Q$ for some $n\}$. It is easily seen that $\sqrt{Q}$ is a completely prime ideal called the associated prime of Q.

For any primary congruence β on S, the zero class Q of β is a primary ideal. Moreover, $\sqrt{Q} = B$ is a completely prime ideal, and so we can form ρ_B. ρ_B is called the associated prime of β. Consequently $B = \{x \in S | x^n \beta 0$ for some integer $n > 0\}$. On occasion ρ_B will be denoted by $\rho_{\sqrt{Q}}$.

THEOREM 5: Let $\beta_1, \beta_2, \ldots, \beta_n$ be a finite set of primary congruences each having the same associated prime congruence ρ_B, then $\bigcap_1^n \beta_i = \beta$ is a primary congruence with associated prime congruence ρ_B.

Proof: Let Q be the zero class of β and Q_i be the zero class of β_i, then $\{x | x^k \beta 0\} = \bigcap_1^n \{x | x^k \beta_i 0\} = B$ so $\sqrt{Q} = B$.

Next let $t \notin B$ so $t^n \not\beta_i 0$ for all i, thus t is cancellative for all i; consequently, t is cancellative for β.

DEFINITION: A congruence γ on S is irreducible if whenever $\gamma = \alpha \cap \pi$ where α and π are congruences on γ, then $\alpha = \gamma$ or $\pi = \gamma$.

Call a semigroup S Noetherian if S has the ascending chain condition on congruences.

LEMMA 6: Every congruence in a Noetherian semigroup S is the finite intersection of irreducible congruences.

Proof: Let F be the collection of congruences on S which are not finite intersections of irreducible congruences. If $F \neq \emptyset$ then F has a maximal element δ. Thus $\delta = \alpha \cap \beta$ where α, β are congruences on S. Since $\alpha = \alpha_1 \cap \ldots \cap \alpha_m$ and $\beta = \beta_1 \cap \ldots \cap \beta_n$ then $\delta = \alpha_n \cap \ldots \cap \alpha_m \cap \beta_1 \cap \ldots \cap \beta_n$ and $\delta \notin F$. Thus $F = \emptyset$.

LEMMA 7: If γ is an irreducible congruence on S, then S/γ is essential over each nonzero subsystem.

Proof: Without loss of generality, let γ be the identity congruence on S. Let $0 \neq N$ be an ideal of S which is not essential in S. Hence there is some non-identity congruence α on S whose restriction to N is the identity. Let β be the rees congruence on S with respect to N - i.e., $\beta = N\times N \cup \{(s,s) | s \notin N\}$. Then $\alpha \cap \beta = id$ but neither α nor β is the identity congruence. This contradicts the irreducibility of id.

THEOREM 8: Every irreducible congruence in a Noetherian semigroup is primary.

Proof: Let $a \in S$. If $a^nS = 0$ then $a^n = 0$. If a is not nilpotent, then $a^nS \neq 0$. For each integer $n \geq 1$, let $\alpha_n = \ker \lambda_{a^n}$ where $\lambda_x: S \to S$ is the mapping defined by $\lambda_x(s) = xs$. By the ascending chain condition there is $N \geq 1$ with $\alpha_n = \alpha_N$ if $n \geq N$. Now the restriction of α_1 to a^NS is the identity since if $(a^Nx, a^Ny) \in \alpha_1$, then $a^{N+1}x = a^{N+1}y$ or $a^Nx = a^Ny$. However, since a^NS is essential in S, $\alpha_1 = id_S$ or a is cancellative.

COROLLARY 9: Every congruence on a Noetherian semigroup is the finite intersection of primary congruences.

DEFINITION: A primary representation of $\gamma = \bigcap_{i=1}^{n} \beta_i$ as primary congruences is irredundant if it satisfies the following two conditions:

(1) $\bigcap_{i\neq j} \beta_i \neq \gamma$ for all j, and

(2) if ρ_{β_i} is the associated prime of β_i, then

$$\rho_{\beta_i} \neq \rho_{\beta_j} \text{ if } i \neq j.$$

LEMMA 10: Let id have an irredundant primary decomposition as id $= \bigcap_{i=1}^{n} \beta_i$. For any minimal congruence ρ_B, $B = \sqrt{Q_i}$ for some i where $Q_i = \{x \mid x^n \beta_i 0 \text{ for } n \geq 1\}$.

Proof: First, there exists some i_o with $\sqrt{Q_{i_o}} \subseteq B$ for otherwise, for each $1 \leq i \leq n$ choose $x_i \in \sqrt{Q_i} \setminus B$. Thus $x_i^{k_i} \in Q_i$ and $\prod_{i=1}^{n} x_i^{k_i} \in \bigcap_{i=1}^{n} Q_i = 0 \in B$, a contradiction.

Secondly, $\rho_{\sqrt{Q_{i_o}}} \subseteq \rho_B$ but since ρ_B is a minimal prime congruence, $\rho_B = \rho_{\sqrt{Q_{i_o}}}$ and so $B = \sqrt{Q_{i_o}}$.

THEOREM 11: A Noetherian semigroup has a finite number of minimal prime congruences.

Proof: By the preceding lemma, each minimal prime congruence is of the form $\rho_{\sqrt{Q_i}}$ where id $= \bigcap_{i=1}^{n} \beta_i$ is an irredundant primary decomposition of id and Q_i is the zero class of β_i. Thus the number of minimal prime congruences is finite.

REFERENCES

1. Allouch, D., Filtre sur un monoide fini, Semigroup Forum 18 (1979), 27-32.
2. Berthiaume, P., The injective envelope of S-sets, Canad. Math. Bull. 10 (1971), 261-273.
3. Goldman, O., Rings and modules of quotients, J. Algebra 13 (1969), 10-47.
4. Hinkle, C. V., Generalized semigroups of quotients, Trans. A.M.S. 183 (1973), 87-117.
5. Hinkle, C. V., Semigroups of right quotients of a semigroup which is a semilattice of groups, J. Algebra 31 (1974), 276-286.
6. Hoehnke, H. J., Zur Definition der Begriffe Primkongruenz und Primäkongruenz in kommutativen Halbgruppen, Monatberichte der Deutschen Akademie der Wissenschaften zu Berlin 6 (1964), 801-804.
7. Luedeman, John K., A generalization of the concept of a ring of quotients, Canad. Math. Bull. 14 (1971), 517-529.

8. McMorris, F. R., The singular congruence and the maximal quotient semigroup, Canad. Math. Bull. 15 (1972), 301-303.

9. Weinert, H. J., S-sets and semigroups of quotients, Semigroup Forum 19 (1980), 1-79.

10. Weinert, H. J., On special right quotient filters of semigroups, Lecture Notes in Mathematics 855 (1981), Springer-Verlag.

11. Weinert, H. J., personal communication.

Department of Mathematical Sciences
Clemson University
Clemson, SC 29631
U.S.A.

REGULAR SEMIGROUPS OF ENDOMORPHISMS OF GROUPS

John David Philip Meldrum

Although semigroups of endomorphisms of groups seem a natural collection of objects to study, little seems to have been done on this subject. Probably the first paper on this theme was Dubreil [4]. This author has written several other relevant papers on semigroups of endomorphisms [4]-[7]. In [5] there are further references. Other papers on the subject are Dolgarev [3], Fuchs and Rangaswamy [10], whose results are summarized in Fuchs [9], §112, a series of papers by Puusemp [14]-[20], and Mahmood, Meldrum and O'Carroll [12]. Baer [1] established the correspondence between idempotent endomorphisms and semidirect decompositions, a key result in our work (lemma 1.1 [12]).

This paper continues the line of investigation begun in [12]. A substantial proportion of the material in this paper was presented by the author at the Semigroup Conference at Oberwolfach in May 1981. The author would like to thank the participants at the conference for their interest, help and advice and the Mathematisches Forschungsinstitut for their hospitality and the support which made the conference possible. Dr. O'Carroll has provided me throughout with a great deal of help and encouragement for which many thanks are due. Professor Schein has provided me with much assitance, particularly in the matter of the list of references, for which I am very grateful.

We consider a group G and $\mathrm{End}\, G$, the semigroup of endomorphisms of G. An endomorphism $a \in \mathrm{End}\, G$ has kernel $\mathrm{Ker}\, a$, image $\mathrm{Im}\, a$. We say that G is a semidirect product of N by H, denoted $G = N \rtimes H$ if $G = NH$, $N \triangleleft G$ (N is a normal subgroup of G), H is a subgroup of G and

$N \cap H = \{1\}$. In this situation we say that H has a normal complement, and N has a complement in G. If $g \in G$, $g = nh$, $n \in N$, $h \in H$, then the map $a \in \text{End } G$ defined by $ga \mapsto h$ is an idempotent endomorphism which we will denote $\pi(N,H)$. Note that $h\pi(N,H) = h$ for all $h \in H$. In Baer [1], it is shown that expressions of G as semidirect products are in 1 - 1 correspondence with idempotent endomorphisms of G (Lemma 1.1 [12]), namely $e^2 = e \in \text{End } G$ corresponds to $G = \text{Ker } e \rtimes \text{Im } e$. This correspondence is a key result in our work and will be used repeatedly without explicit reference. An element a of a semigroup S has an inverse if there exists $b \in S$ such that $aba = a$, $bab = b$. A semigroup is regular if every element of the semigroup has at least one inverse. A semigroup S is inverse if every element of S has a unique inverse. We will follow Howie [11] in our notation for semigroups, and use it as our source of standard results.

1. GENERAL RESULTS ON REGULAR ENDOMORPHISMS

We first note that $aba = a$ forces ab to be an idempotent. It follows that if a and b are inverses, then both ab and ba are idempotents. We now present a result which is really just another version of Lemma 1.3 of [12].

LEMMA 1.1. *Let $a \in \text{End } G$ have inverse a'. Then*

$$\text{Ker } a = \text{Ker } aa', \qquad \text{Im } a = \text{Im } a'a,$$
$$\text{Im } a' = \text{Im } aa', \qquad \text{Ker } a' = \text{Ker } a'a.$$

This result shows that an element of End G has an inverse only if its kernel has a complement and its image has a normal complement. We investigate this relationship further.

THEOREM 1.2. *Let $a \in \text{End } G$. Then a has an inverse if and only if Ker a has a complement and Im a has a normal complement.*

Proof. The implication one way follows immediately from lemma 1.1. So let us assume that Ker a has a complement H, and Im a has a normal complement K. We note that $H \cong G/\text{Ker } a \cong \text{Im } a$. So we can define an isomorphism $\theta : H \to \text{Im } a$ by

$$a = \pi(\text{Ker } a, H)\theta, \tag{1.3}$$

where $h\theta = ha$. Define a' by

$$a' = \pi(K, \text{Im } a)\theta^{-1}. \tag{1.4}$$

We then have

$$aa'a = a\pi(K,\mathrm{Im}\ a)\theta^{-1}\pi(\mathrm{Ker}\ a,H)\theta$$
$$= a\theta^{-1}\theta = a,$$

using the fact that $\pi(K,\mathrm{Im}\ a)$ is the identity map on Im a = Im θ and $\pi(\mathrm{Ker}\ a,H)$ is the identity map on H = Im θ^{-1}. We also have

$$a'aa' = a'\pi(\mathrm{Ker}\ a,H)\theta\pi(K,\mathrm{Im}\ a)\theta^{-1}$$
$$= a'\theta\theta^{-1} = a'$$

similarly. Hence a' is an inverse of a.

The factorisation of a in (1.3) and of its inverse a' in (1.4) is useful. It also follows from the proof that once K and H are chosen, with Ker a' = K, Im a' = H, then a' has to be defined by (1.4). This is because a'a has to be the identity map on Im a, and aa' has to be the identity map on H. From this, the following result is an immediate consequence.

COROLLARY 1.5. Let a ε End G. Then a has a unique inverse if and only if Ker a has a unique complement and Im a has a unique normal complement.

If we look at D_{2n}, the dihedral group of order 2n, with n odd, then we have an example of a group which has n distinct expressions as a semidirect product, namely

$$D_{2n} = Gp < a > \rtimes Gp < a^i b >$$

for $0 \leq i < n$, where $D_{2n} = Gp < a,b;\ a^n = b^2 = abab = 1 >$. So uniqueness of one of the two subgroups in an expression as a semidirect product does not imply uniqueness of the other. In D_{2n}, Gp < a > is the only normal complement, but Gp < a > has n complements.

If $G = C_\infty \times C_2$ is the direct product of a cyclic group of order 2 by an infinite cyclic group, then C_2 is a uniquely defined subgroup of G, but C_∞ is not. As G is abelian, either factor can be considered as the "normal" subgroup in the semidirect decomposition. Hence uniqueness of one member of the pair of subgroups in a semidirect decomposition does not imply uniqueness of the other.

We now consider Green's equivalences for regular semigroups of endomorphisms. We give a quick survey of the notation. If S is a semigroup, S^1 is the semigroup with identity adjoined if necessary. Then

$$a \mathcal{L} b \iff S^1a = S^1b,$$
$$a \mathcal{R} b \iff aS^1 = bS^1,$$
$$\mathcal{D} = \mathcal{L} \circ \mathcal{R} = \mathcal{R} \circ \mathcal{L},$$
$$a \mathcal{J} b \iff S^1aS^1 = S^1bS^1,$$
$$\mathcal{H} = \mathcal{L} \cap \mathcal{R},$$

where $a,b \in S$. In [2] Chapter 2, section 2, Green's equivalences are determined for the semigroup of all mappings from a set X to itself. In particular two mappings are $\mathcal{L}$ equivalent if they have the same image, and $\mathcal{R}$ equivalent if they have the same kernel. Also $\mathcal{D} = \mathcal{J}$ in this situation. Using proposition II.4.5 of [11], this gives us $\mathcal{L}$, $\mathcal{R}$ and $\mathcal{H}$ for regular semigroups of endomorphisms. We will give independent proofs of the characterizations.

Let S be a regular semigroup of endomorphisms of G.

THEOREM 1.6. *$a \mathcal{L} b$ if and only if* Im a = Im b.

Proof. If Im a = Im b, then $ba'a = b$, $ab'b = a$ since $a'a$ is the identity map on Im b = Im a = Im $a'a$, etc. The rest is trivial.

THEOREM 1.7. *$a \mathcal{R} b$ if and only if* Ker a = Ker b.

Proof. Again the result is given by the observation that $aa'b = b$, $bb'a = a$ if Ker a = Ker b.

COROLLARY 1.8. *$a \mathcal{H} b$ if and only if* Ker a = Ker b *and* Im a = Im b.

THEOREM 1.9. *$a \mathcal{D} b$ if and only if some element of* S *induces an isomorphism between* Im a *and* Im b.

Proof. If $a \mathcal{D} b$ then we can find $c \in S$ such that $a \mathcal{L} c$ and $c \mathcal{R} b$. So Im a = Im c, Ker c = Ker b. Thus Im a = Im c $\cong$ G/Ker c = G/Ker b $\cong$ $\cong$ Im b. From (1.3) and (1.4) it is clear that b' induces an isomorphism from Im b onto a complement of Ker b, H say, and c induces an isomorphism from any complement of Ker c onto Im c. As Ker b = Ker c, and Im a = Im c, it follows that $b'c$ induces an isomorphism from Im b onto Im a.

Conversely, let us assume that $c \in S$ induces an isomorphism from Im b onto Im a. Then bc maps G onto Im a, and so Im bc = Im a and $bc \mathcal{L} a$. Obviously Ker b $\leq$ Ker bc. But $g \in$ Ker bc forces $gb \in$ Ker c $\cap$ Im b = {1} by hypothesis. Thus Ker bc $\leq$ Ker b. But then Ker b = Ker bc and $b \mathcal{R} bc$. Hence $b \mathcal{D} a$.

THEOREM 1.10. *$a \mathcal{J} b$ if and only if for some* $c \in S$, Im c $\leq$ Im a *and some element of* S *induces a homomorphism from* Im c *onto* Im b, *and for* some d ε S, Im d ≤ Im b and some element of S induces a homomorphism from Im d onto Im a.

Proof. Suppose a $\mathcal{J}$ b. Then there exists x,y,u,v ε S such that xay = b, ubv = a. So Im xa ≤ Im a and y reduces to a homomorphism from Im xa onto Im b. Similarly Im ub ≤ Im b and v reduces to a homomorphism from Im ub onto Im a.

Conversely suppose Im c ≤ Im a. As a'a is the identity map on Im a'a = Im a, ca'a = c. Let y be the element of S inducing the homomorphism from Im c to Im b. Then ca'ay = b. Similarly we have x ε S such that db'bx = a. This shows that a $\mathcal{J}$ b.

If a ε S, denote by L_a, R_a, J_a the equivalence classes containing a determined by $\mathcal{L}$, $\mathcal{R}$, $\mathcal{J}$ respectively. Then there are natural partial orders on these sets of equivalence classes defined by

$$L_a \leq L_b \text{ if and only if } S^1a \leq S^1b,$$
$$R_a \leq R_b \text{ if and only if } aS^1 \leq bS^1,$$
$$J_a \leq J_b \text{ if and only if } S^1aS^1 \leq S^1bS^1.$$

We can deduce the following results easily.

THEOREM 1.11. $L_a \leq L_b$ if and only if Im a ≤ Im b, $R_a \leq R_b$ if and only if Ker a ≥ Ker b, $J_a \leq J_b$ if and only if for some c ε S with Im c ≤ Im a, some element of S induces a homomorphism from Im c onto Im b.

Proof. $L_a \leq L_b$ if and only if a = xb for some x ε S. But a = xb implies a = ab'b. $R_a \leq R_b$ if and only if a = bx for some x ε S. But a = bx implies Ker a ≥ Ker b and Ker a ≥ Ker b implies a = bb'a. The last part follows from the proof of theorem 1.10.

We finish this section with some general remarks. Every semigroup has a faithful representation as a semigroup of endomorphisms of a suitable group. This result is easy to prove but requires a lot of background. It can be found in Fröhlich [8]. This means that theorem 1.6, theorem 1.7 and corollary 1.8 provide another, somewhat elephantine proof of proposition II.4.5 of [11].

If we compare theorem 1.9 and theorem 1.10, we see that $\mathcal{J} \neq \mathcal{D}$ only if it is possible to find a,b ε S such that there is a homomorphism of a suitable type from a suitable proper subgroup of Im a onto Im b and a suitable homomorphism from some suitable subgroup of Im b onto

Im a. A bit of calculation shows that in that case we end up with an infinite strictly decreasing chain of subgroups of the form {Im a; a ε S}, equivalently with an infinite strictly ascending chain of subgroups of the form {Ker a; a ε S}. So, using theorem 1.11, if either the set of classes $\{L_a;\ a \in S\}$ or the set of classes $\{R_a;\ a \in S\}$ satisfies the minimum condition, then $\mathcal{D} = \mathcal{J}$. Compare proposition II.1.11 of [11].

Clifford and Preston [2] used the term "eggbox" to visualize a $\mathcal{D}$-class: the rows represented an $\mathcal{R}$-class, the columns represented an $\mathcal{L}$-class, the cells an $\mathcal{H}$-class. So in our case the columns are indexed by the individual elements of an isomorphism class of subgroups of the form {Im a; a ε S}, the isomorphisms being induced by elements of S. The rows are indexed by subgroups of the form {Ker a; a ε S} with the property that the images G/Ker a give rise to the isomorphism class mentioned above. Finally the $\mathcal{H}$ classes that are groups are groups of the form $\mathrm{Aut}_S(\mathrm{Im}\ a'a)$, the group of automorphisms of Im a'a induced by elements of S. A special case gives our final result of the section.

THEOREM 1.12. Let G be a group with End G regular. Then the classes that are groups are precisely the groups of the form Aut(Im e) for e an idempotent endomorphism of G.

This follows immediately from the remarks above.

2. REGULAR GROUPS

We consider now groups with the property that End G is regular.

DEFINITION 2.1. A group G with the property that End G is regular is called a regular group.

The first result is an immediate consequence of theorem 1.2.

THEOREM 2.2. A group G is regular if and only if every kernel of an endomorphism has a complement and every image of an endomorphism has a normal complement.

These groups seem to be very hard to pin down, much more so than the inverse groups, those with End G an inverse semigroup ([12]). Of course such groups are regular.

So we put some extra conditions on G. The case of abelian regular groups has already been studied in some depth by Fuchs and Rangaswamy

[10]. The following result is taken from Fuchs [9] and adapted to our notation.

THEOREM 2.3. Let A be an abelian group.

(a) If A is not reduced then A is regular if and only if A is a direct product of a torsion-free divisible group and an elementary group.

(b) If A is a torsion group then A is regular if and only if A is elementary.

(c) If A is reduced and regular then the torsion subgroup T is elementary, A/T is divisible and A lies between the restricted direct product of the primary components and their complete direct product.

The only open question is in part (c), that is to provide the converse when A is reduced. A reduced abelian group is one which has no non-trivial divisible subgroups.

We now consider nilpotent reduced groups. The answer is prehaps a bit disappointing.

THEOREM 2.4. Let G be a nilpotent regular group. Then G is abelian.

Proof. Suppose G is not abelian. Let Z be the centre of G, Z_2/Z be the centre of G/Z. As G is nilpotent, not abelian, then $Z_2 > Z$. Choose $h \in Z_2 - Z$. Then the map

$$g \to [h,g] = h^{-1}g^{-1}hg$$

is a homomorphism from G into Z which is not trivial. See Meldrum [13], lemma 3.2. Call it a. Then $Ga \leq Z \cap G'$, where G' is the derived group of G. So $G = N \rtimes Ga$ since G is regular. But Ga abelian forces $G' \leq N$, and this contradicts $Ga \leq G'$, since G is nilpotent and so $G' < G$. Hence G must be abelian.

The next class of groups to consider are the soluble groups. In this case we only have the very beginnings of a theory, but we present it here in the hope of stimulating further work.

LEMMA 2.5. Let G be a soluble non-abelian regular group. Then G cannot have an infinite cyclic group as homomorphic image.

Proof. Assume the contrary. Then there is an endomorphism of G mapping G to any cyclic subgroup contained in G. If we take such a subgroup in G' then we can use the method at the end of the proof of theorem 2.4 to obtain a contradiction.

LEMMA 2.6. Let G be a soluble non-abelian regular group. If G' contains an element of order p, then G cannot have a non-trivial cyclic group of p-power order as homomorphic image.

Proof. The method of proof is exactly the same as for the previous result.

LEMMA 2.7. Let G be a soluble non-abelian regular group. If $G - G'$ contains an element of order a prime p and G has a non-trivial homomorphic image of p-power order, then all non-trivial elements of order a power of p must lie in $G - G'$ and must be of order p.

Proof. From the hypotheses, it follows that every element of order p generates a subgroup which has a normal complement. By the method of the proof of theorem 2.4, it follows that all elements of order p must be outside G', hence so must all elements of order a power of p. Finally suppose that $g \in G$ has order p^2. Then $G = N \rtimes Gp < g^p >$ and $g = ng^{pr}$ where $n \in N$. So $g^p = n'g^{prp} = n'$, a contradiction. This finishes the proof.

THEOREM 2.8. Let G be a finite regular metabelian group. Then $G = G' \rtimes H$ for some subgroup H, the orders of G' and H are coprime and both groups are elementary.

Proof. From lemma 2.7 it follows that the Sylow subgroups of G are either contained in G' or have trivial intersection with G'. This together with lemma 2.7 gives us everything except that G' is elementary. So assume that G' is not elementary. Choose a prime q such that G' contains an element of order at least q^2. Let q^r be the highest power of q dividing $|G|$, and let $Gp < g_1 > \times \ldots \times Gp < g_s >$ be defined by the condition that $\{g_1,\ldots,g_s\}$ is the maximum number of independent elements of order q^r. Let K be the set of all q^{r-1} powers of elements in G'. Then $K \triangleleft G$ and G/K has order $q^s.|H|$. Consider $g_i^h = g_1^{x(1)} \ldots g_s^{x(s)} g'$ where g' has order at most q^{r-1}. So, writing q^{r-1} as t, we have $(g_i^t)^h = g_1^{tx(1)} \ldots g_s^{tx(s)}$. It follows that $Gp < g_1^t > \times \ldots \times Gp < g_s^t > H = L$ say, is isomorphic to G/K. Hence there exists an endomorphism from G onto L with kernel K, and so L has a normal complement N say. So $g_1 = n\ell$ where $n \in N, \ell \in L$. Since $g_1, n \in G'$, it follows that $\ell \in G'$. Hence $\ell \in Gp < g_1^t, \ldots, g_s^t >$. Now it follows from standard abelian group theory that we have a contradiction. This finishes the proof.

The dihedral groups of order 2n, where n is odd and square-free, satisfy the conditions of theorem 2.8. A bit of calculation which we omit for brevity, shows that such groups are regular. This encourages us to make two conjectures.

Conjecture 1. A necessary condition for a finite soluble group to be regular is that if its order is $q_1 \ldots q_n$ say, where the q_i are powers of distinct primes, for each i, $1 \leq i < n$, $G = N_i \rtimes H_i$, where $|N_i| = q_1 \ldots q_i$, $|H_i| = q_{i+1} \ldots q_n$ and each Sylow subgroup of G is elementary abelian.

Conjecture 2. The above condition is also sufficient.

The first conjecture is, I believe, very likely to be true. I am not so sure about the second.

We finish with a result which came about as the result of a question from M. Petrich at the conference mentioned at the beginning of the paper. In [12], it was shown that if S is an inverse semigroup of endomorphisms of a group G, then it is isomorphic to a semigroup of partial isomorphisms of G. Conversely, a semigroup of partial isomorphisms of G which satisfies some fairly stringent conditions can be "lifted" to an inverse semigroup of endomorphisms of G. The question was: which groups G have the property that they have an inverse semigroup of endomorphisms which gives rise naturally to the semigroup of all partial isomorphisms of G.

THEOREM 2.9. Let G be a group with the property that it has an inverse semigroup of endomorphisms giving rise to the semigroup of all partial isomorphisms of G. Then G is a direct product of elementary p-groups for various primes p, and all such groups satisfy the hypotheses.

Proof. The hypotheses mean that given any partial isomorphism a of G: $H_1 \to H_2$ say, with H_1, H_2 subgroups of G, then there is an endomorphism b of G such that Im b = Im a, $G = \text{Ker } b \rtimes H_1$.

By taking $H_1 = H_2 = \text{Gp} < g >$ for some $g \in G$, it follows that Ker b $\geq G'$, i.e. $g \notin G'$. This is true for all $g \in G$, $g \neq 1$. Hence $G' = \{1\}$ and G is abelian. The hypotheses then state that all non-trivial subgroups are direct factors. Also any subgroup inherits the property of G. So we conclude that all elements have prime order. This shows that the conditions are necessary.

The converse follows from elementary vector space theory and abelian group theory.

REFERENCES

[1] Baer, R. Splitting endomorphisms. Trans. Amer. Math. Soc. 61 (1947), 508-516.

[2] Clifford, A.H. and G.B. Preston, The Algebraic Theory of Semigroups. Vol. 1. Math. Surveys of the American Math. Soc. 7, Providence, R. I., 1961.

[3] Dolgarev, A.I. Certain ideals of the semigroup of endomorphisms of a group and related properties of the group. Modern Algebra (Leningrad) 4 (1976), 67-75 (Russian).

[4] Dubreil, P. Sous-groupes d'un demi-groupe. Demi-groupes des endomorphismes d'un groupe. Sém. Dubreil-Pisot (Algèbre et Théorie des Nombres), 14e. année, 1960/61, no. 16 (13 mars 1961), 15 pp.

[5] Dubreil, P. Endomorphismes. Sém. Dubreil-Pisot (Algèbre et Théorie des Nombres), 18e année, 1964/65, no. 23 (17 mai 1965), 20 pp.

[6] Dubreil, P. Lectures on the algebraic theory of semigroups. Tulane University, New Orleans, 1962.

[7] Dubreil, P. Demi-groupe des endomorphismes d'un algèbre. Eleuteria (Greece). To appear 1982.

[8] Fröhlich, A. On groups over a d. g. near-ring. I. Sum constructions and free R-groups. Quart. J. Math. Oxford. Series 2, 11 (1960), 193-210.

[9] Fuchs, L. Infinite Abelian Groups. Vol. 2. Academic Press. London. 1973.

[10] Fuchs, L. and K.M. Rangaswamy, On generalized regular rings. Math. Z. 107 (1968), 71-81.

[11] Howie, J.M. An introduction to semigroup theory. Academic Press. London. 1976.

[12] Mahmood, S.J., J.D.P. Meldrum and L. O'Carroll, Inverse semigroups and near-rings. J. London Math. Soc. (2), 23 (1981), 45-60.

[13] Meldrum, J.D.P. On nilpotent wreath products. Proc. Cambridge Philos. Soc. 68 (1970), 1-15.

[14] Puusemp, P. Semigroups of endomorphisms of dihedral groups. Student works in algebra and geometry. Tartu (1972), 39-49. (Russian).

[15] Puusemp, P. Semigroups of endomorphisms of two classes of metacyclic groups. Acta et Comm. Univ. Tartu 336 (1974), 100-119. (Russian).

[16] Puusemp, P. The idempotents of endomorphisms semigroups of groups. Acta et Comm. Univ. Tartu 366 (1975), 76-104. (Russian).

[17] Puusemp, P. The endomorphism semigroups of generalized quaternion groups. Acta et Comm. Univ. Tartu 390 (1976), 84-103. (Russian).

[18] Puusemp, P. The endomorphism semigroups of the semidirect product of two cyclic p-groups. Acta et Comm. Univ. Tartu 390 (1976), 104-133. (Russian).

[19] Puusemp, P. On the definability of a periodic abelian group by its semigroup of endomorphisms. Eesti NSV Teaduste Akadeemia Toimetised, Füüsika, Matemaatika 29 (1980), no. 3, 241-245. (Russian).

[20] Puusemp, P. On the definability of a periodic abelian group by its semigroup of endomorphisms in the class of all periodic abelian groups. Eesti NSV Teadusts Akadeemia Toimetised, Füüsika, Matemaatika 29 (1980), no. 3, 246-253. (Russian).

Department of Mathematics,
University of Edinburgh,
King's Buildings,
Edinburgh EH9 3JZ,
Scotland.

ON Γ -DECOMPOSITION OF SEMIGROUPS *)

F. Migliorini - J. Szép

INTRODUCTION

In § 1 we introduce the notations of Γ-decomposition and we prove some general properties of it.

In § 2 we determine the Γ-decompositons of bands and of inverse semigroups.

In § 3 we consider the properties of left $D_L(S)$ and the right $D_R(S)$ decompositions in detail.

1

Let S be a semigroup. Suppose that the only annihilator of S is 0. We recall the following decompositons of S [9]:

(1) $$S = \bigcup_{i=0}^{5} S_i, \qquad S_i \cap S_j = \emptyset \qquad (i \neq j)$$

*) This work was performed in the sphere of G.N.S.A.G.A.(O.N.R.Italia).

(1') $S = \bigcup_{i=0}^{5} D_i , \qquad D_i \cap D_j = \emptyset \qquad (i \neq j)$

where

$S_o = \{a \in S;\ aS \subset S \text{ and } \exists\ x \neq 0 \text{ such that } ax = 0\}$

$S_1 = \{a \in S;\ aS = S \text{ and } \exists\ x \neq 0 \text{ such that } ax = 0\}$

$S_2 = \{a \in S - (S_o \cup S_1);\ aS \subset S \text{ and } \exists\ x_1, x_2 \in S,$
$x_1 \neq x_2 \text{ and } ax_1 = ax_2\}$

$S_3 = \{a \in S - (S_o \cup S_1);\ aS = S \text{ and } \exists\ x_1, x_2 \in S,$
$x_1 \neq x_2 \text{ and } ax_1 = ax_2\}$

$S_4 = \{a \in S - \bigcup_{i=0}^{3} S_i;\ aS \subset S\}$

$S_5 = \{a \in S - \bigcup_{i=0}^{3} S_i;\ aS = S\}$

The subsets D_i $(i=0,1,\ldots,5)$ are defined analogously with multiplication by the element a on the right (instead of left). If the sets S_i $[D_i]$ are nonempty, then they are sub semigroups of S. There are several interrelations concerning the subsemigroups S_i $[D_i]$. For the multiplication of the components with $S_iS_j=S_k$ one obtains:

	S_o	S_2	S_5
S_o	S_o		S_o
S_2	S_o	S_2	S_2
S_5	S_o	S_2	S_5

finite case

	S_o	S_1	S_2	S_3	S_4	S_5
S_o	S_o	S_o		S_o		S_o
S_1		S_1		S_1		S_1
S_2	S_o		S_2	S_3		S_2
S_3		S_1		S_3		S_3
S_4	S_o	S_o	S_2	S_2	S_4	S_4
S_5	S_o	S_1	S_2	S_3	S_4	S_5

infinite case

1.1. The decompositions (1) and (1') of a semigroup S will be called left and right decompositon of S and the following notation will be used:

$$D_L(S) = \{S_i\}, \quad D_R(S) = \{D_i\} \qquad (i=0,1,\dots,5).$$

Let $S_i \cap D_j = C_{ij}$ and $\Gamma = \{C_{ij}\}_{i,j=0,1\dots,5}$

If the subsets C_{ij} are nonempty, then they are subsemigroups of S and

$$(3) \qquad S = \bigcup_{i,j=0}^{5} C_{ij}, \quad C_{ij} \cap C_{hk} = \emptyset \text{ , if } (i,j) \neq (h,k).$$

The decomposition (3) of a semigroup S will be called the fine decomposition or the Γ-decomposition of S.

The Γ-decomposition is a new tool in the structural investigation of semigroups. It is a symmetrical decomposition, that is, the decomposition is independent of the left and right multiplication with the element a. On the other hand, it is possible to study the relation between the $D_L(S)$ and $D_R(S)$ decompositions by considering the subsemigroups C_{ij}. (see e.g. Theorem 1.2).

There are subsemigroups C_{ij} which are empty:

(1i) $(S_1 \cup S_3) \cap (D_1 \cup D_3) = \emptyset$.

The set $S_1 \cup S_3$ is the set of left increasing elements of S, $D_1 \cup D_3$ is the set of right increasing elements of S. It is known, that an element a of S fails to be a left and right increasing element of S. Therefore we get

$$C_{11} = C_{13} = C_{31} = C_{33} = \emptyset.$$

(1ii) $(D_1 \cup D_3) \cap S_5 = \emptyset$, $(S_1 \cup S_3) \cap D_5 = \emptyset$.

Indeed, if $a \in D_1 \cup D_3$, then a is a right increasing element of S, that is, $Sa = S$ and $aS \subset S$ (cf. [1], Theorem VI.2.1). Thus $a \notin S_1 \cup S_3 \cup S_5$, whence it follows that $(D_1 \cup D_3) \cap S_5 = \emptyset$. Similarly we obtein $(S_1 \cup S_3) \cap D_5 = \emptyset$. Therefore

$$C_{15} = C_{35} = C_{51} = C_{53} = \emptyset.$$

(1iii) $S_4 \cap D_5 = \emptyset$, $S_5 \cap D_4 = \emptyset$.

Let $a \in S_4 \cap D_5$. Since D_5 is a left group, there is an idempotent element $e \in D_5$ such that $aa' = e$ ($a' \in D_5$). The element e is a right identity of S and $ae = a$ implies $aec = ac$ for all $c \in S$. But $a \in S_4$ holds and $ec = c$ for every element c of S. Hence it follows that e is a two-sided identity element of S. Thus $a(a'S) = eS = S$ which contradicts $a \in S_4$. Therefore we have $S_4 \cap D_5 = \emptyset$. The assertion $S_5 \cap D_4 = \emptyset$ can be proved similarly. We get

$$C_{45} = C_{54} = \emptyset.$$

These results imply the next theorem.

<u>THEOREM 1.1</u>. <u>For an arbitrary Γ-decomposition of a semigroup</u> S <u>we have</u>

$$C_{11} = C_{13} = C_{31} = C_{33} = C_{15} = C_{51} = C_{35} = C_{53} = C_{45} = C_{54} = \emptyset.$$

1.2. Now we consider some further properties of a Γ-decompositon.

(1.iv) If the only left annihilator of S is 0 (i.e. $xS = \{0\}$ implies $x = 0$), then $C_{10} = C_{30} = C_{50} = \emptyset$.

Let $a \in S_1 \cap D_o$. Then there is a nonzero element x such that $xa = 0$. For an arbitrary $y \in S$ there exists an element v in S so that $y = av$. Thus $xy = xav = 0$. Hence it follows that $xS = \{0\}$, and $x=0$, a contradiction! Therefore $C_{10} = \emptyset$. It can be proved similarly that $C_{30} = \emptyset$, $C_{50} = \emptyset$.

Analogously we have $C_{01} = C_{03} = C_{05} = \emptyset$ provided that the only right annihilator of S is the zero element 0 of S.

(1.v) If $C_{i2} \neq \emptyset$ for i=1,3, or 5 in a semigroup S, then $D_4 = D_5 = \emptyset$. Dually, if $C_{2j} \neq \emptyset$ for j=1,3, or 5 in S, then $S_4 = S_5 = \emptyset$.

Let us consider the case i=1. Let $a \in C_{12} = S_1 \cap D_2$. Then there exist elements $x_1 \neq x_2$ such that $x_1a = x_2a$. Let y be an arbitrary element of S. Then there is an element v such that $y = av$. Hence $x_1y = x_1av = x_2av = x_2y$ and we obtain that $y \in D_o \cup D_1$, or else $y \in D_2 \cup D_5$. Thus we have

$$S = D_o \cup D_1 \cup D_2 \cup D_3,$$

that is, $D_4 \cup D_5 = \emptyset$. The proof is similar for the case i=3 or 5.

(1.vi) If the semigroup S has a right identity element e_1, then $C_{10} = C_{30} = C_{50} = \emptyset$.

Dually, if S contains a left identity element e_2, then $C_{01} = C_{03} = C_{05} = \emptyset$.

Indeed, if $e_1 \in S$ is a right identity of S it is easy to see that S has no nonzero left annihilator. Applying (1.iv) we obtain our assertion.

In the next part of this paper we shall deal with the case of monoids ($1 \in S$). We shall see that at least 22 of the components will be empty.

(1.vii) If $1 \in S$, then $S_i \cap D_j = \emptyset$, $S_j \cap D_i = \emptyset$ ($i=0,2,4$; $j=5$).

Let us apply Theorem VI.2.3. of [1]. Thus, if $a \in S_2 \cap D_5$ (the case $i=2$), then $aS \subset S$ and $Sa = S$. But it follows that $a \in D_1 \cup D_3$, which is a contradiction. Hence $S_2 \cap D_5 = \emptyset$. The proof is similar for the other cases. Therefore $C_{05} = C_{25} = C_{45} = C_{50} = C_{52} = C_{54} = \emptyset$.

(1.viii) If $1 \in S$, then $C_{12} = C_{32} = C_{21} = C_{23} = \emptyset$.

Indeed, in the case of a monoid we have $D_5 \neq \emptyset$, $S_5 \neq \emptyset$ and the assertion follows from (1.v).

(1.ix) If $1 \in S$, then $C_{10} = C_{30} = C_{01} = C_{03} = \emptyset$.

This a direct consequence of (1.vi).

Summing up the assertions (1.vii), (1.viii), (1.ix) we obtain the following result.

THEOREM 1.2. *If $1 \in S$, then the semigroups C_{10}, C_{01}, C_{30}, C_{03}, C_{12}, C_{21}, C_{23}, C_{32}, C_{05}, C_{50}, C_{25}, C_{52} are empty, moreover $C_{14} = S_1$, $C_{34} = S_3$ and dually $C_{41} = D_1$, $C_{43} = D_3$. Furthermore $C_{14} \cup C_{34} \neq \emptyset$ if and only if $C_{41} \cup C_{43} \neq \emptyset$. Finally $C_{55} = S_5 = D_5$.*

Proof. With respect to the second part of the theorem we remark that

$$S_1 = S_1 \cap \left(\bigcup_{i=0}^{5} D_i \right) = C_{14}\ , \quad S_3 = S_3 \cap \left(\bigcup_{i=0}^{5} D_i \right) = C_{34}.$$

Moreover a monoid has a left increasing element ($C_{14} \cup C_{34} \neq \emptyset$) if and only if it has a right increasing element ($D_1 \cup D_3 \neq \emptyset$) (cf. [1], Corollary III.6.12.)

Finally $S_5 = S_5 \cap \left[\bigcup_{i=0}^{5} D_i\right] = C_{55} = D_5$.

To illustrate the Γ-decomposition of a semigroup S we have the following, table of "Cayley-type":

	D_o	D_1	D_2	D_3	D_4	D_5
S_o	C_{00}	$\emptyset$	C_{02}	$\emptyset$	C_{04}	$\emptyset$
S_1	$\emptyset$	$\emptyset$*	$\emptyset$	$\emptyset$*	C_{14}	$\emptyset$*
S_2	C_{20}	$\emptyset$	C_{22}	$\emptyset$	C_{24}	$\emptyset$
S_3	$\emptyset$	$\emptyset$*	$\emptyset$	$\emptyset$*	C_{34}	$\emptyset$*
S_4	C_{40}	C_{41}	C_{42}	C_{43}	C_{44}	$\emptyset$
S_5	$\emptyset$	$\emptyset$*	$\emptyset$	$\emptyset$*	$\emptyset$*	C_{55}

The places denoted by * are empty sets for every semigroup S, while the others are empty only for monoids.

1.3. Some further properties of the Γ-decompositions

(1.x) If $1 \notin S$, then $S_5 \cap D_5 = \emptyset$.

Indeed, if $a \in S_5 \cap D_5$, then $aS = Sa = S$ and thus there are elements $z_1, z_2 \in S$ such that $az_1 = a$, $z_2a = a$. If $v \in S$ then there exist elements, $t, u \in S$ such that $v = at = ua$. Hence it follows

$$vz_1 = uaz_1 = ua = v, \quad z_2v = z_2at = at = v$$

that is z_1 is a right identity and z_2 is a left identity of S. Hence we obtain $z_1 = z_2 = 1$. Therefore $S_5 \cap D_5 \neq \emptyset$ implies $1 \in S$, whence by Theorem 1.2., we have $C_{55} = S_5 = D_5$. In other words:

$$(C_{55} \neq \emptyset) \iff 1 \in S.$$

(1.xi) In a semigroup S we have

$$S_o \cup S_1 = \{0\} \iff D_o \cup D_1 = \{0\}.$$

Indeed, if $a \in S_o \cup S_1$ $(a \neq 0)$ then there is an element $b \neq 0$ such that $ab = 0$ and thus $b \in D_o \cup D_1$. Also the dual statement holds. In the case $S_o \cup S_1 = \{0\}$ we have

$$C_{00} = \{0\}, \; C_{0j} = C_{j0} = \emptyset \quad (j=1,2,3,4,5)$$
$$C_{1j} = C_{j1} = \emptyset$$

(1.xii) If $a \in S_o \cup S_1$, then Sa has no a left identity of S. Dually, if $a \in D_o \cup D_1$ then aS does not contain a right identity of S.

We show the first assertion. In the contrary case, if $a \in S_o \cup S_1$ and Sa contains a nonzero left identity e of S, then there is an element $x \neq 0$ such that $ax = 0$, whence $(Sa)x = S(ax) = \{0\}$. Hence $ex = 0$, which is a contradiction.

(1.xiii) If S is a finite semigroup, then it is easy to see that $C_{ij} = \emptyset$ if $i,j = 1,3$, or 4.

Therefore in the case of a finite semigroup the Γ-decomposition can be characterized by the following table:

	D_o	D_1	D_2	D_3	D_4	D_5
S_o	C_{oo}	$\emptyset$	C_{o2}	$\emptyset$	$\emptyset$	C_{o5}
S_1	$\emptyset$	$\emptyset$*	$\emptyset$	$\emptyset$*	$\emptyset$	$\emptyset$*
S_2	C_{20}	$\emptyset$	C_{22}	$\emptyset$	$\emptyset$	C_{25}
S_3	$\emptyset$	$\emptyset$*	$\emptyset$	$\emptyset$*	$\emptyset$	$\emptyset$*
S_4	$\emptyset$	$\emptyset$	$\emptyset$	$\emptyset$	$\emptyset$	$\emptyset$*
S_5	C_{50}	$\emptyset$*	C_{52}	$\emptyset$*	$\emptyset$*	C_{55}

1.4. In general we know only little information concerning the structure of the semigroups C_{ij} in a semigroup S. However we have the following properties.

(1.xiv) If $S_5 \cap D_5 \neq \emptyset$, then it is a subgroup of S. By (1.i), (1.ii) we have $(S_1 \cup S_3 \cup S_5) \cap (D_1 \cup D_3 \cup D_5) = C_{55}$, i.e. C_{55} contains all the invertible elements of S. Hence, by Theorem VI.1.4 in [1], C_{55} is a subgroup of S and $C_{55} = S_5 = D_5$.

(1.xv) α) C_{i5} [C_{5i}] (i=0,2) is a right [left] cancellative semigroup.

β) C_{i4} [C_{4i}] (i=0,1,2,3) is a right [left] cancellative semigroup without idempotent elements.

γ) C_{44} is a cancellative semigroup without idempotent elements.

The properties α), β), γ) are simple consequences of the definitions of the semigroups D_5 [S_5], D_4 [S_4] and of the property (3.ii), which will be proved later.

1.5. From the relations (2), (2') analogous results may be obtained for the components C_{ij} of the Γ-decomposition.

For instance, we have

(4) $C_{ij}\, C_{k\ell} \subseteq S_i\, S_k \cap D_j\, D_\ell$

Also $C_{5i}\, C_{j5} \subseteq C_{ji}$, $C_{i5}\, C_{5j} \subseteq C_{ij}$ $(i,j = 0,2,5)$ hold.

Other examples:

$$C_{40}\, C_{24} \subseteq C_{20}\ , \quad C_{41}\, C_{14} \subseteq C_{00}\ , \ C_{42}\, C_{24} \subseteq C_{22}$$
$$C_{42}\, C_{34} \subseteq C_{22}\ , \quad C_{41}\, C_{34} \subseteq C_{20}\ , \ \ldots$$
$$C_{20}\, C_{34} \subseteq C_{20}\ , \quad C_{23}\, C_{32} \subseteq C_{22}\ , \ C_{01}\, C_{10} \subseteq C_{00}$$
$$C_{40}\, C_{30} \subseteq C_{20}\ , \quad \ldots$$
$$C_{21}\, C_{50} \subseteq C_{20}\ , \quad C_{05}\, C_{14} \subseteq C_{04}\ , \ C_{05}\, C_{55} \subseteq C_{05}.$$

We remark that $C_{ij}\, C_{hk} \subseteq C_{\ell m}$ implies $\ell, m = 0,2,4,5$ provided that $(i,j) \neq (h,k)$.

2.

In this section we treat the Γ-decompositions of bands as well as of completely regular inverse semigroups.

a) Let S be an idempotent semigroup and let

$$S = \bigcup_{i=0}^{5} S_i = \bigcup_{i=0}^{5} D_i = \bigcup_{i,j=0}^{5} C_{ij}$$

be the left, right and Γ-decomposition of S.

If S is a band, then by [4] , we have

$$S_1 = S_3 = S_4 = \emptyset ,$$

furthermore it can be proved [7] , that

$$S_o = \bigcup_{a \in S_o} L_a , \qquad S_2 = \bigcup_{a \in S_2} L_a ,$$

where L_a is the $\mathcal{L}$-class of a, and it is a rectangular band. Moreover S_5 is a rectangular band as well.

Analogously for the components D_j we have
$D_1 = D_3 = D_4 = \emptyset$, D_5 is a rectangular band; $D_o = \bigcup_{a \in D_o} R_a$,
$D_2 = \bigcup_{a \in D_2} R_a$, where R_a is the $\mathcal{R}$-class of a, which is also a rectangular band.

It is evident that $C_{ij} = \emptyset$, if i or j is 1,3, or 4. Further, if $0 \in S$, then 0 is the only left (right) annihilator of S. Thus by (1.iv), we get $C_{05} = C_{50} = \emptyset$.

Thus the following components are non-empty:
$C_{oo}, C_{o2}, C_{20}, C_{22}, C_{25}, C_{52}, C_{55}$.
We remark that, by (1.v), $C_{55} \neq \emptyset$ implies $C_{25} = C_{52} = \emptyset$.
Finally, in the idempotent semigroup S, the component S_5 is the set of left identity elements, D_5 is the set of all right identity elements. Hence it follows that $C_{55} = S_5 \cap D_5$ is {1} or it is empty, when $1 \notin S$.

Suppose that $1 \notin S$ and $S_5 \neq \emptyset$ (in this case $D_5 = \emptyset$).
We show that $C_{52} = S_5 \cap D_2 = S_5$. Indeed, if S_5 is not contained in D_2, then there is an element $e \in S_5 \setminus D_2$.
Since $S = D_o \cup D_2$, we have $e \in D_o$. There exists an element $y \in S \setminus \{0\}$ such that $ye = 0$. In this case for all $s \in S$:

$$ys = y(es) = (ye)s = 0$$

and hence $s \in D_o$, that is, $S = D_o$. Therefore
$C_{50} = S_5 \cap D_o \supseteq S_5 \neq \emptyset$, which is a contradiction.

Dually, if $D_5 \neq \emptyset$ and $1 \notin S$, then $C_{25} = D_5$, $C_{52} = \emptyset$.

Let us consider the component C_{oo}:

$$C_{oo} = \left[\bigcup_{a\in S_o} L_a\right] \cap \left[\bigcup_{b\in D_o} R_b\right] = \bigcup_{a,b} (L_a \cap R_b),$$

where $L_a \cap R_b$ is a rectangular band. Thus C_{ij} $(i,j = 0,2)$ is the union of rectangular bands. We obtain the following result.

THEOREM 2.1. The Γ-decomposition of an idempotent semigroup S is the following:

1) if $1 \in S$, then $S_5 = D_5 = \{1\}$; $C_{25} = C_{52} = \emptyset$

$S = C_{oo} \cup C_{o2} \cup C_{20} \cup C_{22} \cup \{1\}$;

2) if $1 \notin S$, then

a) if $S_5 \neq \emptyset$: $C_{52} = S_5$, $C_{25} = \emptyset$,

$S = C_{oo}\ \ C_{o2} \cup C_{20} \cup C_{22} \cup S_5$.

b) if $D_5 \neq \emptyset$: $C_{25} = D_5$, $C_{52} = \emptyset$

$S = C_{oo} \cup C_{o2} \cup C_{20} \cup C_{22} \cup D_5$,

where S_5 $[D_5]$ if nonempty, is a right [left] zero band, and C_{ij} $(i,j=0,2)$ is the union of rectangular bands.

It is evident that the above Γ-decomposition is characteristic for an idempotent semigroup.

It is easy to show that $e_1 \in S_o \cup S_2$ and $e \leq e_1$ imply $e \in S_o \cup S_2$. Dually, $e_1 \in D_o \cup D_2$ and $e \leq e_1$ imply $e \in D_o \cup D_2$.

In the Γ-decomposition of S we have: if $e_1 \in \bigcup_{i,j=0}^{2} C_{ij}$ and $e \leq e_1$ then $e \in \bigcup_{i,j=0}^{2} C_{ij}$; especially, if $e_1 \in C_{oo}$ and $e \leq e_1$, then $e \in C_{oo}$. In the case of $C_{55} = \{1\}$ we have $e \leq 1$ for all $e \in S$.

b) Let S be a completely regular inverse semigroup. Let

$$D_L(S) = \{ S_i \}, \quad D_R(S) = \{D_i\} \quad (i=0,1,\dots,5)$$

the left and right decomposition of S, respectively. We shall treat the Γ-decomposition of S. The following result [3] is known:

<u>THEOREM</u>. <u>A regular semigroup</u> S <u>is a completely regular inverse semigroup if and only if</u> $aS = Sa$ <u>holds for all elements</u> $a \in S$.

Furthermore it is known [8], that if S is completely regular, then $S_1 = S_3 = S_4 = \emptyset$, $D_1 = D_3 = D_4 = \emptyset$.

(2.i) $S_5 \cap D_o = S_5 \cap D_2 = \emptyset$. Similarly, $S_o \cap D_5 = S_2 \cap D_5 = \emptyset$.

Indeed, if $a \in S_5$, then $aS = S$. Thus, by the above theorem of S. Lajos, it follows that $Sa = S$, that is, $a \notin D_o \cup D_2$. The proof is similar for the second statement.

Therefore, if S is a completely regular inverse semigroup, then its Γ-decomposition is

$$S = C_{oo} \cup C_{o2} \cup C_{20} \cup C_{22} \cup C_{55}.$$

The component C_{55}, if non-empty, then by (1.xiv) it is a subgroup and $C_{55} = S_5 = D_5$.

(2.ii) a) If $a \in S_i$ $(i=0,2)$, then $L_a \subseteq S_i$.

b) If $a \in D_i$ $(i=0,2)$, then $R_a \subseteq D_i$.

Let us consider the case a). If $a \in S_o$, then there is an element $x \neq 0$ such that $ax = 0$ and $aS \subset S$. Let $c \in L_a$; then $Sc = Sa$, hence $c = ta$, and $cx = (ta)x = 0$. Thus $c \in S_o$ and $L_a \subseteq S_o$ follows. If $a \in S_2$, then there are elements

$x_1, x_2 \in S$ $(x_1 \neq x_2)$ such that $ax_1 = ax_2$ and $aS \subset S$. Let $c \in L_a$; then similarly as above, there is a $v \in S$, so that $c = va$. Hence $cx_1 = (va)x_1 = vax_2 = cx_2$; moreover $c \notin S_o$, because in the contrary case $a = sc$ would belong to S_o. Thus $c \in S_2$ and $L_a \subseteq S_2$.

The proof is similar for the case b).

Let us consider now the components $C_{ij} = S_i \cap D_j$ $(i,j=0,2)$. In this case

$$C_{ij} = \left[\bigcup_{a \in S_i} L_a\right] \cap \left[\bigcup_{b \in D_j} R_b\right] = \bigcup_{\substack{a \in S_i \\ b \in D_j}} (L_a \cap R_b)$$

(2.iii) Each C_{ij} $(i,j=0,2)$ is the union of groups.

Indeed, in a completely regular inverse semigroup [3]

$$\mathcal{L} = \mathcal{R} = \mathcal{H} = \mathcal{D} = \mathcal{J},$$

hence it follows that $L_a = H_a$, $R_b = H_b$, where H_a and H_b are subgroups of S. Hence

$$C_{ij} = \bigcup_{\substack{a \in S_i \\ b \in D_j}} (H_a \cap H_b).$$

The above results will be summarized in the following theorem.

THEOREM 2.2. *A completely regular inverse semigroup S has the following Γ-decomposition:*

$$S = C_{oo} \cup C_{o2} \cup C_{20} \cup C_{22} \cup C_{55},$$

where C_{ij} $(i,j=0,2)$ if non-empty, is the union of subgroups which are H-classes of S and the subgroup C_{55} (if non-empty) is the H-class of 1.

In an inverse semigroup S the "$\leq$" relation is defined as follows:

$a \leq b$ if there is an idempotent e such that a=eb. This is a partial ordering of the elements of S (cf. [2], V.2).

Let $a \leq b$ in S. Then $a = eb$, $e \in E(S)$, thus if $b \in D_o$, then $a \in S_o$. Dually, if $b \in D_o$ and $a \leq b$, then $a \in D_o$. Analogously one can prove that if $b \in S_2$ $[D_2]$ and $a \leq b$ then $a \in S_o \cup S_2$ $[D_o \cup D_2]$. For the Γ-decomposition: if $b \in \bigcup_{i,j=0}^{2} C_{ij}$ and $a \leq b$ then

$a \in \bigcup_{i,j=0}^{2} C_{ij}$ (especially, if $b \in C_{oo}$, then $a \in C_{oo}$).

In C_{55} the ordering is trivial.

3.

In this section we obtain further information concerning the decompositions

$$D_L(S) = \{S_i\}\ , \quad D_r(S) = \{D_i\} \qquad (i=0,1,\ldots,5).$$

(3.i) $D_L(S_5) = S_5^{(5)}$; Dually, $D_R(D_5) = D_5^{(5)}$.

Indeed, S_5 is a right group and for every $s_5 \in S_5$ we have $s_5 S_5 = S_5$ and $s_5 x \neq s_5 y$ if $x \neq y$; $x,y \in S$.

(3.ii) $D_L(S_4) = S_4^{(4)}$; dually $D_R(D_4) = D_4^{(4)}$.

Indeed from $s_4 s = s_4 s'$ it follows that $s = s'$ ($\forall\, s_4 \in S_4$) and it implies that in the decomposition $S_4 = \bigcup_{i=0}^{5} S_i^{(4)}$ there are $S_1^{(4)} = S_2^{(4)} = S_3^{(4)} = \emptyset$.

Furthermore the semigroup S_4 has no idempotent element.

For if $a = a^2 \in S_4$ then for every $s \in S$ from $as = s'$ it follows $as = as'$ that is $s = s'$ and $as = s$. But in this case $a \in S_5$ follows which is a contradiction. It means $S_o^{(4)} = S_5^{(4)} = \emptyset$.

(3.iii). Let us consider the components S_o, S_1 and $\bar{S}_o = S_o \cup S_1$.

We prove the following assertion

$$D_L(\bar{S}_o) = \bar{S}_o \quad \text{dually} \quad D_R(\bar{D}_o) = \bar{D}_o.$$

First of all we prove that for every $\bar{s}_o \in \bar{S}_o$ there is an $x \in S_o$ $(x \neq 0)$ such that $\bar{s}_o\, x = 0$.

Assume that $S_o \neq \{0\}$. If our assertion is not true, then there is an $\bar{s}_o \in \bar{S}_o$ such that $\bar{s}_o x = 0$ $(x \in S_o)$, hence $x=0$. But $\bar{s}_o \in \bar{S}_o$ implies that there exists $s \in S \setminus \{0\}$ so that $\bar{s}_o s=0$ $(s \notin S_o)$. Moreover

1) $s \notin S_5$. For if $s \in S_5$, then $\bar{s}_o\, sS = \bar{s}_o S = \{0\}$; thus $\bar{s}_o S_o = \{0\}$ which is a contradiction. We obtain similarly that $s \notin S_1, S_3$.

2) $s \notin S_4$. Indeed, if $s \in S_4$, then $\bar{s}_o s=0$ implies $\bar{s}_o sS = 0$. But $S_4 S_o \subseteq S_o$ and $\bar{s}_o s\, S_o = \{0\}$ implies $s\, S_o = \{0\}$ which is a contradiction because of $S_o \neq \{0\}$ and $s \in S_o$ would follow.

3) $s \notin S_2$. Indeed, $\bar{s}_o s = 0$ implies $s_o s\, S_o = \{0\}$.

If $s \in S_2$, then $sx = 0$ $(x \in S)$ and it follows that $x=0$. On the other hand, $S_2 S_o \subseteq S_o$ and $s_o s\, S_o = \{0\}$ implies $s\, S_o = \{0\}$ which gives a contradiction.

DEFINITION. Let $D_L(S) = \{S_o, S_1, S_2, S_3, S_4, S_5\}$. We say that S is D_L-<u>irreducible</u>, if $D_L(S_i) = S_i$ $(i=0,1,\ldots,5)$.

By the earlier results, the semigroups $\bar{S}_o$, S_4, S_5 are D_L-irreducible. We define in a similar way the notions of D_R-irreducibility and Γ-irreducibility.

(3.iv) $D_L(S_3) = S_2^{(3)} \cup S_3^{(3)} \cup S_4^{(3)}$. Dually, $D_R(D_3) = D_2^{(3)} \cup D_3^{(3)} \cup D_4^{(3)}$.

Indeed, S_3 does not contain idempotent elements because every element of S_3 is an increasing element of S. Thus $S_o^{(3)} = S_1^{(3)} = S_5^{(3)} = \emptyset$.

(3.v) $D_L(S_1) = S_2^{(1)} \cup S_3^{(1)} \cup S_4^{(1)}$. Dually $D_R(D_1) = D_2^{(1)} \cup D_3^{(1)} \cup D_4^{(1)}$. The proof is similar to (3.iv).

Remark. In connection with (3.iv) and (3.v) we remark, that a semigroup S which has left (right) increasing elements may not have idempotent elements. But the following theorem is true (see [5] , I.1.): "Let $S = aM$, $M \neq S$, M is minimal for a and M is a subsemigroup of S. Then M (and therefore S) contains a subsemigroup which is isomorphic to the bicyclic semigroup $\mathcal{C}(p,q)$."

By the suppositions of the theorem, S has an idempotent element. But a semigroup S every element of which is a left (right) increasing element (such a semigroup was constructed by Teissier [1] , VI.3.14) does not contain an idempotent element. Indeed, $aS = S$ for all $a \in S$, hence S is right simple. If S would contain an idempotent, then it would be a right group, which is left cancellative, a contradiction.

Now we give an application of the above results.

Let S be a semigroup with property $D_L(S) = \{S_o, S_1, S_4, S_5\}$ that is $S = S_o \cup S_1 \cup S_4 \cup S_5$.

Let $I \neq \{0\}$ be a two-sided ideal of S ($I \neq S$). We have $S_1 \cap I = S_5 \cap I = \emptyset$, because I is right ideal of S. Let $S'_o = I \cap S_o$, $S'_4 = I \cap S_4$. Thus $I = S'_o \cup S'_4$. Suppose that S does not contain an element $s \neq 0$ such that $sI = \{0\}$. Then we show that for every $\bar{s}'_o \in \bar{S}'_o$ there exists an $\bar{s} \in I$ ($\bar{s} \neq 0$) such that $\bar{s}'_o\bar{s} = 0$. Since $\bar{s}'_o \in \bar{S}_o$ there is an $s \in S$ ($s \neq 0$) such that $s'_o s = 0$. Let $s'' \in I$. Then $s'_o ss'' = 0$ and $ss'' \in I$. By the assumption $sI \neq \{0\}$ if $s \neq 0$. Hence there is an element $s'' \in I$ such that $ss'' = \bar{s} \neq 0$. This means that $D_L(I) = \{S'_o, S'_4\}$. We have the following

THEOREM 3.6. *Let I be a two-sided ideal of S such that $sI \neq \{0\}$ if $s \neq 0$. Then $D_L(I) = \{S'_o, S'_4\}$, where $S'_o = I \cap S_o$, $S'_4 = I \cap S_4$.*

REFERENCES

[1] Ljapin, E.S. "Semigroups", Transl. Math.Monographs, Vol.3, Providence R.I. (1963).

[2] Howie, J.M. " An introduction to semigroup theory" London, Academic Press (1976).

[3] Lajos, S. "Characterization of completely regular inverse semigroups". Acta Sci. Math. 31 (1970),229-231.

[4] Migliorini,F.-Szép, J. "On a special decomposition of a regular semigroup". Acta Sci.Math. (Szeged), 40, (1978), 121-128.

[5] Migliorini,F. "Studio sui semigruppi con elementi accrescitivi". Rend. Ist.di Matem. Univ. Trieste, 6 (1974), 11-30

[6] Petrich, M. "Introduction to semigroups". C.E.Merrill Publ. Comp. - Columbus, Ohio (1973).

[7] Piochi, B. "Sulla decompsizione di Szép dei semigruppi generati da idempotenti". Ist.Matem.Univ.Siena, R.38. (1981).

[8] Scozzafava, R. "Remarks on Szép's decomposition of semigroups". Quad.Ist.Matem.Univ. Lecce, Q.10 (1978).

[9] Szép, J. "On the structure of finite semigroups, III". Dept. Math., K.Marx.Univ. Econ., Budapest, (1973) No. 3.

F.Migliorini
Universita di Siena
Istituto di Mathematika
Via del Capitano 15
53100 - Siena
Italia

J. Szép
Matematikai és Számitástudományi Intezet
Marx Károly Közgazdaságtudományi Egyetem
1093 Budapest IX
Dimitrov tér 8
Hungary

LANGUAGES OF PRIMITIVE WORDS ASSOCIATED WITH SETS OF NATURAL NUMBERS[1]

A. D. Paradis and G. Thierrin

INTRODUCTION

Let X be a nonempty finite set called an alphabet and let X* be the free monoid generated by X. The elements of X are called words and the identity element of X* is the empty word 1. We use the notation $|x|$ for the length of a word $x \in X^*$. A language L over X is any subset of X*. For any language L over X and any subset A of $N = \{0,1,2,\ldots,n,\ldots\}$, let $L(A) = \{x \mid x \in L \text{ and } |x| \in A\}$. Given a language L over X, the relation P_L is defined by $x \equiv y(P_L)$ if and only if $L..x = L..y$ where $L..x = \{(u,v) \mid u,v \in X^* \text{ and } uxv \in L\}$. P_L is a congruence called the syntactic congruence of L and the quotient monoid $\mathrm{syn}(L) = X^*/P_L$ is called the syntactic monoid of L. The language L is said to be disjunctive if P_L is the identity.

Let $X^+ = X^* - \{1\}$. A word $x \in X^+$ is said to be primitive if $x = f^n$ for some $f \in X^+$ implies $n = 1$. Let Q be the set of all primitive words and, for $i \geq 2$, let $Q^{(i)} = \{f^i \mid f \in Q\}$. It is well known ([3] and [4]) that Q and $Q^{(i)}$, for $i \geq 2$, are disjunctive languages. Also, it has been shown in [5] that the set of all primitive words of even or odd length is disjunctive. Furthermore, in [1] we find that for a prime number $p \geq 3$ and a natural number $0 \leq r \leq p-1$, the sets of words of Q or of $Q^{(i)}$, for $i \geq 2$, having their length congruent to r modulo p are all disjunctive. It is therefore natural to determine for which subset

[1] This research has been supported by grants A9245 and A7877 respectively of the Natural Sciences and Engineering Research Council of Canada.

A of N, the sets Q(A) or $Q^{(i)}(A)$, for $i \geq 2$, are disjunctive. A complete answer to this problem is given in the next section.

If Q(A) is disjunctive, then its syntactic monoid is isomorphic to X*. If Q(A) is not disjunctive, and this happens in infinitely many cases, it is interesting to get some properties of the corresponding syntactic monoid. This is done in the last section of this paper. As a consequence of these results, it is possible to give a formulation of the twin primes conjecture in the language of the semigroup theory.

In this paper, the alphabet X is always assumed to contain at least two letters.

DISJUNCTIVITY OF Q(A) AND $Q^{(i)}(A)$

The subset $A \subseteq N$ is said to be quasi-prime if A is infinite and if there exists $n_0 \in A$ such that for every $n \in A$ with $n \geq n_0$, we have n is prime. A quasi-prime set A is said to be prime if every $n \in A$ is prime.

PROPOSITION 1. Let $A \subseteq N$ be an infinite set. Then Q(A) is disjunctive if and only if A is not quasi-prime.

Proof: (=>) Let $n \in N$ and $a,b \in X$ with $a \neq b$. Since $a^{n+2}b \neq ab^{n+2}$ and Q(A) is disjunctive, then there exist $u,v \in X^*$ such that $ua^{n+2}bv \in Q(A)$ and $uab^{n+2}v \notin Q(A)$ (or vice versa). Let $t = |u|+|v|+n+3$. Then $t > n$ and $t \in A$. Since $uab^{n+2}v \notin Q(A)$ and $t \in A$, then $uab^{n+2}v \notin Q$. Hence $uab^{n+2}v$ is not primitive and since $t = |uab^{n+2}v|$, t cannot be prime. Therefore A is not quasi-prime.

(<=) Let $x,y \in X^*$ with $x \neq y$ and $|x| = |y|$. Let $n \in N$ such that $n > |x|$. Since A is not quasi-prime, there exists $t > (n+|x|)^2$ such that $t = t_1t_2 \in A$ and we can take $t_1 > (n + |x|)$ and $t_2 > 1$. Then, if $a,b \in X$ with $a \neq b$, we have

$$u = (b^n a^{t_1-(n+|x|)})y(b^n a^{t_1-(n+|x|)}y)^{t_2-1} \notin Q(A)$$

and

$$v = (b^n a^{t_1-(n+|x|)})x(b^n a^{t_1-(n+|x|)}y)^{t_2-1} \in Q(A),$$

because clearly u is not primitive and v is primitive with length $t = t_1t_2$. Hence $x \not\equiv y(P_{Q(A)})$ with $x \neq y$ and $|x| = |y|$. It has been shown in [2] that a language L is disjunctive if and only if $x \not\equiv y(P_L)$ for all distinct words x and y of the same length. Therefore Q(A) is disjunctive. #

A language $L \subseteq X^*$ is <u>regular</u> if the syntactic congruence P_L is of finite index.

<u>COROLLARY</u> 2. <u>If</u> A <u>is an infinite regular set of</u> N, <u>then</u> Q(A) <u>is disjunctive</u>.

<u>Proof</u>: A regular set of N being ultimately periodic, since A is infinite and regular, then there exist $m \in N$ with $m > 0$ and $s \in A$ with $s > 1$ such that $\{s + km \mid k \in N\} \subseteq A$. Let $n \in N$. For $t = s+(n+1)sm$, we have $t \in A$, $t > n$ and t is not prime. Therefore A is not quasi-prime and Q(A) is disjunctive. #

For a language L over X and a word $w \in X^*$, let $L.^{\cdot}w = \{x \mid x \in X^*$ and $wx \in L\}$ and $L^{\cdot}.w = \{x \mid x \in X^*$ and $xw \in L\}$.

<u>PROPOSITION</u> 3. Let $A \subseteq N$ be an infinite and not quasi-prime set. For any word $w \in X^*$, we have $Q(A).^{\cdot}w = Q(A)^{\cdot}.w$ and this set is disjunctive.

<u>Proof</u>: It is known ([5]) that for $x,y \in X^*$, if xy is primitive, then yx is also primitive. Hence we get the equality $Q(A).^{\cdot}w = Q(A)^{\cdot}.w$.

We shall now prove that $Q(A).^{\cdot}w$ is disjunctive. Let $x,y \in X^*$ with $x \neq y$ and $|x| = |y|$. Since A is not quasi-prime, there exists $t > (3|w|+2|x|+2)^2$ such that $t = t_1t_2$ and we can take $t_1 > 3|w|+2|x|+2$ and $t_2 > 1$. Then, if $a,b \in X$ with $a \neq b$, we have

$$u = (wba^{t_1-(|w|+|x|+1)})y(wba^{t_1-(|w|+|x|+1)}y)^{t_2-1} \notin Q(A)$$

and

$$v = (wba^{t_1-(|w|+|x|+1)})x(wba^{t_1-(|w|+|x|+1)}y)^{t_2-1} \in Q(A)$$

because clearly u is not primitive and v is primitive with length $t = t_1t_2$. Hence

$$(ba^{t_1-(|w|+|x|+1)})y(wba^{t_1-(|w|+|x|+1)}y)^{t_2-1} \notin Q(A).^{\cdot}w$$

and

$$(ba^{t_1-(|w|+|x|+1)})x(wba^{t_1-(|w|+|x|+1)}y)^{t_2-1} \in Q(A).^{\cdot}w.$$

Thus $x \not\equiv y(P_{Q(A).^{\cdot}w})$ and therefore the set $Q(A).^{\cdot}w$ is disjunctive. #

It is not true in general that for a disjunctive set D and a word $w \in X^*$, that $D.^{\cdot}w$ is also a disjunctive set. By way of example, for any nonempty word $w \in X^*$, the set $D = Q - \{wx \mid x \in X^*\}$ is disjunctive while the set $D.^{\cdot}w = \emptyset$ is not disjunctive.

For $i \geq 2$, the subset $A \subseteq N$ is said to be <u>i-dense</u> if for every $n \in N$, there exists $t \in A$ such that $t > n$ and $t = ki$.

<u>PROPOSITION</u> 4. <u>Let</u> $A \subseteq N$ <u>and</u> $i \geq 2$. <u>Then</u> $Q^{(i)}(A)$ <u>is disjunctive if and only if</u> A <u>is i-dense</u>.

<u>Proof</u>: (=>) Let $n \in N$ and $a,b \in X$ with $a \neq b$. Since $a^{n+1} \neq b^{n+1}$ and $Q^{(i)}(A)$ is disjunctive, then there exist $u,v \in X^*$ such that $ua^{n+1}v \in Q^{(i)}(A)$ and $ub^{n+1}v \notin Q^{(i)}(A)$ (or vice versa). Let $t = |u|+|v|+n+1$. Then $t > n$ and $t \in A$. Also, since $ua^{n+1}v \in Q^{(i)}(A)$, we have $ua^{n+1}v = f^i$ for $f \in Q$ and $t = |ua^{n+1}v| = |f^i| = |f|i$.

(<=) Let $x,y \in X^*$ with $x \neq y$ and $|x| = |y|$. For $n > |x|$, there exists $t \in A$ such that $t > (n + |x|)i$ and $t = ki$. Since $x \neq y$ and $b^n a^{k-(n+|x|)}x$ is primitive with length k, we have

$$(b^n a^{k-(n+|x|)})y(b^n a^{k-(n+|x|)}x)^{i-1} \notin Q^{(i)}(A)$$

and

$$(b^n a^{k-(n+|x|)})x(b^n a^{k-(n+|x|)}x)^{i-1} \in Q^{(i)}(A).$$

Hence $x \not\equiv y(P_{Q^{(i)}(A)})$ and it follows that $Q^{(i)}(A)$ is disjunctive. #

The property stated in Proposition 3 for disjunctive sets of the form Q(A), is also verified for disjunctive sets of the form $Q^{(i)}(A)$.

<u>PROPOSITION</u> 5. <u>Let</u> $i \geq 2$ <u>and let</u> $A \subseteq N$ <u>be an i-dense set. For any word</u> $w \in X^*$, <u>we have</u> $Q^{(i)}(A).^{\cdot}w = Q^{(i)}(A)^{\cdot}.w$ <u>and this set is disjunctive.</u>

Proof: It is known ([5]) that for $i \geq 2$ and $x,y \in X^*$, if $xy \in Q^{(i)}$, then $yx \in Q^{(i)}$. Hence we get the equality $Q^{(i)}(A).^{\cdot}w = Q^{(i)}(A)^{\cdot}.w$.

We shall now prove that $Q^{(i)}(A).^{\cdot}w$ is disjunctive. Let $x,y \in X^*$ with $x \neq y$, $|x| = |y|$ and let $m > |w| + |x|$. Since A is i-dense, there exists $t \in A$ such that $t > (2m+|w|+|x|)i$ and $t = ki$. Then, for $a,b \in X$ with $a \neq b$, we have

$$(wb^m a^{k-(m+|w|+|x|)})y(wb^m a^{k-(m+|w|+|x|)}x)^{i-1} \notin Q^{(i)}(A)$$

and

$$(wb^m a^{k-(m+|w|+|x|)})x(wb^m a^{k-(m+|w|+|x|)}x)^{i-1} \in Q^{(i)}(A)$$

because $x \neq y$ and $wb^m a^{k-(m+|w|+|x|)}x$ is primitive with length k. Hence

$$(b^m a^{k-(m+|w|+|x|)})y(wb^m a^{k-(m+|w|+|x|)}x)^{i-1} \notin Q^{(i)}(A).^{\cdot}w$$

and

$$(b^m a^{k-(m+|w|+|x|)})x(wb^m a^{k-(m+|w|+|x|)}x)^{i-1} \in Q^{(i)}(A).^{\cdot}w.$$

Thus $x \not\equiv y(P_{Q^{(i)}(A).^{\cdot}w})$ and therefore the set $Q^{(i)}(A).^{\cdot}w$ is disjunctive. #

In the following proposition we use $Q^{(1)}(A)$ instead of $Q(A)$, to simplify the presentation, and we are concerned with the product of disjunctive sets of the form $Q^{(i)}(A)$.

PROPOSITION 6. Let $i + j \geq 3$. If $A \subseteq N$ and $B \subseteq N$ are such that $Q^{(i)}(A)$ and $Q^{(j)}(B)$ are disjunctive, then the product

$$Q^{(i)}(A)Q^{(j)}(B) = \{uv \mid u \in Q^{(i)}(A) \text{ and } v \in Q^{(j)}(B)\}$$

is also disjunctive.

Proof: Let $x,y \in X^*$ with $x \neq y$ and $|x| = |y|$. By hypothesis, there exist $s \in B$ such that $s > (1+2|x|)j$ and $s = qj$, and there exists $r \in A$ such that $r > qi$ and $r = pi$. If $i \geq 2$, then we have

$$u_1 = (ba^{p-(1+|x|)})y(ba^{p-(1+|x|)}x)^{i-1}(ba^{q-(1+|x|)}x)^j \notin Q^{(i)}(A)Q^{(j)}(B)$$

and

$$v_1 = (ba^{p-(1+|x|)})x(ba^{p-(1+|x|)}x)^{i-1}(ba^{q-(1+|x|)}x)^j \in Q^{(i)}(A)^{(j)}(B)$$

because u_1 cannot be written as $f^i w$ with f primitive and the words $ba^{p-(1+|x|)}x$ and $ba^{q-(1+|x|)}x$ are primitive with length p and q respectively. If $j \geq 2$, then we have

$$u_2 = (xa^{p-(1+|x|)}b)^i (xa^{q-(1+|x|)}b)^{j-1} y(a^{q-(1+|x|)}b) \notin Q^{(i)}(A)Q^{(j)}(B)$$

and

$$v_2 = (xa^{p-(1+|x|)}b)^i (xa^{q-(1+|x|)}b)^{j-1} x(a^{q-(1+|x|)}b) \in Q^{(i)}(A)Q^{(j)}(B)$$

because u_2 cannot be written as wg^j with g primitive and the words $xa^{p-(1+|x|)}b$ and $xa^{q-(1+|x|)}b$ are primitive with length p and q respectively. In both cases, $x \not\equiv y(P_{Q^{(i)}(A)Q^{(j)}(B)})$ and therefore $Q^{(i)}(A)Q^{(j)}(B)$ is disjunctive. #

Let us point out that in this section, many results contained in [1], [2], [3], [4], and [5] have been generalized using simpler proofs.

THE SYNTACTIC MONOID OF Q(A)

If Q(A) is disjunctive, its syntactic monoid is isomorphic to the free monoid X*. If Q(A) is not disjunctive, then the next proposition shows that its syntactic monoid is quite different from a free monoid.

A monoid M is said to be <u>f-abelian</u> if there exists a finite subset $F \subseteq M$ such that M - F is an abelian subsemigroup of M.

<u>PROPOSITION</u> 7. <u>Let</u> A <u>be an infinite subset of</u> N. <u>Then the following properties are equivalent</u>:

(1) A <u>is quasi-prime</u>;

(2) syn(Q(A)) <u>is f-abelian</u>;

(3) Q(A) <u>is not disjunctive</u>.

<u>Proof</u>: (1) => (2). Since A is quasi-prime, then there exists $n_0 \in A$ such that for every $n \in A$ with $n \geq n_0$, we have n is prime. Let U be the set of classes $\bar{u}$ modulo $P_{Q(A)}$ containing words of length greater or equal to n_0. It is immediate that U is a subsemigroup of syn(Q(A)) and that syn(Q(A)) - U is finite. Now let $\bar{u}, \bar{v} \in U$ with $|u| \geq n_0$ and $|v| \geq n_0$. If $(x,y) \in Q(A)..uv$, then $xuvy \in Q(A)$ and $|xuvy|$ is a prime

number $p \in A$. This implies $|xvuy| = p$ and $xvuy \in Q(A)$. Thus $Q(A)..uv \subseteq Q(A)..vu$. By symmetry we have the inverse inclusion. Hence $uv \equiv vu(P_{Q(A)})$, $\bar{u}\bar{v} = \overline{uv} = \overline{vu} = \bar{v}\bar{u}$ and the subsemigroup U is abelian. Therefore syn(Q(A)) is f-abelian.

(2) => (1) If syn(Q(A)) is f-abelian, then it is clear that it cannot be isomorphic to a free monoid X^* with $|X| \geq 2$. Therefore Q(A) is not disjunctive and, by Proposition 1, A is quasi-prime.

(1) <=> (3) This follows immediately from Proposition 1. #

PROPOSITION 8. Let A be an infinite subset of $N - \{0,1\}$. Then A is prime if and only if syn(Q(A)) is abelian.

Proof: (=>) Let $u,v \in X^*$. If $(x,y) \in Q(A)..uv$, then $xuvy \in Q(A)$ and $|xuvy|$ is a prime number $p \in A$. This implies $|xvuy| = p$ and $xvuy \in Q(A)$. Thus $Q(A)..uv \subseteq Q(A)..vu$. By symmetry, we have the inverse inclusion. Therefore $uv \equiv vu(P_{Q(A)})$ and syn(Q(A)) is abelian.

(<=) By Proposition 7, A is quasi-prime. Assume that A is not prime. Then there exists $m \in A$ such that $m = rs$ with $r > 1$ and $s > 1$. Let $x = a^r b^{r(s-1)}$ with $a,b \in X$ and $a \neq b$. We have $m = |x|$ and it is clear that x is a primitive word. Hence $x \in Q(A)$. Since Q(A) is a union of classes of $P_{Q(A)}$, if $\bar{x}$ is the class of x modulo $P_{Q(A)}$, then $\bar{x} \subseteq Q(A)$. Since $\bar{x} = \bar{a}^r \bar{b}^{r(s-1)}$, then, because syn(Q(A)) is abelian, we have $\bar{x} = (\overline{ab^{(s-1)}})^r$. Therefore $(ab^{s-1})^r \in Q(A)$, a contradiction. Hence A is prime. #

From the above result, it follows then easily that the well known twin primes conjecture that there are infinitely many prime numbers p such that p + 2 is also prime is equivalent to the following conjecture:

> There exists an infinite subset A of $N - \{0,1\}$ containing infinitely many pairs of the form $\{n,n+2\}$ such that syn(Q(A)) is abelian.

REFERENCES

[1] Chien, T. Y., Decomposition of a free monoid into disjunctive languages, Soochow Journal of Mathematics 5(1979), 121-127.

[2] Reis, C. M. and Shyr, H. J., Some properties of disjunctive languages on a free monoid, Information and Control 37(1978), 334-344.

[3] Shyr, H. J., Disjunctive languages on a free monoid, Information and Control 34(1977), 123-129.

[4] Shyr, H. J., Free monoids and languages, Lecture Notes, Soochow University, Taipei (1979).

[5] Shyr, H. J. and Thierrin, G., Disjunctive languages and codes, Proceedings of the 1977 FCT Conference, Poznan, Lecture Notes in Computer Science 56, Springer-Verlag (1977), 171-176.

Département de Mathématiques
Université du Québec à Trois-Rivières
Trois-Rivières, Québec
Canada

Department of Mathematics
The University of Western Ontario
London, Ontario
Canada

EXTENSIONS OF SEMINEARRINGS BY SEMIGROUPS OF RIGHT QUOTIENTS

Hanns Joachim Weinert

§ 1 Introduction

In this paper, we investigate interrelations between (right distributive) seminearrings and different types of semigroups of right quotients, briefly Q_r-semigroups. These interrelations will turn out to be very close, and our results on seminearrings contain and unify (known as well as new) statements on nearrings, semirings[1] and rings.

At first, assume that $(S,+,\cdot)$ is a seminearring and $(T,\cdot)$ a classical Q_r-semigroup $Q_r(S,\Sigma)$ of $(S,\cdot)$ with respect to a subsemigroup Σ of $(S,\cdot)$ (cf. § 3). Then there exists exactly one addition on T such that $(T,+,\cdot)$ becomes a seminearring which contains $(S,+,\cdot)$ as a subseminearring. This result (Thm. 3.2) yields the concept of a classical Q_r-seminearring $(T,+,\cdot) = Q_r(S,\Sigma)$ of a seminearring $(S,+,\cdot)$ with respect to a subsemigroup Σ of $(S,\cdot)$, the subject of § 3. In particular, nearly all supplementary properties of the seminearring S, e.g. to be multiplicatively or additively left [right] cancellative or to be a nearring [semiring], carry over to any classical Q_r-seminearring T of S (Thm. 3.5). Moreover, all classical Q_r-seminearrings of S are contained in a maximal one (cf. Thms. 3.6 and 3.7), and we characterize those seminearrings for which this maximal classical Q_r-seminearring is a seminearfield (Thm. 3.8).

Secondly, assume that $(S,+,\cdot)$ is a seminearring and that $(T',\cdot)$ is a McMorris Q_r-semigroup of $(S,\cdot)$ (cf. § 5). In this

1) For basic definitions and facts concerning these algebras we refer to § 2. We also discuss in § 2 our way of procedure with respect to the situation that all additions which occur throughout this paper need not be associtative (but may be assumed to be if a reader is only interested in this case).

case there is at most one addition on T' such that (T',+,·) is a seminearring containing (S,+,·) as a subseminearring, and (T',·) is contained in a (minimal) McMorris Q_r-semigroup (T,·) of (S,·) which is, with a unique suitable addition, a seminearring (T,+,·) containing (S,+,·) (Thm. 5.4). Correspondingly defined McMorris Q_r-seminearrings are investigated in the second part of § 5, with more or less similar results as above (cf. Thm. 5.6). Note that each classical Q_r-seminearring of S is such a McMorris Q_r-seminearring, but there is less adventage to use this. So we have prefered to deal with the classical case independently, also with respect to its particular interest. Beyond it, McMorris Q_r-seminearrings are closely related to the following much more general concepts, which need similar technical effort, and we have used results of these considerations in our treatment of McMorris Q_r-seminearrings, postponed for this reason.

At third, each (right) S-groupoid M_S^+ with a semigroup (S,·) as operator domain (cf. § 4) defines, depending on two Q_r-filters $\mathcal{F}$ and K of the S-set M_S (cf. Def. 4.1 a)), various Q_r-seminearrings $(Q,+,\cdot) = Q_r(M_S^+,\mathcal{F},K)$ of M_S^+. They arise from a seminearring $(\Lambda_S(\mathcal{F},M_S^+),+,\cdot)$ of certain S-mappings f depending on $\mathcal{F}$ by a congruence κ_K which depends on K (Thm. 4.2, Def. 4.3). The elements of Q are denoted by $[f] = [f]_{\mathcal{F},K}$. The multiplicative semigroup (Q,·) of (Q,+,·) is the Q_r-semigroup $Q_r(M_S,\mathcal{F},K)$ of M_S with respect to $\mathcal{F}$ and K (cf. [39]), and (Q,+) reflects all essential properties of (M,+) (Thm. 4.4). Since each seminearring (S,+,·) determines an S-groupoid S_S^+, all this applies to Q_r-seminearrings $Q = Q_r(S_S^+,\mathcal{F},K)$ of a seminearring S with respect to Q_r-filters $\mathcal{F}$,K of S_S (Def. 5.1), and (Q,·) is now the Q_r-semigroup $Q_r(S_S,\mathcal{F},K)$ of (S,·) with respect to these Q_r-filters, plainly called a Q_r-semigroup of (S,·) today. Surprisingly, the addition of such a Q_r-seminearring $Q_r(S_S^+,\mathcal{F},K)$ of a seminearring S, defined in Thm. 4.2 in a natural way using pointwise addition of mappings, turns out to be uniquely determined by that of S if the Q_r-filter K of S_S is a so-called special one (a far-ranging concept, exclusively considered in various papers). Moreover, each

Q_r-seminearring $(Q_r(S_S^+,\mathcal{F},K),+,\cdot)$ contains a natural homomorphic image of $(S,+,\cdot)$ which is isomorphic iff $\mathcal{F}\subseteq K\subseteq S_S^{\Delta}$, where S_S^{Δ} denotes the (always special) Q_r-filter of S_S which consists of all dense S-subsets (Thm. 5.2). Exactly in this case, a Q_r-seminearring $Q_r(S_S^+,\mathcal{F},K)=Q_r(S_S^+,\mathcal{F},\mathcal{F})=Q_r(S_S^+,\mathcal{F})$ of a seminearring S is a McMorris Q_r-seminearring of S, and the maximal one, $Q_m = Q_r(S_S^+,S_S^{\Delta})$, contains each McMorris Q_r-seminearring T of S as considered above (Thm. 5.4). In general, however, not all of those seminearrings T coincide with a Q_r-seminearring of the form $Q_r(S_S^+,\mathcal{F})$ (Expl. 5.5), whereas all classical ones do.

The last three sections are devoted to semirings. Already in Thms. 3.5, 4.4 and 5.6 we have considered conditions such that certain elements of a Q_r-seminearring Q are also left distributive in Q. In order to obtain subsemirings of Q in this way, we define subcommutative S-groupoids M_S^+ (Prop. 6.1). Moreover, this condition is sufficient to consider, similarly as in § 4, Q_r-semirings $Q_r(M_S^+,\mathcal{F}^+,K^+)$ with respect to Q_r-filters $\mathcal{F}^+$ and K^+ of the S-groupoid M_S^+, all concepts based upon S-homomorphisms on S-subgroupoids (cf. Def. 4.1 b), Thms. 6.2, 6.4). We also compare both kinds of Q_r-filters and investigate the resulting interrelations between Q_r-seminearrings $Q_r(M_S^+,\mathcal{F},K)$ and Q_r-semirings $Q_r(M_S^+,\mathcal{F}^+,K^+)$ (Thms. 6.5, 6.9).

In § 7, we apply these results to S-groupoids S_S^+ obtained from subcommutative semirings S. As in § 5, each Q_r-semiring $(Q_r(S_S^+,\mathcal{F}^+,K^+),+,\cdot)$ contains a natural homomorphic image of $(S,+,\cdot)$, which is isomorphic iff $\mathcal{F}^+\subseteq K^+\subseteq S_S^{+\Delta}$ holds for the dense Q_r-filter $S_S^{+\Delta}$ of S_S^+ (Thm. 7.2). Together with results of § 5, we outline a theory of McMorris Q_r-semirings (Thms. 7.3, 7.4, 7.6), similar to that of McMorris Q_r-seminearrings developed above. In this context we also pose two problems.

Finally, we want to obtain well known ring-theoretical constructions (cf. the introduction of § 8) from our considerations. In general, by the difference between S-submodules and S-subsemimodules of an S-module M_S^+ or S_S^+ for a ring S, the direct way is problematical. We succeeded to introduce, again for any subcommutative S-groupoid M_S^+, restricted concepts of k-closed S-subgroupoids and k-closed Q_r-filters $\mathcal{F}^k$ (Defs. 8.2, 8.4), such that the corresponding Q_r-semirings $Q_r(M_S^+, \mathcal{F}^k, K^k)$ allow direct applications to S-modules and rings. Moreover, interrelations between both kinds of Q_r-filters $\mathcal{F}^+$ and $\mathcal{F}^k$ (Thm. 8.5) yield the main result (Thm. 8.8) that each Q_r-semiring $Q_r(M_S^+, \mathcal{F}^k, K^k)$ is isomorphic to a Q_r-semiring $Q_r(M_S^+, \mathcal{F}^+, K^+)$ of § 6 for a suitable choice of $\mathcal{F}^+$ and K^+.

§ 2 Seminearrings

As far as we know, the notation "seminearring" firstly was introduced in [29] (in a German version "Fasthalbring"). According to [12], [26] and [38], a (right or right distributive) seminearring is defined to be an algebra $(S,+,\cdot)$ such that $(S,+)$ and $(S,\cdot)$ are semigroups and that $(a+b)c = ac+bc$ holds for all $a,b,c \in S$. For examples, investigations and applications we further refer to [13], [14], [40] and to more references given in the latter paper.

Moreover, (right) seminearrings are a common generalization of (right) nearrings and semirings, where $(S,+)$ has to be a group in the first case, and both distributive laws are assumed in the second one. We refer to [26] for nearrings, and to [3], [4], [16], [17], [27], [28], [30], [32] as some of the earliest papers on semirings. We also mention in this context that certain semirings [seminearrings] occur as positive cones of partially ordered rings [nearrings] (cf. [8] and [25], [26]).

As a matter of consequence, we do not need associativity for any addition in the statements of our paper, not even for simplification. (Of course, we prefer examples with associative additions). Moreover, in each case where an addition is used to define another one and the first addition is assumed to be associative, then the latter one will be associative, too. Hence all our concepts and results can be formulated without this assumption, and then they yield corresponding concepts and results restricted to associative additions.

We want to express this generality in our presentation, since then our results also apply to algebras $(S,+,\cdot)$ as for instance neardomains ("Fastbereiche"), which occur in other mathematical contexts (cf. [19], [20] and [33]), and since we intend to give some of these applications in a later paper. In order to avoid a lot of confusing new terms, we proceed as in [40] and generalize the meaning of "seminearring" (and, consequently, of all specialized concepts in the following three definitions) throughout this paper as follows:

DEFINITION 2.1 An algebra $S = (S,+,\cdot)$ is called a (right) seminearring, iff $(S,+)$ is a groupoid, $(S,\cdot)$ a semigroup, and $(a+b)c = ac+bc$ holds for all $a,b,c \in S$. In particular, a seminearring S is called a semiring iff also the left distributive law holds, and a (right) nearring iff $(S,+)$ is a quasigroup.

Moreover, seminearrings and nearrings are tacitly assumed to be right distributive ones in the following.

Since semirings and nearrings have developed as complementary generalizations of rings, it is near by hand to use "ring" for an algebra $(S,+,\cdot)$ which is both, a semiring and a nearring. If $(S,+)$ is associative, this yields the usual concept of (multiplicatively associative) rings and those, whose group $(S,+)$ need not be commutative. (Several papers have investigated "rings" of this kind, cf. [35], also for references.) Without that assumption, according to Def. 2.1, the above use of "ring" refers to a semiring (S,+,·) such that (S,+) is merely a quasigroup. (There are plenty of "rings" in this meaning, even those without a zero.) For this reason, we avoid to use the term "ring" in general statements (except in § 8), but it will be clear that results on seminearrings also yield those on "rings" in each of the above meaning.

By Def. 2.1, seminearrings are algebras with two binary operations. Hence concepts as subseminearring [subsemiring, subnearring], homomorphism, congruence, and congruence-class-seminearring of a seminearring are clear. We call a neutral element o of (S,+) [e of (S,·)] the zero [the identity] of the seminearring S = (S,+,·). If S has a zero o, it may be left absorbing defined by oa = o for all a ∈ S, or briefly oS = o. Dually, o is called right absorbing iff So = o holds, and absorbing iff both is the case. As a consequence of the right distributivity, one easily checks:

Remark 2.2 Let S be a seminearring with zero o. Then oS = o holds if o is the only idempotent of (S,+), or if (S,+) is left or right cancellative. Thus, oS = o holds if (S,+) is a quasigroup (and hence, by the existence of o, a loop).

Note that even in these cases a left absorbing zero of a seminearring S may be at the same time a right identity of (S,·) (cf. [26] for nearrings). As another illustration and for further use we give

Example 2.3 Let (S,·) be the semigroup of all reals $r \in \mathbb{R}$ such that $r \geq r_o \geq 1$ for some fixed $r_o \in \mathbb{R}$, with respect to usual multiplication. Define an addition by $r+s = \max(r,s)$. Then S = (S,+,·) is a semiring with associtative addition, and commutative with respect to both operations. Moreover, r_o is the zero of S, which is cancellable in $(S,\cdot)$, and it coincides with the identity of S if one chooses $r_o = 1$.

On the other hand, S is a subsemiring of the semiring $T = (T,+,\cdot)$, consisting of all positive reals considered in the same way, and T has no zero (in fact, $(T,\cdot)$ is a group and hence $(T,+,\cdot)$ a semifield according to Def. 2.6).

To deal with the following considerations in a convenient way, for each seminearring S we introduce the notation S^* by $S^* = S\setminus\{o\}$ if S has a zero, and by $S^* = S$ otherwise.

DEFINITION 2.4 A seminearring S is called multiplicatively left [right, two-sided] cancellative iff each $a \in S^*$ is left [right, two-sided] cancellable in $(S,\cdot)$. Moreover, S is said to have zero-divisors iff S has a zero o and $a\cdot b = o$ holds for some $a,b \in S^*$.

We need some consequences of this definition, given in the next result, which is proved as a part of Thm. 2.2 in [40]:

Result 2.5 Let S be a seminearring such that $|S| \geqq 2$. Then S is multiplicatively left [right] cancellative iff S satisfies one of the following three statements:

α) S has no zero and $(S,\cdot)$ is a left [right] cancellative semigroup.

β) S has a zero o and $(S,\cdot)$ is a left [right] cancellative semigroup.

γ) S has a zero o, $So = oS = o$ holds and $(S^*,\cdot)$ is a left [right] cancellative subsemigroup of $(S,\cdot)$.

In each of these cases, S has no zero-divisors; moreover, each subseminearring of S is multiplicatively left [right] cancellative, too.

Note that each semiring S of Expl. 2.3 is a multiplicatively cancellative seminearring which belongs to case β). But S is a subsemiring of the semiring T of Expl. 2.3, and T belongs to case α). Further examples for all three cases are near by hand. Note in particular: if $(S,+,\cdot)$ is a multiplicatively left [right] cancellative seminearring without an absorbing zero (i.e. case α) or β)), a multiplicatively left [right] cancellative seminearring $(S\cup\{n\},+,\cdot)$ with an absorbing zero (i.e. case γ)) is obtained by adjoining such an element $n \notin S$ (cf. [40], Lemma 1.3).

It is also known that a nearring S with zero o which is multiplicatively left cancellative is also multiplicatively right cancellative and that the latter holds iff S has no zero-divisors (cf. [40], Prop. 2.4). In general, however, the lack of zero-divisors of a seminearring S means nothing with respect to left [right] cancellativity of S.

<u>DEFINITION 2.6</u> A seminearring [semiring, nearring] S is called a <u>seminearfield</u> [<u>semifield</u>, <u>nearfield</u>] iff $|S| \geq 2$ and $(S^*,\cdot)$ is a group.

This definition, proved to be in fact the most general one which is possible, and various equivalent conditions have been given in [40]. Here we only mention that a seminearfield S such that $|S^*| \geq 2$ is multiplicatively (two-sided) cancellative and satisfies $Sa = aS = S$ for all $a \in S^*$ and, if S has a zero o, also $So = oS = o$ - statements which fail to be true for exactly 5 (of all together 7) non-isomorphic seminearfields with zero of order 2.

In particular, assuming that additions are associative,

the concepts "semifield" and "nearfield" of Def. 2.6 coincide with the usual ones. Moreover, if such a semifield S is at the same time a nearfield, the group (S,+) has to be commutative by [35], § 1, thus S is a field in that usual meaning which does not include commutativity of (S,·).

Without that assumption, according to the above remarks concerning "rings", our statements will also apply to an interpretation of "field" which is a semifield and a nearfield in the meaning of Def. 2.5, hence a semiring (S,+,·) such that (S,+) is a quasigroup, $|S| \geq 2$, and $(S^*,\cdot)$ is a group (cf. Expls. 3.8 and 3.9 of [40]).

§ 3 Classical Q_r-seminearrings $Q_r(S,\Sigma)$

Recall that a semigroup $T = (T,\cdot)$ is defined to be a classical Q_r-semigroup $T = Q_r(S,\Sigma)$ of a semigroup $S = (S,\cdot)$ with respect to a subsemigroup Σ of S iff

i) T contains S as a subsemigroup and T has an identity,

ii) each $\alpha \in \Sigma$ has an inverse $\alpha^{-1} \in T$, and

iii) the subset $S\Sigma^{-1} = \{a\alpha^{-1} \mid a \in S,\ \alpha \in \Sigma\} \subseteq T$ coincides with T.

For given S and Σ, such a semigroup T exists iff 1) each $\alpha \in \Sigma$ is cancellable in S and 2) the (right) Ore-Asano-condition

$$q_r(S,\Sigma):\ \alpha S \cap b\Sigma \neq \emptyset \quad \text{for all } \alpha \in \Sigma,\ b \in S$$

holds, i.e. $\alpha x = b\xi$ has always a solution $(x,\xi) \in S\times\Sigma$. If such a semigroup T exists, it is completely described by

$$(3.1)\quad a\alpha^{-1} = b\beta^{-1} \Leftrightarrow \alpha x = \beta\xi \text{ and } ax = b\xi \text{ for some } (x,\xi) \in S\times\Sigma$$
$$\Leftrightarrow \alpha u = \beta v \Rightarrow au = bv \text{ for all } (u,v) \in S\times S,$$

$$(3.2)\quad a\alpha^{-1}\cdot b\beta^{-1} = (ax)(\beta\xi)^{-1}$$
$$\text{for any } (x,\xi) \in S\times\Sigma \text{ such that } \alpha x = b\xi.$$

Hence T is, up to isomorphisms over S, uniquely determined by S and Σ, which is intended in the notation $T = Q_r(S,\Sigma)$. For a proof and historical references we refer to [37].[1)]

<u>Remark 3.1</u> Any finite number of elements $t_1,\dots,t_n$ of a classical Q_r-semigroup $T = Q_r(S,\Sigma)$ may be written in the form $a_1\delta^{-1},\dots,a_n\delta^{-1}$ with a "common denominator" δ. This follows by induction from

(3.3) $\quad a\alpha^{-1} = (ax)(\beta\xi)^{-1}, \quad b\beta^{-1} = (b\xi)(\beta\xi)^{-1}$

for any $(x,\xi) \in S\times\Sigma$ such that $\alpha x = \beta\xi$.

The next considerations prepare a similar definition of a classical Q_r-seminearring, and we emphasize that extensions by right quotients correspond to right distributive semi-nearrings as considered in this paper.

<u>THEOREM 3.2</u> <u>Let</u> $(T,+,\cdot)$ <u>be a seminearring which contains</u> $(S,+,\cdot)$ <u>as a subseminearring such that</u> $(T,\cdot)$ <u>is a classical</u> <u>Q_r-semigroup of</u> $(S,\cdot)$ <u>with respect to a subsemigroup</u> Σ <u>of</u> $(S,\cdot)$. <u>Then the addition on</u> T <u>is uniquely determined by the addition on</u> S <u>according to</u>

(3.4) $\quad a\alpha^{-1} + b\beta^{-1} = (ax+b\xi)(\beta\xi)^{-1}$

<u>for any</u> $(x,\xi) \in S\times\Sigma$ <u>such that</u> $\alpha x = \beta\xi$.

<u>Moreover, with respect to Remark 3.1, this formula</u> (3.4) <u>is equivalent to the more convenient one</u>

---

1) Note that $Q_r(S,\Sigma)$ and the dual concept $Q_\ell(S,\Sigma)$ coincide for each Σ satisfying 1) which is contained in the centre of S. In this case we have classical Q-semigroups $T = Q(S,\Sigma)$ for which (3.1) reduces to $a\alpha^{-1} = b\beta^{-1} \Leftrightarrow a\beta = b\alpha$ $(x = \beta,\ \xi = \alpha)$ and (3.2) to $a\alpha^{-1}\cdot b\beta^{-1} = (ab)(\beta\alpha)^{-1}$ $(x = b,\ \xi = \alpha)$.

(3.5) $a_1\delta^{-1}+a_2\delta^{-1} = (a_1+a_2)\delta^{-1}$.

<u>Conversely, let</u> $(S,+,\cdot)$ <u>be a seminearring and</u> $(T,\cdot)$ <u>be a classical</u> Q_r<u>-semigroup of</u> $(S,\cdot)$ <u>as above. Then</u> (3.4) <u>defines an addition on</u> T <u>such that</u> $(T,+,\cdot)$ <u>is a seminearring which contains</u> $(S,+,\cdot)$ <u>as a subseminearring.</u>[1)]

<u>Proof</u> Using (3.3) and right distributivity of $(T,+,\cdot)$, we obtain (3.4) and the equivalence to (3.5). For the converse statement, we firstly show that (3.4) defines an operation on T, idepended on the representation of $a\alpha^{-1}$ and $b\beta^{-1}$, and on the choice of $(x,\xi)\in S\times\Sigma$. We do this in a single step and assume $a\alpha^{-1} = a'\alpha'^{-1}$, $b\beta^{-1} = b'\beta'^{-1}$ and $\alpha'x' = \beta'\xi'$ for some $(x',\xi')\in S\times\Sigma$, hence

(3.4') $a'\alpha'^{-1}+b'\beta'^{-1} = (a'x'+b'\xi')(\beta'\xi')^{-1}$.

By $q_r(S,\Sigma)$, there exists a pair $(y,\eta)\in S\times\Sigma$ which satisfies

$$\beta\xi y = \beta'\xi'\eta, \quad \text{thus also} \quad \alpha xy = \alpha'x'\eta$$

(since $\alpha x = \beta\xi$ and $\alpha'x' = \beta'\xi'$). Applying the second equivalence of (3.1) to $b\beta^{-1} = b'\beta'^{-1}$ and $\beta(\xi y) = \beta'(\xi'\eta)$, we get

$$b\xi y = b'\xi'\eta, \quad \text{whereas} \quad axy = a'x'\eta$$

follows from $a\alpha^{-1} = a'\alpha'^{-1}$ in the same way. Hence, by right distributivity of $(S,+,\cdot)$, we have $(ax+b\xi)y = (a'x'+b'\xi')\eta$. By the first equivalence of (3.1), the latter and $(\beta\xi)y = (\beta'\xi')\eta$ imply the equality of the right sides of (3.4) and (3.4'); this proves that (3.4) really defines an addition on T.

That addition on T satisfies also (3.5), since (3.4) for $\beta = \alpha$ and $x = \xi$ yields $a\alpha^{-1}+b\alpha^{-1} = (a\xi+b\xi)(\alpha\xi)^{-1}$, what equals

1) Clearly, we shall write $(S,+,\cdot)$ and $(T,+,\cdot)$ in this context as soon as we have proved Thm. 3.2.

$(a+b)\xi(\alpha\xi)^{-1} = (a+b)\alpha^{-1}$ by calculations in $(S,+,\cdot)$ and $(T,\cdot)$. We use (3.5) and Remark 3.1 to show the right distributive law for $(T,+,\cdot)$ according to

$$
\begin{aligned}
(a\alpha^{-1}+b\alpha^{-1})\cdot c\gamma^{-1} &= (a+b)\alpha^{-1}\cdot c\gamma^{-1} \qquad (\text{and for } \alpha x = c\xi)\\
&= ((a+b)x)(\gamma\xi)^{-1} = (ax+bx)(\gamma\xi)^{-1}\\
&= (ax)(\gamma\xi)^{-1} + (bx)(\gamma\xi)^{-1} = a\alpha^{-1}\cdot c\gamma^{-1} + b\alpha^{-1}\cdot c\gamma^{-1},
\end{aligned}
$$

where $\alpha x = c\xi$ introduced for the second step is also used for both summands in the last step. Since $a+b = a+b$ for all $a,b \in S$ is now clear, again by (3.5), our theorem is proved.

As a consequence of Thm. 3.2 and the above statements on classical Q_r-semigroups we obtain:

COROLLARY 3.3 *Let* $S = (S,+,\cdot)$ *be a seminearring and* Σ *any subsemigroup of* $(S,\cdot)$ *such that* 1) *each* $\alpha \in \Sigma$ *is cancellable in* $(S,\cdot)$ *and* 2) $q_r(S,\Sigma)$ *holds. Then there exists, unique up to isomorphisms over* S, *a seminearring* $T = (T,+,\cdot)$ *which contains* $(S,+,\cdot)$ *as a subseminearring such that* $(T,\cdot)$ *is a classical* Q_r*-semigroup of* $(S,\cdot)$ *with respect to* Σ. *Further, this seminearring* $(T,+,\cdot)$ *is completely described by* (3.1), (3.2), *and* (3.4) *or* (3.5).

DEFINITION 3.4 The seminearring T of Cor. 3.3 is called *the classical* Q_r*-seminearring of* S *with respect to* Σ.

Moreover, according to Thm. 3.2, we may use the notation $T = Q_r(S,\Sigma)$ as for semigroups, if it is clear that we speak about seminearrings. In particular, $T = Q_r(S,\Sigma)$ may be called a *classical* Q_r*-seminearfield*, Q_r*-semiring* or Q_r*-semifield*, Q_r*-nearring* or Q_r*-nearfield of* S *with respect to* Σ, if T happens to be such an algebra.[1] (Footnote see next page.)

We shall see that all these cases for $T = Q_r(S,\Sigma)$ really occur, clearly depending on suitable conditions on S and Σ. Especially, if S is a semiring or a nearring, then each classical Q_r-seminearring $T = Q_r(S,\Sigma)$ of S is likewise a semiring or a nearring, respectively. These statements and various similar ones which include supplementary properties on $(S,+,\cdot)$ and $(T,+,\cdot)$ obviously are contained in the next theorem and will not be formulated explicitely.

THEOREM 3.5 *Let $T = Q_r(S,\Sigma)$ be a classical Q_r-seminearring of a seminearring S with respect to Σ. Then we have:*

a) *If the groupoid $(S,+)$ is commutative, left [right] cancellative, idempotent, a left [right] quasigroup, associative or a group, then $(T,+)$ has the same property.*

b) *There are cases such that S has a zero, whereas T does not have one. But if S has a zero o and*

(3.6) $\quad o\xi = o$ holds for all $\xi \in \Sigma$,

then o is also the zero of T, and conversely.[2)]

c) *If $c \in S$ and $\gamma \in \Sigma$ are left distributive elements of S, then $c\gamma^{-1} \in T$ is a left distributive element of T.*

d) *If the seminearring S is multiplicatively commutative, left [right] cancellative or if S has no zero-divisors, then the same holds for T, and conversely.*

---

1) We avoid to use Q_r-ring and Q_r-field of S with respect to the ambiguity of the terms "ring" and "field" in our context. In fact, all what follows applies also to "rings" and "fields" in each interpretation corresponding to our remarks in § 2.

2) Note that $oS = o$ implies (3.6) and compare Remark 2.2. In particular, if S has a zero which is the only idempotent of $(S,+)$, the same statement holds for T.

<u>Proof</u> a) It is very simple to show each of these assertions if one uses Remark 3.1 and (3.5). For instance, assume that $(S,+)$ is left cancellative and $t_1+t_2 = t_1+t_3$ in $(T,+)$. Then we have $t_1+t_2 = a_1\delta^{-1}+a_2\delta^{-1} = a_1\delta^{-1}+a_3\delta^{-1} = t_1+t_3$, hence $a_1+a_2 = a_1+a_3$ and $a_2 = a_3$ by assumptions on $(S,+)$, i.e. $t_2 = t_3$.

b) For the first statement let S be a semiring of Expl. 2.2 with r_o as zero. Then $(T,\cdot) = Q_r(S,S) = Q(S,S)$, considered in $(\mathbb{R},\cdot)$, is the group of all positive reals, already regarded as a semifield $(T,+,\cdot)$ with respect to $t_1+t_2 = \max(t_1,t_2)$ in the same example. (Note that by Thm. 3.2 the latter has to be the unique addition which extends that of S, and that clearly (3.5) defines this addition on $(T,\cdot)$.) Hence the classical Q_r-semiring T of S has no zero.

For the remaining equivalence, we firstly assume that S has a zero o. To calculate t+o and o+t for any $t = x\xi^{-1} \in T$ by (3.5), we have to write $o \in T$ with the same denominator. We use (3.6) to do this by $o = o(\xi\xi^{-1}) = (o\xi)\xi^{-1} = o\xi^{-1}$. Now (3.5) implies $t+o = t = o+t$ for all $t \in T$. Conversely, if T has a zero o, this element has at least one representation $o = a\alpha^{-1} \in S\Sigma^{-1}$. Moreover, for each $t \in T$ and each $\xi \in \Sigma$, from

$$t\xi^{-1}+o = o+t\xi^{-1} = t\xi^{-1} \Rightarrow t+o\xi = o\xi+t = t$$

we obtain $o\xi = o$, i.e. (3.6) for the zero o of T. But this yields $o = o\alpha = (a\alpha^{-1})\alpha = a \in S$, hence o is also the zero of S and satisfies (3.6).

c) In order to calculate $c\gamma^{-1}\cdot a\alpha^{-1}+c\gamma^{-1}\cdot b\alpha^{-1}$ for given elements $a\alpha^{-1}$ and $b\alpha^{-1}$ of T, we apply Remark 3.1 a second time to $\gamma^{-1}a$ and $\gamma^{-1}b$. Hence $\gamma^{-1}a = d_1\delta^{-1}$ and $\gamma^{-1}b = d_2\delta^{-1}$ hold for suitable $d_1, d_2 \in S$, $\delta \in \Sigma$. From $\gamma d_1 = a\delta$ and $\gamma d_2 = b\delta$ we get $\gamma(d_1+d_2) = (a+b)\delta$ by the assumption on γ. Since we also have $cd_1 + cd_2 = c(d_1+d_2)$, we obtain by (3.5) and (3.2)

$$c\gamma^{-1}\cdot a\alpha^{-1} + c\gamma^{-1}\cdot b\alpha^{-1} = (cd_1)(\alpha\delta)^{-1} + (cd_2)(\alpha\delta)^{-1}$$

$$= (cd_1+cd_2)(\alpha\delta)^{-1} = (c(d_1+d_2))(\alpha\delta)^{-1}$$

$$= c\gamma^{-1}\cdot(a+b)\alpha^{-1} = c\gamma^{-1}\cdot(a\alpha^{-1}+b\alpha^{-1}).$$

d) The converse statements (and both concerning commutativity) are trivial with respect to b). Now let $(T,\cdot) = Q_r(S,\Sigma)$ be a Q_r-semigroup of a semigroup $(S,\cdot)$. Then it is known (cf. [34], § 4) that $(T,\cdot)$ is left [right] cancellative if $(S,\cdot)$ has this property. This proves the concerning statements for seminearrings if $(S,+,\cdot)$ belongs to the cases α) or β) of Result 2.5. If $(S,+,\cdot)$ belongs to case γ), then by $So = oS = o$ it follows from part b) that o is also the zero of $(T,+,\cdot)$. Further, $(T^*,\cdot)$ for $T^* = T\setminus\{o\}$, is obviously the classical Q_r-semigroup $(T^*,\cdot) = Q_r(S^*,S^*)$ of $(S^*,\cdot)$. Since $(S^*,\cdot)$ is left [right] cancellative in case γ), the result cited above yields the same for the semigroup $(T^*,\cdot)$. Moreover, $So = oS = o$ implies $To = oT = o$. Hence the seminearring $(T,+,\cdot)$ satisfies γ) of Result 2.5, too, and therefore it is multiplicatively left [right] cancellative.

For the last statement, let S have no zero-divisors and suppose $a\alpha^{-1}\cdot b\beta^{-1} = o$ for some elements of T. Again by b), the zero o of T belongs to S and (3.6) holds. Thus we obtain $a\alpha^{-1}\cdot b = o$ and $ax = o$ for $\alpha^{-1}b = x\xi^{-1}$. By assumption on S, this implies $a = o$ and hence $a\alpha^{-1} = o$, or $x = o$ and hence $\alpha^{-1}b = o$. Thus it remains to show that $\alpha o = b$ yields $b = o$.

Applying $q_r(S,\Sigma)$ to α and o, we obtain $\alpha y = o\eta = o$ for some $(y,\eta) \in S\times\Sigma$. Since S has no zero-divisors, the latter yields $y = o$ and therefore $o = \alpha o = b$, as we were to show.

We now turn to some other questions in this context:

<u>THEOREM 3.6</u> <u>Assume that the seminearring</u> S <u>has at least one classical</u> Q_r<u>-seminearring. Then there exists, unique up to isomorphisms over</u> S, <u>a maximal classical</u> Q_r<u>-seminearring</u> $T_m = Q_r(S,\Sigma_m)$ <u>of</u> S. <u>More precisely, for a suitable choice of</u> Σ_m, <u>each</u> Q_r<u>-seminearring</u> $T = Q_r(S,\Sigma)$ <u>satisfies</u> $\Sigma \subseteq \Sigma_m$, <u>and</u> T_m <u>contains (an isomorphic copy of)</u> T <u>as a subseminearring.</u>

<u>Proof</u> Recall that each subsemigroup Σ of $(S,\cdot)$ which consists of cancellable elements of $(S,\cdot)$ and satisfies $q_r(S,\Sigma)$ determines a classical Q_r-semigroup $(T,\cdot) = Q_r(S,\Sigma)$ of $(S,\cdot)$ and hence also a classical Q_r-seminearring $(T,+,\cdot) = Q_r(S,\Sigma)$ of $(S,+,\cdot)$. But the same T may be such a Q_r-extension of S with respect to different subsemigroups $\Sigma,\Sigma',\ldots$ of S, all contained in a maximal one, say Σ^*, which obviously consists of all elements $\tau \in S$, such that $\tau^{-1} \in T$ exists. In [34] and [37], those "relatively maximal subsemigroups Σ^* of (right) denominators of $(S,\cdot)$" have been considered, which clearly correspond to all classical Q_r-semigroups [Q_r-seminearrings] in a bijective way. Moreover, all subsemigroups of (right) denominators of a semigroup $(S,\cdot)$ generate a subsemigroup Σ_m, which is again such a subsemigroup. The proof follows by induction on the number of generating elements for each element of Σ_m (cf. [34], § 4, Satz 3). The existence of this "absolutely maximal subsemigroup Σ_m of (right) denominators of $(S,\cdot)$" and the above considerations imply Thm. 3.6.

In the same way, using again the interrelation between classical Q_r-semigroups and Q_r-seminearrings due to Thm. 3.2, we obtain from a corresponding theorem in [36] directly:

THEOREM 3.7 Let $T = Q_r(S,\Sigma)$ be a classical Q_r-seminearring of a seminearring S and let $R = Q_r(T,\theta)$ be such a Q_r-seminearring of T. Then R is also a classical Q_r-seminearring $Q_r(S,\Sigma_R)$ of S, choosing Σ_R to be the set of all elements of S which have an inverse in R. In particular, the maximal classical Q_r-seminearring T_m of S has no proper extension as a classical Q_r-seminearring.

Finally, we characterize seminearrings S which are even embeddable into seminearfields by a classical Q_r-extension, which clearly concerns the maximal Q_r-seminearring of S.

THEOREM 3.8 Let S be a seminearring [semiring, nearring] such that $|S^*| \geq 2$. Then there exists a maximal classical Q_r-seminearring T_m of S, which is a seminearfield [semifield, nearfield] iff S is multiplicatively cancellative and the Ore-Asano-condition $q_r(S,\Sigma_m)$ holds for the set Σ_m of all cancellable elements of $(S,\cdot)$, where we have $\Sigma_m = S$ if $(S,+,\cdot)$ satisfies α) or β) of Result 2.5, and $\Sigma_m = S^*$ in the case γ). In particular, the Q_r-seminearfield $T_m = Q_r(S,\Sigma_m)$ of S has a zero exactly in the latter case.

Proof If S is embeddable into any seminearfield T, $|S^*| \geq 2$ yields $|T^*| \geq 2$. Hence, by the remarks behind Def. 2.6, the seminearfield T is multiplicatively cancellative, which implies the same for the semiring S by Result 2.5. Hence we may restrict ourselves to multiplicatively cancellative seminearrings S in the following.[1] (Footnote see next page.)

Let T be a Q_r-seminearfield of S. If T has a zero o, it is also the zero of S by Thm. 3.5 b), and (3.6) and $|S^*| \geq 2$ imply that o is not cancellable in $(S,\cdot)$. Hence both, T and S, belong to case γ) and one has $To = oT = o$. This means that $(T^*,\cdot)$ is a group, hence all elements of $\Sigma = S^*$ are invertible in T, which implies $q_r(S,S^*)$. Conversely, if S satisfies γ) and $q_r(S,S^*)$, the Q_r-seminearring $T = Q_r(S,S^*)$ exists, it belongs also to γ) by Thm. 3.5 b), hence $(T^*,\cdot)$ is a group and $(T,+,\cdot)$ a seminearfield.

If $T = Q_r(S,\Sigma)$ is a Q_r-seminearfield of S without a zero, $(T,\cdot)$ is a group and S satisfies α) or β) by Thm. 3.5 b). In both cases, all elements of $\Sigma = S$ are invertible in T, which implies $q_r(S,S)$. The converse statement is shown as above.

The statements concerning semirings and nearrings (and similar ones including supplementary properties) follow now by Thm. 3.5. An example for case β), a seminearring S with a cancellable zero o and a Q_r-seminearfield $T_m = Q_r(S,S)$ which has no zero, has already occured (in fact for semirings) in Expl. 2.3, cf. the proof of Thm. 3.5 b).

§ 4 Q_r-seminearrings $Q_r(M_S^+,\mathcal{F},K)$ of an S-groupoid M_S^+

Let $S = (S,\cdot)$ be a semigroup, and $(M,+)$ a groupoid. Then the set M together with a mapping $(m,s) \to ms$ from $M\times S$ into M is called a (<u>right</u>) <u>S-set</u>, denoted by M_S, iff

(4.1) $\quad (ms_1)s_2 = m(s_1s_2)$ for all $m \in M$; $s_1,s_2 \in S$.

As in [39], § 3, we assume nothing about $\Theta(M_S) = \{\delta \in M \mid \delta S = \delta\}$,

1) If such a seminearring S is finite, it clearly is a semi-nearfield itself, and $q_r(S,\Sigma_m)$ holds trivially.

the set of all fixed elements of M_S; in particular, $\Theta(M_S)$ may be empty. Likewise, M_S need neither be centered ($S = S^o$ and $|\Theta(M_S)| = 1$) nor unitary ($S = S^1$ and $m1 = m$ for all $m \in M$).

Taking the addition on M into consideration, we call an S-set M_S a (<u>right</u>) <u>S-groupoid</u> and denote it by M_S^+ iff also

(4.2) $(m_1+m_2)s = m_1s+m_2s$ for all $m_1, m_2 \in M$; $s \in S$.

Since both, S-sets as well a S-groupoids, may be considered as universal algebras with unary operations for each $s \in S$, the concepts of an <u>S-subset</u> A_S of M_S and an <u>S-subgroupoid</u> A_S^+ of M_S^+ are clear. For any S-groupoid M_S^+, we denote by $\mathfrak{M}_S$ the set of all S-subsets of M_S, whereas $\mathfrak{M}_S^+$ means the set of all S-subgroupoids of M_S^+; obviously, we have $M \in \mathfrak{M}_S^+ \subseteq \mathfrak{M}_S$ and $\Theta(M_S) \in \mathfrak{M}_S^+$ iff $\Theta(M_S) \neq \emptyset$.

Note that "S" in this context always refers to a certain semigroup of (right) operators. On the other hand, starting with a semigroup $(S,\cdot)$ or a seminearring $(S,+,\cdot)$, the semigroup $(S,\cdot)$ operates on $M = S$ by its multiplication. In both cases, the resulting S-set will be denoted by S_S, and the S-groupoid corresponding to a seminearring by S_S^+. We also use $\mathfrak{S}_S$ and $\mathfrak{S}_S^+$ instead of $\mathfrak{M}_S$ and $\mathfrak{M}_S^+$ in this context, clearly now the set of all right ideals of the semigroup $(S,\cdot)$ or of the seminearring $(S,+,\cdot)$, respectively[1]. The application of the following considerations on general S-groupoids M_S^+ to S-groupoids S_S^+ obtained from seminearrings will be of

---

1) In this general context, it is meaningful to define left, right and two-sided ideals of a seminearring $(S,+,\cdot)$ as the corresponding ideals of $(S,\cdot)$ which are additively closed. This definition implies the most common concepts for semirings, but applied to nearrings or rings, the resulting concepts need modification to coincide with the usual ones.

special interest and the subject of § 5.

For any S-sets N_S and M_S, a mapping $f: N \to M$ such that

(4.3) $\quad f(ns) = f(n)s \quad$ for all $n \in N,\ s \in S$

holds is called a (<u>right</u>) <u>S-mapping</u>. We denote the set of all these mappings by $\Lambda_S(N_S, M_S)$, or shortly by $\Lambda_S(N,M)$. If N_S^+ and M_S^+ are S-groupoids, an S-mapping $f \in \Lambda_S(N_S, M_S)$ which is also a homomorphism of $(N,+)$ into $(M,+)$ according to

(4.4) $\quad f(n_1+n_2) = f(n_1)+f(n_2) \quad$ for all $n_1, n_2 \in N$

is called a (<u>right</u>) <u>S-homomorphism</u>. We denote the set of all these mappings by $\mathrm{Hom}_S(N_S^+, M_S^+)$, or shortly by $\mathrm{Hom}_S(N,M)$. As a matter of fact, also the S-mappings of M_S will be important in dealing with S-groupoids M_S^+ and seminearrings.

Henceforth, all concepts introduced so far are tacitly assumed to be right ones, which will correspond to our consideration of right distributive seminearrings. For the first part of the next definition compare [11] and [39], Def. 3.1, and we refer to the latter paper for more details:

<u>DEFINITION 4.1</u> a) A <u>right quotient filter</u> (<u>Q_r-filter</u>) $\mathcal{F}$ <u>of an S-set</u> M_S is defined by $\emptyset \neq \mathcal{F} \subseteq \mathfrak{M}_S$ and the conditions

(4.5) $\quad A \in \mathcal{F},\ A \subseteq B \in \mathfrak{M}_S \Rightarrow B \in \mathcal{F}$

(4.6) $\quad A, B \in \mathcal{F} \Rightarrow A \cap B \in \mathcal{F}$

(4.7) $\quad A, B \in \mathcal{F},\ f \in \Lambda_S(A,M) \Rightarrow f^{-1}(B) \in \mathcal{F}$.

b) A <u>right quotient filter</u> (<u>Q_r-filter</u>) $\mathcal{F}^+$ <u>of an S-groupoid</u> M_S^+ is defined by $\emptyset \neq \mathcal{F}^+ \subseteq \mathfrak{M}_S^+$ and conditions (4.5^+) - (4.7^+) as above, but $\mathfrak{M}_S$ replaced by $\mathfrak{M}_S^+$ and $\Lambda_S(A,M)$ by $\mathrm{Hom}_S(A,M)$.

Since we later have to use both concepts simultaneously,

terms as Λ_S-filter $\mathcal{F}$ of M_S for a) and Hom_S-filter $\mathcal{F}^+$ of M_S^+ for b) would be more significant, but we did not like to replace the notation of a right [or left] quotient filter, used by several authors in the last years in the context of S-sets and semigroups (see e.g. [10], [23]).

We further recall that an S-subset A_S of an S-set M_S is defined to be dense in M_S iff

(4.8) for each $m_1 \neq m_2$ and m in M there exists an $s \in S$ such that $m_1 s \neq m_2 s$ and $ms \in A$.

Note that for each S-groupoid M_S^+ the set M_S^Δ, consisting of all S-subsets which are dense in M_S, is either empty, or, clearly iff M_S is dense in M_S, a Q_r-filter of M_S. In the latter case, the set $M_S^{+\Delta}$ of all S-subgroupoids which are dense in M_S is not empty, too, and $M_S^{+\Delta}$ is a Q_r-filter of M_S^+.

THEOREM 4.2 Let M_S^+ be an S-groupoid, $\mathcal{F}$ a Q_r-filter of the S-set M_S, and denote by $\Lambda_S(\mathcal{F},M_S^+) = \Lambda_S(\mathcal{F},M_S) = \cup\{\Lambda_S(A,M) \mid A \in \mathcal{F}\}$ the set of all S-mappings f defined on an S-subset $A = D_f \in \mathcal{F}$, the domain of f, into M_S. Then we have:

a) In a natural way, $(\Lambda_S(\mathcal{F},M_S^+),+,\cdot)$ is a seminearring with respect to operations $+,\cdot$ defined by

(4.9) $(f+g)(x) = f(x)+g(x)$ for all $x \in D_f \cap D_g \in \mathcal{F}$,

(4.10) $(f\cdot g)(x) = f(g(x))$ for all $x \in g^{-1}(D_f) \in \mathcal{F}$.

Moreover, $1_M \in \Lambda_S(M,M)$ is the identity of $(\Lambda_S(\mathcal{F},M_S^+),+,\cdot)$.

b) If $(M,+)$ is commutative, associative or idempotent, then the same holds for $(\Lambda_S(\mathcal{F},M_S^+),+)$. Further, if $(M,+)$ happens to have a neutral o which satisfies $oS = o$, then 0 defined by $0(x) = o$ for all $x \in M$ is the zero of $(\Lambda_S(\mathcal{F},M_S^+),+,\cdot)$.

c) Each Q_r-filter $\mathcal{K}$ of M_S such that $\mathcal{F} \subseteq \mathcal{K}$ defines by

(4.11) $\quad f \equiv g\ (\kappa_{\mathcal{K}})$ iff $C_{f,g} = \{x \in D_f \cap D_g \mid f(x) = g(x)\} \in \mathcal{K}$

a relation $\kappa_{\mathcal{K}}$ on $\Lambda_S(\mathcal{F}, M_S^+)$, which is a congruence with respect to both operations (4.9) and (4.10). In the important case that $\mathcal{F} \subseteq \mathcal{K} \subseteq M_S^{\Delta}$, one has

(4.12) $\quad f \equiv g\ (\kappa_{\mathcal{K}}) \Leftrightarrow C_{f,g} = D_f \cap D_g \Leftrightarrow f \equiv g\ (\kappa_{\mathcal{F}})$.

Proof All statements except those in which the addition is in fact involved are known for the semigroup $(\Lambda_S(\mathcal{F}, M_S), \cdot)$ (cf. [39], Thm. 3.5 and Thm. 5.4 i)). Moreover, (4.9) defines an addition on $\Lambda_S(\mathcal{F}, M_S) = \Lambda_S(\mathcal{F}, M_S^+)$ since, for each $s \in S$,

$$(f+g)(xs) = f(x)s + g(x)s = (f(x)+g(x))s = ((f+g)(x))s$$

holds by (4.1) and (4.2) for all $x \in D_{f+g} = D_f \cap D_g \in \mathcal{F}$. Thus the proof of a) is completed if $(f+g)\cdot h = f\cdot h + g\cdot h$ is shown for all $f,g,h \in \Lambda_S(\mathcal{F}, M_S^+)$. Using (4.9) and (4.10), we see by

$$D_{(f+g)h} = h^{-1}(D_f \cap D_g) = h^{-1}D_f \cap h^{-1}D_g = D_{fh+gh}$$

that the domains of $(f+g)h$ and $fh+gh$ coincide, and for all x contained in that common domain we have

$$((f+g)h)x = (f+g)(hx) = f(hx) + g(hx) = (fh)x + (gh)x = (fh+gh)x.$$

The proof of b) is straightforward. For c) it is known that $\kappa_{\mathcal{K}}$ is a congruence on $(\Lambda_S(\mathcal{F}, M_S^+), \cdot)$ and that (4.12) is true if $\mathcal{F} \subseteq \mathcal{K} \subseteq M_S^{\Delta}$. Thus is remains to show that $\kappa_{\mathcal{K}}$ is also compatible with respect to the addition (4.9). So we assume (4.11) and similarly $f' \equiv g'\ (\kappa_{\mathcal{K}})$, i.e. $C_{f',g'} \in \mathcal{K}$. Then we have $C_{f,g} \cap C_{f',g'} \subseteq C_{f+f',g+g'}$, which implies $C_{f+f',g+g'} \in \mathcal{K}$ by (4.6) and (4.5) for $\mathcal{K}$ and therefore $f+f' \equiv g+g'\ (\kappa_{\mathcal{K}})$.

DEFINITION 4.3 Let M_S^+ be an S-groupoid, $\mathcal{F} \subseteq \mathcal{K}$ Q_r-filters of the S-set M_S, and $(\Lambda_S(\mathcal{F},M_S^+),+,\cdot)$ the seminearring with the congruence κ_K of Thm. 4.2. Then the seminearring

$$Q_r(M_S^+,\mathcal{F},K) = (Q_r(M_S^+,\mathcal{F},K),+,\cdot) = (\Lambda_S(\mathcal{F},M_S^+)/\kappa_K,+,\cdot),$$

consisting of all κ_K-classes $[f] = [f]_{\mathcal{F},K}$ of $\Lambda_S(\mathcal{F},M_S^+)$, is called the Q_r-seminearring of M_S^+ with respect to $\mathcal{F}$ and K.

In particular, if $\mathcal{F} = K$, we write $Q_r(M_S^+,\mathcal{F},\mathcal{F}) = Q_r(M_S^+,\mathcal{F})$ and speak of the Q_r-seminearring of M_S^+ with respect to $\mathcal{F}$.[1)]

By our way of proceeding it is clear, that the multiplicative semigroup of the Q_r-seminearring $(Q,+,\cdot) = Q_r(M_S^+,\mathcal{F},K)$ of M_S^+ is the Q_r-semigroup $(Q,\cdot) = Q_r(M_S,\mathcal{F},K)$ of the S-set M_S with respect to $\mathcal{F}$ and K as defined in [39], Def. 3.6. However, in general there are Q_r-seminearrings $(Q,+,\cdot) = Q_r(M_S^+,\mathcal{F},K)$ which have different additions, but the same Q_r-semigroup $(Q,\cdot) = Q_r(M_S,\mathcal{F},K)$, for instance those corresponding to the same S-set M_S, but established with different additions to make it an S-groupoid M_S^+ (but cf. Thm. 5.2 c)).

THEOREM 4.4 Let $Q = Q_r(M_S^+,\mathcal{F},K)$ be the Q_r-seminearring of an S-groupoid M_S^+ with respect to Q_r-filters $\mathcal{F}$ and K of M_S. Then, clearly, $[1_M]$ is the identity of Q, and we have:

a) If the groupoid $(M,+)$ is commutative, left [right] cancellative, idempotent, a left [right] quasigroup, associative or a group, then $(Q,+)$ has the same property.

b) If $(M,+)$ has a neutral o which satisfies $oS = o$, then $[0]$ such that $0(x) = o$ for all $x \in M$ is the zero of $(Q,+,\cdot)$.

---

1) Note in this context that for $\mathcal{F} \subseteq K \subseteq M_S^\Delta$ we always have $Q_r(M_S^+,\mathcal{F},K) = Q_r(M_S^+,\mathcal{F},M_S^\Delta) = Q_r(M_S^+,\mathcal{F},\mathcal{F}) = Q_r(M_S^+,\mathcal{F})$ by (4.12).

c) <u>If</u> [d] $\in$ Q <u>contains</u> <u>an</u> <u>S-mapping</u> $d \in \Lambda_S(D_d, M)$ <u>which</u> <u>is</u> <u>in</u> <u>fact</u> <u>an</u> <u>S-homomorphism</u> <u>defined</u> <u>on</u> <u>an</u> <u>S-subgroupoid</u> D_d, <u>then</u> [d] <u>is</u> <u>a</u> <u>left</u> <u>distributive</u> <u>element</u> <u>of</u> $(Q,+,\cdot)$.

<u>Proof</u> a) Assume first that (M,+) is left cancellative and that $[f]+[g] = [f]+[h]$ holds in Q. Then, by (4.9) and (4.11), $f(x)+g(x) = f(x)+h(x)$ holds for all $x \in C_{f+g,f+h} \in \mathcal{K}$, which implies $g(x) = h(x)$ for those x and hence $C_{f+g,f+h} \subseteq C_{g,h}$. This yields $C_{g,h} \in \mathcal{K}$ by (4.5), hence $g \equiv h\ (\kappa_{\mathcal{K}})$ or $[g] = [h]$. Thus (Q,+) is left cancellative, too. To see that (Q,+) is a left quasigroup if the same holds for (M,+), it suffices to show that for each $[g], [h] \in Q$ there is an $[f] \in Q$ which satisfies $[f]+[g] = [h]$. For each $x \in D_g \cap D_h \in \mathcal{F}$, we define $f(x) = y \in M$ according to $y+g(x) = h(x)$. Due to (4.2), we have $f \in \Lambda_S(D_g \cap D_h, M)$; thus $[f] = [f]_{\mathcal{F},\mathcal{K}}$ defines an element of Q and $[f]+[g] = [h]$ holds by $C_{f+g,h} = D_g \cap D_h \in \mathcal{F} \subseteq \mathcal{K}$ according to the definition of f. Hence (Q,+) is a left quasigroup.

Using left-right duality with respect to the additions, and that b) of Thm. 4.2 clearly transfers to the epimorphic image $(Q,+,\cdot)$ of $(\Lambda_S(\mathcal{F}, M_S^+),+,\cdot)$, we obtain all statements of a) and also b) of Thm. 4.4.

c) We are going to show that $[d]([f]+[g]) = [d][f]+[d][g]$ holds for all $[f],[g] \in Q$. With d,f and g as representatives for these classes, d(f+g) is defined on

$$D_{d(f+g)} = (f+g)^{-1}D_d = \{x \in D_f \cap D_g \mid f(x)+g(x) \in D_d\},$$

which contains the domain of df+dg,

$$D_{df+dg} = f^{-1}D_d \cap g^{-1}D_d = \{x \in D_f \cap D_g \mid f(x) \in D_d \text{ and } g(x) \in D_d\},$$

since $f(x), g(x) \in D_d \in \mathfrak{M}_S^+$ implies $f(x)+g(x) \in D_d$. Using now the assumption $d \in \mathrm{Hom}_S(D_d, M)$, we obtain for all $x \in D_{df+dg}$

$$(d(f+g))(x) = d(f(x)+g(x)) = d(f(x))+d(g(x)) = (df+dg)(x).$$

Hence $D_{df+dg} \in \mathcal{F} \subseteq K$ is contained in $C_{d(f+g),df+dg}$, and so the latter is in K by (4.5), proving our statement by (4.11).

Clearly, there are now two directions to go on. On the one hand, we will apply these considerations to S-groupoids S_S^+ obtained by seminearrings S as announced above (§ 5). On the other hand, one would like to introduce conditions such that left distributive elements as in c) of Thm. 4.4 form a subsemiring of a seminearring $Q_r(M_S^+, \mathcal{F}, K)$ (§ 6). We conclude this section with the following statement and leave out the proof which is straightforward.

PROPOSITION 4.5 Let $\mathcal{F}_1 \subseteq \mathcal{F}_2 \subseteq K$ be Q_r-filters of the S-set M_S of an S-groupoid M_S^+. Then there is a monomorphism of the Q_r-seminearring $Q_1 = Q_r(M_S^+, \mathcal{F}_1, K)$ into the Q_r-seminearring $Q_2 = Q_r(M_S^+, \mathcal{F}_2, K)$, given by

$$[f]_{\mathcal{F}_1,K} \to [f]_{\mathcal{F}_2,K} \in Q_2 \quad \text{for all} \quad [f]_{\mathcal{F}_1,K} \in Q_1.$$

§ 5 Q_r-seminearrings $Q_r(S_S^+, \mathcal{F}, K)$ of a seminearring S and McMorris Q_r-seminearrings

DEFINITION 5.1 Let S be a seminearring, S_S^+ the S-groupoid determined by S and $\mathcal{F} \subseteq K$ Q_r-filters of the S-set S_S. Then the Q_r-seminearring $Q_r(S_S^+, \mathcal{F}, K)$ of S_S^+ according to Def. 4.3 is called the Q_r-seminearring of the seminearring S with respect to $\mathcal{F}$ and K.

Recall that $Q_r(S_S^+,\mathcal{F},\mathcal{K})$ consists of the $\kappa_{\mathcal{K}}$-classes $[f]_{\mathcal{F},\mathcal{K}}$, where now $f \in \Lambda_S(\mathcal{F},S_S^+)$ is an S-mapping on some $D_f \in \mathcal{F} \subseteq \mathcal{S}_S$ into the S-set S_S. Moreover, the operations on $Q_r(S_S^+,\mathcal{F},\mathcal{K})$ are defined by (4.9) and (4.10) for any representatives.

Clearly, all statements of § 4 apply to this special case, and the multiplicative semigroup of such a Q_r-seminearring $Q_r(S_S^+,\mathcal{F},\mathcal{K})$ of $(S,+,\cdot)$ is now the Q_r-semigroup $Q_r(S_S,\mathcal{F},\mathcal{K})$ of the semigroup $(S,\cdot)$ (denoted by $Q_r(S,\mathcal{F},\mathcal{K})$ in [39], Def. 3.6). Moreover, we have the following interrelations between S and $Q_r(S_S^+,\mathcal{F},\mathcal{K})$ as seminearrings:

THEOREM 5.2 *Let $Q = Q_r(S_S^+,\mathcal{F},\mathcal{K})$ be the Q_r-seminearring of the seminearring S with respect to the Q_r-filters $\mathcal{F} \subseteq \mathcal{K}$ of S_S, and denote by λ_a the inner left translation $x \to \lambda_a(x) = ax$ of $(S,\cdot)$. Then we have:*

a) *There is a homomorphism $\sigma = \sigma_{\mathcal{F},\mathcal{K}}$ of $(S,+,\cdot)$ into $(Q,+,\cdot)$, which is defined by*

$$(5.1) \qquad a \to \lambda_a \to [\lambda_a]_{\mathcal{F},\mathcal{K}} = \sigma(a) = a^* \in Q,$$

and for each representative f of an element $[f] \in Q$ one has

$$(5.2) \qquad [f]\cdot[\lambda_a] = [\lambda_{f(a)}], \text{ or shorter } [f]\cdot a^* = f(a)^*,$$

always for all $a \in D_f$.

b) *The congruence $\psi_{\mathcal{K}}$ on $(S,+,\cdot)$ which corresponds to the homomorphism $\sigma = \sigma_{\mathcal{F},\mathcal{K}}$ above is characterized by*

$$(5.3) \qquad a \equiv b\ (\psi_{\mathcal{K}}) \Leftrightarrow \lambda_a \equiv \lambda_b\ (\kappa_{\mathcal{K}}) \Leftrightarrow C_{\lambda_a,\lambda_b} \in \mathcal{K}$$
$$\Leftrightarrow ax = bx \text{ for all } x \in K \text{ for some } K \in \mathcal{K}.$$

In particular, $\sigma_{\mathcal{F},\mathcal{K}}$ is injective iff $\mathcal{F} \subseteq \mathcal{K} \subseteq S_S^{\Delta}$ holds. In this case, we have $Q = Q_r(S_S^+,\mathcal{F},\mathcal{K}) = Q_r(S_S^+,\mathcal{F},S_S^{\Delta}) = Q_r(S_S^+,\mathcal{F})$, and

$(S,+,\cdot)$ *may be considered as a subseminearring of* $(Q,+,\cdot)$ *if one identifies each* $a \in S$ *with* $\sigma(a) = a^* = [\lambda_a] \in Q$.

c) *If the* Q_r*-filter* $\mathcal{K}$ *equals* S_S^Δ, *more generally, if* $\mathcal{K}$ *is a special* Q_r*-filter of* S_S 1), *then the addition of the* Q_r*-seminearring* $Q = Q_r(S_S^+, \mathcal{F}, \mathcal{K})$ *is uniquely determined by the addition of the seminearring* $(S,+,\cdot)$ *if one merely demands that* (5.1) *is a homomorphism.*

Proof a) and b) By the right distributivity of S, the first step in (5.1) defines a homomorphism into the seminearring $\Lambda_S(\mathcal{F}, S_S^+)$ of Thm. 4.2, and the second one corresponds to the congruence $\kappa_{\mathcal{K}}$ defined there. This yields that $\sigma_{\mathcal{F},\mathcal{K}}$ is a seminearring homomorphism and that (5.3) holds. But (5.3) and the remaining statements are now the same as for the Q_r-semigroup $Q_r(S_S, \mathcal{F}, \mathcal{K})$ and proved in [39], Thm. 5.4 and Lemma 5.5.

c) Assume that the Q_r-semigroup $(Q,\cdot) = Q_r(S_S, \mathcal{F}, \mathcal{K})$ is established with an addition $\texttt{+}$ such that (5.1) is a homomorphism of $(S,+,\cdot)$ into $(Q,\texttt{+},\cdot)$ and that all elements $[\lambda_a]_{\mathcal{F},\mathcal{K}}$ are right distributive in $(Q,\texttt{+},\cdot)$. This will be enough to show that this addition $\texttt{+}$ coincides with the addition + of the Q_r-seminearring $Q_r(S_S^+, \mathcal{F}, \mathcal{K})$, which was defined by (4.9) and $[f]+[g] = [f+g]$. Toward this end, we consider $[f]\texttt{+}[g] = [h]$ for any $[f],[g] \in Q$. Multiplying the latter from the right side by $[\lambda_a]$ for all $a \in D_f \cap D_g \cap D_h = D \in \mathcal{F}$, we obtain from (5.2) $[\lambda_{f(a)}]\texttt{+}[\lambda_{g(a)}] = [\lambda_{h(a)}]$. Now we use that (5.1) is assumed to

---

1) A Q_r-filter $\mathcal{K}$ of a semigroup $(S,\cdot)$ is called special iff for each $C \in \mathcal{K}$ and each mapping $c \to K_c$ of C into $\mathcal{K}$ the S-set $\cup\{cK_c \mid c \in C\}$ is again in $\mathcal{K}$ (cf. [39], § 11 and [10]). Special Q_r-filters of semigroups are the most considered ones (cf. e.g. [23]), and S_S^Δ is easily checked to have this property.

be a homomorphism with respect to +, what yields

$[\lambda_{f(a)}] + [\lambda_{g(a)}] = [\lambda_{f(a)+g(a)}]$, or $\lambda_{f(a)+g(a)} \equiv \lambda_{h(a)}\ (\kappa_{\mathcal{K}})$,

again for all $a \in D \in \mathcal{F}$. Hence, by (5.3), for each such $a \in D$ there exists some $K_a \in \mathcal{K}$ such that $(f(a)+g(a))x = h(a)x$ holds in $(S,+,\cdot)$ for all $x \in K_a$. All together, we have

$$f(ax)+g(ax) = h(ax) \quad \text{for all } ax \in \cup\{aK_a \mid a \in D\} = K.$$

Since $a \in D \in \mathcal{F} \subseteq \mathcal{K}$, $K_a \in \mathcal{K}$ and $\mathcal{K}$ is assumed to be special, we obtain $K \in \mathcal{K}$ (cf. the last footnote) and

$$(f+g)(y) = f(y)+g(y) = h(y) \quad \text{for all } y \in K \in \mathcal{K}$$

(note that the first equation, corresponding to (4.9), holds for all $y \in D_f \cap D_g$ and hence for all $y \in K$). According to (4.11), the last formula implies $K \subseteq C_{f+g,h}$ or $f+g \equiv h\ (\kappa_{\mathcal{K}})$, as we were to show.

We now turn to Q_r-seminearrings $Q = Q_r(S_S^+,\mathcal{F},\mathcal{K})$ of a seminearring S which, in a natural way, may be considered as extensions of S. By Thm. 5.2 b), those Q_r-seminearrings of S are characterized by $\mathcal{F} \subseteq \mathcal{K} \subseteq S_S^\Delta$ and do not depend on the choice of $\mathcal{K}$ between $\mathcal{F}$ and S_S^Δ (cf. (4.12)). Hence such a Q_r-seminearring $Q_r(S_S^+,\mathcal{F},\mathcal{K}) = Q_r(S_S^+,\mathcal{F})$ of S exists for each Q_r-filter $\mathcal{F}$ of S_S contained in S_S^Δ, presumed that $S_S^\Delta \neq \emptyset$.

<u>In the following, we always consider</u> S <u>as a subseminearring of</u> $Q_r(S_S^+,\mathcal{F})$ <u>by the identification</u> $a = [\lambda_a]$. With this interpretation, each classical Q_r-seminearring $Q_r(S,\Sigma)$ of S is such a Q_r-seminearring $Q_r(S_S^+,\mathcal{F}_\Sigma)$, where the corresponding Q_r-filter $\mathcal{F}_\Sigma \subseteq S_S^\Delta$ is given by $\mathcal{F}_\Sigma = \{A \in \mathcal{S}_S \mid A \cap \Sigma \neq \emptyset\}$ (cf. [39], Thm. 9.1). On the other hand, these Q_r-seminearrings $Q_r(S_S^+,\mathcal{F})$

with the above identification will turn out to be particular cases of McMorris Q_r-seminearrings of S, according to the following definitions:

Let $(T,\cdot)$ be a semigroup, $(S,\cdot)$ a subsemigroup of $(T,\cdot)$, and T_S the S-set obtained from T by right multiplication in $(T,\cdot)$. Then T is called a <u>McMorris Q_r-semigroup of</u> S iff S_S is dense in T_S according to (4.8) (cf. [24] and [39], § 2).

As usual, we denote this situation by $S \leqq T$. One easily checks that, for each chain $(S,\cdot) \subseteq (T,\cdot) \subseteq (R,\cdot)$ of subsemigroups, $S \leqq T \leqq R$ is equivalent to $S \leqq R$. As a consequence, a semigroup S has a McMorris Q_r-semigroup iff $S \leqq S$; by (4.8), this means that S is a right reductive semigroup (cf. [6], § 1.3), which is in turn equivalent to $S_S^{\Delta} \neq \emptyset$.

<u>DEFINITION 5.3</u> A seminearring $(T,+,\cdot)$ containing $(S,+,\cdot)$ as a subseminearring is called a <u>McMorris Q_r-seminearring of</u> S iff $(T,\cdot)$ is a McMorris Q_r-semigroup of $(S,\cdot)$. It will cause no confusion to use then $S \leqq T$ also for seminearrings.

By the above remarks on semigroups, $S_S^{\Delta} \neq \emptyset$ is necessary for the existence of McMorris Q_r-seminearrings of S, which we shall investigate now similarly to our considerations in § 3 on classical Q_r-seminearrings. The next theorem concerns the uniqueness of addition (cf. Thm. 3.2) and the interrelations to the above Q_r-seminearrings $Q_r(S_S^{+},\mathcal{F})$, $\mathcal{F} \subseteq S_S^{\Delta}$, where the latter provide the existence of a maximal McMorris Q_r-seminearring (cf. Thm. 3.6). Note in this context that for each chain $S \subseteq T \subseteq R$ of subseminearrings $S \leqq T \leqq R$ holds iff $S \leqq R$ (in fact a stronger statement than the corresponding one in Thm. 3.7).

THEOREM 5.4 For each seminearring $S=(S,+,\cdot)$ we have:

a) If $S_S^\Delta \neq \emptyset$, each Q_r-seminearring $Q=Q_r(S_S^+,\mathcal{F})=Q_r(S_S^+,\mathcal{F},S_S^\Delta)$ of S with respect to a Q_r-filter $\mathcal{F}$ of S_S contained in S_S^Δ is a McMorris Q_r-seminearring of S, and the addition defined on Q is the only one such that the Q_r-semigroup $(Q,\cdot)=Q_r(S_S,\mathcal{F})$ of $(S,\cdot)$ becomes a seminearring which contains S as a subseminearring.

b) If T is any McMorris Q_r-seminearring of S, then the addition on T is uniquely determined by the addition on S. Moreover, there is a smallest Q_r-filter $\mathcal{F}_o \subseteq S_S^\Delta$ of S_S such that T is (isomorphic over S) embeddable into the Q_r-seminearring $Q_o=Q_r(S_S^+,\mathcal{F}_o)$. Using suitable identifications, we obtain a chain of seminearrings

$$(5.4) \qquad S \leqq T \leqq Q_r(S_S^+,\mathcal{F}_o)=Q_o \leqq Q_r(S_S^+,S_S^\Delta)=Q_m,$$

where each step is a McMorris extension, determining the addition uniquely from the left to the right, and each step may be a proper one. In particular, T is contained in the maximal McMorris Q_r-seminearring $Q_m=Q_r(S_S^+,S_S^\Delta)$ of S, which is uniquely determined by S up to isomorphisms over S.

c) If $(T',\cdot)$ is any McMorris Q_r-semigroup of $(S,\cdot)$, there need not be an addition on T' such that $(T',+,\cdot)$ becomes a seminearring containing S as a subseminearring. More precisely, $(T',\cdot)$ is embeddable into a smallest McMorris Q_r-seminearring T of S according to b), but T' is not closed with respect to the unique addition on T if $T' \subset T$.

Proof a) It is known that the Q_r-semigroup $(Q,\cdot)=Q_r(S_S,\mathcal{F})$ of $(S,\cdot)$ is a McMorris Q_r-semigroup of $(S,\cdot)$ by the identification $a = [\lambda_a]$ (cf. [39], Thm. 5.4). Since the latter corresponds to (5.1) as a monomorphism, the uniqueness of the addition follows from Thm. 5.2 c).

b) At first we recall statements on the embedding of $(T,\cdot)$ (cf. [39], Prop. 7.1, Thms. 7.2 and 7.5). From $S \leqq T$ we get $S_S^\Delta \neq \emptyset$, thus S_S^Δ is a Q_r-filter of S_S. By $\mu_t(s) = t\cdot s$ for each $t \in T$ one defines an S-mapping $\mu_t \in \Lambda_S(S_S, T_S)$, and there is a smallest Q_r-filter $\mathcal{F}_o \subseteq S_S^\Delta$ of S containing the S-subsets

$$(5.5) \qquad \mu_t^{-1}(S) = \{s \in S \mid ts \in S\} = A_t$$

for all $t \in T$. Hence for the restrictions μ'_t of μ_t on $\mu_t^{-1}(S)$ we have $\mu'_t \in \Lambda_S(\mathcal{F}_o, S_S)$, and $t \to [\mu'_t]_{\mathcal{F}_o}$ defines a monomorphism of $(T,\cdot)$ into $(Q_o,\cdot) = (Q_r(S_S,\mathcal{F}_o),\cdot)$. Moreover, $\mathcal{F}_o$ is the smallest Q_r-filter of S_S which allows such an embedding.

We now state that $t \to [\mu'_t]$ is also a homomorphism of $(T,+)$ into $(Q_o,+)$, where $(Q_o,+,\cdot) = Q_r(S_S^+,\mathcal{F}_o)$ is a seminearring as considered in a). This follows since $t_1+t_2 = t_3$ and the right distributive law in T imply $\mu'_{t_1} + \mu'_{t_2} = \mu'_{t_3}$ in $\Lambda_S(\mathcal{F}_o, S_S^+)$, and yields the embedding of T into Q_o in (5.4) as seminearrings. The inclusion can be proper, as we shall see in Expl. 5.5. The next step in (5.4), which is written more explicitely $Q_o = Q_r(S_S^+,\mathcal{F}_o,S_S^\Delta) \leq Q_r(S_S^+,S_S^\Delta,S_S^\Delta) = Q_m$, follows from Prop. 4.5. It can be a proper one for semigroups (cf. [39], Expl. 9.4), and hence also for seminearrings[1]. The maximality of Q_m as a McMorris Q_r-seminearring of S may be defined by the property that each McMorris Q_r-seminearring T of S is embeddable into Q_m as just proved. For two other characterizations in this

---

1) Each semigroup $(S,\cdot)$ turns even into a semiring $(S,+,\cdot)$ by trivial additions as $a+b = a$ or $a+b = b$ for all $a,b \in S$.

context, which obviously transfer from semigroups to seminearrings, see [39], Thm. 7.7.

For the first statement of b), still unproved till now, we note that all considerations above apply to any addition on a McMorris Q_r-semigroup $(T,\cdot)$ of $(S,\cdot)$ such that $(T,+,\cdot)$ is a seminearring containing $(S,+,\cdot)$ as a subseminearring. But this implies the uniqueness of such an addition on T, since $t \to [\mu'_t]$ is a monomorphism of $(T,+,\cdot)$ into $(Q_o,+,\cdot)$ and the addition of the latter seminearring is uniquely determined by that of S according to a).

A direct proof of this statement, independent on part a) and hence on Thm. 5.2, for which one merely needs that the S-subset A_t defined in (5.5) belongs to S_S^Δ for each $t \in T$, is as follows: Assume that there are two additions of this kind, say $+_1$ and $+_2$, and that

$$t_1 +_1 t_2 = t_3 \neq t_4 = t_1 +_2 t_2 \quad \text{holds for some } t_1, t_2 \in T.$$

Since $A_{t_1} \cap A_{t_2} \cap A_{t_3} \cap A_{t_4} = A \in S_S^\Delta$, we obtain in $(S,+,\cdot)$

$$t_1 s + t_2 s = t_3 s = t_4 s = t_1 s + t_2 s \quad \text{for all } s \in A,$$

which contradicts $t_3 s \neq t_4 s$ for at least one $s \in A$ by (4.8).

c) By the same considerations as in part b) of this proof, the McMorris Q_r-semigroup $(T',\cdot)$ of $(S,\cdot)$ is also embeddable into a Q_r-semigroup $(Q_o,\cdot) = (Q_r(S_S,\mathcal{F}_o),\cdot)$. The latter is a seminearring $(Q_o,+,\cdot)$ which contains $(S,+,\cdot)$ only for the addition defined on $Q_o = Q_r(S_S^+,\mathcal{F}_o)$. Hence any addition on T' such that $S \leq T' \leq Q_o$ would be a chain of subseminearring has to be the restriction of that addition on Q_o, by the first statement of b) applied to T' and Q_o. So our proof will be completed by an example such that a semigroup $(T',\cdot)$ contained in $(Q_o,+,\cdot)$ is not additively closed in Q_o.

<u>Example 5.5</u> Let $(S,+,\cdot)$ be the ring defined by the tables

+	o	a	b	c
o	o	a	b	c
a	a	o	c	b
b	b	c	o	a
c	c	b	a	o

·	o	a	b	c
o	o	o	o	o
a	o	a	a	o
b	o	b	b	o
c	o	c	c	o

.

Obviously, the S-set S_S of $(S,\cdot)$ is dense in S_S, whereas each proper S-subset, for instance $\{o,a,b\}$, is not. Hence we have $S_S^\Delta = \{S\}$, which implies that the maximal McMorris Q_r-semigroup $(Q_m,\cdot) = Q_r(S_S,S_S^\Delta)$ of $(S,\cdot)$ coincides in this example with the semigroup $\Lambda_S(S_S^\Delta,S) = \Lambda_S(S,S)$ of all left translations of $(S,\cdot)$, and the congruence κ_K for $K = S_S^\Delta$ is the equality. One easily checks that Q_m consists of 4^3 mappings $f = [f]$, where $f(o) = o$ holds and $f(a), f(b), f(c)$ may be chosen arbitrarily. According to our procedure above, we identify each $x \in S$ with $[\lambda_x] = \lambda_x$.

Moreover, $(Q_m,+,\cdot) = Q_r(S_S^+,S_S^\Delta)$ is the only Q_r-seminearring of S according to Thm. 5.4 a), clearly the maximal McMorris Q_r-seminearring of S; its unique addition is just given by (4.9) in this case. Due to Thm. 4.4, $(Q_m,+)$ is a commutative group, $o = \lambda_o$ is the zero, and $1_S = 1$ is the identity of the seminearring $(Q_m,+,\cdot)$, which is in fact a nearring. As one will suppose, the 4^2 mappings $f \in Q_m$ which are endomorphisms of $(S,+)$ are the only left distributive elements of Q_m (see Thm. 5.6 c) in this context).

To complete the above proof, consider for instance the

subsemigroup $T' = S^1 = S \cup \{1\}$ of $(Q_m, \cdot)$. Then we have $S \leq T' \leq Q_m$, but T' cannot be closed with respect to the addition of Q_m, since T' consists of 5 elements, and the order of a subsemigroup of the finite group $(Q_m, +)$ has to divide 4^3. One also checks that T' generates additively a McMorris Q_r-seminearring T of S which consists of S and the 4 elements 1, 1+a, 1+b, 1+c, in fact again a nearring. Since there is no proper Q_r-filter $\mathcal{F}_o \subseteq S_S^\Delta$ of S_S, we also have obtained an example for the fact that the step $T \leq Q_o = Q_m$ in (5.4) may well be a proper one. Let us note that the same considerations work, for instance, if one replaces the element 1 above by the mapping $f \in Q_m$ such that $f(o) = f(a) = f(b) = o$, $f(c) = c$, which clearly is not a left distributive element of Q_m.

The next theorem corresponds to Thm. 3.5 for the classical case, and it should be also compared with Thm. 4.4 for Q_r-seminearrings $Q_r(M_S^+, \mathcal{F}, K)$ of an S-groupoid M_S^+. According to the situation described in Thm. 5.4, each Q_r-seminearring $Q_r(S_S^+, \mathcal{F})$ of S such that $\mathcal{F} \subseteq S_S^\Delta$ is a McMorris Q_r-seminearring of S, whereas an arbitrary McMorris Q_r-seminearring T of S is in general only (isomorphic to) a subseminearring of a former one. More precisely, if $\mathcal{F}_o$ is the Q_r-filter corresponding to T by (5.5), for each Q_r-filter $\mathcal{F}$ of S_S such that $\mathcal{F}_o \subseteq \mathcal{F} \subseteq S_S^\Delta$ holds, we have $T \leq Q_r(S_S, \mathcal{F}_o) \leq Q_r(S_S^+, \mathcal{F})$ according to Prop. 4.5 (clearly again with suitable identifications).

<u>THEOREM 5.6</u> <u>For a seminearring</u> S, <u>let</u> T <u>be any McMorris</u> <u>Q_r-seminearring of</u> S, <u>and</u> $Q = Q_r(S_S^+, \mathcal{F})$ <u>a McMorris Q_r-seminearring of</u> S <u>given by a Q_r-filter</u> $\mathcal{F} \subseteq S_S^\Delta$, <u>in particular the maximal McMorris Q_r-seminearring</u> $Q_r(S_S^+, S_S^\Delta)$ <u>of</u> S.

a) If the groupoid (S,+) is commutative, left [right] cancellative, idempotent or associative, then (T,+) has the same property. If (S,+) is a left [right] quasigroup or a group, then (Q,+) has the same property, but this need not be true for (T,+).

b) If S has a zero o such that oS = o holds, then o is also the zero of T and satisfies oT = o.

c) An element $t = [\mu'_t]$ of $T \subseteq Q_o = Q_r(S_S^+, \mathcal{F}_o)$ is left distributive in T iff the S-mapping $\mu'_t \in \Lambda_S(A_t, S_S)$ is itself an S-homomorphism and A_t, defined by (5.5), an S-subgroupoid.

d) If S is multiplicatively commutative, left [right] cancellative or if S has no zero-divisors, then only the first property transfers to T, whereas the others need not even be true for Q.

Proof a) According to Thm. 4.4 a), each of these properties of (S,+) transfers to any Q_r-seminearring $Q = Q_r(S_S^+, \mathcal{F})$ of S_S^+. Since T is contained in such a seminearring by Thm. 5.4 b), it remains to prove that (T,+) need not be a group if (S,+) is one. Again we give an example for a ring $(S,+,\cdot)$, defined to be the ring ideal generated by x in the polynomial ring $\mathbb{R}[x]$. Then the set T of all polynomials $f(x) = a_o + a_1 x + \ldots$ of $\mathbb{R}[x]$ such that $a_o \geq 0$ obviously is a subsemiring $(T,+,\cdot)$ of $\mathbb{R}[x]$ which satisfies $(S,\cdot) \leq (T,\cdot)$. Hence T is a McMorris Q_r-seminearring of S, but (T,+) is not a group.

b) Due to Thm. 4.4 b), [0] such that $0(x) = o$ holds for all $x \in S$ is the zero of Q. By our embedding of S into Q, we have identified $[0] = [\lambda_o]$ with o. To prove $oQ = o$ by way of contradiction, assume $oq \neq o$ for some $q \in Q$. Since S_S is dense in Q_S, there is an $s \in S$ such that $oqs \neq os$ and $qs \in S$, which contradicts $oS = o$. Thus b) is proved for Q and hence for T, which is contained in such a $Q_o = Q_r(S_S^+, \mathcal{F}_o)$ by Thm. 5.4 b).

c) We refer to the embedding of T just mentioned and assume that t is left distributive in T. Then $A_t = \mu_t^{-1}(S)$ as defined in (5.5) is clearly an S-subgroupoid of S_S^+, and μ'_t is an S-homomorphism on A_t. Conversely, if these both statements hold, then $t = [\mu'_t]$ is a left distributive element even in $Q_o = Q_r(S_S^+, \mathcal{F}_o)$ (and also in $Q_m = Q_r(S_S^+, S_S^\Delta)$) by Thm. 4.4 c).

d) It is known that a McMorris Q_r-semigroup $(T, \cdot)$ of $(S, \cdot)$ is commutative if $(S, \cdot)$ is commutative (cf. [24] or [39], Cor. 8.2). For the negative statements, let $(S^*, \cdot)$ be the free semigroup generated by two elements a and b, and define an addition by $s_1 + s_2 = s_1$ for all $s_i \in S^*$. Then $(S^*, +, \cdot)$ is a (two-sided) cancellative semiring which belongs to case α) of Result 2.5; we adjoin an absorbing zero $n = o$ and obtain a seminearring $(S, +, \cdot)$ which satisfies γ) (cf. the remark behind Result 2.5). Since $S = aS^1 \cup bS^1$ and $aS^1 \cap bS^1 = \{o\}$ hold, each choice of $f(a), f(b) \in S$ defines an S-mapping $f \in \Lambda_S(S,S)$ by [39], Lemma 3.9 b). Moreover, we clearly have $S \in S_S^\Delta$. Now consider $[f], [g] \in Q_m = Q_r(S_S^+, S_S^\Delta)$ for $f(a) = a$, $f(b) = o$ and $g(a) = o$, $g(b) = b$. Using (4.12), we have $f \not\equiv \lambda_o \not\equiv g\ (\kappa_K)$, $K = S_S^\Delta$, or $[f] \neq o \neq [g]$ in Q_m. But $f \cdot g = \lambda_o$, i.e. $[f][g] = o$, shows that Q_m has zero-divisors, hence Q_m is also neither left nor right cancellative by Result 2.5.

Comparing again our results on McMorris Q_r-seminearrings with those in the classical case, it remains to give a statement corresponding to Thm. 3.8. But this will be in fact Thm. 3.8 itself according to the following simple

PROPOSITION 5.7 Let S be a seminearring such that $|S^*| \geq 2$ and T a McMorris Q_r-seminearring of S which is a seminearfield. Then T is a classical Q_r-seminearfield of S.

Proof Again we have $|T^*| \geq 2$, which implies $To = oT = o$ iff the seminearfield T has a zero o. In any case, $(T^*, \cdot)$ is a group. By (4.8), for each $t_1 \neq t_2$ and t of T^*, there is an $s \in S$ such that $t_1 s \neq t_2 s$ and $ts = s' \in S$ holds. The first implies that s cannot be the zero of T (if there is one). Hence $s \in T^*$ has an inverse $s^{-1} \in T^*$ and we obtain $t = s's^{-1}$, as we were to show.

§ 6 Q_r-semirings $Q_r(M_S^+, \mathcal{F}^+, K^+)$ of a subcommutative S-groupoid M_S^+

Let us start with a Q_r-seminearring $Q = Q_r(M_S^+, \mathcal{F}, K)$ of an S-groupoid M_S^+ with respect to Q_r-filters $\mathcal{F} \subseteq K$ of M_S (Def. 4.3). It is clear that the product of left distributive elements of $(Q, +, \cdot)$ is again such an element. We are going to discuss sufficient conditions (applicable to each Q_r-seminearring Q of this kind) which imply that this also holds for the sum $[f] + [g] = [f+g]$ of left distributive elements $[f], [g] \in Q$.

According to Thm. 5.6 c), we may use Thm. 4.4 c) as a criterion for such a test. Hence restrictions of f, g and f+g to suitable S-subgroupoids of M_S^+ have to be S-homomorphisms into M_S^+, in particular

(6.1) $\quad (f+g)(x+y) = f(x+y)+g(x+y) = (f(x)+f(y))+(g(x)+g(y))$

(6.2) $(f+g)(x)+(f+g)(y) = (f(x)+g(x))+(f(y)+g(y))$

have to coincide for all x,y of a suitable S-subset $K \in \mathcal{K}$. For the latter it is clearly sufficient that

(6.3) $(a+b)+(c+d) = (a+c)+(b+d)$ for all $a,b,c,d \in M$

is satisfied, and we call a groupoid (M,+) subcommutative iff (6.3) holds. We note that this concept was already used in the context of (additively associative) semirings (cf. [9], [38], [42]), and that for instance each rectangular band (M,+) is subcommutative[1]. In fact, subcommutativity will prove to be a condition, convenient and sufficient for all statements for the rest of our paper, and it clearly applies to all associative and commutative additions. But we emphasize that most of the following statements are also true for ad hoc constructed examples or special types of algebras under consideration for which the addition is not subcommutative.

PROPOSITION 6.1 *Let M_S^+ be a subcommutative S-groupoid, $\mathcal{F} \subseteq \mathcal{K}$ Q_r-filters of the S-set M_S, and $Q = Q_r(M_S^+, \mathcal{F}, \mathcal{K})$ the Q_r-seminearring with respect to $\mathcal{F}$ and $\mathcal{K}$. Let U be the set of all elements $[f] \in Q$ for which at least one representative, say f itself, is an S-homomorphism defined on an S-subgroupoid $D_f \in \mathfrak{M}_S^+$, e.g. $[1_M]$. Then $(U,+,\cdot)$ is a subsemiring of $(Q,+,\cdot)$. Moreover, $(Q,+)$ and hence $(U,+)$ are subcommutative.*

Proof For $[f],[g] \in U$ with representatives f,g as described above, $D_f \cap D_g$ is also an S-subgroupoid of M_S^+. Thus (6.1) and (6.2) are meaningful for all $x,y \in D_f \cap D_g$, and they coincide by (6.3). Hence f+g is an S-homomorphism on $D_f \cap D_g \in \mathfrak{M}_S^+$ and

1) For seminearfields and semifields S such that (S,+) is a rectangular band we refer to [41].

$[f]+[g]=[f+g]\in U$ holds. One also checks that $g^{-1}(D_f)$ is an S-subgroupoid, and $f\cdot g$ is then clearly an S-homomorphism on $g^{-1}(D_f)\in\mathfrak{M}_S^+$. Hence U is a subseminearring of Q, and each element $[f]\in U$ is left distributive by Thm. 4.4 c), even in Q. The last statement follows from

$$(f_1+f_2)(x)+(f_3+f_4)(x) = (f_1+f_3)(x)+(f_2+f_4)(x)$$

for arbitrary $f_i\in\Lambda_S(\mathcal{F},M_S^+)$ and all $x\in D_{f_1}\cap\ldots\cap D_{f_4}\in\mathcal{F}$, which is checked straightforward due to (6.3).

In order to obtain semirings in such a way, it is near by hand to deal from the first merely with S-subgroupoids and S-homomorphisms of an S-groupoid M_S^+, and so with Q_r-filters $\mathcal{F}^+$ of M_S^+ according to Def. 4.1 b). For these considerations, which are parallel to those in § 4, we need as above the (only sufficient) assumption that M_S^+ is subcommutative[1].

THEOREM 6.2 *Let M_S^+ be a subcommutative S-groupoid, $\mathcal{F}^+$ a Q_r-filter of M_S^+, and $Hom_S(\mathcal{F}^+,M_S^+)=\cup\{Hom_S(A,M)\mid A\in\mathcal{F}^+\}$ the set of all S-homomorphisms f, defined on an S-subgroupoid $A_S^+=D_f\in\mathcal{F}^+$, into M_S^+. Then we have:*

a) *In a natural way, $(Hom_S(\mathcal{F}^+,M_S^+),+,\cdot)$ is a seminearring with 1_M as identity for operations $+,\cdot$ defined by*

(4.9^+) $\quad (f+g)(x) = f(x)+g(x)$ *for all* $x\in D_f\cap D_g\in\mathcal{F}^+$,

(4.10^+) $\quad (f\cdot g)(x) = f(g(x))$ *for all* $x\in g^{-1}(D_f)\in\mathcal{F}^+$.

b) *If (S,+) is commutative, associative or idempotent, then the same holds for $(Hom_S(\mathcal{F}^+,M_S^+),+)$, which is clearly always*

1) According to § 8, the following generalizes in fact the well known constructions of Johnson [18] and of Utumi [31] for rings S, clearly based on S-modules. An attempt [5] to transfer the latter to nearrings deals in both cases with $Q_r(S_S^+,S_S^\Delta)$ in our notation of § 5, cf. MR 46 # 1851.

subcommutative like $(M,+)$. Further, if $(M,+)$ has a neutral o satisfying $oS = o$, then 0 defined by $0(x) = o$ for all $x \in M$ is the zero of $(\mathrm{Hom}_S(\mathcal{F}^+,M_S^+),+,\cdot)$.

c) Each Q_r-filter $\mathcal{K}^+$ of M_S^+ such that $\mathcal{F}^+ \subseteq \mathcal{K}^+$ defines by

$$(4.11^+) \quad f \equiv g\ (\kappa_{\mathcal{K}^+}) \ \text{ iff } \ C_{f,g} = \{x \in D_f \cap D_g \mid f(x) = g(x)\} \in \mathcal{K}^+$$

a congruence on $(\mathrm{Hom}_S(\mathcal{F}^+,M_S^+),+,\cdot)$. If $\mathcal{F}^+ \subseteq \mathcal{K}^+ \subseteq M_S^{+\Delta}$, one has

$$(4.12^+) \quad f \equiv g\ (\kappa_{\mathcal{K}^+}) \iff C_{f,g} = D_f \cap D_g \iff f \equiv g\ (\kappa_{\mathcal{F}^+}).$$

d) Whereas $\mathrm{Hom}_S(\mathcal{F}^+,M_S^+)$ is in general only a seminearring, we obtain a semiring modulo $\kappa_{\mathcal{K}^+}$ and denote it by

$$Q_r(M_S^+,\mathcal{F}^+,\mathcal{K}^+) = (\mathrm{Hom}_S(\mathcal{F}^+,M_S^+)/\kappa_{\mathcal{K}^+},+,\cdot).$$

Proof Appart from a few supplements, the proof is essentially the same as for Thm. 4.2, including the used references to conclusions in [39]. The latter, concerning the semigroup $(\Lambda_S(\mathcal{F},M_S),\cdot)$, transfer to the semigroup $(\mathrm{Hom}_S(\mathcal{F}^+,M_S^+),\cdot)$ due to the correspondence of both kinds of Q_r-filters defined in Def. 4.1, of (4.11) and (4.11^+), and of M_S^Δ and $M_S^{+\Delta}$, both defined by (4.8). Additionally, one has to check that S-mappings defined by (4.9^+) and (4.10^+) are S-homomorphisms, and that $(\mathrm{Hom}_S(\mathcal{F}^+,M_S^+),+)$ is subcommutative, considerations which are the same as those in the proof of Prop. 6.1, and use (6.3). For d) we state that $(h(f+g))(x)$ and $(hf+hg)(x)$ coincide on the domain $D_{hf+hg} \in \mathcal{F}^+$ of $hf+hg$, and $D_{hf+hg} \subseteq D_{h(f+g)} \in \mathcal{F}^+$ (cf. the proof of Thm. 4.4 c)). Hence $hf+hg$ and $h(f+g)$ coincide on $D_{hf+hg} \in \mathcal{F}^+ \subseteq \mathcal{K}^+$ which proves the left distributivity modulo $\kappa_{\mathcal{K}^+}$. On the other hand, $D_{hf+hg} \subset D_{h(f+g)}$ happens frequently and implies $hf+hg \neq h(f+g)$ in $\mathrm{Hom}_S(\mathcal{F}^+,M_S^+)$. For example, consider the semiring $(\mathbb{N},+,\cdot)$ of natural numbers with the usual operations as an S-groupoid $M_S^+ = \mathbb{N}_{\mathbb{N}}^+$ and the S-homomorphisms $h = \frac{1}{2}$, $f = g = 1$ defined on $D_h = 2\mathbb{N}$, $D_f = D_g = \mathbb{N}$ in the obvious way. Then we obtain $D_{hf+hg} = 2\mathbb{N}$ and $D_{h(f+g)} = \mathbb{N}$.

DEFINITION 6.3 Let M_S^+ be a subcommutative S-groupoid and $\mathcal{F}^+ \subseteq \mathcal{K}^+$ Q_r-filters of M_S^+. Then the semiring $Q_r(M_S^+,\mathcal{F}^+,\mathcal{K}^+)$ introduced in Thm. 6.2 d) is called the *Q_r-semiring of M_S^+ with respect to $\mathcal{F}^+$ and $\mathcal{K}^+$*. Again we write $[f] = [f]_{\mathcal{F}^+,\mathcal{K}^+}$ for its elements and use $Q_r(M_S^+,\mathcal{F}^+,\mathcal{F}^+) = Q_r(M_S^+,\mathcal{F}^+)$ for the *Q_r-semiring of M_S^+ which respect to $\mathcal{F}^+$* iff $\mathcal{F}^+ = \mathcal{K}^+$.

THEOREM 6.4 *Let $Q = Q_r(M_S^+,\mathcal{F}^+,\mathcal{K}^+)$ be the Q_r-semiring of a subcommutative S-groupoid M_S^+ with respect to Q_r-filters $\mathcal{F}^+$ and $\mathcal{K}^+$ of M_S^+. Then $[1_M]$ is the identity of Q, and we have:*

a) *If the groupoid $(M,+)$ is commutative, left [right] cancellative, idempotent, a left [right] quasigroup, associative or a group, then $(Q,+)$ has the same property.*

b) *If $(M,+)$ has a neutral o which satisfies $oS = o$, then $[0]$ such that $0(x) = o$ for all $x \in M$ is the zero of $(Q,+,\cdot)$.*

Proof One can transfer the proof for Thm. 4.4 step by step with the appropriate interpretation of concepts and references. The only thing one has to add concerns the proof that $(Q,+)$ is a left quasigroup if $(M,+)$ is one, namely to show that the mapping f defined by $f(x) = y \in M$ for $y+g(x) = h(x)$ is also an additive homomorphism on $D_g \cap D_h \in \mathcal{F}^+$. To do this, we consider $f(x) = y$ as above and $f(u) = v$, $v+g(u) = h(u)$ for $x,u \in D_g \cap D_h$. Then we obtain by (6.3)

$$(y+v)+(g(x)+g(u)) = (y+g(x))+(v+g(u)) = h(x)+h(u)$$

or $(y+v)+g(x+u) = h(x+u)$, which yields $f(x+u) = y+v = f(x)+f(u)$.

It is also clear that a statement corresponding to Prop. 4.5 holds for Q_r-semirings according to Def. 6.3. Though we need it later, we may forgo its explicite formulation.

As already announced in § 1, we are now going to compare semirings U obtained from a subcommutative S-groupoid M_S^+ by Prop. 6.1 with Q_r-semirings $Q_r(M_S^+, \mathcal{F}^+, K^+)$ according to Def. 6.3. In this direction we start with the following theorem, true for arbitrary S-groupoids M_S^+, and note that its restrictive statements (the first one of b) and that the inclusions (6.4) and (6.5) may be proper) cannot be improved for subcommutative S-groupoids M_S^+, in fact not even for those with associative and commutative addition[1].

THEOREM 6.5 Let M_S^+ be an S-groupoid. Then we have:

a) Each Q_r-filter $\mathcal{F}$ of the S-set M_S determines a Q_r-filter $\mathcal{F}^{\oplus} = \mathcal{F} \cap \mathfrak{M}_S^+$ of the S-groupoid M_S^+ such that $\mathcal{F}^{\oplus} \subseteq \mathcal{F}$, the largest one contained in $\mathcal{F}$.

b) For a Q_r-filter $\mathcal{F}^+$ of M_S^+ there need not be a Q_r-filter $\mathcal{F}$ of M_S such that $\mathcal{F}^+ \subseteq \mathcal{F}$. If those Q_r-filters $\mathcal{F}$ exist, their intersection $\mathcal{F}^{+\#} = \cap \mathcal{F}$ is the smallest Q_r-filter of M_S which contains $\mathcal{F}^+$. The latter is always the case if M_S^+ is given by $M_S^+ = S_S^+$ for any seminearring $(S,+,\cdot)$.

c) Applying both procedures one after the other, we get

(6.4) $\qquad \mathcal{F} \supseteq \mathcal{F}^{\oplus\#} = (\mathcal{F} \cap \mathfrak{M}_S^+)^{\#}$

for each Q_r-filter $\mathcal{F}$ of M_S, and

1) An S-groupoid M_S^+ such that $(M,+)$ is associtative clearly is called an S-semigroup. In the following examples, we deal in fact with S-semigroups and S-subsemigroups. But as these examples concern general statements for S-groupoids, we use there also the more general terms with "groupoid" to facilitate the comparison with our general theorems.

(6.5) $\mathcal{F}^+ \subseteq (\mathcal{F}^+)^{\#\oplus} = \mathcal{F}^{+\#} \cap \mathfrak{M}_S^+$

for all Q_r-filters $\mathcal{F}^+$ of M_S^+ such that $\mathcal{F}^{+\#}$ exists. In both cases, the inclusions may be proper, even for S-groupoids $M_S^+ = S_S^+$ obtained from a semiring S whose both operations are associative and commutative. Finally, for three consecutive steps (if $\mathcal{F}^{+\#}$ exists) one has

(6.6) $\mathcal{F}^{\oplus\#\oplus} = \mathcal{F}^{\oplus}$ and $(\mathcal{F}^+)^{\#\oplus\#} = (\mathcal{F}^+)^{\#}$.

Proof a) Since (4.5^+) and (4.6^+) for $\mathcal{F}^{\oplus}$ are obvious[1], we only show (4.7^+) and assume $A,B \in \mathcal{F}^{\oplus}$ and $f \in Hom_S(A,M)$. This yields $A,B \in \mathcal{F}$ and $f \in \Lambda_S(A,M)$, hence $f^{-1}(B) \in \mathcal{F}$ by (4.7) for $\mathcal{F}$ (and, in particular $f^{-1}(B) \neq \emptyset$). For each $x,y \in f^{-1}(B)$, we obtain $x,y \in A$ and $f(x),f(y) \in B$, therefore $x+y \in A$ and $f(x+y) = f(x)+f(y) \in B$ and thus $x+y \in f^{-1}(B)$, i.e. $f^{-1}(B) \in \mathfrak{M}_S^+$. The latter and $f^{-1}(B) \in \mathcal{F}$ imply $f^{-1}(B) \in \mathcal{F} \cap \mathfrak{M}_S^+ = \mathcal{F}^{\oplus}$.

b) The first statement will be shown in Expl. 6.10. The next follows from the fact, that any intersection of Q_r-filters of M_S is again one (cf. [39], Lemma 3.3; the same holds, by the way, for Q_r-filters of any S-groupoid M_S^+). For the last assertion we use that for each semigroup $(S,\cdot)$ the S-set S_S has a unique maximal Q_r-filter $\mathcal{F} = \overline{S}_S$, which consists of all S-subsets A such that $xS \cap A \neq \emptyset$ holds for all $x \in S$ (cf. [39], Def. 5.7, Thm. 5.9 ii)). Hence it is enough to show that each Q_r-filter $\mathcal{F}^+$ of S_S^+ is contained in that $\mathcal{F} = \overline{S}_S$. But for any $A \in \mathcal{F}^+$ and all $x \in S$, we have $\lambda_x^{-1}(A) = \{s \in S \mid xs \in A\} \in \mathcal{F}$ by (4.7^+), which proves $xS \cap A \neq \emptyset$ and $A \in \mathcal{F} = \overline{S}_S$.

---

1) We mention in this context that (4.6^+) [(4.6)] is a consequence of (4.7^+) [(4.7)] due to $A \cap B = (1_A)^{-1}(B)$, or that (4.7^+) [(4.7)] can be weakened by $A = B$ (cf. [39], Lemma 3.2).

c) By the meaning of the operations $^{\oplus}$ and $^{\#}$ the inclusions (6.4) and (6.5) are obvious. That both may be proper, even for S-groupoids $M_S^+ = S_S^+$ as described above, will be shown in Expls. 6.7 and 6.8. The equalities (6.6), familiar in similar situations, are easily seen by a simple sketch.

Remark 6.6 By our definitions (cf. (4.8)), the Q_r-filter $M_S^{+\Delta}$ of all dense S-subgroupoids of M_S^+ is obtained from the Q_r-filter M_S^{Δ} of all dense S-subsets of M_S just as described in Thm. 6.5 a), i.e. $M_S^{+\Delta} = M_S^{\Delta\oplus} \subseteq M_S^{\Delta}$. Consequently, by (6.6), we obtain (6.5) in this case with equality, i.e.

$$(6.5^{\Delta}) \qquad M_S^{+\Delta} = (M_S^{+\Delta})^{\#\oplus} \quad \text{iff } M_S \text{ is dense in } M_S.$$

On the other hand, concerning (6.4), we merely have

$$(6.4^{\Delta}) \qquad M_S^{\Delta} \supseteq (M_S^{\Delta})^{\oplus\#} \quad \text{iff } M_S \text{ is dense in } M_S,$$

and we shall see that <u>also this inclusion may be proper, even for S-groupoids $M_S^+ = S_S^+$ obtained from a semiring S for which both operations are associative and commutative.</u>

Example 6.7 Let S be the set of all subsets of {a,b} in the notation $o = \emptyset$, $a = \{a\}$, $b = \{b\}$ and $1 = \{a,b\}$. Using now the union [intersection] as addition [multiplication], we obtain a semiring $(S,+,\cdot)$ such that both operations are associative, commutative and idempotent. Let S_S^+ be the corresponding S-groupoid. One easily checks that $B = \{o,a,b\}$ is the smallest S-subset of S_S which is dense in S_S. Hence the Q_r-filter S_S^{Δ} of S_S consists of B and S. Since B is not an S-subgroupoid, the Q_r-filter $S_S^{\Delta\oplus} = S_S^{+\Delta}$ of all dense S-subgroupoids of S_S^+ coincides with {S}. Since {S} is also a Q_r-filter of the S-set S_S, we obtain

$$\{B,S\} = S_S^{\Delta} \supset (S_S^{\Delta})^{\oplus\#} = (S_S^{+\Delta})^{\#} = \{S\}^{\#} = \{S\}$$

and thus our assertion on the inclusions (6.4^{Δ}) and (6.4).

Example 6.8 We use even a ring $(S,+,\cdot) = \{a,2a,\dots,8a = o\}$, additively generated by an element a of order 8, define its multiplication by $a\cdot a = 4a$, and consider the S-groupoid S_S^+. We state that the S-subgroupoids $B = \{o,4a\}$, $\{o,2a,4a,6a\}$, S form a Q_r-filter $\mathcal{F}^+$ of S_S^+. One only has to check (4.7^+), which depends on the fact that each S-homomorphism f, since $f(4a)+f(4a) = f(o) = o$, maps 4a either onto 4a or onto o. Next we claim that the smallest Q_r-filter $\mathcal{F}$ of the S-set S_S which contains B consists of all S-subsets of S_S, i.e. $\mathcal{F} = \mathcal{S}_S$. For this purpose we use that $g(4a) = 2a$, $g(o) = o$ is clearly an S-mapping $g \in \Lambda_S(B,S)$. Then it follows from $B \in \mathcal{F}$ by (4.7), that $g^{-1}(B) = \{x \in B \mid g(x) \in B\} = \{o\} \in \mathcal{F}$ holds. Now the latter yields $\mathcal{F} = \mathcal{S}_S$ by (4.5) since the absorbing zero o of $(S,+,\cdot)$ is clearly contained in each S-subset of S_S. Hence we have $(\mathcal{F}^+)^{\#} = \mathcal{F} = \mathcal{S}_S$ and $(\mathcal{F}^+)^{\#\oplus} = \mathcal{S}_S^{\oplus} = \mathcal{S}_S^+$, the set of S-subgroupoids of S_S^+. Because $\{o\}$ is not contained in $\mathcal{F}^+$ (in fact we have $\mathcal{F}^+ \cup \{o\} = \mathcal{S}_S^{\oplus}$), we obtain $\mathcal{F}^+ \subset (\mathcal{F}^+)^{\#\oplus}$ for the inclusion (6.5).

THEOREM 6.9 For any subcommutative S-groupoid M_S^+ we have:

a) Let $\mathcal{F} \subseteq \mathcal{K}$ be Q_r-filters of the S-set M_S and $Q = Q_r(M_S^+,\mathcal{F},\mathcal{K})$ the corresponding Q_r-seminearring. Then the subsemiring U of Q as described in Prop. 6.1 is isomorphic to the Q_r-semiring $Q^+ = Q_r(M_S^+,\mathcal{F}^{\oplus},\mathcal{K}^{\oplus})$ of M_S^+ with respect to the Q_r-filters $\mathcal{F}^{\oplus} = \mathcal{F} \cap \mathfrak{M}_S^+$ and $\mathcal{K}^{\oplus} = \mathcal{K} \cap \mathfrak{M}_S^+$ of the S-groupoid M_S^+. Again with representatives $f \in [f] \in U$ which are S-homomorphisms on $D_f \in \mathfrak{M}_S^+$, this isomorphism is given in a natural way by

(6.7) $[f]_{\mathcal{F},K} \to [f]_{\mathcal{F}^\oplus, K^\oplus}$ for all $[f]_{\mathcal{F},K} \in U$.

b) To check converse possibilities, let $\mathcal{F}^+ \subseteq K^+$ be Q_r-filters of the S-groupoid M_S^+. Then, in general, the corresponding Q_r-semiring $Q^+ = Q_r(M_S^+, \mathcal{F}^+, K^+)$ cannot be obtained as a subsemiring U as described in a), even not if Q_r-filters $\mathcal{F}$ and K of the S-set M_S exist such that $\mathcal{F}^+ \subseteq \mathcal{F}$ and $K^+ \subseteq K$ hold.[1)]

Proof a) We use (4.11) and (4.11$^+$) to show that (6.7) really defines a mapping, say φ, and that φ is injective. For the first, assume $f \equiv g\ (\kappa_K)$ for S-homomorphisms f and g. This implies $C_{f,g} \in K$ and $C_{f,g} \in \mathfrak{M}_S^+$, the latter since $f(x) = g(x)$ and $f(y) = g(y)$ of course yield $f(x+y) = g(x+y)$ for S-homomorphisms. Hence we have $C_{f,g} \in K \cap \mathfrak{M}_S^+ = K^\oplus$, i.e. $f \equiv g\ (\kappa_{K^\oplus})$, and see that (6.7) does not depend on the choice of the admitted representatives of $[f]_{\mathcal{F},K} \in U$. The converse conclusion, proving injectivity, is trivial by $K^\oplus \subseteq K$. Further, φ is surjective, since for each $[f]_{\mathcal{F}^\oplus, K^\oplus} \in Q^+$ each representative f is an S-homomorphism defined on an S-subgroupoid $D_f \in \mathcal{F}^\oplus \subseteq \mathcal{F}$. Therefore $[f]_{\mathcal{F},K}$ defines an element of Q which belongs to U. By the corresponding definitions of operations, φ is now clearly an isomorphism of $(U,+,\cdot)$ onto $(Q^+,+,\cdot)$.

b) We shall prove this negative statement even for Q_r-semirings $Q^+ = (M_S^+, \mathcal{F}^+, K^+)$ such that $\mathcal{F}^+ = K^+$. Clearly, part a) is not applicable if there is no Q_r-filter $\mathcal{F}$ of M_S containing $\mathcal{F}^+$, according to the first assertion of Thm. 6.5 b), still open

1) We only remark in this context, that for each Q_r-semiring $Q^+ = Q_r(M_S^+, \mathcal{F}^+, K^+)$ of M_S^+ such that $\mathcal{F}^+ \subseteq K^+ \subseteq M_S^{+\Delta}$ the Q_r-seminearring $Q = Q_r(M_S^+, \mathcal{F}^{+\#}, K^{+\#})$ exists and that Q^+ is isomorphic embeddable into the subsemiring U of Q as considered in a).

till now. By the following example, we shall prove the latter as well as Thm. 6.9 b), even in the case $\mathcal{F}^+ = \mathcal{K}^+ \subseteq \mathcal{F} = \mathcal{K}$.

<u>Example 6.10</u> Let $(M,+) = (\mathbb{N},+)$ be the additive semigroup of natural numbers, $(S,\cdot) = (2\mathbb{N}+1,\cdot)$ the multiplicative one of odd numbers, both with respect to the usual operations, and let S operate on M by $ms = m\cdot s$, the product in $(\mathbb{N},\cdot)$. Then M_S^+ is an S-groupoid whose addition is associative and commutative. Since each subsemigroup A of (M,+) is also an S-subgroupoid of M_S^+, the set of all subsemigroups of (M,+), say $\mathfrak{M}^+$, coincides with the set $\mathfrak{M}_S^+$ of all S-subgroupoids. Now we use the obvious fact that for each S-groupoid M_S^+ the set $\mathfrak{M}_S^+$ is itself a Q_r-filter of M_S^+ iff $\mathfrak{M}_S^+$ is closed with respect to intersections (i.e. (4.6^+) for $\mathfrak{M}_S^+$), which is clearly the case for the S-groupoid M_S^+ introduced above.

We use this Q_r-filter $\mathcal{F}^+ = \mathfrak{M}_S^+ = \mathfrak{M}^+$ of M_S^+ to show the last remaining statement of Thm. 6.5, the first one of part b). For this purpose, we define a mapping g on M by

$$(6.8) \qquad g(m) = 2m \quad \text{for } 2\nmid m \text{ and}$$
$$g(m) = m' \quad \text{for } m = 2^{\nu}m',\ \nu \geq 1,\ 2\nmid m'.$$

For all $s \in S = 2\mathbb{N}+1$, we obtain $g(ms) = 2ms = g(m)s$ for $2\nmid m$, and $g(ms) = (ms)' = m's = g(m)s$ for $2\mid m$. This proves $g \in \Lambda_S(M,M)$. By way of contradiction, assume that there is a Q_r-filter $\mathcal{F}$ of M_S which contains $\mathcal{F}^+$. Using $2\mathbb{N} \in \mathcal{F}^+ \subseteq \mathcal{F}$ and (4.7), we get

$$g^{-1}(2\mathbb{N}) = \{m \in M \mid g(m) \in 2\mathbb{N}\} \in \mathcal{F}.$$

But $g(m) = m' \notin 2\mathbb{N}$ for $2\mid m$ and $g(m) = 2m \in 2\mathbb{N}$ for $2\nmid m$ yield $g^{-1}(2\mathbb{N}) = 2\mathbb{N}+1 \in \mathcal{F}$, hence $g^{-1}(2\mathbb{N}) \cap 2\mathbb{N} = \emptyset$ which contradicts that $\mathcal{F}$ satisfies (4.6).

To prove Thm. 6.9 b) in the case that the Q_r-filter $\mathcal{F}^+ = \mathcal{K}^+$ of M_S^+ is contained in a Q_r-filter $\mathcal{F} = \mathcal{K}$ of M_S, we modify our example and consider M and S including the number 0. Then each S-subgroupoid A of M_S^+ is a subsemigroup of $(M,+)$ which contains 0, and conversely. By a similar argument as above, the set $\mathcal{F}^+$ of all S-subgroupoids $A \neq \{0\}$ is a Q_r-filter of M_S^+, and (6.8) defines now an S-mapping $g \in \Lambda_S(M,M)$ including $g(0) = 0$. Then, for each Q_r-filter $\mathcal{F}$ of M_S such that $\mathcal{F}^+ \subseteq \mathcal{F}$, we obtain $g^{-1}(2\mathbb{N}_0) \cap 2\mathbb{N}_0 = \{0\} \in \mathcal{F}$ similarly as above.

Hence the corresponding Q_r-seminearring $Q = Q_r(M_S^+,\mathcal{F},\mathcal{F})$ of M_S^+ has merely one element, since all S-mappings $f \in \Lambda_S(\mathcal{F},M)$ coincide on $\{0\} \in \mathcal{F}$ due to $f(0) = 0$.

<u>But the Q_r-semiring</u> $Q^+ = Q_r(M_S^+,\mathcal{F}^+,\mathcal{F}^+)$ <u>of</u> M_S^+ <u>is isomorphic to the classical Q_r-semifield</u> $Q_r(\mathbb{N}_0,\mathbb{N})$ <u>of all non-negative rationals</u>. This is not so surprising as it may seem at the first glance. Note that the special choice of the operator domain S of M_S^+ was essential for the above considerations, but it has in fact no influence on the Q_r-filter $\mathcal{F}^+$ of M_S^+ and on the construction of Q^+. Both remain the same if we replace the operator domain S by $\mathbb{N}_0$ itself. But then one considers, for $(S,+,\cdot) = (\mathbb{N}_0,+,\cdot)$, the S-groupoid S_S^+ and the Q_r-semiring $Q^+ = Q_r(S_S^+,\mathcal{F}^+)$ with respect to the Q_r-filter $\mathcal{F}^+$ of S_S^+ containing all S-subgroupoids $A \neq \{0\}$. Now it is easy to check that Q^+ is isomorphic to $Q_r(\mathbb{N}_0,\mathbb{N})$, but this will be a special case of Thm. 7.6 b). - (Concerning our example in the original version without 0, one obtains likewise that the Q_r-semiring $Q^+ = Q_r(M_S^+,\mathcal{F}^+,\mathcal{F}^+)$ is then isomorphic to the classical Q_r-semifield $Q_r(\mathbb{N},\mathbb{N})$ of all positive rationals.)

§ 7 Q_r-semirings $Q_r(S_S^+,\mathcal{F}^+,K^+)$ of a subcommutative semiring S and McMorris Q_r-semirings

According to § 5, we are going to apply our considerations on Q_r-semirings $Q_r(M_S^+,\mathcal{F}^+,K^+)$ to S-groupoids S_S^+ obtained from a seminearring $(S,+,\cdot)$. By our assumption in § 6, of course we need that $(S,+)$ is subcommutative, and we shall use the abbreviation <u>subcommutative</u> <u>seminearring</u> in this case. Moreover, we want to consider interrelations between $(S,+,\cdot)$ and $(Q_r(S_S^+,\mathcal{F}^+,K^+),+,\cdot)$ due to the homomorphism (7.1); for this reason $(S,+,\cdot)$ has to be also left distributive:

<u>DEFINITION 7.1</u> Let S be a subcommutative semiring, S_S^+ the S-groupoid determined by S, and $\mathcal{F}^+ \subseteq K^+$ Q_r-filters of the S-groupoid S_S^+. Then the Q_r-semiring $Q_r(S_S^+,\mathcal{F}^+,K^+)$ of S_S^+ according to Def. 6.3 is called the <u>Q_r-semiring of the semiring S with respect to $\mathcal{F}^+$ and K^+</u>.

<u>THEOREM 7.2</u> <u>Let $Q = Q_r(S_S^+,\mathcal{F}^+,K^+)$ be the Q_r-semiring of the subcommutative semiring S with respect to the Q_r-filters $\mathcal{F}^+ \subseteq K^+$ of S_S^+, and denote by λ_a the inner left translation $x \to \lambda_a(x) = ax$ of $(S,\cdot)$. Then we have:</u>

a) <u>There exists a homomorphism $\sigma = \sigma_{\mathcal{F}^+,K^+}$ of $(S,+,\cdot)$ into $(Q,+,\cdot)$ which is defined by</u>

(7.1) $\quad a \to \lambda_a \to [\lambda_a]_{\mathcal{F}^+,K^+} = \sigma(a) \in Q$,

<u>and for each representative f of an element $[f] \in Q$ one has</u>

(7.2) $\quad [f]\cdot[\lambda_a] = [\lambda_{f(a)}]$ <u>always for all</u> $a \in D_f$.

b) <u>The congruence ψ_{K^+} on $(S,+,\cdot)$ which corresponds to the</u>

homomorphism $\sigma = \sigma_{\mathcal{F}^+,\mathcal{K}^+}$ *above is characterized by*

$$(7.3) \qquad a \equiv b\ (\psi_{\mathcal{K}^+}) \ \Leftrightarrow\ \lambda_a \equiv \lambda_b\ (\kappa_{\mathcal{K}^+}) \ \Leftrightarrow\ C_{\lambda_a,\lambda_b} \in \mathcal{K}^+$$
$$\Leftrightarrow\ ax = bx \ \text{for all}\ x \in K \ \text{for some}\ K \in \mathcal{K}^+.$$

In particular, $\sigma_{\mathcal{F}^+,\mathcal{K}^+}$ *is injective iff* $\mathcal{F}^+ \subseteq \mathcal{K}^+ \subseteq S_S^{+\Delta}$. *In this case, we have* $Q = Q_r(S_S^+,\mathcal{F}^+,\mathcal{K}^+) = Q_r(S_S^+,\mathcal{F}^+,S_S^{+\Delta}) = Q_r(S_S^+,\mathcal{F}^+)$, *and* $(S,+,\cdot)$ *may be considered as a subsemiring of* $(Q,+,\cdot)$ *if one identifies each* $a \in S$ *with* $\sigma(a) = [\lambda_a] \in Q$.

c) *If the* Q_r*-filter* $\mathcal{K}^+$ *equals* $S_S^{+\Delta}$, *more generally, if* $\mathcal{K}^+$ *is a special* Q_r*-filter of* S_S^+ [1], *then the addition of the* Q_r*-semiring* $Q = Q_r(S_S^+,\mathcal{F}^+,\mathcal{K}^+)$ *is uniquely determined by the addition of the semiring* $(S,+,\cdot)$ *if one merely demands that* (7.1) *is a homomorphism.*

Proof a) and b) By the left distributivity of S, λ_a is an S-homomorphism defined on S_S^+; hence the first step in (7.1) is a homomorphism of S into the seminearring $\mathrm{Hom}_S(\mathcal{F}^+,S_S^+)$ of Thm. 6.2, and the second one corresponds to the congruence $\kappa_{\mathcal{K}^+}$ defined there. This shows that $\sigma_{\mathcal{F}^+,\mathcal{K}^+}$ is a semiring homomorphism and (7.3). For (7.2), we state that for each $a \in D_f$ we have $(f\cdot\lambda_a)(x) = f(ax) = f(a)\cdot x$ for all $x \in S$. Moreover, $\sigma_{\mathcal{F}^+,\mathcal{K}^+}$ is injective, by (7.3), iff for each $K \in \mathcal{K}^+$ and any elements $a \neq b$ of S there exists a $k \in K$ such that $ak \neq bk$.

---

1) A Q_r-filter $\mathcal{K}^+$ of a semiring $(S,+,\cdot)$ is called special iff for each $C \in \mathcal{K}^+$ and each mapping $c \to K_c$ of C into $\mathcal{K}^+$ every semiring ideal $A \in \mathcal{S}_S^+$ such that $\cup\{cK_c \mid c \in C\} \subseteq A$ holds is also in $\mathcal{K}^+$. This concept will prove to be the adequate one to special Q_r-filters of semigroups and will be investigated in a forthcoming paper by U.Hebisch. In particular, $\mathcal{K}^+ = S_S^{+\Delta}$ is a special Q_r-filter of S_S^+ in this meaning, but we do not need this explicitely in our proof.

This is clearly the case if $K^+ \subseteq S_S^{+\Delta}$, since then each $K \in K^+$ is dense in S_S. For the converse, we assume that K^+ has the above property and consider any elements $a \neq b$ and c of S. Since $\lambda_c \in Hom_S(S,S)$ holds, for each $K \in K^+$ we obtain

$$\lambda_c^{-1}(K) \cap K = \{k_o \in K \mid ck_o \in K\} = K_o \in K^+$$

by (4.7^+) and (4.6^+). By our assumption on K^+, there exists a $k_o \in K_o \subseteq K$ satisfying $ak_o \neq bk_o$. Together with $ck_o \in K$, this proves that each $K \in K^+$ is dense in S_S, i.e. $K^+ \subseteq S_S^{+\Delta}$. The remaining statement of b) is a consequence of (4.12^+).

c) Assume that the semigroup $(Q,\cdot)$ of $Q = (S_S^+, \mathcal{F}^+, K^+)$ is established with an addition $\oplus$ such that (7.1) is a homomorphism of $(S,+,\cdot)$ into $(Q,\oplus,\cdot)$ and that all elements $[\lambda_a]_{\mathcal{F}^+,K^+}$ are right distributive in $(Q,\oplus,\cdot)$. Again this will be sufficient to show that this addition $\oplus$ coincides with the addition $+$ of the Q_r-semiring $Q_r(S_S^+, \mathcal{F}^+, K^+)$, which was defined by (4.9^+) and $[f]+[g] = [f+g]$. By the corresponding conclusions as in the proof of Thm. 5.2 c), we assume $[f] \oplus [g] = [h]$ and obtain

$$\lambda_{f(a)+g(a)} \equiv \lambda_{h(a)} \quad (\kappa_{K^+})$$

for all $a \in D_f \cap D_g \cap D_h = D \in \mathcal{F}^+$. If $\mathcal{F}^+ \subseteq K^+ \subseteq S_S^{+\Delta}$ holds, we have proved above that $\sigma_{\mathcal{F}^+,K^+}$ is injective, and we get in this case $f(a)+g(a) = h(a)$ for all $a \in D \in \mathcal{F}^+$. The latter implies $f+g \equiv h$ modulo $\kappa_{\mathcal{F}^+}$, and hence also modulo κ_{K^+} according to (4.12^+), which proves our statement for $\mathcal{F}^+ \subseteq K^+ \subseteq S_S^{+\Delta}$.

If K^+ is assumed to be special, we obtain as in the corresponding part of the proof of Thm. 5.2 c) that

$$f(ax) + g(ax) = h(ax) \quad \text{for all} \quad ax \in \cup\{aK_a \mid a \in D\}$$

holds for suitable $K_a \in K^+$. Now we define a subset A of S by

$A=\{y\in D_f\cap D_g \mid f(y)+g(y)=h(y)\}$, which clearly contains the above union. It remains to show $A\in\mathcal{S}_S^+$, since then A belongs to K^+ (cf. the last footnote) and

$$(f+g)(y) = f(y)+g(y) = h(y) \quad \text{for all} \quad y\in A\in K^+$$

implies $C_{f+g,h}\in K^+$ by (4.5^+), i.e. $f+g\equiv h\,(\kappa_{K^+})$ by (4.11^+). To prove $A\in\mathcal{S}_S^+$, it is obvious that A is an S-subset of S_S. To show that A is also closed with respect to the addition of $(S,+,\cdot)$, we assume $y,z\in A$. Then we obtain

$$\begin{aligned} f(y+z)+g(y+z) &= (f(y)+f(z))+(g(y)+g(z)) \\ &= (f(y)+g(y))+(f(z)+g(z)) = h(y)+h(z) = h(y+z), \end{aligned}$$

using that $(S,+)$ is subcommutative, which yields $y+z\in A$.

Again we turn to Q_r-semirings $Q=Q_r(S_S^+,\mathcal{F}^+,K^+)$ of a subcommutative semiring S which, in a natural way, may be considered as extensions of S. By Thm. 7.2 b), those Q_r-semirings are characterized by $\mathcal{F}^+\subseteq K^+\subseteq S_S^{+\Delta}$ and do not depend on the choice of K^+ between $\mathcal{F}^+$ and $S_S^{+\Delta}$ (cf. (4.12^+)). Hence such a Q_r-semiring $Q_r(S_S^+,\mathcal{F}^+,K^+)=Q_r(S_S^+,\mathcal{F}^+)$ of S exists for each Q_r-filter $\mathcal{F}^+$ of S_S^+ contained in $S_S^{+\Delta}$, presumed that $S_S^{+\Delta}\neq\emptyset$, which is the same as $S_S^{\Delta}\neq\emptyset$.[1)]

<u>In the following</u>, for $\mathcal{F}^+\subseteq S_S^{+\Delta}$, <u>we always consider</u> S <u>as a subsemiring of</u> $Q=Q_r(S_S^+,\mathcal{F}^+)$ <u>by the identification</u> $a=[\lambda_a]$. In this interpretation, each of these Q_r-semirings Q will turn out to be a subcommutative McMorris Q_r-semiring of S according to Def. 5.3, which means that $(Q,\cdot)$ is a McMorris Q_r-semigroup of $(S,\cdot)$. On the other hand, Thms. 5.4 and 5.6

1) The Q_r-semiring $Q_r(S_S^+,S_S^{+\Delta})$ for a semiring $(S,+,\cdot)$ whose both operations are associative and commutative was directly constructed in [15].

provide many results on McMorris Q_r-semirings T of an arbitrary semiring S. So at first we state those general results in Thm. 7.3, and deal subsequently with the subcommutative case, in particular with the concept of a maximal McMorris Q_r-semiring of S.

THEOREM 7.3 Let S be any semiring and T a McMorris Q_r-semiring of S according to Def. 5.3. Then we have:

a) The addition on T is uniquely determined by the addition on S, whereas for an arbitrary McMorris Q_r-semigroup $(T',\cdot)$ of $(S,\cdot)$ there need not exist an addition on T' such that $(T',+,\cdot)$ becomes a semiring containing S as a subsemiring.

b) All statements of Thm. 5.6 a),b) are verbatim valid for semirings S and T as above.

c) For each $t \in T$, the set $\mu_t^{-1}(S) = \{s \in S \mid ts \in S\} = A_t$ is an S-subgroupoid of S_S^+ contained in $S_S^{+\Delta}$, and μ'_t defined by $\mu'_t(s) = ts$ for all $s \in A_t$ satisfies $\mu'_t \in \mathrm{Hom}_S(A_t, S)$.

d) If S is multiplicatively commutative, left [right] cancellative or if S has no zero-divisors, then only the first property transfers to T, whereas the others do not and may be violated even for a maximal McMorris Q_r-semiring T of S.

Proof a) The first statement holds by Thm. 5.4 b) even for seminearrings S and T. For the second one, we can use the proof of Thm. 5.4 c) and Expl. 5.5, since the seminearring S considered in this example is in fact a ring.

b) Again we may regard S and T as seminearrings and apply Thm. 5.6 a) and b), since the counter-example given in the proof of part a) deals in fact with semirings S and T.

c) As already cited at the outset of the proof of Thm. 5.4 b), from $(S,\cdot) \leq (T,\cdot)$ it follows that $A_t \in S_S^{\Delta}$. Because T is a semiring, A_t is additively closed and so contained in $S_S^{+\Delta}$, and obviously μ'_t is an S-homomorphism.

d) The statement on commutativity is known for $(S,\cdot) \leq (T,\cdot)$. The other ones will be shown in Expls. 7.7 and 7.8.

THEOREM 7.4 For a subcommutative semiring S we have:

a) If $S_S^{+\Delta} \neq \emptyset$, each Q_r-semiring $Q^+ = Q_r(S_S^+, \mathcal{F}^+) = Q_r(S_S^+, \mathcal{F}^+, S_S^{+\Delta})$ of S with respect to a Q_r-filter $\mathcal{F}^+$ of S_S^+ contained in $S_S^{+\Delta}$ is a subcommutative McMorris Q_r-semiring of S.

b) If T is any McMorris Q_r-semiring of S, then the (uniquely determined) addition on T is subcommutative. Moreover, for the Q_r-filter $\mathcal{F}_o^{\oplus} = \mathcal{F}_o \cap \mathcal{S}_S^+ \subseteq S_S^{+\Delta}$ of S_S^+ (see the supplement), T is (isomorphic over S) embeddable into the Q_r-semiring $Q_o^+ = Q_r(S_S^+, \mathcal{F}_o^{\oplus})$. Using suitable identifications, we obtain a chain of semirings

$$(7.4)\qquad S \leqq T \leqq Q_r(S_S^+, \mathcal{F}_o^{\oplus}) = Q_o^+ \leqq Q_r(S_S^+, S_S^{+\Delta}) = Q_m^+ ,$$

where each step is a McMorris extension and may be proper. In particular, T is contained in the maximal McMorris Q_r-semiring $Q_m^+ = Q_r(S_S^+, S_S^{+\Delta})$ of S, which is uniquely determined by S up to isomorphisms over S.

Supplement Applying Thm. 5.4 b) to S and T as seminearrings,

$$(5.4)\qquad S \leqq T \leqq Q_r(S_S^+, \mathcal{F}_o) = Q_o \leqq Q_r(S_S^+, S_S^{\Delta}) = Q_m$$

holds for the McMorris Q_r-seminearrings Q_o and Q_m of S with respect to Q_r-filters $\mathcal{F}_o$ and S_S^{Δ} of S_S,[1] where $\mathcal{F}_o$ was the smallest one suitable for the embedding (5.4). We also state

1) Footnote see next page.

that $Q_m^+ \leq Q_o$ and $Q_m^+ \leq Q_m$ hold, and refer to Problem 7.5.

<u>Proof</u> a) For any elements $[f] \neq [g]$, $[h]$ of Q^+, all given representatives f, g and h are defined on $D = D_f \cap D_g \cap D_h \in \mathcal{F}^+$. From $[f] \neq [g]$ it follows $f(a) \neq g(a)$ for at least one element $a \in D$ by (4.12^+). Then we obtain for $[\lambda_a] = a \in Q^+$ by (7.2)

$$[f][\lambda_a] = [\lambda_{f(a)}] = f(a) \neq g(a) = [\lambda_{g(a)}] = [g][\lambda_a]$$

and $[h][\lambda_a] = [\lambda_{h(a)}] = h(a) \in S$. Hence $(S,\cdot)$ is dense in $(Q^+,\cdot)$ according to (4.8), as we were to show.

b) We start with the situation sketched in the supplement. In this context, $\mathcal{F}_o$ was defined as the smallest Q_r-filter of S_S containing the S-subsets $A_t = \{s \in S \mid ts \in S\}$ for all $t \in T$ (actually, $\mathcal{F}_o$ is the intersection of all those Q_r-filters of S_S), and t was identified with $[\mu'_t] \in Q_o$. By Thm. 7.3 c), each A_t is now an S-subgroupoid of S_S^+ and $\mu'_t \in Hom_S(A_t, S)$. Hence applying Prop. 6.1, T is a subsemiring of the subsemiring U_o of $Q_o = Q_r(S_S^+, \mathcal{F}_o, S_S^\Delta)$, and we get by the way that all these seminearrings are subcommutative. Due to Thm. 6.9 a), U_o is isomorphic to the Q_r-semiring

$$Q_o^+ = Q_r(S_S^+, \mathcal{F}_o^\oplus, S_S^{\Delta\oplus}) = Q_r(S_S^+, \mathcal{F}_o^\oplus, S_S^{+\Delta}) = Q_r(S_S^+, \mathcal{F}_o^\oplus),$$

and we have obtained $T \leq Q_o^+$ in (7.4) up to an isomorphism over S. The last step, $Q_r(S_S^+, \mathcal{F}_o^\oplus, S_S^{+\Delta}) \leq Q_r(S_S^+, S_S^{+\Delta}, S_S^{+\Delta}) = Q_m^+$, and thus the maximality of Q_m^+, follows from the version of

1) Recall that a Q_r-semiring $Q_r(S_S^+, \mathcal{F}^+)$ of S with respect to a Q_r-filter $\mathcal{F}^+$ of S_S^+ is a special case of the Q_r-semirings constructed in Thm. 6.2 (with S-homomorphisms and S-subgroupoids), whereas a Q_r-seminearring $Q_r(S_S^+, \mathcal{F})$ of S with respect to a Q_r-filter $\mathcal{F}$ of S_S is a special case of the Q_r-seminearrings constructed in Thm. 4.2 (with S-mappings and S-subsets), regardless whether the latter may be a semiring, too.

Prop. 4.5 for Q_r-semirings according to Def. 6.3, mentioned behind Thm. 6.4. Further, $Q_o^+ \simeq U_o \subseteq Q_o$ and $(S,\cdot) \leq (Q_o,\cdot)$ yield $Q_o^+ \leq Q_o$ by the identification $Q_o^+ = U_o$, and similarly we get $Q_m^+ \leq Q_m$ as stated in the supplement.

Thus it remains to prove that these steps may be proper. For $T < Q_o^+$, in fact $T < Q_o^+ = Q_m^+$, we can use Expl. 5.5. Since the ring S satisfies $\{S\} = S_S^\Delta = S_S^{+\Delta}$, there is only one Q_r-semiring according to Thm. 7.4 a), $Q_o^+ = Q_m^+ = Q_r(S_S^+, S_S^{+\Delta})$. It consists of the 4^2 endomorphisms of $(S,+)$, and it is in fact a ring (in the usual meaning). The McMorris Q_r-semiring T considered in Expl. 5.5 (consisting of S and the 4 elements 1, 1+a, 1+b, 1+c), also in fact a ring, is properly contained in $Q_o^+ = Q_m^+$. Note in this context that S has no classical Q_r-extensions.

Examples such that $Q_o^+ < Q_m^+$ holds in (7.4) abound even for classical Q_r-semirings, and we shall quote some trivial ones in connection with Thm. 7.6.

We like to point out the following difference between Thm. 5.4 b) and Thm. 7.4 b). Whereas in the first case we really have used the smallest Q_r-filter $\mathcal{F}_o$ of S_S for which (5.4) holds, we have used the Q_r-filter $\mathcal{F}_o \cap \mathcal{S}_S^+ = \mathcal{F}_o^\oplus$ of S_S^+ for (7.4). <u>In fact, we know that for each McMorris Q_r-semiring</u> T <u>of</u> S <u>there is a unique smallest Q_r-filter $\mathcal{F}_o^+$ of S_S^+ such that</u> (7.4) <u>holds with</u> $Q_r(S_S^+,\mathcal{F}_o^+)$ <u>instead of</u> $Q_r(S_S^+,\mathcal{F}_o^\oplus)$. Moreover, $\mathcal{F}_o^+$ is the intersection of all Q_r-filters of S_S^+ which contain the S-subgroupoids A_t of S_S^+ for all $t \in T$, such that $\mathcal{F}_o^+ \subseteq \mathcal{F}_o^\oplus$ holds. Since we could not clear up this situation, we pose the

<u>Problem 7.5</u> For all McMorris Q_r-semirings T of a subcommu-

tative semiring S, decide whether $\mathcal{F}_o^{\oplus} = \mathcal{F}_o \cap \mathfrak{S}_S^+$ coincides with the smallest Q_r-filter $\mathcal{F}_o^+$ of S_S^+ as just described.

As a supplement to these considerations we state

THEOREM 7.6 For each subcommutative semiring S we have:

a) Let $T = Q_r(S_S^+, \mathcal{F})$ be the Q_r-seminearring of S with respect to a Q_r-filter $\mathcal{F} \subseteq S_S^{\Delta}$ of S_S and $Q_r(S_S^+, \mathcal{F}^{\oplus})$ the Q_r-semiring with respect to the Q_r-filter $\mathcal{F}^{\oplus} = \mathcal{F} \cap \mathfrak{S}_S^+$ of S_S^+. Then $T = Q_r(S_S^+, \mathcal{F})$ is a semiring iff $Q_r(S_S^+, \mathcal{F}) \simeq Q_r(S_S^+, \mathcal{F}^{\oplus})$ holds.

b) Each classical Q_r-seminearring $T = Q_r(S, \Sigma)$ of S satisfies

$$(7.5) \qquad T = Q_r(S,\Sigma) = Q_r(S_S^+, \mathcal{F}_\Sigma) \simeq Q_r(S_S^+, \mathcal{F}_\Sigma^{\oplus})$$

with respect to the Q_r-filter $\mathcal{F}_\Sigma = \{A \in \mathfrak{S}_S \mid A \cap \Sigma \neq \emptyset\}$ of S_S.

Proof a) If $T = Q_r(S_S^+, \mathcal{F})$ is a semiring, then by Thm. 5.6 c) (or by Thms. 7.4 a) and 7.3 c)) each $t = [\mu_t'] \in T = Q_r(S_S^+, \mathcal{F})$ is represented by an S-homomorphism defined on an S-subgroupoid of S_S^+. Thus $T = Q_r(S_S^+, \mathcal{F}, S^{\Delta})$ coincides with its subsemiring U as defined in Prop. 6.1, and $U \simeq Q_r(S_S^+, \mathcal{F}^{\oplus}, S^{\Delta\oplus}) = Q_r(S_S^+, \mathcal{F}^{\oplus})$ follows from Thm. 6.9 a). The converse statement is trivial.

b) As already mentioned behind Thm. 5.2, each classical Q_r-semigroup $(T,\cdot) = Q_r(S,\Sigma)$ of $(S,\cdot)$ may be identified with the Q_r-semigroup $Q_r(S_S, \mathcal{F}_\Sigma)$ for $\mathcal{F}_\Sigma$ as defined above (cf. [39], Thm. 9.1). Due to Thm. 3.2, this transfers to the classical Q_r-seminearring $(T,+,\cdot) = Q_r(S,\Sigma)$ of $(S,+,\cdot)$, clearly with respect to the Q_r-seminearring $Q_r(S_S^+, \mathcal{F}_\Sigma)$. Moreover, if S is a semiring, the same holds for $T = Q_r(S,\Sigma)$ by Thm. 3.5. Hence (7.5) follows from part a).

We already have refered to Thm. 7.6 b) in Expl. 6.10: For $(S,+,\cdot) = (\mathbb{N}_0,+,\cdot)$ and $Q_r(\mathbb{N}_0, \mathbb{N}) = Q_r(S,\Sigma)$, the Q_r-filter

$\mathcal{F}_\Sigma = \{A \in \mathcal{S}_S \mid A \cap \mathbb{N} \neq \emptyset\}$ of S_S consists of all S-subsets $A \neq \{0\}$ of $S = \mathbb{N}_0$ and coincides (in this case) with $\mathcal{F}_\Sigma^\oplus$, denoted by $\mathcal{F}^+$ in the second version of Expl. 6.10. Obviously, we have $\mathcal{F}_\Sigma = \mathcal{F}_\Sigma^\oplus = S_S^\Delta = S_S^{+\Delta}$, and $Q_r(\mathbb{N}_0,\mathbb{N}) \simeq Q_r(S_S^+, S_S^{+\Delta}) = Q_m^+ = Q_m$ is the maximal McMorris extension of $S = \mathbb{N}_0$. Considering $S = \mathbb{N}_0$ in the same way for suitable subsemigroups Σ of $(\mathbb{N},\cdot)$, for instance $\Sigma = \{2^\nu \mid \nu \in \mathbb{N}_0\}$, we clearly obtain McMorris Q_r-semirings $Q_r(\mathbb{N}_0,\Sigma) \simeq Q_r(S_S^+,\mathcal{F}_\Sigma^\oplus)$ which are properly contained in the maximal one. Examples of this kind (e.g. with $\mathbb{Z}$ instead of $\mathbb{N}_0$ and many other ones) have been mentioned concerning the last step in (7.4) in the proof of Thm. 7.4 b).

We close this section with examples of maximal McMorris Q_r-semirings of subcommutative as well as non-subcommutative semirings S and a corresponding problem in the latter case.

Example 7.7 Let S be the additively commutative semiring given by the tables

+	o	a	b
o	o	a	b
a	a	a	b
b	b	b	b

·	o	a	b
o	o	o	o
a	o	a	a
b	o	b	b

.

Clearly, o is the zero of S, but S has no zero-divisors and is right cancellative. Moreover, we have $\{S\} = S_S^\Delta = S_S^{+\Delta}$, and the maximal McMorris Q_r-seminearring $Q_m = Q_r(S_S^+, S_S^\Delta)$ consists of the 9 S-mappings on S_S given by the table

	λ_o	λ_a	λ_b	1	f	g	u	v	w
o	o	o	o	o	o	o	o	o	o
a	o	a	b	a	o	o	b	a	b
b	o	a	b	b	b	a	a	o	o

.

One easily checks that $o = \lambda_o$, $a = \lambda_a$, $b = \lambda_b$ as well as 1, f and g are also S-homomorphisms, hence the elements of the maximal McMorris Q_r-semiring $Q_m^+ = Q_r(S_S^+, S_S^{+\Delta})$. With respect to Thm. 7.3 d) we state $f \cdot g = o$; hence Q_m^+ has zero-divisors and is not right cancellative.

Example 7.8 a) The semiring $(S^*, +, \cdot)$ used in the proof of Thm. 5.6 d) is subcommutative, and we write in this part a) $S^* = S$. By $s_1 + s_2 = s_1$ for all $s_i \in S$, each S-subset A of S_S is also an S-subgroupoid of S_S^+, and oviously dense in S_S. Thus we have $S_S^\Delta = S_S^{+\Delta}$ and also $\Lambda_S(S_S^\Delta, S) = \mathrm{Hom}_S(S_S^{+\Delta}, S)$. Hence, the maximal McMorris Q_r-seminearring $Q_m = Q_r(S_S^+, S_S^\Delta)$ and the maximal McMorris Q_r-semiring $Q_m^+ = Q_r(S_S^+, S_S^{+\Delta})$ coincide. To complete the proof of Thm. 7.3 d) concerning the left cancellativity, we state that S is (even two-sided) cancellative. Now we define an S-homomorphism f on S by $f(a) = f(b) = a$. Then also $f \cdot f = f$ is defined on $S \in S_S^\Delta = S_S^{+\Delta}$, which yields by (4.12) or (4.12⁺) that $[f][f] = [f][1_S] = [1_S][f]$ holds in $Q_m = Q_m^+$, disproving left and right cancellativity.

b) Turning back to the notation $S = S^*$ and $S = S^* \cup \{o\}$ used in the proof of Thm. 5.6 d), the semiring S is not subcommutative. Now we also have $S_S^\Delta = S_S^{+\Delta}$, but $\Lambda_S(S_S^\Delta, S) \supset \mathrm{Hom}_S(S_S^{+\Delta}, S)$, where the latter consists of all S-mappings f which satisfy $f(x) = o \Leftrightarrow x = o$ for each $x \in D_f$. Moreover, again there is a unique maximal McMorris Q_r-semiring Q_m^+ of S, which consists of all S-homomorphisms contained in $Q_m = Q_r(S_S^+, S_S^\Delta)$. So we may write $Q_m^+ = Q_r(S_S^+, S_S^{+\Delta})$ - but note that this notation is only defined for subcommutative semirings S. One also can prove that Q_m^+ is neither left nor right cancellative (take f as above with $f(o) = o$), but Q_m^+ has no zero-divisors (whereas Q_m has those according to the proof of Thm. 5.6 d)).

Example 7.9 a) A much simpler example of a non-subcommutative seminearring $S = \{o,a,b\}$ which also has a unique maximal McMorris Q_r-semiring is given by the tables

+	o	a	b
o	o	a	b
a	a	a	a
b	b	b	b

·	o	a	b
o	o	o	o
a	o	a	a
b	o	b	b

·	o	a	b
o	o	o	o
a	a	a	a
b	b	b	b

with respect to the left hand multiplication. As in Expl. 7.7 we have $\{S\} = S_S^{\Delta} = S_S^{+\Delta}$, and the maximal McMorris Q_r-seminearring $Q_m = Q_r(S_S^+, S_S^{\Delta})$ consists of the same 9 S-mappings as in Expl. 7.7 (in both examples, $(Q_m, \cdot)$ is the maximal McMorris Q_r-semigroup of the same semigroup $(S,\cdot)$, but the additions on S and Q_m differ essentially). The following table gives all 7 endomorphisms of $(S,+)$

(7.6)

	λ_o	λ_a	λ_b	1	u	λ'_a	λ'_b
o	o	o	o	o	o	a	b
a	o	a	b	a	b	a	b
b	o	a	b	b	a	a	b

(cf. [42], § 3). But the last both are no S-mappings, and one easily checks that S, T_1 and T_2 according to

$$S = \{o=\lambda_o, a=\lambda_a, b=\lambda_b\} < S \cup \{1\} < T_2 = S \cup \{1,u\} < Q_r(S_S^+, S_S^{\Delta})$$

are all McMorris Q_r-semirings of S. Hence, $T_2 = Q_m^+$ is the unique maximal one, which consists as in Expl. 7.8 b) of all S-homomorphisms contained in the maximal McMorris Q_r-seminearring Q_m of S. Again one is tempted to extend the notation $Q_m^+ = Q_r(S_S^+, S_S^{+\Delta})$ for all those S-homomorphisms also to non-subcommutative semirings S, but the next part b) shows that there is no corresponding concept behind that.

b) We consider the semiring $S = \{o,a,b\}$ with the right hand multiplication. Since now each element of S_S is fixed, we have $\{S\} = S_S^{\Delta} = S_S^{+\Delta}$ and the maximal McMorris Q_r-seminearring $Q_m = Q_r(S_S^+, S_S^{\Delta})$ consists of all 3^3 mappings of S into S (i.e., $(Q_m, \cdot)$ is the full transformation semigroup on S, cf. [39], Expl. 4.6). The embedding of S by (7.1) into Q_m yields now $o = \lambda_o$, $a = \lambda'_a$ and $b = \lambda'_b$. But in this case the S-homomorphisms contained in Q_m, clearly those listed in (7.6), do not form a subsemiring of Q_m. More than that, one checks that it is not even possible to extend S to a semiring adjoining some of 4 S-homomorphisms λ_a, λ_b, 1 and u. Thus S itself is the maximal McMorris Q_r-semiring Q_m of S.

After this and other experiences we pose the following

Problem 7.10 Let S be any semiring such that $S_S^{\Delta} \neq \emptyset$. Does there exist maximal McMorris Q_r-semirings of S, especially a unique one, say Q_m^+, which contains (up to isomorphisms over S) all McMorris Q_r-semirings T of S corresponding to Q_m^+ in Thm. 7.4 b) for the subcommutative case?
(Note that all these semirings have to be contained in the maximal McMorris Q_r-seminearring $Q_r(S_S^+, S_S^{\Delta})$ by Thm. 5.4 b).)

§ 8 Background for applications to rings

Clearly, all our previous constructions yield rings by suitable assumptions, and each result on semirings is, all the more, applicable to rings. But in this way we consider "rings as semirings", and the central question concerns the interrelation between our constructions and various types of "quotient rings" well known in ring theory (cf. e.g. [2], Chap. II, § 2, [7], [18], [22], [31], and the references there).

To be more precise, let us agree to use "ring" in this last section in the common meaning (i.e. for a semiring S such that (S,+) is a commutative group). Accordingly, we shall speak about a (right) S-module M_S^+ including that S is a ring, (M,+) an abelian group and that

(8.1) $\qquad m(s_1+s_2) = ms_1+ms_2 \quad$ for all $m \in M$, $s_i \in S$

holds (but we do not assume M_S^+ to be unitary, not even if S should have an identity). Then each "quotient ring" of a ring S mentioned above, considered as a right one in the following, has a construction similar to those of Thm. 6.2 as follows. For some S-module M_S^+ (in particular for $M_S^+ = S_S^+$) a set $\mathcal{F}^{\pm}$ of S-submodules of M_S^+ is defined or considered, which satisfies (at least) the conditions (4.5^+) - (4.7^+) for a right quotient filter, clearly with respect to the set of all S-submodules of M_S^+ and to S-homomorphisms. Since the latter concept for S-modules is the same as for S-groupoids (where $S = (S,\cdot)$ is the multiplicative semigroup of the ring S), we may call $\mathcal{F}^{\pm}$ a subtractively closed Q_r-filter of M_S^+, and the procedure described by (4.9^+), (4.10^+) and (4.11^+) as in Thm. 6.2 yields a ring, mostly called quotient ring with some supplementary marking. Of course, we speak here about the Q_r-ring $Q_r(M_S^+,\mathcal{F}^{\pm},K^{\pm})$ of the S-module M_S^+ with respect to $\mathcal{F}^{\pm} \subseteq K^{\pm}$. Fortunately, it is not necessary to check the corresponding proofs; we shall see in Thm. 8.8 that even more general constructions are a consequence of Thm. 6.2. We also anticipate the following answer to the above question to mark the scope of the considerations in this section:[1)]

<u>THEOREM 8.1</u> <u>We assume that</u> S <u>is a ring</u>, M_S^+ <u>an S-module, and</u> $Q_r(M_S^+,\mathcal{F}^\pm,K^\pm)$ <u>the</u> Q_r<u>-ring of the S-module</u> M_S^+ <u>with respect to subtractively closed</u> Q_r<u>-filters</u> $\mathcal{F}^\pm \subseteq K^\pm$ <u>of the S-module</u> M_S^+ <u>as introduced above</u>. <u>Then, considering</u> S <u>as a semiring and</u> M_S^+ <u>as an S-semigroup, there are</u> Q_r<u>-filters</u> $\mathcal{F}^+ \subseteq K^+$ <u>of the S-groupoid</u> M_S^+ <u>such that the related</u> Q_r<u>-semiring</u> $Q_r(M_S^+,\mathcal{F}^+,K^+)$ <u>is, in a natural way, isomorphic to the ring</u> $Q_r(M_S^\pm,\mathcal{F}^\pm,K^\pm)$.

According to the above, the crucial point is the difference between (right) semiring ideals and (right) ring ideals of a ring S, more general, between S-subsemigroups and S-subgroups of an S-module M_S^+.[2)] For this reason we introduce the following restriction for S-subsemigroups, in fact for subgroupoids of an arbitrary groupoid, which generalizes a concept due to [4] (cf. also [17],[21]):

<u>DEFINITION 8.2</u> Let A be a subgroupoid of a groupoid $(M,+)$. Then A is called <u>k-closed</u> (with respect to M) iff

(8.2) $a+x \in A$ or $x+a \in A$ for some $a \in A$, $x \in M$ implies $x \in A$.

---

1) To our best knowledge, all ring-theoretical constructions deal with $\mathcal{F}^\pm = K^\pm$. Defining the latter as the set of $\cap$-large (equivalently: essential) S-submodules of M_S^+, Johnson [18] introduced $Q_r(M_S^+,\mathcal{F}^\pm)$ as the extended centralizer of the ring S over M_S^+. Utumi's quotient ring [31] conforms to $Q_r(S_S^+,\mathcal{F}^\pm)$ with respect to the set $\mathcal{F}^\pm$ of all dense right ideals of S.

2) A sufficient condition such that both kinds of (right) ideals [S-subsemigroups and S-subgroups] coincide is that S has a (right) identity [M_S^+ is unitary].

We further introduce the set $\mathfrak{m}^k$ of all k-closed subgroupoids of (M,+) which contains M by (8.2), and for an S-groupoid M_S^+ the set $\mathfrak{m}_S^k$ of all k-closed S-subgroupoids of M_S^+.

Since $\mathfrak{m}_S^k = \mathfrak{m}_S^+ \cap \mathfrak{m}^k$ holds, the next lemma applies also to k-closed S-subgroupoids of each S-groupoid M_S^+ (where S denotes any semigroup (S,·)), hence in particular to k-closed right ideals of any semiring (S,+,·).

Lemma 8.3 a) The intersection $A \cap B$ of k-closed subgroupoids A, B of a groupoid (M,+) is either empty or a k-closed subgroupoid of (M,+).

b) Let A and B be k-closed subgroupoids of (M,+) and let f be a homomorphism of (A,+) into (M,+). Then $f^{-1}(B)$ is either empty or a k-closed subgroupoid of (M,+).

c) If (M,+) is a group, then a subsemigroup A of (M,+) is k-closed iff it is a subgroup of (M,+).[1)]

Proof We omit to show a) and assume $f^{-1}(B) \neq \emptyset$ for b). Then $f^{-1}(B)$ is a subgroupoid of (M,+). To show $f^{-1}(B) \in \mathfrak{m}^k$, assume $a+x \in f^{-1}(B)$ for some $a \in f^{-1}(B)$ and $x \in M$. Then it follows

$$a+x \in A,\ a \in A \ \text{ and } \ f(a)+f(x) = f(a+x) \in B,\ f(a) \in B,$$

hence $x \in A$ and $f(x) \in B$, since A and B are k-closed, what means $x \in f^{-1}(B)$. The conclusion for $x+a \in f^{-1}(B)$ is the same. Concerning c), let A be a k-closed subsemigroup of a group (M,+). Then A contains the neutral o of (M,+) by $a+o \in A$ for

1) In this context we note the following without proof: Let (M,+) be a semigroup generating a commutative group (G,+). Then for each subgroup U of (G,+), the intersection $U \cap M$ is either empty or a k-closed subsemigroup A of (M,+). Conversely, each k-closed subsemigroup A of (M,+) is obtained as such an intersection $A = U \cap M$ (but does, in general, not determine U in this way uniquely).

some $a \in A$, and $-a \in A$ by $a+(-a) = o \in A$ for all $a \in A$. Conversely, each subgroup A of a group is obviously k-closed.

Recall that Lemma 8.3 also applies to S-groupoids and to seminearrings. We only formulate the following applications of part c), which are instructive for our procedure:

c') For any S-groupoid M_S^+ such that $(M,+)$ is a group, the concepts "k-closed S-subgroupoid of M_S^+" and "S-subgroup of M_S^+" coincide.

c") For a ring S, the concepts "k-closed (right) semiring ideal of S" and "(right) ring ideal of S" coincide.

Consequently, considerations e.g. on semirings S dealing only with k-closed (right) ideals of S will be immediately applicable to rings. In fact, this will be our way to go, for instance to obtain Thm. 8.1, and so we have to include into our considerations of § 6 and § 7 the possibility to restrict them to k-closed S-subgroupoids.

<u>DEFINITION 8.4</u> Let M_S^+ be any S-groupoid and $\mathfrak{M}_S^k = \mathfrak{M}_S^+ \cap \mathfrak{M}^k$ the set of all k-closed S-subgroupoids of M_S^+. Then a <u>k-closed Q_r-filter</u> $\mathcal{F}^k$ <u>of the S-groupoid</u> M_S^+ is defined by $\emptyset \neq \mathcal{F}^k \subseteq \mathfrak{M}_S^k$ and the conditions

(4.5^k) $\quad A \in \mathcal{F}^k,\ A \subseteq B \in \mathfrak{M}_S^k \Rightarrow B \in \mathcal{F}^k$

(4.6^k) $\quad A,B \in \mathcal{F}^k \Rightarrow A \cap B \in \mathcal{F}^k$

(4.7^k) $\quad A,B \in \mathcal{F}^k,\ f \in \mathrm{Hom}_S(A,M) \Rightarrow f^{-1}(B) \in \mathcal{F}^k.$

In particular, if S is a ring and M_S^+ an S-module, then the k-closed Q_r-filters $\mathcal{F}^k$ of M_S^+ are, due to Lemma 8.3c), just the "subtractively closed" Q_r-filters $\mathcal{F}^\pm$ of M_S we have used above to describe the ring-theoretical constructions and to formulate Thm. 8.1. For the next theorem, compare the parallelisms and the differences to Thm. 6.5.

THEOREM 8.5 Let M_S^+ be an S-groupoid. Then we have:

a) Each Q_r-filter $\mathcal{F}^+$ of the S-groupoid M_S^+ determines a k-closed Q_r-filter $\mathcal{F}^{+(k)} = \mathcal{F}^+ \cap \mathfrak{M}_S^k$ of M_S^+ such that $\mathcal{F}^{+(k)} \subseteq \mathcal{F}^+$, the largest one contained in $\mathcal{F}^+$.

b) For each k-closed Q_r-filter $\mathcal{F}^+$ of the S-groupoid M_S^+ there always exists a unique samllest Q_r-filter $\mathcal{F}^{k\#}$ of M_S^+ which contains $\mathcal{F}^k$. It is given by

$$(8.3) \qquad \mathcal{F}^{k\#} = \{ A^+ \in \mathfrak{M}_S^+ \mid A^k \subseteq A^+ \text{ for some } A^k \in \mathcal{F}^k \},$$

and hence clearly the intersection of all Q_r-filters $\mathcal{F}^+$ of M_S^+ satisfying $\mathcal{F}^k \subseteq \mathcal{F}^+$.

c) Applying both procedures one after the other, we get

$$(8.4) \qquad \mathcal{F}^+ \supseteq (\mathcal{F}^+)^{(k)\#} = (\mathcal{F}^+ \cap \mathfrak{M}_S^k)^\#$$

for each Q_r-filter $\mathcal{F}^+$ of M_S^+, and

$$(8.5) \qquad \mathcal{F}^k = (\mathcal{F}^k)^{\#(k)} = \mathcal{F}^{k\#} \cap \mathfrak{M}_S^k$$

for each k-closed Q_r-filter $\mathcal{F}^k$ of M_S^+. The inclusion (8.4) may be proper, even for S-groupoids $M_S^+ = S_S^+$ obtained from a semiring S with identity whose both operations are associative and commutative. Finally, for three steps one has

$$(8.6) \qquad (\mathcal{F}^+)^{(k)\#(k)} = (\mathcal{F}^+)^{(k)} \text{ and } (\mathcal{F}^k)^{\#(k)\#} = (\mathcal{F}^k)^\#.$$

d) Let M_S^+ be an S-groupoid such that $(M,+)$ is a commutative group and that $oS = o$ holds for the neutral o of $(M,+)$, in particular an S-module M_S over a ring S. Then (8.4) holds with equality for each Q_r-filter $\mathcal{F}^+$ of M_S^+.

Proof a) In general, we have $\mathcal{F}^{+(k)} = \mathcal{F}^{+} \cap \mathfrak{M}_S^k = \mathcal{F}^{+} \cap \mathfrak{M}^k$. Then (4.5^k) and (4.6^k) for $\mathcal{F}^{+(k)}$ are obvious from the corresponding properties of $\mathcal{F}^{+}$ and Lemma 8.3 a).[1] To show (4.7^k), we assume $A,B \in \mathcal{F}^{+(k)}$ and $f \in \mathrm{Hom}_S(A,M)$. Then $f^{-1}(B) \in \mathcal{F}^{+}$ holds by (4.7^{+}), and $f^{-1}(B) \neq \emptyset$ and Lemma 8.3 b) imply $f^{-1}(B) \in \mathfrak{M}^k$, hence $f^{-1}(B) \in \mathcal{F}^{+(k)}$.

b) The set $\mathcal{F}^{k\#}$ defined by (8.3) consists at least of all $A^k \in \mathcal{F}^k$, and it obviously satisfies (4.5^{+}) and (4.6^{+}). To show (4.7^{+}), we assume $A^{+},B^{+} \in \mathcal{F}^{k\#}$ and $f \in \mathrm{Hom}_S(A^{+},M)$. Then there are $A^k, B^k \in \mathcal{F}^k$ such that $A^k \subseteq A^{+}$ and $B^k \subseteq B^{+}$, and the restriction $f|A^k$ of f to A^k satisfies $f|A^k \in \mathrm{Hom}_S(A^k,M)$. The latter yields $(f|A^k)^{-1}(B^k) \in \mathcal{F}^k$ by (4.7^k) for $\mathcal{F}^k$, hence

$$f^{-1}(B^{+}) \supseteq f^{-1}(B^k) \supseteq (f|A^k)^{-1}(B^k) \in \mathcal{F}^k$$

and $f^{-1}(B^{+}) \in \mathfrak{M}_S^{+}$ imply $f^{-1}(B^{+}) \in \mathcal{F}^{k\#}$ by (8.3). Thus $\mathcal{F}^{k\#}$ is a Q_r-filter of the S-groupoid M_S^{+} which contains $\mathcal{F}^k$, and by (8.3) and (4.5^{+}) it is the smallest one of this kind.

c) The inclusion (8.4) and $\mathcal{F}^k \subseteq (\mathcal{F}^k)^{\#(k)}$ of (8.5) are clear by the statements a) and b). For the converse inclusion of (8.5), assume $A^{+} \in (\mathcal{F}^k)^{\#(k)} = \mathcal{F}^{k\#} \cap \mathfrak{M}_S^k$. By (8.3) and $A^{+} \in \mathcal{F}^{k\#}$, there exists some $A^k \in \mathcal{F}^k$ such that $A^k \subseteq A^{+}$. Using $A^{+} \in \mathfrak{M}_S^k$ and (4.5^k) for $\mathcal{F}^k$, we obtain $A^{+} \in \mathcal{F}^k$, the other inclusion of (8.5). That (8.4) may be proper, even for S-groupoids $M_S^{+} = S_S^{+}$ as described above, will be shown in Expl. 8.7. Finally, the right equality of (8.6) is now immediate by (8.5), whereas the check of the left one is routine.

---

1) Again one may use that (4.6^k) is a consequence of (4.7^k), due to $A \cap B = (1_A)^{-1}(B)$. On the other hand, (4.5^k), (4.6^k) and (4.7^k) weakened by $A = B$ define the same concept as Def. 8.4.

d) It is clear that an S-module satisfies $oS = o$ by (8.1). Thus it remains to show $\mathcal{F}^+ \subseteq (\mathcal{F}^+)^{(k)\#}$ for an S-groupoid M_S^+, using only that $(M,+)$ is an abelian group and $oS = o$. But these assumptions are sufficient to check that $f(x) = -x$ for all $x \in M$ defines an S-homomorphism $f \in \mathrm{Hom}_S(M,M)$. Further, for each $A^+ \in \mathcal{F}^+$ one has

$$f^{-1}(A^+) = \{y \in M \mid f(y) = -y = x \in A^+\} = -A^+ \in \mathcal{F}^+,$$

and hence $B^+ = A^+ \cap -A^+ \in \mathcal{F}^+$ by (4.7^+) and (4.6^+). Now $B^+ = B^k$ is obviously k-closed (a subgroup of $(M,+)$ by Lemma 8.3 c)), which implies $B^k \in \mathcal{F}^{+(k)} = \mathcal{F}^+ \cap \mathfrak{M}_S^k$. Applying (8.3) to $\mathcal{F}^{+(k)}$, from $B^k \subseteq A^+ \in \mathfrak{M}_S^+$ one obtains $A^+ \in (\mathcal{F}^{+(k)})^\# = (\mathcal{F}^+)^{(k)\#}$ as we were to show.

<u>Remark 8.6</u> Applying Thm. 8.5 a) to the Q_r-filter $\mathcal{F}^+ = M_S^{+\Delta}$ of all dense S-subgroupoids of any S-groupoid M_S^+ (provided $M_S^\Delta \neq \emptyset$ and hence $M_S^{+\Delta} \neq \emptyset$), we obtain the k-closed Q_r-filter $(M_S^{+\Delta})^{(k)}$ of all k-closed dense S-subgroupoids of M_S which we denote by $M_S^{k\Delta}$. Then we have $M_S^{k\Delta} = (M_S^{+\Delta})^{(k)} \subseteq M_S^{+\Delta}$ and

(8.5^Δ) $\quad M_S^{k\Delta} = (M_S^{k\Delta})^{\#(k)}$ iff M_S is dense in M_S.

On the other hand, concerning (8.4), one merely can state

(8.4^Δ) $\quad M_S^{+\Delta} \supseteq (M_S^{+\Delta})^{(k)\#} = (M_S^{k\Delta})^\#$ iff M_S is dense in M_S,

and we shall prove that <u>also this inclusion may be proper, even for S-groupoids $M_S^+ = S_S^+$ obtained from a semiring S with identity whose both operations are associative and commutative</u>. On the other hand, we clearly have <u>equality in</u> (8.4^Δ) <u>if M_S^+ satisfies the assumptions of Thm. 8.5 d)</u>.

<u>Example 8.7</u> For simplification, we use a subsemiring S of Expl. 2.3, given by $S = \{2^n \mid n \in \mathbb{N}_0\}$ with the usual multipli-

cation and $2^n+2^m = 2^{\max(n,m)}$. Clearly, both operations of $(S,+,\cdot)$ are associative and commutative, 2^0 is the identity (and the zero) of $(S,+,\cdot)$, and all ideals of $(S,\cdot)$ are given by $A_m = \{ 2^{m+x} \mid x \in \mathbb{N}_0 \}$ for all $m \in \mathbb{N}_0$. Further, each A_m is dense in S_S, and also a subsemigroup of $(S,+)$. Thus, for the S-groupoid S_S^+, the Q_r-filter $S_S^{+\Delta}$ of all dense S-subsemigroups is the set $\mathcal{S}_S^+$ of all S-subsemigroups of S_S^+ (all semiring ideals of $(S,+,\cdot)$). Now suppose that A_m is k-closed. Then $2^m+2^0 = 2^m \in A_m$ and $2^0 \in S$ imply $2^0 \in A_m$, i.e. $A_m = A_0 = S$. Therefore, the k-closed Q_r-filter $S_S^{k\Delta} = (S_S^{+\Delta})^{(k)}$ coincides with $\{S\}$, which implies $\{S\}^\# = \{S\}$ by (8.3). Hence (8.4^Δ) reads in this case

$$\mathcal{S}_S^+ = S_S^{+\Delta} \supset (S_S^{+\Delta})^{(k)\#} = (S_S^{k\Delta})^\# = \{S\}^\# = \{S\},$$

which also proves that the inclusion (8.4) may be proper.

Now we come to the main result of this section, which means the following for any subcommutative S-groupoid M_S^+:

1) The construction of the Q_r-semiring $Q_r(M_S^+,\mathcal{F}^+,\mathcal{K}^+)$ given in Thm. 6.2 can be repeated with the restriction to k-closed Q_r-filters $\mathcal{F}^k \subseteq \mathcal{K}^k$ instead of $\mathcal{F}^+ \subseteq \mathcal{K}^+$.

2) Each Q_r-semiring $Q_r(M_S^+,\mathcal{F}^k,\mathcal{K}^k)$ obtained in that way, in particular any "quotient ring" quoted above, is isomorphic to the Q_r-semiring $Q_r(M_S^+,\mathcal{F}^{k\#},\mathcal{K}^{k\#})$ already given by Thm. 6.2.

3) The proof of these statements reduces essentially to the proof that there is such an isomorphism as claimed in 2).

In this context we note that for any S-homomorphisms f,g the set $C_{f,g} = \{ x \in D_f \cap D_g \mid f(x) = g(x) \}$ is always an S-subgroupoid of M_S^+, but in general not a k-closed one (not even if

$D_f, D_g \in \mathfrak{M}_S^k$). By the first statement and by (4.5$^+$) for $\mathcal{K}^+$, we may replace (4.11$^+$) in Thm. 6.2 by the equivalent definition

$\triangleq$ (4.11$^+$) $\quad f \equiv g\ (\kappa_{\mathcal{K}^+})$ iff $f(x) = g(x)$ holds for all $x \in K^+$ for some $K^+ \in \mathcal{K}^+$,

and we shall use this formulation in the following.

THEOREM 8.8 Let M_S^+ be a subcommutative S-groupoid, $\mathcal{F}^k \subseteq \mathcal{K}^k$ k-closed Q_r-filters of M_S^+, and $\mathrm{Hom}_S(\mathcal{F}^k, M_S^+)$ the set of all S-homomorphisms f, defined on some k-closed S-subgroupoid $D_f \in \mathcal{F}^k$, into M_S^+. Then $(\mathrm{Hom}_S(\mathcal{F}^k, M_S^+), +, \cdot)$ becomes an algebra with operations $+, \cdot$ defined by

(4.9^k) $\quad (f+g)(x) = f(x) + g(x)$ for all $x \in D_f \cap D_g \in \mathcal{F}^k$,

(4.10^k) $\quad (f \cdot g)(x) = f(g(x))$ for all $x \in g^{-1}D_f \in \mathcal{F}^k$,

and we obtain a congruence on $(\mathrm{Hom}_S(\mathcal{F}^k, M_S^+), +, \cdot)$ by

(4.11^k) $\quad f \equiv g\ (\kappa_{\mathcal{K}^k})$ iff $f(x) = g(x)$ holds for all $x \in K^k$ for some $K^k \in \mathcal{K}^k$.

We denote by $Q^k = Q_r(M_S^+, \mathcal{F}^k, \mathcal{K}^k)$ the resulting congruence-class-algebra, which consists of the classes $[f] = [f]_{\mathcal{F}^k, \mathcal{K}^k}$ and state that $(Q^k, +, \cdot)$ is a semiring which is isomorphic to the Q_r-semiring $(Q, +, \cdot) = Q_r(M_S^+, \mathcal{F}^{k\#}, \mathcal{K}^{k\#})$ with respect to the Q_r-filters $\mathcal{F}^{k\#} \subseteq \mathcal{K}^{k\#}$ of M_S^+ (cf. Def. 6.3 and Thm. 8.5 b). This isomorphism is defined by

(8.7) $\quad [f]_{\mathcal{F}^{k\#}, \mathcal{K}^{k\#}} \to [f]_{\mathcal{F}^k, \mathcal{K}^k}$ for all $[f] \in Q$

and means that Q^k consists essentially of the same classes as Q, but without those representatives f whose domain D_f is not k-closed, i.e. $D_f \in \mathcal{F}^{k\#} \setminus \mathcal{F}^k$.

Proof Since $\mathcal{F}^k \subseteq \mathcal{F}^{k\#}$ (cf. (8.3)), it is known by Thm. 6.2 a) that $f+g$ and $f \cdot g$ as defined in (4.9^k) and (4.10^k) are S-homomorphisms, and their domains are contained in $\mathcal{F}^k$ by (4.6^k) and (4.7^k). Hence $(\mathrm{Hom}_S(\mathcal{F}^k, M_S^+), +, \cdot)$, shorted by $(H, +, \cdot)$ for the moment, is an algebra. All we have to do now is to show that the mapping φ defined by

$$(8.8) \qquad f \to \varphi(f) = [f] = [f]_{\mathcal{F}^{k\#}, \mathcal{K}^{k\#}} \quad \text{for all } f \in H$$

is a homomorphism of $(H, +, \cdot)$ onto the semiring $(Q, +, \cdot)$, and that $\varphi(f) = \varphi(g)$ is equivalent to (4.11^k). To see that φ is surjective, let $[f]_{\mathcal{F}^{k\#}, \mathcal{K}^{k\#}}$ be any element of Q and $D_f \in \mathcal{F}^{k\#}$. By (8.3), there exists some $A^k \in \mathcal{F}^k$ such that $A^k \subseteq D_f$, and the restriction $\mathfrak{f} = f \mid A^k$ of f to $A^k = D_{\mathfrak{f}} \in \mathcal{F}^k \subseteq \mathcal{F}^{k\#}$ satisfies $\mathfrak{f} \equiv f\ (\kappa_{\mathcal{K}^{k\#}})$, since both coincide on $D_{\mathfrak{f}} \in \mathcal{F}^{k\#}$ and $\mathcal{F}^k \subseteq \mathcal{K}^k$ implies $\mathcal{F}^{k\#} \subseteq \mathcal{K}^{k\#}$ by (8.3). Thus $\mathfrak{f}$ is mapped to $[\mathfrak{f}] = [f] \in Q$ and φ is onto. Moreover, φ is compatible with the operations defined by (4.9^k) and (4.10^k) on $(H, +, \cdot)$ and the operations on $(Q, +, \cdot)$, defined by representatives and (4.9^+) and (4.10^+).

So we obtain (8.7) and all remaining statements of Thm. 8.8 if we show that $\varphi(f) = \varphi(g)$ defines the relation $f \equiv g\ (\kappa_{\mathcal{K}^k})$ according to (4.11^k). Assuming the latter, f and g coincide on some $K^k \in \mathcal{K}^k \subseteq \mathcal{K}^{k\#}$ which yields $f \equiv g\ (\kappa_{\mathcal{K}^{k\#}})$, i.e. $[f] = [g]$ in Q or $\varphi(f) = \varphi(g)$ by (8.8). Conversely, $\varphi(f) = \varphi(g)$ implies $[f] = [g]$ in Q and hence $f(x) = g(x)$ for all $x \in K^+$ for some $K^+ \in \mathcal{K}^{k\#}$. Again by (8.3), for $K^+ \in \mathcal{K}^{k\#}$ there is some $K^k \in \mathcal{K}^k$ such that $K^k \subseteq K^+$. All the more, we have $f(x) = g(x)$ for all $x \in K^k \in \mathcal{K}^k$, which is (4.11^k).

To sum up, the concept of k-closed Q_r-filters of M_S^+ and Thm. 8.8 yield, corresponding to the Thms. 4.2 and 6.2, the third construction of this kind. According to the Defs. 4.3

and 6.3, we give the

DEFINITION 8.9 Let M_S^+ be a subcommutative S-groupoid and $\mathcal{F}^k \subseteq \mathcal{K}^k$ k-closed Q_r-filters of M_S^+. Then we call the semiring $Q^k = Q_r(M_S^+, \mathcal{F}^k, \mathcal{K}^k)$ obtained in Thm. 8.8 the Q_r-semiring of M_S^+ with respect to $\mathcal{F}^k$ and $\mathcal{K}^k$.

In particular, if $\mathcal{F}^k = \mathcal{K}^k$, we write $Q_r(M_S^+, \mathcal{F}^k, \mathcal{F}^k) = Q_r(M_S^+, \mathcal{F}^k)$ and speak of the Q_r-semiring of M_S^+ with respect to $\mathcal{F}^k$.

There is no doubt that also this third concept has its own interest - it is just the one which contains the first constructions appeared in this context by Johnson [18] and Utumi [31] as special cases. On the other hand, we must not investigate these Q_r-semirings $Q^k = Q_r(M_S^+, \mathcal{F}^k, \mathcal{K}^k)$ in the same way as we did with the Q_r-semirings $Q_r(M_S^+, \mathcal{F}^+, \mathcal{K}^+)$, since the natural isomorphism (8.7) transfers all results to Q^k which are applicable to $Q_r(M_S^+, \mathcal{F}^{k\#}, \mathcal{K}^{k\#})$ (Thm. 6.4 and most of § 7). So it merely remains to complete Thm. 8.8, according to the Thms. 4.2 and 6.2, for the case of dense k-closed Q_r-filters:

COROLLARY 8.10 Let M_S^+ be a subcommutative S-groupoid such that $M_S^\Delta \neq \emptyset$ holds, $\mathcal{F}^k \subseteq \mathcal{K}^k \subseteq M_S^{k\Delta}$ k-closed Q_r-filters of M_S^+ contained in $M_S^{k\Delta} = M_S^\Delta \cap \mathfrak{m}^k$, and $Q^k = Q_r(M_S^+, \mathcal{F}^k, \mathcal{K}^k)$ the corresponding Q_r-semiring. Then Q^k does not depend on the choice of $\mathcal{K}^k$ between $\mathcal{F}^k$ and $M_S^{k\Delta}$ according to

$$(4.12^k) \qquad f \equiv g \ (\kappa_{\mathcal{K}^k}) \iff f \equiv g \ (\kappa_{\mathcal{F}^k}) \iff C_{f,g} = D_f \cap D_g \in \mathfrak{m}_S^k,$$

and we have in particular

$$Q^k = Q_r(M_S^+, \mathcal{F}^k, M_S^{k\Delta}) = Q_r(M_S^+, \mathcal{F}^k, \mathcal{F}^k) = Q_r(M_S^+, \mathcal{F}^k).$$

Proof Note that $\mathcal{F}^{k\#} \subseteq \mathcal{K}^{k\#} \subseteq (M_S^{k\Delta})^\# \subseteq M_S^{+\Delta}$ follows from the assumption $\mathcal{F}^k \subseteq \mathcal{K}^k \subseteq M_S^{k\Delta}$ and (8.4^Δ) in Remark 8.6. For each k-closed Q_r-filter K^k between $\mathcal{F}^k$ and $M_S^{k\Delta}$, including $\mathcal{F}^k = K^k$,

$$f \equiv g\,(\kappa_{K^k}) \Leftrightarrow f \equiv g\,(\kappa_{K^{k\#}}) \quad \text{for all } f,g \in \mathrm{Hom}_S(\mathcal{F}^k, M_S^+)$$

holds by Thm. 8.8. Since the right side is, due to (4.12^+), also equivalent to $f \equiv g\,(\kappa_{\mathcal{F}^{k\#}})$, we obtain the first equivalence in (4.12^k). Also by (4.12^+), the latter is equivalent to $C_{f,g} = D_f \cap D_g$ (even for $f,g \in \mathrm{Hom}_S(\mathcal{F}^{k\#}, M_S^+)$). Thus $\mathcal{F}^k \subseteq \mathcal{F}^{k\#}$ and Lemma 8.3 a) imply the second equivalence in (4.12^k).

REFERENCES

[1] Botero de Meza, M.M., Zur Theorie allgemeincr Rechtsquotientenhalbgruppen und Rechtsquotientenhalbringe, Dissertation, Technische Universität Clausthal (1975).

[2] Bourbaki, N., Eléments de Mathématique, 27, Algèbre Commutative, Chap. I, II, Paris (1961).

[3] Bourne, S., The Jacobson radical of a semiring, Proc. Nat. Acad. Sci. USA 37 (1951), 163-170.

[4] Bourne, S. and H. Zassenhaus, On the semiradical of a semiring, Proc. Nat. Acad. Sci. USA 44 (1958), 907-914.

[5] Chew, K.L. and G.H. Chan, On extensions of near-rings, Nanta Math. 5 (1971), 12-21.

[6] Clifford, A.H. and G.B. Preston, The algebraic theory of semigroups, Amer. Math. Soc. Vol.I (1961), Vol.II (1967).

[7] Faith, C., Lectures on injective modules and quotient rings, Lecture Notes in Mathematics 49 (1967).

[8] Fuchs, L., Teilweise geordnete algebraische Strukturen, Vandenhoeck & Ruprecht (1966).

[9] Grillet, M.P., Embedding of a semiring into a semiring with identity, Acta Math. Acad. Sci. Hungar. 20 (1969), 121-128.

[10] Hebisch, U., On right quotient filters of semigroups, this Lecture Notes.

[11] Hinkle, C.V., Jr., Semigroups of right quotients of a semigroup which is a semilattice of groups, Semigroup Forum 5 (1972), 167-173.

[12] Hogewijs, H., Semi-nearrings-embedding, Med. Konink. Acad. Wetensch. Lett. Schone Kunst, België Kl. Wetensch. 32 (1970), 3-11.

[13] van Hoorn, W.G., Some generalizations of the Jacobson radical for seminearrings and semirings, Math. Z. 118 (1970), 69-82.

[14] van Hoorn, W.G. and G. van Rootselaar, Fundamental notions in the theory of seminearrings, Comp. Math. 18 (1967), 65-78.

[15] Huq, S.A., The semiring of quotients of commutative semirings, Colloq. Math. 28 (1973), 185-188.

[16] Iizuka, K., On the Jacobson radical of a semiring, Tohoku Math. J. 2 (1959), 409-421.

[17] Iséki, K. and Y. Miyanaga, Notes on topological spaces III. On space of maximal ideals of semiring, Proc. Japan Acad. 32 (1956), 325-328.

[18] Johnson, R.E., The extended centralizer of a ring over a modul, Proc. Amer. Math. Soc. 2 (1951), 891-895.

[19] Karzel, H., Zusammenhänge zwischen Fastbereichen, scharf zweifach transitiven Permutationsgruppen und 2-Strukturen mit Rechtecksaxiom, Abh. Math. Sem. Univ. Hamburg 32 (1969), 191-206.

[20] Kerby, W. and H. Wefelscheid, Bemerkungen über Fastbereiche und scharf zweifach transitive Gruppen, Abh. Math. Sem. Univ. Hamburg 37 (1972), 20-29.

[21] LaTorre, D.R., On h-ideals and k-ideals in hemirings, Publ. Math. Debrecen 12 (1965), 219-226.

[22] Luedeman, J.K., A generalization of the concept of a ring of quotients, Canad. Math. Bull. 14 (1971) 517-529.

[23] Luedeman, J.K., Torsion theories and semigroups of quotients, this Lecture Notes.

[24] McMorris, F.R., The maximal quotient semigroup, Semigroup Forum 4 (1972), 360-364.

[25] Pilz, G., Geordnete Fastringe, Abh. Math. Sem. Univ. Hamburg 35 (1970), 83-88.

[26] Pilz, G., Near-Rings, North-Holland Publ. Comp. (1977).

[27] Rédei, L., Die Verallgemeinerung der Schreierschen Erweiterungstheorie, Acta Sci. Math. (Szeged) 14 (1952), 252-273.

[28] Rédei, L., Algebra, Budapest (1954), Leipzig (1959), London (1967).

[29] van Rootselaar, B., Algebraische Kennzeichnung freier Wortarithmetiken, Comp. Math. 15 (1963), 156-186.

[30] Steinfeld, O., Über die Struktursätze der Semiringe, Acta Math. Acad. Sci. Hungar. 10 (1959), 149-155.

[31] Utumi, Y., On quotient rings, Osaka Math. J. 8 (1956), 1-18.

[32] Vandiver, H.S., Note on a simple type of algebra in which the cancellation law of addition does not hold, Bull. Amer. Math. Soc. 40 (1934), 914-920.

[33] Wefelscheid, H., ZT-subgroups of sharply 3-transitive groups, Proc. Edinburgh Math. Soc. (2) 23 (1980), 9-14.

[34] Weinert, H.J., Über Halbringe und Halbkörper II, Acta Math. Acad. Sci. Hungar. 14 (1963), 209-227.

[35] Weinert, H.J., Ringe mit nichtkommutativer Addition I, Jber. Deutsch. Math.-Verein. 77 (1975), 10-27.

[36] Weinert, H.J., Zur Erweiterung algebraischer Strukturen durch Rechtsquotientenbildung, Acta Math. Acad. Sci. Hungar. 16 (1965), 213-214.

[37] Weinert, H.J., On the extension of partial orders on semigroups of right quotients, Trans. Amer. Math. Soc. 142 (1969), 345-353.

[38] Weinert, H.J., Related representation theorems for rings, semirings, near-rings and semi-near-rings by partial transformations and partial endomorphisms, Proc. Edinburgh Math. Soc. 20 (1976-77), 307-315.

[39] Weinert, H.J., S-sets and semigroups of quotients, Semigroup Forum 19 (1980), 1-78.

[40] Weinert, H.J., Seminearrings, seminearfields and their semigroup-theoretical background, Semigroup Forum 24 (1982), 231-254.

[41] Weinert, H.J., Zur Theorie der Halbfastkörper, to appear in Studia Sci. Math. Hungar.

[42] Weinert, H.J. and R.D. Griepentrog, Embedding semirings by translational hulls, Semigroup Forum 14 (1977), 235-246.

Institut für Mathematik
Technische Universität Clausthal
Erzstraße 1
D-3392 Clausthal-Zellerfeld

Vol. 845: A. Tannenbaum, Invariance and System Theory: Algebraic and Geometric Aspects. X, 161 pages. 1981.

Vol. 846: Ordinary and Partial Differential Equations, Proceedings. Edited by W. N. Everitt and B. D. Sleeman. XIV, 384 pages. 1981.

Vol. 847: U. Koschorke, Vector Fields and Other Vector Bundle Morphisms – A Singularity Approach. IV, 304 pages. 1981.

Vol. 848: Algebra, Carbondale 1980. Proceedings. Ed. by R. K. Amayo. VI, 298 pages. 1981.

Vol. 849: P. Major, Multiple Wiener-Itô Integrals. VII, 127 pages. 1981.

Vol. 850: Séminaire de Probabilités XV. 1979/80. Avec table générale des exposés de 1966/67 à 1978/79. Edited by J. Azéma and M. Yor. IV, 704 pages. 1981.

Vol. 851: Stochastic Integrals. Proceedings, 1980. Edited by D. Williams. IX, 540 pages. 1981.

Vol. 852: L. Schwartz, Geometry and Probability in Banach Spaces. X, 101 pages. 1981.

Vol. 853: N. Boboc, G. Bucur, A. Cornea, Order and Convexity in Potential Theory: H-Cones. IV, 286 pages. 1981.

Vol. 854: Algebraic K-Theory. Evanston 1980. Proceedings. Edited by E. M. Friedlander and M. R. Stein. V, 517 pages. 1981.

Vol. 855: Semigroups. Proceedings 1978. Edited by H. Jürgensen, M. Petrich and H. J. Weinert. V, 221 pages. 1981.

Vol. 856: R. Lascar, Propagation des Singularités des Solutions d'Equations Pseudo-Différentielles à Caractéristiques de Multiplicités Variables. VIII, 237 pages. 1981.

Vol. 857: M. Miyanishi. Non-complete Algebraic Surfaces. XVIII, 244 pages. 1981.

Vol. 858: E. A. Coddington, H. S. V. de Snoo: Regular Boundary Value Problems Associated with Pairs of Ordinary Differential Expressions. V, 225 pages. 1981.

Vol. 859: Logic Year 1979-80. Proceedings. Edited by M. Lerman, J. Schmerl and R. Soare. VIII, 326 pages. 1981.

Vol. 860: Probability in Banach Spaces III. Proceedings, 1980. Edited by A. Beck. VI, 329 pages. 1981.

Vol. 861: Analytical Methods in Probability Theory. Proceedings 1980. Edited by D. Dugué, E. Lukacs, V. K. Rohatgi. X, 183 pages. 1981.

Vol. 862: Algebraic Geometry. Proceedings 1980. Edited by A. Libgober and P. Wagreich. V, 281 pages. 1981.

Vol. 863: Processus Aléatoires à Deux Indices. Proceedings, 1980. Edited by H. Korezlioglu, G. Mazziotto and J. Szpirglas. V, 274 pages. 1981.

Vol. 864: Complex Analysis and Spectral Theory. Proceedings, 1979/80. Edited by V. P. Havin and N. K. Nikol'skii, VI, 480 pages. 1981.

Vol. 865: R. W. Bruggeman, Fourier Coefficients of Automorphic Forms. III, 201 pages. 1981.

Vol. 866: J.-M. Bismut, Mécanique Aléatoire. XVI, 563 pages. 1981.

Vol. 867: Séminaire d'Algèbre Paul Dubreil et Marie-Paule Malliavin. Proceedings, 1980. Edited by M.-P. Malliavin. V, 476 pages. 1981.

Vol. 868: Surfaces Algébriques. Proceedings 1976-78. Edited by J. Giraud, L. Illusie et M. Raynaud. V, 314 pages. 1981.

Vol. 869: A. V. Zelevinsky, Representations of Finite Classical Groups. IV, 184 pages. 1981.

Vol. 870: Shape Theory and Geometric Topology. Proceedings, 1981. Edited by S. Mardešić and J. Segal. V, 265 pages. 1981.

Vol. 871: Continuous Lattices. Proceedings, 1979. Edited by B. Banaschewski and R.-E. Hoffmann. X, 413 pages. 1981.

Vol. 872: Set Theory and Model Theory. Proceedings, 1979. Edited by R. B. Jensen and A. Prestel. V, 174 pages. 1981.

Vol. 873: Constructive Mathematics, Proceedings, 1980. Edited by F. Richman. VII, 347 pages. 1981.

Vol. 874: Abelian Group Theory. Proceedings, 1981. Edited by R. Göbel and E. Walker. XXI, 447 pages. 1981.

Vol. 875: H. Zieschang, Finite Groups of Mapping Classes of Surfaces. VIII, 340 pages. 1981.

Vol. 876: J. P. Bickel, N. El Karoui and M. Yor. Ecole d'Eté de Probabilités de Saint-Flour IX – 1979. Edited by P. L. Hennequin. XI, 280 pages. 1981.

Vol. 877: J. Erven, B.-J. Falkowski, Low Order Cohomology and Applications. VI, 126 pages. 1981.

Vol. 878: Numerical Solution of Nonlinear Equations. Proceedings, 1980. Edited by E. L. Allgower, K. Glashoff, and H.-O. Peitgen. XIV, 440 pages. 1981.

Vol. 879: V. V. Sazonov, Normal Approximation – Some Recent Advances. VII, 105 pages. 1981.

Vol. 880: Non Commutative Harmonic Analysis and Lie Groups. Proceedings, 1980. Edited by J. Carmona and M. Vergne. IV, 553 pages. 1981.

Vol. 881: R. Lutz, M. Goze, Nonstandard Analysis. XIV, 261 pages. 1981.

Vol. 882: Integral Representations and Applications. Proceedings, 1980. Edited by K. Roggenkamp. XII, 479 pages. 1981.

Vol. 883: Cylindric Set Algebras. By L. Henkin, J. D. Monk, A. Tarski, H. Andréka, and I. Németi. VII, 323 pages. 1981.

Vol. 884: Combinatorial Mathematics VIII. Proceedings, 1980. Edited by K. L. McAvaney. XIII, 359 pages. 1981.

Vol. 885: Combinatorics and Graph Theory. Edited by S. B. Rao. Proceedings, 1980. VII, 500 pages. 1981.

Vol. 886: Fixed Point Theory. Proceedings, 1980. Edited by E. Fadell and G. Fournier. XII, 511 pages. 1981.

Vol. 887: F. van Oystaeyen, A. Verschoren, Non-commutative Algebraic Geometry, VI, 404 pages. 1981.

Vol. 888: Padé Approximation and its Applications. Proceedings, 1980. Edited by M. G. de Bruin and H. van Rossum. VI, 383 pages. 1981.

Vol. 889: J. Bourgain, New Classes of $\mathcal{L}^p$-Spaces. V, 143 pages. 1981.

Vol. 890: Model Theory and Arithmetic. Proceedings, 1979/80. Edited by C. Berline, K. McAloon, and J.-P. Ressayre. VI, 306 pages. 1981.

Vol. 891: Logic Symposia, Hakone, 1979, 1980. Proceedings, 1979, 1980. Edited by G. H. Müller, G. Takeuti, and T. Tugué. XI, 394 pages. 1981.

Vol. 892: H. Cajar, Billingsley Dimension in Probability Spaces. III, 106 pages. 1981.

Vol. 893: Geometries and Groups. Proceedings. Edited by M. Aigner and D. Jungnickel. X, 250 pages. 1981.

Vol. 894: Geometry Symposium. Utrecht 1980, Proceedings. Edited by E. Looijenga, D. Siersma, and F. Takens. V, 153 pages. 1981.

Vol. 895: J.A. Hillman, Alexander Ideals of Links. V, 178 pages. 1981.

Vol. 896: B. Angéniol, Familles de Cycles Algébriques – Schéma de Chow. VI, 140 pages. 1981.

Vol. 897: W. Buchholz, S. Feferman, W. Pohlers, W. Sieg, Iterated Inductive Definitions and Subsystems of Analysis: Recent Proof-Theoretical Studies. V, 383 pages. 1981.

Vol. 898: Dynamical Systems and Turbulence, Warwick, 1980. Proceedings. Edited by D. Rand and L.-S. Young. VI, 390 pages. 1981.

Vol. 899: Analytic Number Theory. Proceedings, 1980. Edited by M.I. Knopp. X, 478 pages. 1981.

Vol. 900: P. Deligne, J. S. Milne, A. Ogus, and K.-Y. Shih, Hodge Cycles, Motives, and Shimura Varieties. V, 414 pages. 1982.

Vol. 901: Séminaire Bourbaki vol. 1980/81 Exposés 561–578. III, 299 pages. 1981.

Vol. 902: F. Dumortier, P.R. Rodrigues, and R. Roussarie, Germs of Diffeomorphisms in the Plane. IV, 197 pages. 1981.

Vol. 903: Representations of Algebras. Proceedings, 1980. Edited by M. Auslander and E. Lluis. XV, 371 pages. 1981.

Vol. 904: K. Donner, Extension of Positive Operators and Korovkin Theorems. XII, 182 pages. 1982.

Vol. 905: Differential Geometric Methods in Mathematical Physics. Proceedings, 1980. Edited by H.-D. Doebner, S.J. Andersson, and H.R. Petry. VI, 309 pages. 1982.

Vol. 906: Séminaire de Théorie du Potentiel, Paris, No. 6. Proceedings. Edité par F. Hirsch et G. Mokobodzki. IV, 328 pages. 1982.

Vol. 907: P. Schenzel, Dualisierende Komplexe in der lokalen Algebra und Buchsbaum-Ringe. VII, 161 Seiten. 1982.

Vol. 908: Harmonic Analysis. Proceedings, 1981. Edited by F. Ricci and G. Weiss. V, 325 pages. 1982.

Vol. 909: Numerical Analysis. Proceedings, 1981. Edited by J.P. Hennart. VII, 247 pages. 1982.

Vol. 910: S.S. Abhyankar, Weighted Expansions for Canonical Desingularization. VII, 236 pages. 1982.

Vol. 911: O.G. Jørsboe, L. Mejlbro, The Carleson-Hunt Theorem on Fourier Series. IV, 123 pages. 1982.

Vol. 912: Numerical Analysis. Proceedings, 1981. Edited by G. A. Watson. XIII, 245 pages. 1982.

Vol. 913: O. Tammi, Extremum Problems for Bounded Univalent Functions II. VI, 168 pages. 1982.

Vol. 914: M. L. Warshauer, The Witt Group of Degree k Maps and Asymmetric Inner Product Spaces. IV, 269 pages. 1982.

Vol. 915: Categorical Aspects of Topology and Analysis. Proceedings, 1981. Edited by B. Banaschewski. XI, 385 pages. 1982.

Vol. 916: K.-U. Grusa, Zweidimensionale, interpolierende Lg-Splines und ihre Anwendungen. VIII, 238 Seiten. 1982.

Vol. 917: Brauer Groups in Ring Theory and Algebraic Geometry. Proceedings, 1981. Edited by F. van Oystaeyen and A. Verschoren. VIII, 300 pages. 1982.

Vol. 918: Z. Semadeni, Schauder Bases in Banach Spaces of Continuous Functions. V, 136 pages. 1982.

Vol. 919: Séminaire Pierre Lelong – Henri Skoda (Analyse) Années 1980/81 et Colloque de Wimereux, Mai 1981. Proceedings. Edité par P. Lelong et H. Skoda. VII, 383 pages. 1982.

Vol. 920: Séminaire de Probabilités XVI, 1980/81. Proceedings. Edité par J. Azéma et M. Yor. V, 622 pages. 1982.

Vol. 921: Séminaire de Probabilités XVI, 1980/81. Supplément: Géométrie Différentielle Stochastique. Proceedings. Edité par J. Azéma et M. Yor. III, 285 pages. 1982.

Vol. 922: B. Dacorogna, Weak Continuity and Weak Lower Semicontinuity of Non-Linear Functionals. V, 120 pages. 1982.

Vol. 923: Functional Analysis in Markov Processes. Proceedings, 1981. Edited by M. Fukushima. V, 307 pages. 1982.

Vol. 924: Séminaire d'Algèbre Paul Dubreil et Marie-Paule Malliavin. Proceedings, 1981. Edité par M.-P. Malliavin. V, 461 pages. 1982.

Vol. 925: The Riemann Problem, Complete Integrability and Arithmetic Applications. Proceedings, 1979-1980. Edited by D. Chudnovsky and G. Chudnovsky. VI, 373 pages. 1982.

Vol. 926: Geometric Techniques in Gauge Theories. Proceedings, 1981. Edited by R. Martini and E.M. de Jager. IX, 219 pages. 1982.

Vol. 927: Y. Z. Flicker, The Trace Formula and Base Change for GL (3). XII, 204 pages. 1982.

Vol. 928: Probability Measures on Groups. Proceedings 1981. Edited by H. Heyer. X, 477 pages. 1982.

Vol. 929: Ecole d'Eté de Probabilités de Saint-Flour X – 1980. Proceedings, 1980. Edited by P.L. Hennequin. X, 313 pages. 1982.

Vol. 930: P. Berthelot, L. Breen, et W. Messing, Théorie de Dieudonné Cristalline II. XI, 261 pages. 1982.

Vol. 931: D.M. Arnold, Finite Rank Torsion Free Abelian Groups and Rings. VII, 191 pages. 1982.

Vol. 932: Analytic Theory of Continued Fractions. Proceedings, 1981. Edited by W.B. Jones, W.J. Thron, and H. Waadeland. VI, 240 pages. 1982.

Vol. 933: Lie Algebras and Related Topics. Proceedings, 1981. Edited by D. Winter. VI, 236 pages. 1982.

Vol. 934: M. Sakai, Quadrature Domains. IV, 133 pages. 1982.

Vol. 935: R. Sot, Simple Morphisms in Algebraic Geometry. IV, 146 pages. 1982.

Vol. 936: S.M. Khaleelulla, Counterexamples in Topological Vector Spaces. XXI, 179 pages. 1982.

Vol. 937: E. Combet, Intégrales Exponentielles. VIII, 114 pages. 1982.

Vol. 938: Number Theory. Proceedings, 1981. Edited by K. Alladi. IX, 177 pages. 1982.

Vol. 939: Martingale Theory in Harmonic Analysis and Banach Spaces. Proceedings, 1981. Edited by J.-A. Chao and W.A. Woyczyński. VIII, 225 pages. 1982.

Vol. 940: S. Shelah, Proper Forcing. XXIX, 496 pages. 1982.

Vol. 941: A. Legrand, Homotopie des Espaces de Sections. VII, 132 pages. 1982.

Vol. 942: Theory and Applications of Singular Perturbations. Proceedings, 1981. Edited by W. Eckhaus and E.M. de Jager. V, 363 pages. 1982.

Vol. 943: V. Ancona, G. Tomassini, Modifications Analytiques. IV, 120 pages. 1982.

Vol. 944: Representations of Algebras. Workshop Proceedings, 1980. Edited by M. Auslander and E. Lluis. V, 258 pages. 1982.

Vol. 945: Measure Theory. Oberwolfach 1981, Proceedings. Edited by D. Kölzow and D. Maharam-Stone. XV, 431 pages. 1982.

Vol. 946: N. Spaltenstein, Classes Unipotentes et Sous-groupes de Borel. IX, 259 pages. 1982.

Vol. 947: Algebraic Threefolds. Proceedings, 1981. Edited by A. Conte. VII, 315 pages. 1982.

Vol. 948: Functional Analysis. Proceedings, 1981. Edited by D. Butković, H. Kraljević, and S. Kurepa. X, 239 pages. 1982.

Vol. 949: Harmonic Maps. Proceedings, 1980. Edited by R.J. Knill, M. Kalka and H.C.J. Sealey. V, 158 pages. 1982.

Vol. 950: Complex Analysis. Proceedings, 1980. Edited by J. Eells. IV, 428 pages. 1982.

Vol. 951: Advances in Non-Commutative Ring Theory. Proceedings, 1981. Edited by P.J. Fleury. V, 142 pages. 1982.

Vol. 952: Combinatorial Mathematics IX. Proceedings, 1981. Edited by E. Billington, S. Oates-Williams, and A.P. Street. XI, 443 pages. 1982.

Vol. 953: Iterative Solution of Nonlinear Systems of Equations. Proceedings, 1982. Edited by R. Ansorge, Th. Meis, and W. Törnig. VII, 202 pages. 1982.